Edited by
Fabrizio Cavani, Stefania Albonetti,
Francesco Basile, and Alessandro Gandini

Chemicals and Fuels from Bio-Based Building Blocks

Volume 1

Edited by Fabrizio Cavani, Stefania Albonetti, Francesco Basile, and Alessandro Gandini

Chemicals and Fuels from Bio-Based Building Blocks

Volume 1

Editors

Prof. Fabrizio Cavani
Dipartimento di Chimica Industriale
Viale Risorgimento 4
40136 Bologna
Italy

Prof. Stefania Albonetti
Dipartimento di Chimica Industriale
Viale Risorgimento 4
40136 Bologna
Italy

Prof. Francesco Basile
Dipartimento di Chimica Industriale
Viale Risorgimento 4
40136 Bologna
Italy

Prof. Alessandro Gandini
Universidade de Sao Paulo
PB 780
13560-970 Sao Carlos, SP
Brazil

Library of Congress Card No.: applied for

British Library Cataloguing-in-Publication Data
A catalogue record for this book is available from the British Library.

Bibliographic information published by the Deutsche Nationalbibliothek
The Deutsche Nationalbibliothek lists this publication in the Deutsche Nationalbibliografie; detailed bibliographic data are available on the Internet at <http://dnb.d-nb.de>.

Print ISBN: 978-3-527-33897-9
ePDF ISBN: 978-3-527-69819-6
ePub ISBN: 978-3-527-69822-6
Mobi ISBN: 978-3-527-69821-9
oBook ISBN: 978-3-527-69820-2

Cover Design Formgeber, Mannheim, Germany
Typesetting SPi Global, Chennai, India
Printing and Binding Markono Print Media Pte Ltd, Singapore

Printed on acid-free paper

Contents

List of Contributors

Stefania Albonetti
Alma Mater
Studiorum – University of
Bologna
Dipartimento di Chimica
Industriale "Toso Montanari"
Viale Risorgimento 4
40136 Bologna
Italy

Aristos Aristidou
Cargill Biotechnology R&D
2500 Shadywood Road
Excelsior, MN 55331
USA

Udo Armbruster
Leibniz-Institut für Katalys e.V.
Albert-Einstein-Street 29a
18059 Rostock
Germany

Mehrdad Arshadi
Swedish University of
Agricultural Sciences
Department of Forest
Biomaterials and Technology
Umea
Sweden

Francesco Basile
Alma Mater
Studiorum – University of
Bologna
Dipartimento di Chimica
Industriale "Toso Montanari"
Viale Risorgimento 4
40136 Bologna
Italy

Anna Katharina Beine
RWTH Aachen University
Institut für Technische und
Makromolekulare Chemie
Lehrstuhl für Heterogene
Katalyse und Technische Chemie
Worringerweg 2
52074 Aachen
Germany

Giuseppe Bellussi
ENI S.p.A
Downstream R&D Development
Operations and Technology
Via Maritano 26
20097 S. Donato Milanese
Italy

Daniele Bianchi
Eni S.p.A.
Renewable Energy and
Environmental R&D
Center – Istituto eni Donegani
Via G. Fauser 4
28100 Novara
Italy

Eric Black
Cargill Corn Milling North
America
15407 McGinty Road West
Wayzata, MN 55391
USA

Thomas Bonnotte
Univ. Lille
CNRS, Centrale Lille, ENSCL
Univ. Artois
UMR 8181 – UCCS – Unité de
Catalyse et Chimie du Solide
F-59000 Lille
France

Thomas R. Boussie
Rennovia Inc.
3040 Oakmead Village Drive
Santa Clara California 95051
USA

Massimo Bregola
Cargill Starches & Sweeteners
Europe
Divisione Amidi
Via Cerestar 1
Castelmassa
RO 45035
Italy

Pieter C.A. Bruijnincx
Utrecht University
Faculty of Science
Debye Institute for
Nanomaterials Science
Inorganic Chemistry and
Catalysis
Universiteitsweg 99
3584 CG Utrecht
The Netherlands

Tuong V. Bui
University of Oklahoma
Chemical, Biological, and
Materials Engineering
100 East Boyd Street
Norman, OK 73019
USA

Vincenzo Calemma
ENI S.p.A
Downstream R&D Development
Operations and Technology
Via Maritano 26
20097 S. Donato Milanese
Italy

Federico Capuano
Eni S.p.A.
Refining and Marketing and
Chemicals
Via Laurentina 449
00142 Roma
Italy

Alfred Carlson
BioAmber, Inc.
3850 Annapolis Lane North
Plymouth, MN 55447
USA

Roberto Werneck do Carmo
BRASKEMS.A.
Chemical Processes from
Renewable Raw Materials
Renewable Technologies
Rua Lemos Monteiro 120
05501-050 São Paulo, SP
Brazil

Fabrizio Cavani
Alma Mater
Studiorum – University of
Bologna
Dipartimento di Chimica
Industriale "Toso Montanari"
Viale Risorgimento 4
40136 Bologna
Italy

Annamaria Celli
University of Bologna
Department of Civil, Chemical,
Environmental and Materials
Engineering
Via Terracini 28
40131 Bologna
Italy

Gabriele Centi
University of Messina
ERIC aisbl and CASPE/INSTM
Department DIECII
Section Industrial Chemistry
Viale F. Stagno D'Alcontras 31
98166 Messina
Italy

Sanjay Charati
Solvay R&I
Centre de Lyon
Saint Fons 69190
France

Alessandro Chieregato
Alma Mater
Studiorum – Università di
Bologna
Dipartimento di Chimica
Industriale "Toso Montanari"
Viale Risorgimento 4
40136 Bologna
Italy

James H. Clark
University of York
Green Chemistry Centre of
Excellence
York
YO10 5DD
UK

Corine Cochennec
Solvay R&I
Centre de Lyon
Saint Fons 69190
France

Bill Coggio
BioAmber, Inc.
3850 Annapolis Lane North
Plymouth, MN 55447
USA

Martino Colonna
University of Bologna
Department of Civil, Chemical,
Environmental and Materials
Engineering
Via Terracini 28
40131 Bologna
Italy

Steven Crossley
University of Oklahoma
Chemical, Biological, and
Materials Engineering
100 East Boyd Street
Norman, OK 73019
USA

Manilal Dahanayake
Solvay R&I
Centre de Lyon
Saint Fons 69190
France

Paulo Luiz de Andrade Coutinho
BRASKEM S.A.
Knowledge Management
Intellectual Property and
Renewables
Corporative Innovation
Rua Lemos Monteiro 120
05501-050 São Paulo, SP
Brazil

Jean-Claude de Troostembergh
Cargill Biotechnology R&D
Havenstraat 84
Vilvoorde 1800
Belgium

Gary M. Diamond
Rennovia Inc.
3040 Oakmead Village Drive
Santa Clara California 95051
USA

Eric Dias
Rennovia Inc.
3040 Oakmead Village Drive
Santa Clara California 95051
USA

Jean-Luc Dubois
ARKEMA France
420 Rue d'Estienne d'Orves
92705 Colombes
France

Franck Dumeignil
Univ. Lille
CNRS, Centrale Lille
ENSCL, Univ. Artois
UMR 8181 – UCCS – Unité de
Catalyse et Chimie du Solide
F-59000 Lille
France

and

Maison des Universités
Institut Universitaire de France
IUF
103 Bd St-Michel
75005 Paris
France

Alan Barbagelata El-Assad
BRASKEM S.A.
Innovation in Renewable
Technologies
Corporative Innovation
Rua Lemos Monteiro 120
05501-050 São Paulo, SP
Brazil

Mihaela Florea
University of Bucharest
Department of Organic
Chemistry
Biochemistry and Catalysis
4-12 Regina Elisabeta Boulevard
030016 Bucharest
Romania

Alessandro Gandini
University of São Paulo
São Carlos Institute of Chemistry
Avenida Trabalhador
São-carlense 400
CEP 13466-590
São Carlos, SP
Brazil

Nicholas Gathergood
Tallinn University of Technology
Department of Chemistry
Tallinn
Estonia

Patrick Gilbeau
Solvay R&I
Centre de Lyon
Saint Fons 69190
France

Claudio Gioia
University of Bologna
Department of Civil, Chemical,
Environmental and Materials
Engineering
Via Terracini 28
40131 Bologna
Italy

Gianni Girotti
Versalis S.p.A.
Green Chemistry R&D Centre
Via G. Fauser 4
28100 Novara
Italy

Peter J.C. Hausoul
RWTH Aachen University
Institut für Technische und
Makromolekulare Chemie
Lehrstuhl für Heterogene
Katalyse und Technische Chemie
Worringerweg 2
52074 Aachen
Germany

Cheng-Jyun Huang
Industrial Technology Research
Institute
Material and Chemical Research
Laboratories
321 Kuang Fu Road
Hsinchu 30011
Taiwan

Ying-Ting Huang
Industrial Technology Research
Institute
Material and Chemical Research
Laboratories
321 Kuang Fu Road
Hsinchu 30011
Taiwan

Thuan Minh Huynh
Leibniz-Institut für Katalys e.V.
Albert-Einstein-Street 29a
18059 Rostock
Germany

and

4 Nguyen Thong Street
District 3
Ho Chi Minh City
Vietnam

Alberto Iaconi
SPIGA BD S.r.l.
Via Pontevecchio 55
16042 Carasco, GE
Italy

Selma Barbosa Jaconis
BRASKEM S.A.
Knowledge Management and Technology Intelligence
Corporative Innovation
Rua Lemos Monteiro 120
05501-050 São Paulo, SP
Brazil

Guang-Way Bill Jang
Industrial Technology Research Institute
Material and Chemical Research Laboratories
321 Kuang Fu Road
Hsinchu 30011
Taiwan

Ruben Jolie
Cargill Biotechnology R&D
Havenstraat 84
Vilvoorde 1800
Belgium

Benjamin Katryniok
Univ. Lille
CNRS
Centrale Lille
ENSCL, Univ. Artois
UMR 8181 – UCCS – Unité de Catalyse et Chimie du Solide
F-59000 Lille
France

and

Ecole Centrale de Lille
ECLille 59655
Villeneuve d'Ascq
France

Apostolis Koutinas
Agricultural University of Athens
Department of Food Science and Human Nutrition
Athens
Greece

Marie-Pierre Labeau
Solvay R&I
Centre de Lyon
Saint Fons 69190
France

Talita M. Lacerda
University of São Paulo
São Carlos Institute of Chemistry
Avenida Trabalhador
São-carlense 400
CEP 13466-590
São Carlos, SP
Brazil

Paola Lanzafame
University of Messina
ERIC aisbl and CASPE/INSTM
Department DIECII
Section Industrial Chemistry
Viale F. Stagno D'Alcontras 31
98166 Messina
Italy

Philippe Lapersonne
Solvay R&I
Centre de Lyon
Saint Fons 69190
France

Susanna Larocca
SO.G.I.S. S.p.A.
Via Giuseppina 132
26048 Sospiro, CR
Italy

Kit Lau
BioAmber, Inc.
3850 Annapolis Lane North
Plymouth, MN 55447
USA

Chia-Ling Li
Industrial Technology Research Institute
Material and Chemical Research Laboratories
321 Kuang Fu Road
Hsinchu 30011
Taiwan

Alice Lolli
Alma Mater Studiorum – Università di Bologna
Dipartimento di Chimica Industriale "Toso Montanari"
Viale Risorgimento 4
40136 Bologna
Italy

Erica Lombardi
Alma Mater Studiorum – Università di Bologna
Dipartimento di Chimica Industriale "Toso Montanari"
Viale Risorgimento 4
40136 Bologna
Italy

Mateus SchreinerGarcez Lopes
BRASKEM S.A.
Innovation in Renewable Technologies
Corporative Innovation
Rua Lemos Monteiro 120
05501-050 São Paulo, SP
Brazil

Rafael Luque
University of Cordoba
Department of Organic Chemistry
E14014 Cordoba
Spain

Rodolfo Mafessanti
Alma Mater Studiorum – Università di Bologna
Dipartimento di Chimica Industriale "Toso Montanari"
Viale Risorgimento 4
40136 Bologna
Italy

Philippe Marion
Solvay R&I
Centre de Lyon
Saint Fons 69190
France

Andreas Martin
Leibniz-Institut für Katalys e.V.
Albert-Einstein-Street 29a
18059 Rostock
Germany

Sergio Martins
Solvay R&I
Centre de Lyon
Saint Fons 69190
France

Christopher Mercogliano
BioAmber, Inc.
3850 Annapolis Lane North
Plymouth, MN 55447
USA

Jim Millis
BioAmber, Inc.
3850 Annapolis Lane North
Plymouth, MN 55447
USA

François Monnet
Solvay R&I
Centre de Lyon
Saint Fons 69190
France

Piergiuseppe Morone
University of Rome
Department of Law and Economics
Unitelma-Sapienza
Rome
Italy

Vince Murphy
Rennovia Inc.
3040 Oakmead Village Drive
Santa Clara California 95051
USA

Ronaldo Nascimento
Solvay R&I
Centre de Lyon
Saint Fons 69190
France

Juliana Velasquez Ochoa
Alma Mater Studiorum – Università di Bologna
Dipartimento di Chimica Industriale "Toso Montanari"
Viale Risorgimento 4
40136 Bologna
Italy

Regina Palkovits
RWTH Aachen University
Institut für Technische und Makromolekulare Chemie
Lehrstuhl für Heterogene Katalyse und Technische Chemie
Worringerweg 2
52074 Aachen
Germany

Vasile I. Parvulescu
University of Bucharest
Department of Organic Chemistry
Biochemistry and Catalysis
4-12 Regina Elisabeta Boulevard
030016 Bucharest
Romania

Sébastien Paul
Univ. Lille
CNRS
Centrale Lille
ENSCL, Univ. Artois
UMR 8181 – UCCS – Unité de Catalyse et Chimie du Solide
F-59000 Lille
France

and

Ecole Centrale de Lille
ECLille
59655 Villeneuve d'Ascq
France

Siglinda Perathoner
University of Messina
ERIC aisbl and CASPE/INSTM
Department DIECII
Section Industrial Chemistry
Viale F. Stagno D'Alcontras 31
98166 Messina
Italy

Carlo Perego
Eni S.p.A.
Renewable Energy and
Environmental R&D
Center – Istituto eni Donegani
Via G. Fauser 4
28100 Novara
Italy

Paolo Pollesel
ENI S.p.A
Downstream R&D Development
Operations and Technology
Via Maritano 26
20097 S. Donato Milanese
Italy

Katie Privett
University of York
Green Chemistry Centre of
Excellence
York
YO10 5DD
UK

Daniel E. Resasco
University of Oklahoma
Chemical, Biological, and
Materials Engineering
100 East Boyd Street
Norman, OK 73019
USA

Marco Ricci
Versalis S.p.A.
Green Chemistry R&D Centre
Via G. Fauser 4
28100 Novara
Italy

Giacomo Rispoli
Eni S.p.A.
Refining and Marketing and
Chemicals
Via Laurentina 449
00142 Roma
Italy

Matthias N. Schneider
Baerlocher GmbH
Freisinger Straße 1
85716 Unterschleissheim
Germany

Franco Speroni
Solvay R&I
Centre de Lyon
Saint Fons 69190
France

Everton Simões Van-Dal
BRASKEM S.A.
Innovation in Renewable
Technologies
Corporative Innovation
Rua Lemos Monteiro 120
05501-050 São Paulo, SP
Brazil

Micaela Vannini
University of Bologna
Department of Civil, Chemical,
Environmental and Materials
Engineering
Via Terracini 28
40131 Bologna
Italy

Bert M. Weckhuysen
Utrecht University
Faculty of Science
Debye Institute for
Nanomaterials Science
Inorganic Chemistry and
Catalysis
Universiteitsweg 99
3584 CG Utrecht
The Netherlands

Sophie C.C. Wiedemann
Utrecht University
Faculty of Science
Debye Institute for
Nanomaterials Science
Inorganic Chemistry and
Catalysis
Universiteitsweg 99
3584 CG Utrecht
The Netherlands

and

Croda Nederland
B.V. Buurtje 1
2800 BE Gouda
The Netherlands

Jinn-Jong Wong
Industrial Technology Research
Institute
Material and Chemical Research
Laboratories
321 Kuang Fu Road
Hsinchu 30011
Taiwan

Yu Zhang
Alma Mater
Studiorum – Università di
Bologna
Dipartimento di Chimica
Industriale "Toso Montanari"
Viale Risorgimento 4
40136 Bologna
Italy

Preface

Today the biorefinery concept is being applied to integrate the production of chemicals, fuels, and materials from renewable resources and wastes. But the concept is not new, since the industry of oils and fats, among others, has already for some time been transforming by-products or co-products received from other industries into chemicals for several diverse sectors while combining this production with that of fuels, such as biogas, obtained from organic residues.

However, the sector is rapidly evolving, and new concepts and ideas are setting the scene. It is impressive to see how the scientific and technical advancements in this field have been, and still are, changing the scenario at an unprecedented pace.

This book was conceived with the ambitious aim of preparing something different from the information already available. In other words, for the authors it was not only a matter of updating the panorama with the latest developments in technologies and transformation processes but also of offering readers the possibility to view the "world of renewables" from a more rational perspective. Instead of contributions focusing on how a certain biomass or bio-based building block can be transformed into specific products, we decided to offer an overview from the standpoints of *reaction* and *products*. This led to the organization of the book into different sections in which the classes of reactions (oxidation, hydrodeoxygenation, C–C bond formation) are examined or the different types of monomers (succinic acid, adipic acid, furandicarboxylic acid, glycols, and acrylic acid), polymers, and drop-in chemicals (olefins, aromatics, and syngas, produced from renewables) are discussed. A closing section of the book contains several contributions from chemical industries operating in the field of biorefinery development. Throughout the book it is possible to find points of intersection between the chapters on products and those on reactions that are finding a place in biorefinery models. This approach gives a three-dimensional perspective on the production of bio-based building blocks and, looking at the future, facilitates the development of new processes and placement of new products in an increasingly integrated context.

This book is therefore organized into three main sections. The first section (15 chapters) is devoted to a discussion of the main products attainable from renewable raw materials. The different types of products have, in turn, been further separated into three sections: (i) chemicals and fuels from bio-based building blocks (B^4), (ii) B^4 monomers, and (iii) polymers from B^4. To provide a reliable description of the state of research and development and of industrial implementations of the production of different molecules, we asked several experts from universities and industry to present the most recent results obtained in their respective field, together with their ideas on the topic. Given the structure of the book, the chapters on polymers concentrate exclusively on the use of bio-based building blocks as potential monomers and on the most recent studies dealing with their polymerizations and copolymerizations, as well as on the properties of the ensuing materials.

The second part of the book focuses on the class of reactions needed to obtain chemicals and fuels from bio-based building blocks. Hydrogenation, deoxygenation, C–C bond formation, and oxidation are key reactions in the upgrading of these compounds, and Chapters 14–17 provide the reader with a basic understanding, offering an overview of the possibilities offered by these tools using different raw materials. Several examples of homogeneous and heterogeneous catalysis are discussed, with an emphasis on the industrial aspects, providing a comprehensive picture and addressing the main issues associated with biomass transformations.

The book concludes with several contributions from scientists working in the industry. This part is introduced by Centi's chapter on "A vision for future biorefineries." In recent years, numerous bio-based companies, seeing new technologies as ways to gain market positions through innovation, have been created all over the world. However, they have often appeared to be driven primarily by the need to raise funds, rather than by a true desire for innovation and a clear analysis of the market perspectives. For example, the US government's push to develop hydrocarbons from biomass led to the creation of over 100 venture-capital companies, most of which have now closed or markedly changed their strategies. Hence there is the trend to reconsider the biorefinery model and analyze the new directions and scenarios to evaluate possibilities and anticipate needs for research. Companies themselves are involved in this process, and the last part of the book presents the different models of biorefineries they have developed in recent years.

SOGIS SpA and Baerlocher GmbH (Chapter 19) analyze the possibility of fitting biorefinery concepts into the well-established oleochemistry value chains. Indeed, oleochemistry has always used renewable raw materials (plant oils) or side streams from food production (animal fats) and may therefore be considered a model for green chemistry. Specifically, the emerging opportunities of food supply chain waste form the topic covered by R. Luque and coworker (Chapter 26). Researchers around the world are working to find innovative solutions to the food waste problem through valorization, and interdisciplinary collaborative approaches are crucial for creating a viable set of solutions in chemical, material, and fuel production.

Large chemical companies, such as Arkema (Chapter 20) and Eni (Chapter 25), highlight the necessity to facilitate the transition to bio-based raw materials, creating existing fossil-based molecules from renewables (benefiting from the existing market and regulations) and using existing infrastructure to lower capital costs. In 2014, renewable products generated around €700 million of the turnover of Arkema; today this company has several plant-based factories operating worldwide, demonstrating the economic feasibility of bio-based production when the products have a technical advantage over fossil-based materials. Similar considerations have also been reported by Solvay (Chapter 24), describing several case studies that provide a good picture of the issue of introducing new sustainable chemical processes into the various sectors of chemical production. Moreover, they have shown the wide range of benefits that the use of renewable raw materials can bring to the chemical industry, taking into account the real sustainability of the processes applied.

The keys to the current and future success of a bioeconomy are analyzed by Cargill in Chapter 21, describing the concept of colocation as a model for the production of bio-based chemicals from starch, while Braskem reports on the economic viability of sugarcane biorefineries in Brazil (Chapter 22), highlighting the importance of integration in value chain models as a key element for the future of chemical production from renewables. Moreover, they underscore the importance of industrial biotechnology as the basis of these industries in the future.

Versalis, the major Italian chemical company belonging to the Eni group (Chapter 23), discusses the company's strategy in the field, describing several projects mainly related to the synthetic elastomer business. Particular attention is focused on the production of natural rubber from guayule plants and the development of a process for the production of butadiene from bio-based raw materials.

This book is aimed at R&D engineers and chemists in chemical and related industries, chemical engineering and chemistry students, and chemical, refining, pharmaceutical, and biotechnological industry managers. Moreover, it can be useful for decision-making managers in funding institutions, providing them with an overview of new trends in the bio-based economy.

In closing, we would like to express our sincere thanks to all the authors who contributed to the compilation of this book, without whom this project would never have seen the light. We are also grateful to all the people at Wiley for their patience and professionalism.

1 December 2015
Bologna

Fabrizio Cavani, Stefania Albonetti,
Francesco Basile, and Alessandro Gandini

Part I
Drop-in Bio-Based Chemicals

Chemicals and Fuels from Bio-Based Building Blocks, First Edition.
Edited by Fabrizio Cavani, Stefania Albonetti, Francesco Basile, and Alessandro Gandini.

1 Olefins from Biomass

Alessandro Chieregato, Juliana Velasquez Ochoa, and Fabrizio Cavani

1.1 Introduction

The depletion of oil, the related environmental and economic concerns, together with the opportunities seen in renewable building blocks have paved the way for a significant reorganization of the chemical industry. Almost as a revival of the early twentieth century chemical industry, todays' oil refineries are in the process of being redesigned coupling petrochemical processes with bio-based productions and fermentation technologies. In this context, the invention of new (or reconsidered) processes for the synthesis of C_2–C_4 olefins from renewables is of crucial importance, since these molecules are fundamental building blocks for the chemical industry. Indeed, mainly due to a switch from naphtha to natural gas – primarily ethane – as a feedstock for steam crackers, there may be a shortage of C_3–C_5 olefins in the near future, making necessary the recourse to alternative feedstocks [1].

Light olefins are key building blocks for the production of strategic bulk chemical products; ethylene is used primarily to manufacture polyethylene, ethylene chloride, and ethylene oxide, which are used for packaging, plastic processing, construction, textiles, and so on. Propylene is used to make polypropylene, but it is also a basic product necessary for producing propylene oxide, acrylic acid, and many other derivatives; not only the plastic processing, packaging industry, and furnishing sector but the automotive industry too are users of propylene and its derivatives. Butenes, 1,3-butadiene (BD), isobutene, and isoprene are important monomers, comonomers, or intermediates for the production of synthetic rubber (mainly for tires and automobile components), lubricants, fuels, and fuel additives [2].

In this chapter the production of C_2–C_4 olefins from renewable sources is reviewed, highlighting the technologies involved, the best-performing catalysts, and the optimal engineering parameters but also discussing the reaction mechanisms. Among the viable options, particular focus is given to the more environmentally benign and sustainable routes, that is, the syntheses involving the least possible number of steps and relatively mild reaction conditions. Indeed,

Chemicals and Fuels from Bio-Based Building Blocks, First Edition.
Edited by Fabrizio Cavani, Stefania Albonetti, Francesco Basile, and Alessandro Gandini.

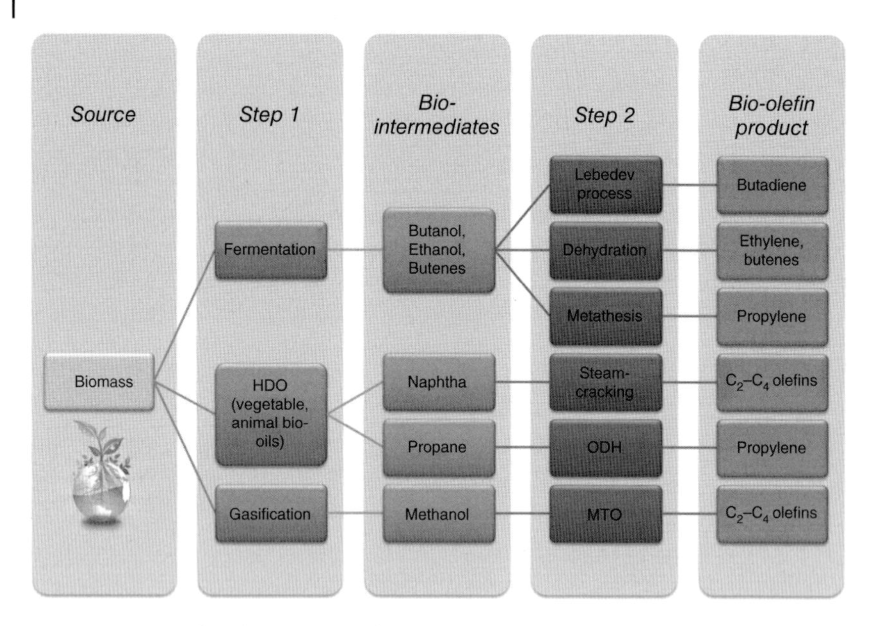

Figure 1.1 Routes from biomass to olefins.

processes such as oxidative dehydrogenation (ODH) or methanol to olefin (MTO) using methanol derived, for instance, from biosyngas produced by pyrolysis/gasification of biomass are of interest today [3]; however, they are energy-intensive processes and require several steps for the production of biochemicals.

The production of olefins from renewable sources can be carried out starting directly from the biofeedstock or certain intermediates (platform molecules) which are already available commercially, such as bioethanol and butanediols (BDOs) (Figure 1.1). In this chapter, the various routes starting from bioalcohols and that starting from bio-oils via hydrodeoxygenation (HDO) and cracking will be considered and described in detail.

1.2 Olefins from Bioalcohols

1.2.1 Ethanol to Ethylene

Bioethanol can be produced from biomass by fermentation processes; typically, engineered yeasts are used to transform C_5 and C_6 sugars into C_2 alcohol. Although this is a consolidated pathway that makes it possible to obtain high selectivity, low accumulation of by-products, high ethanol yield, and high fermentation rates [4], its economy is greatly affected by sugar production prices. A more attractive approach would be the direct use of chemically more complex biomass, such as cellulose and hemicellulose, or even lignocellulose, due to their significantly higher availability and lower cost. Nevertheless, nowadays the technologies for the direct transformation of lignocellulosic biomass to ethanol present – with only a few reported exceptions – unsatisfactory performances for industrial applications [5]. For these reasons, so far it has been possible to put the synthesis of bioethylene from bioethanol into practice only in regions where the cost of sugars is very low, for example, in Brazil [6, 7]. In spite of these economic issues, this reaction has been the subject of a vast scientific production [8]. The ethylene market is growing continuously (US$ 1.3 billion market at 35% growth between 2006 and 2011), as is the demand for renewable polyethylene. It has been estimated that the demand for the bio-based olefin corresponds to 10% of the global polyethylene market, whereas its supply presently totals only <1% [7]. Nevertheless, the possibility of synthesizing ethylene from the steam cracking (or oxydehydrogenation (ODH)) of ethane, available at cheap prices from natural (shale) gas, might represent a significant economic obstacle for further developments of bioethylene production. A niche production of bioethylene, however, might be possible in those markets looking for small-scale volumes, that is, where full-scale crackers (using either naphtha or natural gas) would not be commercially viable. In order to meet this demand, several companies, such as BP, Total Petrochemicals, and Solvay, have been researching and have patented various technologies for the dehydration of ethanol into ethylene [9–11].

Despite the simple chemistry that one might expect from the dehydration of ethanol into ethylene, the careful tuning of the acid–base and redox properties of the catalysts used, as well as their time-on-stream stability, are mandatory requirements; however, these goals are not so straightforward to reach. Indeed, although a great number of acid catalysts can perform ethanol dehydration with selectivity and conversion >95%, only a few of them are able to resist coke deactivation for long periods of time, thus making the periodic regeneration of the catalyst compulsory [6, 8]. The problem might be solved using fluidized bed reactors that present a more uniform temperature profile (which limits the by-product formation) than fixed-bed technologies and that makes it possible to regenerate the spent catalyst under continuous conditions. Still, friction and collision problems between catalyst particles remain general issues with this kind of reactor to the advantage of the more conventional fixed-bed approach.

The first catalysts used industrially for this process were based on immobilized phosphoric acid used at temperatures as high as 500 °C; however, the significant coke formation and the high temperatures required encouraged the development of more efficient systems. Generally speaking, all the acid catalysts developed in the last decades have almost complete initial activity and ethylene selectivity [6]. For instance, in the 1980s, alumina-based catalysts were developed; among them, the SynDol® catalyst from Halcon/Scientific Design Company (Al_2O_3–MgO/SiO_2) was claimed to lead to conversion and selectivity >97% at 99% conversion, with regeneration intervals of 8–12 months [8]. This might be the catalyst currently used by India Glycols to produce bioethylene glycol through ethanol dehydration. Recently, BP has also developed new efficient catalysts based on modified tungsten-based heteropolyacids supported on various substrates (e.g., porous silica), which operate at 180–250 °C and present very low selectivities into by-products (e.g., ethane) which are difficult to separate [11]. Many other acid catalysts, such as zeolites (e.g., ZSM-5) and SAPO, are active for ethanol dehydration at low temperature (~200 °C); their relatively fast deactivation is generally considered to limit their applicability, although Solvay has recently patented the production of bioethylene using these kinds of materials as preferred catalysts [10].

As far as the reaction mechanism is concerned, ethanol dehydration is usually mentioned to be an acid–base concerted mechanism with the formation of intermediate ethoxy and hydroxy species on the catalyst surface, with consecutive water desorption as the rate-determining step (Scheme 1.1) [1, 6].

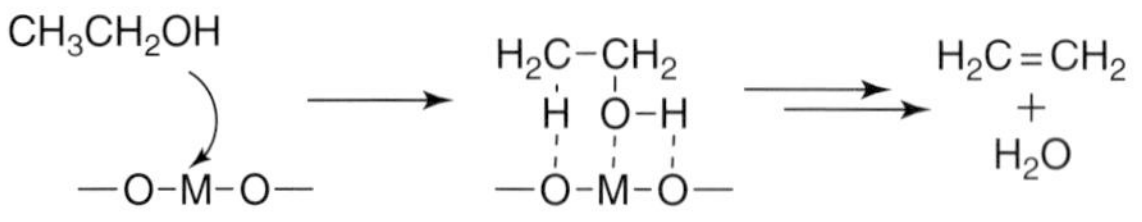

CH_3CH_2OH + —O–M–O— ⟶ $H_2C{-}CH_2$ (H, O–H) on —O–M–O— ⟶ $H_2C{=}CH_2$ + H_2O

Scheme 1.1 Commonly accepted general mechanism for ethanol dehydration on solid catalysts.

However, as highlighted by the high number of papers recently published on this topic [12–17], the reaction mechanism seems to be more complicated than the general mechanism proposed with formation of intermediate ethoxide species; particularly, reaction temperature, nature of the acid–base sites, and ethanol partial pressure are fundamental variables that govern that phenomenon at the molecular level. New insights obtained into the reaction mechanism for ethanol transformation into 1-butanol (Guerbet reaction) and BD (Lebedev reaction) on oxide catalysts with basic features [18] (*vide infra*) might be extended to acid catalysts: Indeed, basic oxides such as MgO or CaO form ethylene and other by-products (e.g., acetaldehyde, hydrocarbons, methane, CO_x, H_2, etc.) that are also observed with acid systems [8]. Nevertheless, the final product distribution is clearly a function of the acid–base properties of the catalyst surface, which facilitate or hamper the different parallel and consecutive pathways that are thermodynamically possible.

1.2.2
Ethanol to Butadiene

Because of the forecasted decrease in BD production by means of the conventional naphtha cracking and extraction from C_4 fraction, as well as of the possible increase of the biosourced rubber requested by future legislation, several alternative routes are currently under investigation for the synthesis of bio-BD, starting from various renewable sources (Figure 1.2). Among these, the transformation of bioethanol into BD is the route raising the greatest expectations, also due to the fact that these technologies were already being practiced at an industrial level during the period 1930–1970s, before the advent of naphtha cracking, which made all the synthetic routes less economically convenient.

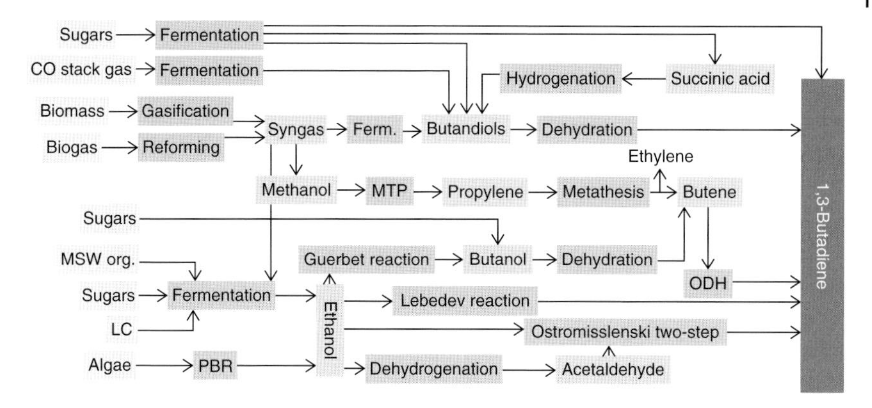

Figure 1.2 Several alternative routes investigated for the synthesis of biobutadiene.

Ethanol conversion into BD is a reaction with a very long history in the chemical industry and has encountered renewed interest within the biorefinery context today. The first accounts on this transformation date back to the beginning of the twentieth century, whereas it became an industrial process starting in the 1920–1930s [19]. From then up until the end of World War II, BD production from ethanol represented the main route for the manufacture of synthetic rubber, and the main players in this field were undoubtedly Russia (i.e., USSR) and the United States.

Although both the reaction mechanism and the catalysts' composition have always been a matter of debate since the very early stages of its history (*vide infra*), the ethanol upgrading into BD can be summarized as a dehydrogenation, dehydration, and condensation reaction (Scheme 1.2).

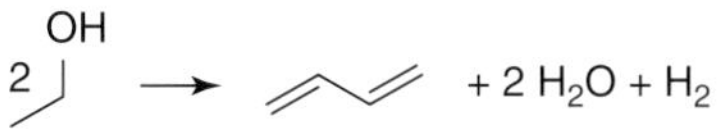

OH

2 ⟶ $+ 2\ H_2O + H_2$

Scheme 1.2 Overall reaction stoichiometry from ethanol to butadiene.

To carry out this transformation, the USSR opted for a single-step approach where ethanol was directly made to react on a multifunctional catalyst, whereas the United States found more convenience in a two-step synthesis, where the dehydrogenation step was separated from the condensation and dehydration ones. The two processes are also called the Lebedev (one-step) and Ostromisslenski (two-step) reactions, respectively, being named after their original inventors [19].

From the 1920–1940 period, the most abundant details on the kind of technology used at an industrial level can be found in the patent literature for the American two-step approach, which was operated by the Carbide and Carbon Chemicals Corporation [20]. A simplified flow sheet of the chemical plant is reported in Figure 1.3. Ethanol was dehydrogenated to acetaldehyde in a first reactor of the shell-and-tube type (R1 in Figure 1.3); the catalyst used in the first step contained copper, and the reaction was conducted at 280 °C with 10% water vapor in the inlet feed. Unconverted ethanol and reaction products such as hydrogen and acetaldehyde were sent to a preliminary separation zone 1 (SZ1) along with some products (acetaldehyde, BD, diethyl ether, ethanol, mono-olefins, hydrogen, and saturated hydrocarbon gases) coming from the second reactor (R2). In SZ1, by means of in-series scrubbing towers, pure hydrogen was separated from all the other molecules. In the same zone of the chemical plant, a stream of light gases was vented (and likely used as industrial fuel); this stream, labeled as "gases" in Figure 1.3, was composed of a mixture of ethylene, propylene, saturated hydrocarbons, carbon dioxide, and carbon monoxide; no BD should be present in this stream since it was most likely all absorbed in the first scrubber and sent to separation zone 2 (SZ2). In the R2 of the shell-and-tube type, ethanol was made to react with acetaldehyde in a molar ratio of about 3 to 1 so as to produce BD. The catalyst used in this second step was claimed to be composed of 2.4% Ta_2O_5 on doped silica gel, operating between 300 and 350 °C, even if other kinds of silica-doped catalysts could have been applied [21, 22] (*vide infra*). The outstream from R2 was sent to SZ2 along with the BD-rich flow derived from the first scrubbing tower of the SZ1. SZ2 was composed of a distillation column and a scrubbing tower, both necessary to obtain a pure stream of BD. The bottom fractions from both rectification columns, containing water, acetaldehyde, diethyl ether, ethanol, and by-products, were collected and mixed with the bottom fractions coming from SZ1. The inlet feed of separation zone 3 (SZ3) was thus composed of a mixture containing approximately 15% acetaldehyde, 5% diethyl ether, 40% ethanol, 35% water, and 5% by-products. SZ3 was made of three distillation columns that had to separate both water and by-products from acetaldehyde and ethanol, which were recycled to both R1 and R2. The by-products withdrawn from SZ3 were composed of acetaldehyde, diethyl ether, ethyl acetate, butyraldehyde, methyl ethyl ketone (MEK), and other minor impurities. Importantly, both outstreams leaving SZ3 were not pure streams of ethanol and acetaldehyde, respectively: the acetaldehyde-rich stream was actually composed of 75% acetaldehyde, 20% diethyl ether, and 5% by-products, whereas the ethanol-rich stream was a mixture of 85% ethanol, 10% water, and 5% by-products.

It should be mentioned that during the same decade, the Carbide and Carbon Chemicals Corporation submitted other patents discussing alternative plant configurations with the aim of improving the overall process economy [20, 23]; thus, compared to the reaction scheme just discussed, some differences might have been used in the actual industrial plant. Nevertheless, these alternative configurations are only minor variations concerning the separation zones and the technology used for product recovery, which, in the end, do not alter significantly the general plant configuration as reported in Figure 1.3.

With regard to the Lebedev (one-step) process, details on the chemical plant configuration used in the USSR are more difficult to find, even if the general approach should be similar to the American technology. However, due to the lower BD purity known to be obtained through the one-step approach, one might expect a more complex separation procedure so as to finally gain the high-purity BD required for an efficient polymerization of the olefin. Nowadays, as previously mentioned, both the catalyst composition and the reaction mechanism are still an important subject of debate, whether talking about the one- or two-step

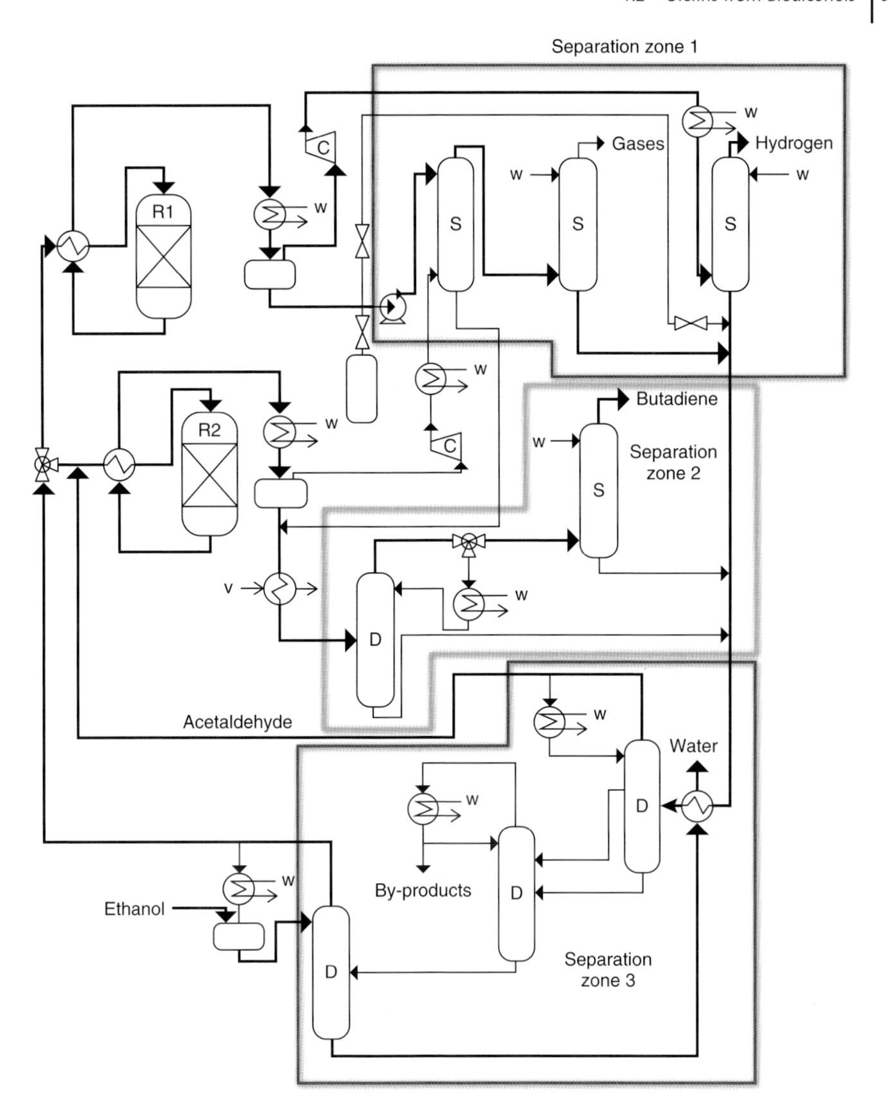

Figure 1.3 Schematic flow diagram of the two-step process for making butadiene from ethanol and acetaldehyde, as inferred from [20]. Symbols: R1: reactor to convert ethanol into acetaldehyde, R2: reactor to convert ethanol and acetaldehyde to butadiene, D: distillation column, S: scrubber, C: compressor, W: water, and V: vapor.

approach. In recent years, a number of papers and reviews have been published on ethanol transformation into BD, with particular attention to the Lebedev approach, due to the less demanding economic and engineering requirements theoretically needed by a one-pot synthesis. A very exhaustive review on both the catalyst and reaction mechanisms was published in 2014 by Sels and coworkers [24]. The catalysts for the Lebedev one-step process can be divided into three main families: (i) doped alumina catalysts, (ii) magnesia–silica catalysts, and (iii) other catalysts. Some of the most interesting results obtained for catalysts belonging to each category are reported in Figure 1.4 and the associated Table 1.1. From there it is possible to see that most of them are in the 20–40% yield range, and only few of them overcome the 60% including the ones reported by Ohnishi *et al.* whose values were taken during the first 10 min of reaction and thus are not representative of the steady state.

As far as the Ostromisslenski two-step approach is concerned, the main difference to the Lebedev one is obviously the separation of the rate-determining dehydrogenation step (i.e., ethanol dehydrogenation into acetaldehyde) from an *in situ* reaction to a dedicated and separate unit. Provided this diversification, the remaining catalysts' features for both approaches stay the same. Acetaldehyde can be produced from ethanol with or without oxygen in the feed [19], leading to water or hydrogen as a coproduct, respectively. Some of the most efficient catalysts are summarized in Table 1.2.

The other hot topic in the conversion of ethanol into BD is undoubtedly the reaction mechanism. Although various routes have been proposed [24], the most generally accepted key step in the reaction mechanism is believed to be the aldol condensation of acetaldehyde. Remarkably, this was also supposed to be the key step for the synthesis of 1-butanol from ethanol, that is, the Guerbet synthesis. Nevertheless, since the downing of gas-phase Guerbet and Lebedev syntheses, the aldol route has often been criticized [20].

First of all, the intermediate acetaldol has never been detected among reaction products, but this detail is not sufficient for ruling out this route. Additionally, already in 1949, the engineers working for the Carbide and Carbon Chemicals Corporation published a paper in which they affirmed that if acetaldol was fed

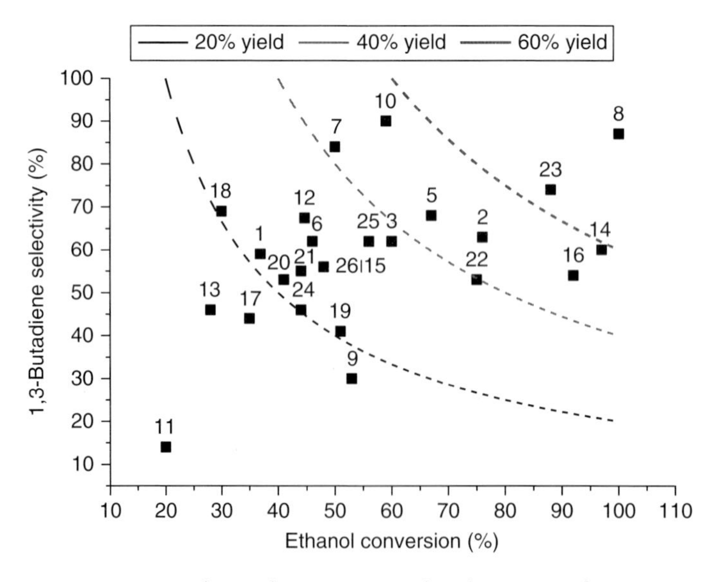

Figure 1.4 1,3-Butadiene selectivity versus ethanol conversion for representative catalysts.

Table 1.1 Active systems in the transformation of ethanol to 1,3-butadiene.

Number	Catalyst	*T* (°C)	Ethanol conversion (%)	1,3-BD selectivity (%)	1,3-BD yield (%)	References (year)
1	Mg/Si (2/3) wet kneading with acetic acid	380	36.9	59.0	21.8	[25] (1947)
2	$2\%Cr_2O_3-76\%MgO-11\%SiO_2-11\%$ kaolin	435	76.0	63.0	47.9	[26] (1954)
3	$2\%Cr_2O_3-79\%MgO-19\%SiO_2$	400	60.0	62.0	37.2	[27] (1960)
4	Al/Zn 6/4 (fluidized bed)	425	100.0	73.0	3.1	[28] (1963)
5	$ZnO-MgO-SiO_2-$kaolin	410	67.0	68.0	45.6	[29] (1966)
6	$Mg(OH)_2$ + colloidal SiO_2	380	46.0	62.0	28.5	[30] (1972)
7	Mg/Si wet kneading 1/1	350	50.0	84.0	42.0	[31] (1985)
8	Mg/Si wet kneading 1/1 + 0.1% Na	350	100.0	87.0	87.0	[31] (1985)
9	MgO/SiO_2 (0.83:1)	350	53.0	30.0	15.9	[32] (1988)
10	$10\%NiO-28\%MgO-62\%SiO_2$	280	59.0	90.0	53.1	[33] (1996)
11	(Ca/P = 1.62) hydroxyapatite	350	20.0	14.0	2.8	[34] (2008)
12	$Cu(1\%)Zr(1\%)Zn(1\%)/SiO_2$	375	44.6	67.4	30.1	[35] (2011)
13	Mg/Si 2/1 mechanical mixing (with $Mg(OH)_2$)	350	28.0	46.0	12.9	[36] (2012)
14	Mg/Si 2/1 wet kneading + Cu 5%	350	97.0	60.0	58.2	[36] (2012)
15	Mg/Si 2/1 wet kneading + Cu 5%	350	46.6	56.0	26.1	[36] (2012)
16	$4\%Ag-55\%MgO-41\%SiO_2$	400	92.0	54.0	49.7	[36] (2012)
17	Mg/Si 3/1	325	35.0	44.0	15.4	[37] (2014)
18	Mg/Si 95/5 + Zr 1.5% + Zn 0.5%	325	30.0	69.0	20.7	[37] (2014)
19	Mg/Si 1/1 + Zr 1.5% + Zn 0.5%	325	51.0	41.0	20.9	[37] (2014)
20	Al and Zn nitrate impregnated on γ-Al_2O_3	400	41.0	53.0	21.7	[38] (2014)
21	Al and Zn nitrate impregnated on γ-Al_2O_3 + H_2O_2 in the feed	395	44.0	55.0	24.2	[38] (2014)
22	CuO/SiO_2-MgO	425	75.0	53.0	39.8	[39] (2014)
23	$Ag/ZrO_2/SiO_2$	314	88.0	74.0	65.1	[40] (2014)
24	$1.0Ag/MgO-SiO_2$	400	44.0	46.0	20.2	[41] (2015)
25	$4\%ZnO/MgO-SiO_2$(1:1)	375	56.0	62.0	34.7	[42] (2015)
26	Ag/ZrBEA	327	48.0	56.0	26.9	[43] (2015)

Table 1.2 Catalysts for the conversion of ethanol into acetaldehyde.

Catalyst[a)]	T (°C)	EtOH conversion (%)	C_2H_4O selectivity (%)	References
0.2%Au–0.2%Cu/SiO_2 (yes)	250	100	100	[44]
Au(6.7)/MCM-41 (yes)	200	20	90	[45]
5% WO_3–95% V_2O_5 (yes)	280	48	98	[46]
5.75 wt%Cu supported on rice husk ash (no)	275	80	100	[47]
Au(4.9)/SBA-15 and Au(5.8)/SBA-16 (no)	350	95	90–100	[45]

a) In parenthesis the presence or not of oxygen in the feed.

on the industrial catalyst to make BD, it was reversed to acetaldehyde and not dehydrated to crotonaldehyde, thus making clear that acetaldol cannot be the key intermediate for the production of BD from ethanol [48]. More recently, Meunier and coworkers [49] also definitively ruled out acetaldehyde self-aldolization as a main reaction pathway for the gas-phase transformation of ethanol into 1-butanol over hydroxyapatite, that is, the best catalysts so far reported in literature for this reaction. Therefore, neither the Lebedev nor the gas-phase Guerbet synthesis can have acetaldol as the key reaction intermediate, as summarized in Scheme 1.3.

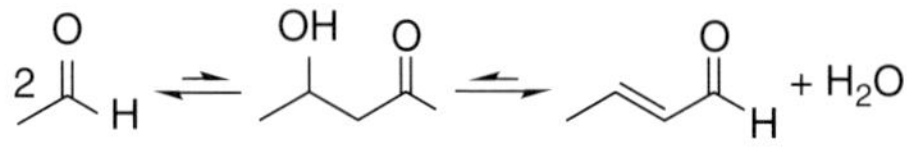

$$2\,CH_3CHO \rightleftharpoons CH_3CH(OH)CH_2CHO \rightleftharpoons CH_3CH{=}CHCHO + H_2O$$

Scheme 1.3 Under the reaction conditions used for the gas-phase Lebedev and Guerbet processes, acetaldol is mainly reversed to acetaldehyde and not upgraded to crotonaldehyde.

Once the thermodynamically hampered aldol condensation is ruled out, the formation of C_4 compounds from ethanol in the gas phase must go through alternative pathways. Very recently an unconventional route has been proposed that avoids aldol condensation and does not require the Meerwein–Ponndorf–Verley (MPV) reaction to justify the formation of the aforementioned products, at least on basic oxide catalysts [18]. By means of a multifaceted approach using catalytic tests, DRIFTS analyses, and thermodynamic and Density Functional Theory (DFT) calculations, it was possible to assign a carbanionic species of ethanol as the common key intermediate for the formation of C_4 products (and ethylene [*vide supra*]). Once this species is formed, the carbanion can (i) dehydrate to form ethylene, (ii) react with another molecule of ethanol, or (iii) react with a molecule of acetaldehyde previously produced by ethanol dehydrogenation (either *in situ*, e.g., by reduced metals, or separately with a dedicated catalyst). In the first case, 1-butanol is directly produced from two molecules of ethanol as a primary product, without the need for acetaldehyde. In the latter case, the reaction of the C_2 carbanion with an adsorbed molecule of acetaldehyde determines the formation of an adsorbed intermediate similar to 1,3-butandiol, which desorbs in the gas phase as crotyl alcohol. The latter can be finally dehydrated into BD, thanks to the high temperature of the Lebedev process (300–450 °C) and/or the acid sites always present in typical Lebedev catalysts. The overall reaction mechanism proposed is shown in Scheme 1.4.

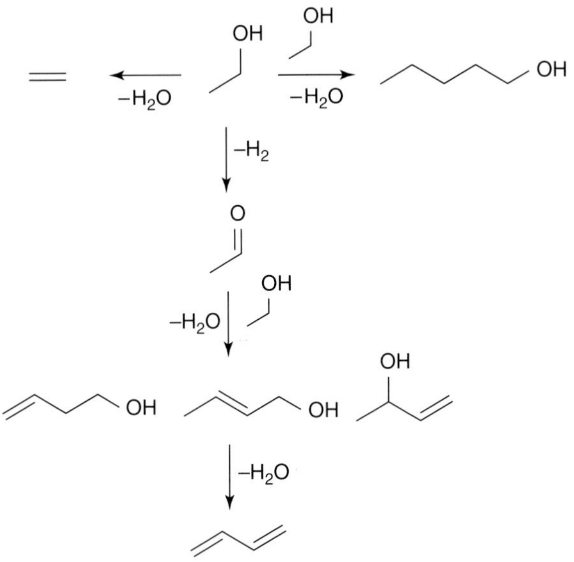

Scheme 1.4 General reaction network for the Lebedev and Guerbet processes in the gas phase on oxide catalysts with basic features. (Adapted from [18].)

Isobutene has also been recently demonstrated to be obtainable directly from ethanol [50, 51]. Nanosized Zn–Zr mixed oxides showed selectivity up to 83% at complete ethanol conversion (reaction temperature 450 °C).

Lastly, ethanol can be also converted into propylene on single multifunctional catalysts, mainly by the dimerization of intermediately formed ethylene (via acid-catalyzed dehydration) and subsequent metathesis of the resulting butenes with unreacted ethylene. Sc/In_2O_3 allowed obtaining propylene yields up to 62% at total ethanol conversion, if hydrogen and water were co-fed with ethanol at 550 °C [52].

1.2.3 C_3 Alcohols to Olefins

As far as the synthesis of C_3 olefins and particularly propylene is concerned, dehydration of C_3 alcohols might be an interesting option. Although in principle various alcohols could be used as raw materials, only a few of them are actually economically viable solutions. Indeed, at least so far, the prices of many C_3 alcohols are higher than or comparable to that of propylene. However, an attractive and sustainable route might be the transformation of glycerol to olefin (GTO) and, particularly, to propylene, in great demand. Indeed, glycerol is produced worldwide in massive amounts as a coproduct of biodiesel synthesis, and it is currently more of a burden to treat as a waste than an economic opportunity enhancing the value of the whole biofuel production chain. Several chemicals and/or biological routes have been proposed in literature to upgrade glycerol into valuable molecules, although only few of them may be economically and technically viable for commercial application [5]. Nevertheless, the conversion of this polyol to olefins might both improve the supply for short-chain olefins and add economic value to biodiesel production.

GTO is a relatively new field of research, with only a few examples in literature and satisfactory catalytic results reported [53]. Indeed, the removal of three atoms of oxygen from such a highly hydrophilic molecule is a difficult task. One of the first reports published in literature on this topic [54] used a fluid catalytic cracking (FCC) approach, in which glycerol was made to react in a microdowner and in fixed microactivity test reactors using zeolites as catalysts. In spite of the high temperatures used (290–720 °C), C_3 and C_2 oxygenated species were always the main products, mainly as acrolein and acetaldehyde. Moreover, carbon monoxide also formed in large quantities at high temperatures (selectivity up to 51%), along with a wide range of by-products. Nevertheless, at temperatures higher than 500 °C, C_2–C_4 olefins formed in considerable amounts along with methane, ethane, and higher hydrocarbons. At 700 °C, ethylene selectivity was 21.8% whereas that of propylene was 7.8%. More recently, other groups have investigated metal-modified zeolites for analogous FCC–GTO [55, 56]; however the yields in light olefin have never reached 20%.

Recently, both in patent and open literatures, alternative processes to the direct cracking approach have been proposed in order to convert glycerol into light olefins. Schmidt and coworkers [57] reported a three-step conversion of glycerol into (i) acrolein, (ii) propanal, and (iii) propylene + ethylene. Both the first and last steps used zeolites (HZSM-5 and HBEA) as dehydration and deoxygenation (i.e., cracking) catalysts, respectively, whereas Pd/α-Al_2O_3 was used for the hydrogenation step.

In defiance of the direct cracking of glycerol previously discussed, the cracking of the intermediately formed propanal makes it possible to obtain much higher yields into light olefins at relatively low temperatures (400–500 °C); at 86% conversion of propanal, fed as pure compound in a preliminary catalytic test, some yields to ethylene and propylene as high as 70% were registered. However, the fast deactivation of the HBEA was observed. Another approach for the production of propylene from glycerol was also proposed by Cao and coworkers [58]; the polyol was first hydrodeoxygenated to *n*-propanol and finally dehydrated to propene. Working in a continuous flow reactor with two in-series catalytic beds, that is, Ir/ZrO_2 and HZSM-5-30, under optimized working conditions (250 °C, P H_2 = 1 MPa), propylene selectivity as high as 85% at total glycerol conversion was achieved.

Lastly, an interesting and original approach for the direct conversion of glycerol into propylene has recently been patented by Dow Global Technologies [59]. In this case, glycerol is made to react in a batch reactor with hydroiodic acid (HI) (HI-to-glycerol ratio ~1:10) at 210 °C under mild hydrogen pressure (~4 bar). Selectivity to propylene as high as 96% at 24% glycerol conversion is claimed after a 6 h reaction. During the process, HI is oxidized to I_2, and in order to act as a catalyst, it must be reduced back to HI by the molecular hydrogen present in the reaction media.

Overall, the economic sense of such processes must be seriously evaluated not only in terms of viability of these multistep approaches but also in terms of product prices: added-value molecules such as acrolein, propanal, and 1-propanol are transformed into propylene which has a lower price. For instance, compared on molar bases, the price of 1-propanol is around 20 ctUSD/mol while that of propylene is still approximately 7 ctUSD/mol.

1.2.4
C_4 Alcohols to Olefins

A number of C_4 alcohols/diols can be used to produce olefins. Among alcohols, *n*-butanol is a valuable choice. It can be synthesized through both chemical and biotechnological processes. Among chemical processes, propene hydroformylation (also called *oxo-synthesis*) [60] is the preferred route today at an industrial level, and theoretically, it could also be applied to produce biobutanol from biopropylene and biosyngas. However, a less energy-intensive and more direct synthesis would be highly desirable, especially for small-size/on-purpose plants. Until the mid-1950s, *n*-butanol was synthesized by acetaldehyde aldol condensation followed by hydrogenation; thus this process could be used to produce biobutanol using the dehydrogenation of bioethanol as a route to bioacetaldehyde.

More conveniently, bioethanol could be directly transformed into *n*-butanol through the Guerbet process. The latter is an established chemical route mainly followed for the production of highly branched and saturated alcohols through the condensation of two primary alcohols. The higher alcohols produced are important intermediates for the synthesis of surfactants [61]; however, of great interest would also be the formation of short-chain C_4 alcohols, particularly *n*-butanol, to be used both as fuels and as chemical building blocks. Several catalysts have been reported in literature for the conversion of ethanol into butanol, and the most representative ones are listed in Table 1.3. MgO is generally recognized as a reference material for this reaction, mainly due to its simplicity and reproducibility; selectivity of about 30–35% can be obtained in the gas phase at low ethanol conversions (~10%) [18]. However, more efficient catalysts have been reported in the last decade; selectivities up to 75% and 85% can be reached for gas-phase [66] and liquid-phase processes [69], respectively. In the first case, basic oxides (e.g., MgO, hydroxyapatites, or hydrotalcites) may be used as catalysts to carry out the reaction, continuously, at atmospheric pressure and temperatures of 200–400 °C. Conversely, for the liquid process, Ru complexes make it possible to obtain much higher ethanol conversions (up to 45% vs. ~15% for the gas-phase route) with high *n*-butanol selectivities; the process, however, is discontinuous and requires long reaction times to achieve high conversions.

As previously discussed for the synthesis of BD from ethanol, the mechanism for the formation of the C–C bond in the gas-phase process is still controversial.

Table 1.3 Catalytic systems for the conversion of ethanol into *n*-butanol.

Catalyst[a)]	*T* (°C)	EtOH conversion (%)	1-BuOH selectivity (%)	References
MgO	250	10	35	[18]
CuMgAlO (5:1:3)	260	9	80	[62]
MgAlO (3:1)	350	35	37	[63]
MgAlO (3:1)	350	23	17	[64]
MgFeO (2.9:1.1)	380	60	16	[65]
CaPO (1.64:1)	300	15	75	[66]
SrPO (1.7:1)	300	11	86	[67]
20.7% Ni/Al_2O_3[b)]	250	25	80	[68]
trans-$[RuCl_2(3)_2]$[b)]	150	45	85	[69]

a) The atomic ratio between metals is shown in parenthesis.
b) Liquid-phase process.

Although aldol condensation is still generally mentioned as the key step, it is worthy of note that the best-performing catalyst, that is, hydroxyapatite, has significantly fewer dehydrogenation features than transition metal oxide-based catalysts. Indeed, Kozlowski and Davis [61] have stressed that the dehydrogenation rate of copper-containing material published by Gines and Iglesia [70] was about 370 times greater than that of hydroxyapatite with a Ca/P ratio of 1.67 [34]. Considering the fundamental role of acetaldehyde formation in the aldol scheme and the lower overall *n*-butanol formation on the Cu-containing oxide than on hydroxyapatite, it seems likely that acetaldehyde (and therefore aldol condensation) is not the key step for producing *n*-butanol from ethanol in the gas phase. This is a remarkable observation that supports the recent hypothesis that two molecules of ethanol react together directly through a carbanionic intermediate to form *n*-butanol as a primary product [18].

n-Butanol, as well as other valuable alcohols and diols, can also be produced directly by fermentation processes. One of the oldest technologies is the acetone/butanol/ethanol (ABE) process in which these molecules are produced using genetically modified bacteria for the fermentation of carbohydrates; however, due to low productivity and difficult product separation, its economy is not sustainable when compared to petrochemical routes, at least with current oil prices. The biological synthesis of C_4 diols seems to be more promising, and a number of companies such as LanzaTech, Versalis/Genomatica, Genecor/Goodyear, and Global Bioenergies are developing their own biochemical routes to BDOs [1].

Whether the bio-C_4 alcohols/diols are obtained from direct or indirect (bio-) chemical routes, they can be either used as such or finally dehydrated into olefins. *n*-Butanol can be easily and efficiently dehydrated into 1-butene, with a mechanism analogous to ethanol dehydration, by using low-to-medium-strength acid catalysts; however, it can also be upgraded to nonlinear C_4 olefins as a one-pot reaction. To do so, strong acid catalysts and higher reaction temperatures must be used so as to combine the dehydration step with skeletal isomerization. Zeolite catalysts such as Theta-1 and ZSM-23 gave high yields of isobutene (~60%)

and stable catalytic behavior [131]. In 2009, BP also filed a patent to skeletally isomerize *n*-butanol into linear and nonlinear hydrocarbons (mainly isobutene) on acid zeolite catalysts with unidirectional, nonintersecting channels [71].

The double dehydration of several BDOs can lead to the formation of BD, although its formation is often associated with several monodehydrated by-products, particularly unsaturated alcohols (alkenols). Among BDOs, 1,3-BDO is one of the most promising intermediates for the production of BD; for instance, on $SiO_2-Al_2O_3$ catalysts, BD is produced directly with selectivity up to 36%, but unsaturated alcohols are also major products (e.g., 3-buten-1-ol) [72, 73]. BD yields as high as 90% were claimed to be obtained in old patent literature through 1,3-BDO dehydration on doped phosphoric acid heterogeneous catalysts [24].

Theoretically, 1,4-BDO dehydration would be another option for producing BD; however, the monodehydration forms tetrahydrofuran (THF) – as the main reaction product, and the following deoxygenation of furan is very difficult. Nevertheless, the reaction is possible, and recycling the unreacted THF to the reactor has made it possible to obtain BD yields up to 95% [74]. Recent results show an easier and more promising BD formation if 2,3-BDO is used as the starting reagent [75], which can be produced via glucose fermentation. Scandium oxide (Sc_2O_3) calcined at high temperature (800 °C) showed BD yields as high as 88% at 411 °C, when H_2 was used as the carrier gas. Also, if a first catalytic bed of Sc_2O_3 was coupled with a consecutive bed of alumina (in the same reactor), a stable 94% BD selectivity was obtained; indeed, the intermediately formed 1-buten-3-ol was more efficiently dehydrated into BD. These results are a considerable step forward toward the direct double dehydration of 2,3-BDO, since in previous literature mainly monodehydration occurred and formed unsaturated alcohols or MEK [76, 77]. An alternative way to produce BD from BDOs is a two-step approach in which each hydroxyl group is eliminated using two different catalysts. For instance, various examples have been reported by Sato *et al.* [76] (and references therein) for the selective dehydration of BDOs into unsaturated alcohols. A summary of the best results obtained is shown in Table 1.4.

Table 1.4 Catalytic systems used for (butane)diol(s) dehydration.

Catalyst	BDO[a]	*T* (°C)	Conversion (%)	Selectivity (%)[a]	References
$SiO_2-Al_2O_3$	1.3	250	74	36 (BD); 28 (3B1OL)	[73]
Supported H_3PO_4	1.4	380	100	95 (BD)[b]	[74]
Sc_2O_3	2.3	411	100	88 (BD)	[75]
CeO_2	1.3	325	43.9	58.1 (3B2OL); 41.1 (2B1OL)	[78]
CeO_2	1.3	325	82.2	60.0 (3B2OL); 36.5 (2B1OL)	[79]
CeO_2	1.4	400	40.5	69.4 (3B1OL); 8.5 (2B1OL)	[80]
ZrO_2	1.4	350	86.4	48.0 (3B1OL); 44.9 (THF)	[81]
1.5 $Na-ZrO_2$	1.4	325	18.7	71.8 (3B1OL); 20.8 (THF)	[82]

a) BDO: butanediol; 3B2OL: 3-buten-2-ol; 2B1OL: 2-buten-1-ol; 3B1OL: 3-buten-1-ol; and THF: tetrahydrofuran.

b) Recycling the intermediately formed THF to the reactor.

Considering the efficient production of unsaturated alcohols from BDOs, the former can be dehydrated into BD on silica- and/or alumina-based catalysts.

Lastly, some patents have also been recently filed concerning the dehydration of BDOs for the production of olefins, particularly BD [83–85]. Rare earth (mixed) phosphate and hydroxyapatite–alumina catalysts are claimed to lead to high BD yields and diol conversions as well as to long-term stability.

1.3 Alternative Routes to Bio-Olefins

1.3.1 Catalytic Cracking

The catalytic cracking of low-value fats, greases, oils, and other renewable sources is one of the most potentially useful methods to obtain olefins from bio-oil upgrading. There are two types of process: Fluid catalytic cracking (FCC) and steam cracking. The former has already been studied for the transformation of vegetable oils (or their blend with vacuum gas oil) into a mixture of gasoline and cracking gas containing propylene [86]. However, the main products in FCC are liquid fuels. On the other hand, steam cracking might be more suitable for obtaining olefins. This process involves two separate steps: (i) Hydrodeoxygenation (HDO) to remove oxygen content from triglycerides and fatty acids in the feedstock – to obtain hydrocarbon chains in the diesel range and renewable naphtha and (ii) cracking of the naphtha to obtain olefins and some gasoline. The advantage of this approach is that existing conversion and production units can be used, thus eliminating the cost of building new "on-purpose" facilities [87]. A simplified diagram of the integrated process is presented in Figure 1.5.

Step 1: The first step of the process (HDO) is usually carried out using traditional hydrotreating catalysts such as (Ni)Co–MoS_2/Al_2O_3. This process includes the treatment of the feedstock at moderate temperatures (280–400 °C) and high hydrogen pressure (20–300 bar): this hydrogen should be preferentially produced with renewable energy sources (by steam reforming of ethanol, e.g., but preferably from water thermolysis and photolysis using solar energy) [88].

The problem of using transition metal sulfided catalysts for the HDO of bio-oils is that they might deactivate during a prolonged operation time due to sulfur stripping and surface oxidation caused by the low content of this heteroatom in the biofeedstock compared to fossil fuel oils. One suggested alternative for avoiding this problem is the co-feeding of H_2S to the system in order to regenerate the sulfide sites. The use of H_2S, however, also has some drawbacks, such as the formation of sulfur-containing products, and also the fact that H_2S could block the adsorption over active sites [89].

Alternative catalytic systems for this first step include noble metals like Pd, Pt, Ru, Rh, or even Ni and Co supported on C, ZrO_2, SiO_2, MgO, or zeolites (Al_2O_3 has been shown to catalyze coke deposition). Transition metal carbides, nitrides,

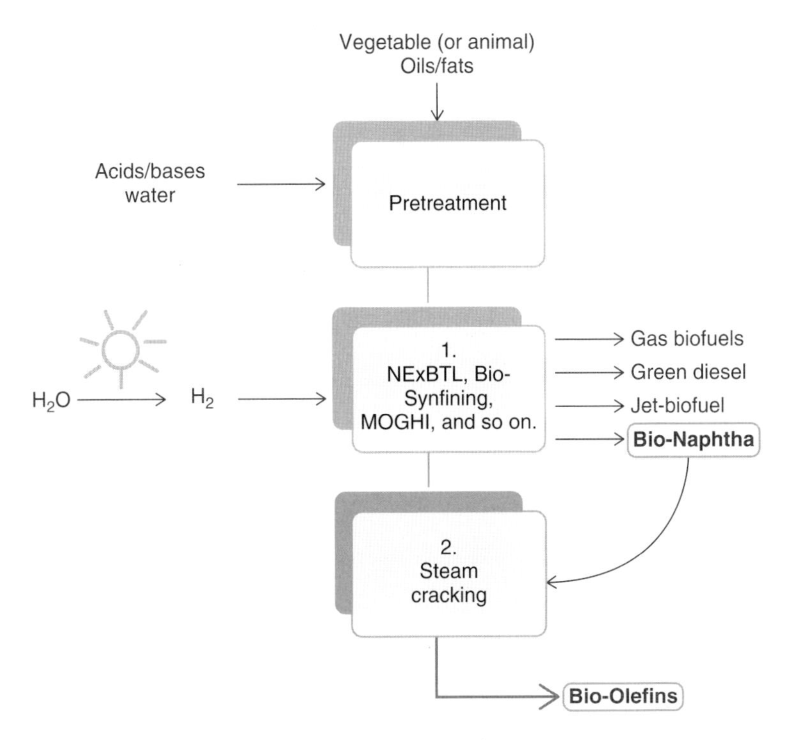

Figure 1.5 Two-step process for biomass/oil upgrading.

and phosphides have also shown promising performances in the HDO of bio-oils (or model compounds) [90]. All these catalytic systems feature advantages and challenges that require further investigation in order to develop more efficient processes. For instance, the know-how in sulfided catalyst synthesis and commercialization held by many large industries makes it worthwhile to continue investigating these materials. On the other hand, noble metal catalysts have the ability to activate H_2 under low-pressure conditions and, moreover, can operate in acidic or aqueous environments.

With regard to the mechanism involved, HDO of biomass entails a complex reaction network that includes decarbonylation, decarboxylation, hydrocracking, hydrogenolysis, and hydrogenation. When using transition metal sulfides, the pathway suggested resembles that for conventional oil HDO; oxygen from the biomolecule adsorbs on a vacancy of the MoS_2 matrix. Simultaneously, the H_2 from the feed dissociatively adsorbs on the catalyst surface, forming S–H (and Mo–H) species. The addition of a proton to the adsorbed oxygenated molecule and the elimination of water produce the deoxygenated product [89]. In this type of catalyst, Mo serves as an active element, while Co and Ni act as promoters [91].

The HDO is a process with high carbon efficiency and therefore a high production potential [88]. There are industrial processes such as Bio-Synfining, property of Syntroleum Corporation, currently available which are able to transform vegetable or animal oils, fats, and greases into renewable synthetic fuels that include diesel, naphtha, and propane. The renewable distillate produced from a plant could be separated into its components and then be used for more profitable applications such as olefin production. In fact, they have already patented the specific process aimed at maximizing naphtha production [92].

Another company having a similar technology is Neste Oil. Branded as NExBTL, this process uses tallow to produce fuels, mainly in the diesel range, but it produces also jet fuel, propane, and renewable naphtha which, as in the former case, could be hydrocracked to obtain olefins. Further competitors/producers of biofuels (including bio-naphtha) from biomass (or waste) include Total Petrochemicals [93], Biochemtex (MOGHI process), Eni/Honeywell UOP (Ecofining), Solena fuels (GreenSky), Rentech, the Energy and Environmental Research Center (EERC) in collaboration with the refiner Tesoro, Finland's UPM, and the Renewable Energy Group (REG) that recently acquired Syntroleum. On the other hand, Sasol and Shell have developed biomethane routes to obtain bionaphtha.

The three main reactions that occur during the first step are:

HDO:

$$(C_nCOO)_3C_3H_5 \xrightarrow[\text{Catalyst}]{H_2} nC_{n+1}H_{2n+4} + 6H_2O + C_3H_8$$

Decarboxylation:

$$(C_nCOO)_3C_3H_5 \xrightarrow[\text{Catalyst}]{H_2} nC_nH_{2n+2} + 3CO_2 + C_3H_8$$

$$CO_2 + H_2 \xrightarrow[\text{Catalyst}]{} CO + H_2O \quad \text{Reverse WGS}$$

Isomerization:

$$n\text{-}C_{n+1}H_{2n+4} + n\text{-}C_nH_{2n+2} \xrightarrow[\text{Catalyst}]{} i\text{-}C_{n+1}H_{2n+4} + i\text{-}C_nH_{2n+2}$$

During this step, both the reaction temperature and the type of catalyst determine the products distribution. A severe hydrotreatment would lead to a high production of naphtha (C_5–C_{10}), whereas mild hydrotreatment conditions promote the production of green diesel.

Step 2: Once the renewable naphtha is obtained, the steam cracking step yields olefins and other compounds such as hydrogen, methane, ethylene, and aromatics (the latter in a lower content since the bionaphtha is expected to be highly paraffinic). However in this case, also, the exact composition of the outlet stream depends on several factors. Generally, propylene production is higher when using mild reaction conditions, whereas the yield to ethylene and aromatics increases at higher temperatures. In general, the process is carried out at atmospheric pressure and at temperatures of around 800 °C with approximately 0.2–1.0 kg of steam per kilogram of feedstock.

Studies on a pilot plant scale of the complete process (HDO and steam cracking) have already been performed using wood-derived tall oil [94] or a bio-oil blend (mainly fat and grease from prepared foods) [95], achieving olefin yields of over 50%; these yields were higher than those obtained by cracking fossil-based naphtha under similar conditions. Moreover, other advantages were the lower optimum temperature needed to maximize light olefins (entailing less energy input and fewer by-products such as pyrolysis fuel and gasoline) and lesser coke formation that occurs in longer length runs.

There are also related approaches such as the direct cracking of the bio-oil. For instance, Gong *et al.* [96] proposed the use of a modified La–HZSM-5 which, under optimized conditions (600 °C, 6%La, and weight hourly space velocity (WHSV) of $0.4\,h^{-1}$), produced $0.28\,kg_{olefins}\,kg_{bio\text{-}oil}{}^{-1}$. The same group also studied the production of olefins by mixing the catalyst (La–HZSM-5) and the dry biomass directly, thus obtaining $0.12\,kg_{olefins}\,kg_{dry\ biomass}{}^{-1}$ when using sugarcane [97].

In order to obtain higher yields of olefins by HDO + steam cracking, the key points which need further research and development are:

- Limiting the coke formation during the HDO step: the high amount of cyclic (and aromatic) products formed affects the catalyst lifetime considerably and make extremely high H_2 pressure necessary to attain better results. Up to now, lifetimes of much more than 200 h have not been achieved with any current catalyst due to carbon deposition [88].
- The control of the reaction heat in the HDO: this is extremely important since the highly exothermic nature of the reactions involved may cause unwanted side reactions, such as cracking, polymerization, ketonization, cyclization, aromatization, and coking of the catalyst.
- For steam cracking, there is a higher formation of CO_x, probably due to the absence of sulfur on the feedstock which, in the case of fossil fuel-based feedstock, is present and interacts with Ni in reactor walls, avoiding oxidation. Thus further studies of the interaction of S-free feedstock with industrial catalysts and the reactor are necessary.

In conclusion, the understanding of these mechanisms and the subsequent optimization of operating conditions and catalysts are still needed for the HDO and steam reforming of renewables, in order to bring them to an industrial-scale usage. Nevertheless, if this can be achieved, CO_2-neutral fuels can be produced via biomass transformation in a sustainable manner.

1.3.2 Metathesis

The metathesis reaction involves the exchange of a bond (or bonds) between similar species so that the bonding associations in products are similar to those in reactants. In the case of olefins, it can be used to produce propylene from more abundant (or cheaper) alkenes such as ethylene and butenes (see Scheme 1.5).

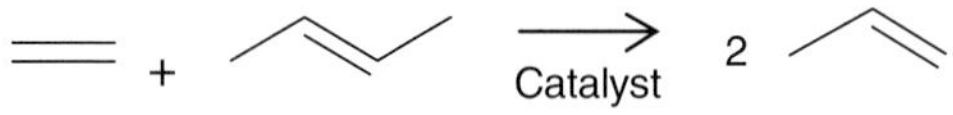

Scheme 1.5 Ethylene and 2-butene metathesis to produce propylene.

This kind of process is not new, but its "greener" version is still being researched. There are two approaches for carrying out such a process: (i) dimerization of the bioethylene produced by the dehydration of bioethanol and the reaction of the obtained butenes with the remaining ethylene and (ii) directly reacting biobutene (produced from biobutanol dehydration) and bioethylene.

The dimerization of ethylene is a technology that is already available. It is generally carried out under mild to moderate conditions (0–100 °C and 50–300 psig) [98]. Common metals in dimerization catalysts are Ti and Ni. Axens' *AlphaButol* is a commercial example of such a method.

Figure 1.6 shows a simplified flow diagram of the second type of process (similar to Lummus' Olefin Conversion Technology (*OCT*)). Here fresh and recycled biobutene streams are mixed with bioethylene and its recycled part. Reagents are sent first into a guard bed (GB) to remove all traces of impurities. Afterwards, the feed is heated before entering the reactor (R1), where the metathesis reaction occurs. Lastly comes the separation step (S) which is generally performed by distillation.

Catalytic systems for this process can be either homogeneous or heterogeneous. The homogeneous ones are usually in the form of organometallic complexes. However, heterogeneous systems are preferred due to their easier handling and low separation problems. Typical solid catalysts for olefin metathesis consist of a group VI A or VII A metal oxide (such as W, Mo, or Re oxides) supported on silica or alumina. Tungsten-based catalysts work at relative higher reaction temperature (>300 °C) to obtain the equilibrium conversion [99–103]. The OCT process by

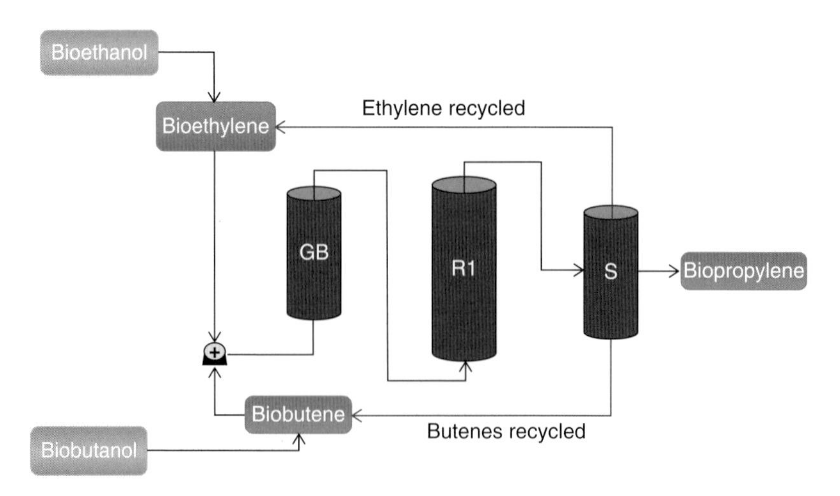

Figure 1.6 Bio-olefin metathesis flow diagram. GB: guard bed, R: isomerization/metathesis reactor, and S: separation unit.

ABB Lummus Global, for instance, is based on WO_3/SiO_2 as the metathesis catalyst and takes place in a fixed-bed reactor at T >260 °C and 30–35 bar. The conversion of butene is above 60% per pass, and the selectivity to propene is >90% [104]. However, the modification of the support (using zeolites) and thus the acidity and degree of reduction of W have been shown to promote the metathesis activity at lower reaction temperatures (~180 °C) for WO_3 catalysts [105]. Recently, W directly incorporated into mesoporous frameworks has shown a potential for this reaction, by making the isolation of the active sites possible [106, 107].

On the other hand, Re oxide catalysts operate at milder conditions and usually have a selectivity close to 100%; their drawback, however, is the fast deactivation (even by H_2O traces), which makes necessary an intensive feed pretreatment and constant catalyst regeneration [108]. Re_2O_7/Al_2O_3 was the catalyst chosen for the *Meta-4* process for propylene production which was developed by the Institut Français du Pétrole (IFP) and the Chinese Petroleum Corporation (Taiwan). However, due to the high cost of the catalyst and the deactivation issue, today this process is not being commercialized. A further disadvantage of Re-based catalysts is linked to the volatility of Re_2O_7 – which sublimates at 262 °C – which may lead to the loss of rhenium during the synthesis of the catalyst (calcination).

As for Mo oxide catalysts, they are recently receiving much attention [109–112], since they offer a good compromise between activity and robustness, while they are already used at an industrial scale in the Shell higher olefin process (*SHOP*) for the production of linear alpha olefins via ethylene oligomerization and olefin metathesis [113]. One advantage of the metathesis process is that it does not produce propane, thus making it unnecessary to install a P–P splitter (which is typically used in steam crackers and FCC units) and consequently reducing the investment cost.

Regarding the feed, it is generally observed that the reactions of isobutene or 1-butene with ethylene are nonproductive. Therefore it is common to use a double-bond isomerization catalyst (Mg, Ca, or Ba oxide) directly mixed with the metathesis catalyst in order to obtain the more reactive 2-butene [114]. However, if the available (bio-) source is rich in isobutene, the skeletal isomerization to obtain *n*-butene might require a separate step such as catalytic distillation (*CDIsis* technology, also licensed by ABB Lummus). The ethylene-to-butene ratio fed to the reactor must be controlled in order to minimize C_{5+} olefins and other by-products. Typical 2-butene conversions range between 60% and 75%, with a selectivity to propylene around 90% [104].

Another option that consists of the use of 1-butene (coming from biobutanol dehydration) alone as a feedstock is being investigated in order to overcome the dependence on the more costly ethylene [115–117]. This process is performed in the presence of an acidic cocatalyst via the isomerization to 2-butenes followed by cross metathesis between 1-butene and 2-butenes resulting in the formation of propylene and 2-pentenes [118]. The obtained 2-pentene can react further with 1-butene to produce more propylene and 3-hexene, and the cycle may continue until the C_4 conversion achieves around 65% and a propylene yield of 30% per pass [119]. Table 1.5 summarizes the most recent results in heterogeneous systems dealing with the metathesis of ethylene and 2-butene. This general picture shows that there is still margin for improvement in this type of process, since alternative systems (different active metals, supports, feeds) and studies with real biofeedstock have received little attention.

Concerning the metathesis mechanism, the generally accepted one resembles that proposed by Chauvin, which is based on results for homogeneous systems [127]. It includes the initial formation of a metal-carbene species and its further propagation by a cyclic reaction with olefins (Scheme 1.6).

The initial carbene formation is believed to occur through a pseudo-Wittig reaction of the metal-oxo precursor with an olefin (Scheme 1.7). In the case of crossed metathesis of ethylene and 1-butene, the precursor is believed to react first with butene [128].

Many factors are reported to affect the structure–activity relationship, such as the degree of polymerization of surface MO_x species and the acidic properties of catalysts. For instance, for $Mo/Al_2O_3-SiO_2$, Hahn *et al.* [129] suggested that polymerized octahedral MoO_x entities play a key role in directing the reactions toward propylene, since they hinder the isomerization of 2-butene to 1-butene, thus avoiding the further nonselective metathesis as compared to tetrahedral sites in which the production of 1-butene and its isomerization are faster than the reaction with ethylene. A recent review of Lwin and Wachs offers a comprehensive summary of the literature on this topic. They conclude, however, that there is still a need for more direct observation measurements (*in situ* and operando) which are able to confirm the catalytic active sites and intermediates proposed [130].

Today, in a related approach, the one-pot process from ethanol to propylene is being investigated as well and – while in the case of oxides such as Sc/In_2O_3 and $Y_2O_3-CeO_2$, the pathway goes through the formation of acetaldehyde, acetone, and lastly propene – in the case of Ni/MCM-41 catalyst, according to them, the key step is the metathesis of the produced ethylene. In fact, ethylene is first dimerized on Ni sites to 1-butene, while the latter can be isomerized to 2-butene on acidic sites. Then the formed 2-butene is supposed to react via cross metathesis with unconverted ethylene. The mechanism claimed implies that Ni is able to perform a metathesis reaction by the formation of a nickel carbene intermediate. An exclusively acidic mechanism was ruled out because of the large pore size of MCM-41 [52, 123].

The metathesis reaction also finds application at an industrial level as a way to enhance the more abundant olefins. An example can be seen in the partnership of the companies Elevance and Versalis for the scale-up of a metathesis reaction of natural oil esters and olefins, particularly ethylene. Other important commercial processes include the SHOP for producing linear higher olefins from ethylene. Furthermore, the production of hexane and neohexene and many other polymers, agrochemicals, pheromones, and fragrances have reached the commercialization stage using the relatively new metathesis chemistry [119]. Therefore, the metathesis process, even if still under development, may become an important source of propylene.

Table 1.5 Recent results for propylene production by the metathesis of ethylene with butenes.

Catalyst	Conditions[a)]	1,3-Butadiene yield (or selectivity) (%)	Reference (year)
Molybdenum based (Mo)			
$MoO_3/SiO_2-Al_2O_3$ (one-pot aerosol)	Ethylene/2-butene = 1/1 $T = 40\,°C$	Propylene selectivity above 99%	[109] (2012)
$MoO_3/Al_2O_3-C+HNO_3$	Ethylene/2-butene = 1/1 $WHSV_{(C_2H_4+C_4H_{82})} = 1\,h^{-1}$ $T = 130\,°C$ and $P = 1.0\,MPa$	2-Butene conversion 69% Propylene yield (with respect to ethylene) 77.4%	[111] (2014)
3.0Mo/(MCM-22-30%Al_2O_3)	Ethylene/2-butene = 4/1 $WHSV_{(C_2H_4+C_4H_{82})} = 0.5-2\,h^{-1}$ $T = 125\,°C$ and $P = 1.0\,MPa$	2-Butene conversion 86% Propylene selectivity 95%	[112] (2010)
Tungsten based (W)			
WO_3/SiO_2 calcined at 550 °C	Ethylene/2-butene = 2/1 $T = 400\,°C$ and $P = 0.1\,MPa$	2-Butene conversion 86% Propylene selectivity 81%	[100] (2012)
WH/Al_2O_3	Ethylene/2-butene = 0.35/1 Total flow rate = 20 ml min^{-1} $T = 150\,°C$ and $P = 1\,bar$	Total C conversion 43% 142.8 $mmol_{c3=g.cat\ h-1}$	[99] (2013)
$WO_3/SiO_2-(nano)TiO_2$	Ethylene/2-butene = 2/1 $WHSV = 2.34\,h^{-1}$ $T = 400\,°C$ and $P = 0.1\,MPa$	2-Butene conversion ~83% Propylene selectivity ~67%	[103] (2013)
$WO_3/SiO_2-Al_2O_3$ (nonhydrolytic sol–gel synthesis)	Ethylene/2-butene = 2/1 Total flow rate 20 $cm^3\,min^{-1}$ $T = 450\,°C$ and $P = 0.1\,MPa$	2-Butene conversion 92% Propylene selectivity 60%	[101] (2014)

(continued overleaf)

Table 1.5 *(Continued)*

Catalyst	Conditions[a)]	1,3-Butadiene yield (or selectivity) (%)	Reference (year)
$WO_3/SiO_2-Al_2O_3$ (one-pot aerosol)	Ethylene/2-butene = 1/1 $T = 250\,°C$	($65\ mmol\ g^{-1}\ h^{-1}$) Propylene yield 39%	[102] (2014)
WO_3/SiO_2 (NaOH modified)	Ethylene/2-butene = 2/1 $T = 400\,°C$ and $P = 0.1\ MPa$	2-Butene conversion 81% Propylene selectivity 77%	[120] (2014)
W-FDU-12 and MgO (doped vs. supported)	Ethylene/1-butene = 2 $T = 450\,°C$ and $P = 0.1\ MPa$ $WHSV = 0.9\ h^{-1}$	1-Butene conversion: 79% Propylene selectivity 90%	[106] (2015)
W-KIT-6 (+MgO/Al_2O_3) One-pot hydrothermal synthesis (Si/W = 25)	Ethylene/1-butene = 2 $T = 350\,°C$ and $P = 0.1\ MPa$ $WHSV = 0.9\ h^{-1}$	1-Butene conversion: 52% Propylene selectivity 74%	[107] (2013)
Rhenium based (Re)			
$Re/SiO_2-Al_2O_3$ (nonhydrolytic sol–gel synthesis)	Ethylene/2-butene = 1/1 $T = 40\,°C$ and atmospheric pressure	2-Butene conversion 60% Propylene selectivity 77%	[108] (2013)
10%Re/Al_2O_3 (high-throughput preparation and screening)	Ethylene/2-butene = 1/1 $T = 75\,°C$ and atmospheric pressure WHSV $1.3\ h^{-1}$	Total C conversion: 61% Propylene selectivity 87% Propylene yield around 55–67%	[121] (2008)

Alternative systems			
WO_3/MTS-9 (titanium silica sieve)	2-Butene $T = 320\,°C$ and $P = 0.8$ MPa, WHSV $= 6.4\,h^{-1}$	2-Butene conversion around 42% Propylene selectivity 45–50%	[122] (2011)
Ni/MCM-41 The template-ion exchange (TIE) method	Ethylene 10% $T = 400\,°C$ and $P = 0.8$ MPa, WHSV $= 6.4\,h^{-1}$	Ethylene conversion 57% Propylene selectivity 24%	[123] (2008)
Ni/Al-MCM-41	Ethylene 10% $T = 375–450\,°C$ and $P = 0.8$ MPa, WHSV $= 1.4\,h^{-1}$	Maximum propylene selectivity 18–22% Maximum propylene yield 26%	[124] (2013)
$10WO_3/Al_2O_3$-70HY	1-Butene $T = 180\,°C$ and $P = 0.1$ MPa, WHSV $= 1.5\,h^{-1}$	1-Butene conversion: 95% Propylene yield 21%	[117] (2009)
$Re/SiO_2-Al_2O_3$	Ethylene/1-pentene = 3 GHSV of $500\,h^{-1}$ T = 35 °C	Propylene yield of 88% 2-Pentene conversion 100% 455 min on stream	[125] (2012)
MoO_x/γ-Al_2O_3 one-dimensional support, synthesized via oleylamine-assisted hydrothermal method	1-Butene $T = 120\,°C$ and $P = 0.1$ MPa, WHSV $= 2.4\,h^{-1}$	1-Butene conversion: 82% Propylene yield 30%	[115] (2014)
WO_3/SiO_2	Ethylene/decene = 15 $T = 360–400\,°C$	Decene conversion >90% Propylene selectivity 70–75%	[126] (2013)

a) Molar reagent ratio, WHSV: weight hourly space velocity (or GHSV: gas hourly space velocity), T: temperature, and P: pressure.

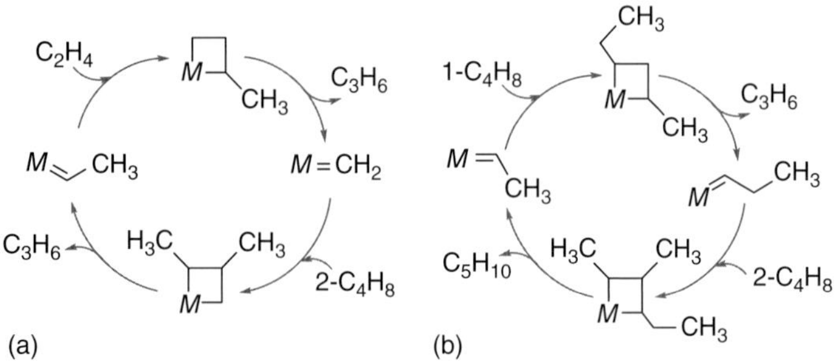

(a) (b)

Scheme 1.6 Mechanism of the metathesis of ethylene with 2-butene (a) and 1-butene with 2-butene (b).

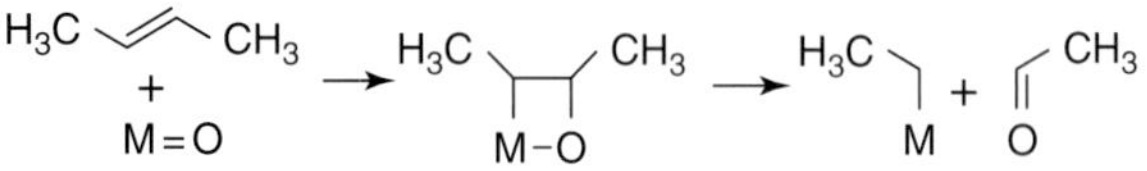

Scheme 1.7 Formation of the initial M-carbene species.

1.4 Conclusions

The field of bio-olefin production is growing extremely fast. Bioalcohol dehydration or chain-length increase routes (i.e. Lebedev and Guerbet reactions) are already practicable at an industrial level. Indeed, excellent yields and stable catalytic behavior have been demonstrated. Last but not least, the practicability of these processes is more likely compared to several other bioased productions proposed in literatre, showing better economic return. The most striking example is the synthesis of biobutadiene; several joint ventures and industrial alliances have been signed during the latest 10 years, and a possible future shortage of this olefin, coupled with the incentives deriving from the "green label" assigned to tires and other products obtained from biorubber, might lead to a quick construction of new plants in the next years.

References

1. Lanzafame, P., Centi, G., and Perathoner, S. (2014) *Catal. Today*, **234**, 2–12.
2. Geilen, F.M.A., Stochniol, G., Peitz, S., and Schulte-Koerne, E. (2013) Butenes, in *Ullmann's Encyclopedia of Industrial Chemistry*, Wiley-VCH Verlag GmbH & Co. KGaA.
3. Lanzafame, P., Centi, G., and Perathoner, S. (2014) *Chem. Soc. Rev.*, **43**, 7562–7580.
4. Kosaric, N., Duvnjak, Z., Farkas, A., Sahm, H., Bringer-Meyer, S., Goebel, O., and Mayer, D. (2000) Ethanol, in *Ullmann's Encyclopedia of Industrial Chemistry*, Wiley-VCH Verlag GmbH & Co. KGaA.
5. Sheldon, R.A. (2014) *Green Chem.*, **16**, 950–963.
6. Fan, D., Dai, D.-J., and Wu, H.-S. (2012) *Materials (Basel)*, **6**, 101–115.

7. Althoff, J., Biesheuvel, K., De Kok, A., Pelt, H., Ruitenbeek, M., Spork, G., Tange, J., and Wevers, R. (2013) *ChemSusChem*, **6**, 1625–1630.
8. Zhang, M. and Yu, Y. (2013) *Ind. Eng. Chem. Res.*, **52**, 9505–9514.
9. Minoux, D., Nesterenko, N., Vermeiren, W., and Sander Van, D. (2009) WO Patent 2009098268A1, Dehydration of alcohols in the presence of an inert component. Total Petrochemicals Research Feluy.
10. Fukumoto, M. and Kimura, A. (2013) Process for the manufacture of ethylene by dehydration of ethanol. EP Patent 2594546A1, Solvay SA.
11. Patrick Gracey, B. and Partington, S.R. (2009) Process for preparing ethene. WO Patent 2009050433A1, BP plc.
12. Phung, T.K. and Busca, G. (2015) *Chem. Eng. J.*, **272**, 92–101.
13. Kang, M., DeWilde, J.F., and Bhan, A. (2015) *ACS Catal.*, **5**, 602–612.
14. Hou, T., Zhang, S., Chen, Y., Wang, D., and Cai, W. (2015) *Renew. Sustain. Energy Rev.*, **44**, 132–148.
15. Christiansen, M.A., Mpourmpakis, G., and Vlachos, D.G. (2015) *J. Catal.*, **323**, 121–131.
16. Potter, M.E., Cholerton, M.E., Kezina, J., Bounds, R., Carravetta, M., Manzoli, M., Gianotti, E., Lefenfeld, M., and Raja, R. (2014) *ACS Catal.*, **4**, 4161–4169.
17. DeWilde, J.F., Czopinski, C.J., and Bhan, A. (2014) *ACS Catal.*, **4**, 4425–4433.
18. Chieregato, A., Velasquez Ochoa, J., Bandinelli, C., Fornasari, G., Cavani, F., and Mella, M. (2015) *ChemSusChem*, **8**, 377–388.
19. Angelici, C., Weckhuysen, B.M., and Bruijnincx, P. (2013) *ChemSusChem*, **6**, 1595–1614.
20. Marsh, J.L., Murray, I.L., and Smith, J.S.P. (1946) Process for making butadiene. US Patent 2403742 A, Carbide & Carbon Chem Corp, pp. 203–209.
21. Toussaint, W.J., Dunn, J.T., and Jackson, D.R. (1947) *Ind. Eng. Chem.*, **39**, 120–125.
22. Toussaint, W.J. and Dunn, J.T. (1947) Process for making diolefins. Carbide & Carbon Chem Corp. US Patent 2421361.
23. Murray, I.L. (1941) Recovery method for cyclic vapor phase reaction products. US Patent 2249847A, Carbide & Carbon Chem Corp.
24. Makshina, E.V., Dusselier, M., Janssens, W., Degrève, J., Jacobs, P.A., and Sels, B.F. (2014) *Chem. Soc. Rev.*, 7917–7953.
25. Natta, G. and Rigamonti, G. (1947) *Chim. Ind. (Milan)*, **29**, 239–244.
26. László, I.R., Falkay, B., and Hegyessy, L. (1954) *Magy. Kem. Foly.*, **60**, 54–74.
27. Kovarik, B. (1960) *Collect. Czech. Chem. Commun.*, **26**, 1918–1924.
28. Bhattacharyya, S.K. and Avasthi, B.N. (1963) *Ind. Eng. Chem. Process Des. Dev.*, **2**, 45–51.
29. Berak, J.M., Guczalski, R., and Wojcik, J. (1966) *Acta Chim. Acad. Sci. Hung.*, **50**, 163–166.
30. Niiyama, H., Morii, S., and Echigoya, E. (1972) *Bull. Chem. Soc. Jpn.*, **45**, 655–659.
31. Ohnishi, R., Akimoto, T., and Tanabe, K. (1985) *J. Chem. Soc., Chem. Commun.*, 1613–1614.
32. Kvisle, S., Aguero, A., and Sneeden, R.P.A. (1988) *Appl. Catal.*, **43**, 117–131.
33. Kitayama, Y., Satoh, M., and Kodama, T. (1996) *Catal. Lett.*, **36**, 95–97.
34. Tsuchida, T., Kubo, J., Yoshioka, T., Sakuma, S., Takeguchi, T., and Ueda, W. (2008) *J. Catal.*, **259**, 183–189.
35. Jones, M.D., Keir, C.G., Di Iulio, C., Robertson, R.A.M., Williams, C.V., and Apperley, D.C. (2011) *Catal. Sci. Technol.*, **1**, 267–272.
36. Makshina, E.V.V., Janssens, W., Sels, B.F.F., and Jacobs, P.A. (2012) *Catal. Today*, **198**, 338–344.
37. Lewandowski, M., Babu, G.S., Vezzoli, M., Jones, M.D., Owen, R.E., Mattia, D., Plucinski, P., Mikolajska, E., Ochenduszko, A., and Apperley, D.C. (2014) *Catal. Commun.*, **49**, 25–28.
38. Ezinkwo, G.O., Tretjakov, V.F., Talyshinky, R.M., Ilolov, A.M., and Mutombo, T.A. (2014) *Catal. Commun.*, **43**, 207–212.
39. Angelici, C., Velthoen, M.E.Z., Weckhuysen, B.M., and Bruijnincx, P.C.A. (2014) *ChemSusChem*, **7**, 2505–2515.

40. Sushkevich, V.L., Ivanova, I.I., Ordomsky, V.V., and Taarning, E. (2014) *ChemSusChem*, **7**, 2527–2536.
41. Janssens, W., Makshina, E.V., Vanelderen, P., De Clippel, F., Houthoofd, K., Kerkhofs, S., Martens, J.A., Jacobs, P.A., and Sels, B.F. (2015) *ChemSusChem*, **8**, 994–1088.
42. Larina, O., Kyriienko, P., and Soloviev, S. (2015) *Catal. Lett.*, **145**, 1162–1168.
43. Sushkevich, V., Ivanova, I.I., and Taarning, E. (2015) *Green Chem.*, **17**, 2552–2559.
44. Redina, E.A., Greish, A.A., Mishin, I.V., Kapustin, G.I., Tkachenko, O.P., Kirichenko, O.A., and Kustov, L.M. (2015) *Catal. Today*, **241**, 246–254.
45. Guan, Y. and Hensen, E.J.M. (2009) *Appl. Catal., A: Gen.*, **361**, 49–56.
46. Kim, D.-W., Kim, H., Jung, Y.-S., Kyu Song, I., and Baeck, S.-H. (2008) *J. Phys. Chem. Solids*, **69**, 1513–1517.
47. Chang, F.-W., Yang, H.-C., Roselin, L.S., and Kuo, W.-Y. (2006) *Appl. Catal., A: Gen.*, **304**, 30–39.
48. Jones, H., Stahly, E., and Corson, B. (1949) *J. Am. Chem. Soc.*, **767**, 1822–1828.
49. Scalbert, J., Thibault-Starzyk, F., Jacquot, R., Morvan, D., and Meunier, F. (2014) *J. Catal.*, **311**, 28–32.
50. Sun, J., Zhu, K., Gao, F., Wang, C., Liu, J., Peden, C.H.F., and Wang, Y. (2011) *J. Am. Chem. Soc.*, **133**, 11096–11099.
51. Liu, C., Sun, J., Smith, C., and Wang, Y. (2013) *Appl. Catal., A: Gen.*, **467**, 91–97.
52. Iwamoto, M. (2015) *Catal. Today*, **242**, 243–248.
53. Zacharopoulou, V., Vasiliadou, E.S., and Lemonidou, A.A. (2015) *Green Chem.*, **17**, 903–912.
54. Corma, A., Huber, G.W., Sauvanaud, L., and O'Connor, P. (2008) *J. Catal.*, **257**, 163–171.
55. Zakaria, Z.Y., Linnekoski, J., and Amin, N.A.S. (2012) *Chem. Eng. J.*, **207–208**, 803–813.
56. Zakaria, Z.Y., Amin, N.A.S., and Linnekoski, J. (2013) *Biomass Bioenergy*, **55**, 370–385.
57. Blass, S.D., Hermann, R.J., Persson, N.E., Bhan, A., and Schmidt, L.D. (2014) *Appl. Catal., A: Gen.*, **475**, 10–15.
58. Yu, L., Yuan, J., Zhang, Q., Liu, Y.-M., He, H.-Y., Fan, K.-N., and Cao, Y. (2014) *ChemSusChem*, **7**, 743–747.
59. Deshpande, R., Davis, P., Pandey, V., and Kore, N. (2013) Dehydroxylation of crude alcohol streams using a halogen-based catalyst. WO Patent 2013090076A1, Dow Global.
60. Hahn, H.-D., Dämbkes, G., and Rupprich, N. (2000) Butanols, in *Ullmann's Encyclopedia of Industrial Chemistry*, Wiley-VCH Verlag GmbH & Co. KGaA.
61. Kozlowski, J.T. and Davis, R.J. (2013) *ACS Catal.*, **3**, 1588–1600.
62. Marcu, I.-C., Tichit, D., Fajula, F., and Tanchoux, N. (2009) *Catal. Today*, **147**, 231–238.
63. Carvalho, D.L., de Avillez, R.R., Rodrigues, M.T., Borges, L.E.P., and Appel, L.G. (2012) *Appl. Catal., A: Gen.*, **415–416**, 96–100.
64. León, M., Díaz, E., and Ordóñez, S. (2011) *Catal. Today*, **164**, 436–442.
65. León, M., Díaz, E., Vega, A., Ordóñez, S., and Auroux, A. (2011) *Appl. Catal., B: Environ.*, **102**, 590–599.
66. Tsuchida, T., Sakuma, S., Takeguchi, T., and Ueda, W. (2006) *Ind. Eng. Chem. Res.*, **45**, 8634–8642.
67. Ogo, S., Onda, A., Iwasa, Y., Hara, K., Fukuoka, A., and Yanagisawa, K. (2012) *J. Catal.*, **296**, 24–30.
68. Riittonen, T., Toukoniitty, E., Madnani, D.K., Leino, A.-R., Kordas, K., Szabo, M., Sapi, A., Arve, K., Wärnå, J., and Mikkola, J.-P. (2012) *Catalysts*, **2**, 68–84.
69. Dowson, G.R.M., Haddow, M.F., Lee, J., Wingad, R.L., and Wass, D.F. (2013) *Angew. Chem. Int. Ed.*, **52**, 9005–9008.
70. Gines, M.J.L. and Iglesia, E. (1998) *J. Catal.*, **176**, 155–172.
71. Philip Atkins, M., Chadwick, D., Ibrahim Barri, S.A., and Al-Hajri, R. (2009) A process for the conversion of *n*-butanol to di-isobutene and pentene and/or di-pentene. WO Patent 2009074804A1, BP plc.
72. Díez, V.K., Torresi, P.A., Luggren, P.J., Ferretti, C.A., and Di Cosimo, J.I. (2013) *Catal. Today*, **213**, 18–24.

73. Ichikawa, N., Sato, S., Takahashi, R., and Sodesawa, T. (2006) *J. Mol. Catal. A: Chem.*, **256**, 106–112.
74. Reppe, W., Steinhofer, A., and Daumiller, G. (1953) Verfahren zur Herstellung von Diolefinen. DE Patent 899350, BASF AG.
75. Duan, H., Yamada, Y., and Sato, S. (2015) *Appl. Catal., A: Gen.*, **491**, 163–169.
76. Duan, H., Yamada, Y., and Sato, S. (2014) *Appl. Catal., A: Gen.*, **487**, 226–233.
77. Zhang, W., Yu, D., Ji, X., and Huang, H. (2012) *Green Chem.*, **14**, 3441–3450.
78. Sato, S., Takahashi, R., Sodesawa, T., Honda, N., and Shimizu, H. (2003) *Catal. Commun.*, **4**, 77–81.
79. Kobune, M., Sato, S., and Takahashi, R. (2008) *J. Mol. Catal. A: Chem.*, **279**, 10–19.
80. Igarashi, A., Ichikawa, N., Sato, S., Takahashi, R., and Sodesawa, T. (2006) *Appl. Catal., A: Gen.*, **300**, 50–57.
81. Yamamoto, N., Sato, S., Takahashi, R., and Inui, K. (2005) *Catal. Commun.*, **6**, 480–484.
82. Yamamoto, N., Sato, S., Takahashi, R., and Inui, K. (2006) *J. Mol. Catal. A: Chem.*, **243**, 52–59.
83. Horiuchi, N., Ikenaga, H., and Shohji, T. (2014) Method of Producing Conjugated Diene. (Mitsui chemicals inc.). JP Patent 2014172883, MITSUI che.
84. Millet, J.-M., Belliere-Baca, V., Nguyen, T.T.N., Huet, R., Rey, P., and Afanasiev, P. (2014) Procede de preparation d'une olefine par conversion catalytique d'au moins un alcool. WO Patent 2014118484A1, Adisseo Fr.
85. Han, Y.H. and Kim, R.H. (2012) Commercial preparation of 1,3 Butadiene from 2,3 Butanediol by catalytic dehydration reactions using hydroxyapatite-alumina catalyst. KR Patent 2012107353.
86. Bielansky, P., Weinert, A., Schönberger, C., and Reichhold, A. (2011) *Fuel Process. Technol.*, **92**, 2305–2311.
87. Pyl, S.P., Schietekat, C.M., Reyniers, M.-F., Abhari, R., Marin, G.B., and Van Geem, K.M. (2011) *Chem. Eng. J.*, **176–177**, 178–187.
88. Mortensen, P.M., Grunwaldt, J.-D., Jensen, P.A., Knudsen, K.G., and Jensen, A.D. (2011) *Appl. Catal., A: Gen.*, **407**, 1–19.
89. Gandarias, I. and Arias, P.L. (2013) *Hydrotreating Catalytic Processes for Oxygen Removal in the Upgrading of Bio-Oils and Bio-Chemicals, Liquid, Gaseous and Solid Biofuels – Conversion Techniques*, Prof. Zhen Fang (Ed.), ISBN: 978-953-51-1050-7, InTech, DOI: 10.5772/52581. pp. 327–356.
90. Ruddy, D.A., Schaidle, J.A., Ferrell, J.R. III, Wang, J., Moens, L., and Hensley, J.E. (2014) *Green Chem.*, **16**, 454–490.
91. He, Z. and Wang, X. (2012) *Catal. Sustainable Energy*, **1**, 28–52.
92. Abhari, R., Tomlinson, H.L., and Roth, G. (2013) Biorenewable naphtha. US Patent 8558042B2, Syntroleum.
93. Vermeiren, W. and Van Gyseghem, N. (2011) A process for the production of bio-naphtha from complex mixtures of natural occurring fats & oils. WO Patent 2011012439A1, Total Petrochemicals Research Feluy.
94. Pyl, S.P., Dijkmans, T., Antonykutty, J.M., Reyniers, M.-F., Harlin, A., Van Geem, K.M., and Marin, G.B. (2012) *Bioresour. Technol.*, **126**, 48–55.
95. Dijkmans, T., Pyl, S.P., Reyniers, M.-F., Abhari, R., Van Geem, K.M., and Marin, G.B. (2013) *Green Chem.*, **15**, 3064.
96. Gong, F., Yang, Z., Hong, C., Huang, W., Ning, S., Zhang, Z., Xu, Y., and Li, Q. (2011) *Bioresour. Technol.*, **102**, 9247–9254.
97. Huang, W., Gong, F., Fan, M., Zhai, Q., Hong, C., and Li, Q. (2012) *Bioresour. Technol.*, **121**, 248–255.
98. Hood, A.D. and Bridges, R.S. (2014) Staged propylene production process. WO Patent 2014110125A1, Lyondell C.
99. Mazoyer, E., Szeto, K.C., Merle, N., Norsic, S., Boyron, O., Basset, J.M., Taoufik, M., and Nicholas, C.P. (2013) *J. Catal.*, **301**, 1–7.
100. Chaemchuen, S., Phatanasri, S., Verpoort, F., Sae-ma, N., and Suriye, K. (2012) *Kinet. Catal.*, **53**, 247–252.
101. Maksasithorn, S., Praserthdam, P., Suriye, K., Devillers, M., and

Debecker, D.P. (2014) *Appl. Catal., A: Gen.*, **488**, 200–207.
102. Debecker, D.P., Stoyanova, M., Rodemerck, U., Colbeau-Justinc, F., Boissère, C., Chaumonnot, A., Bonduelle, A., and Sanchez, C. (2014) *Appl. Catal., A: Gen.*, **470**, 458–466.
103. Limsangkass, W., Phatanasri, S., Praserthdam, P., Panpranot, J., Jareewatchara, W., Na Ayudhya, S.K., and Suriye, K. (2013) *Catal. Lett.*, **143**, 919–925.
104. Mol, J.C. (2004) *J. Mol. Catal. A: Chem.*, **213**, 39–45.
105. Huang, S., Liu, S., Xin, W., Bai, J., Xie, S., Wang, Q., and Xu, L. (2005) *J. Mol. Catal. A: Chem.*, **226**, 61–68.
106. Xu, W., Lin, C., Liu, H., Yu, H., Tao, K., and Zhou, S. (2015) *RSC Adv.*, **5**, 23981–23989.
107. Hu, B., Liu, H., Tao, K., Xiong, C., and Zhou, S. (2013) *J. Phys. Chem. C*, **117**, 26385–26395.
108. Bouchmella, K., Hubert Mutin, P., Stoyanova, M., Poleunis, C., Eloy, P., Rodemerck, U., Gaigneaux, E.M., and Debecker, D.P. (2013) *J. Catal.*, **301**, 233–241.
109. Debecker, D.P., Stoyanova, M., Colbeau-Justin, F., Rodemerck, U., Boissiâre, C., Gaigneaux, E.M., and Sanchez, C. (2012) *Angew. Chem. Int. Ed.*, **51**, 2129–2131.
110. Liu, S., Huang, S., Xin, W., Bai, J., Xie, S., and Xu, L. (2004) *Catal. Today*, **93–95**, 471–476.
111. Gordeev, A.V. and Vodyankina, O.V. (2014) *Pet. Chem.*, **54**, 452–458.
112. Liu, S., Li, X., Xin, W., Xie, S., Zeng, P., Zhang, L., and Xu, L. (2010) *J. Nat. Gas Chem.*, **19**, 482–486.
113. Busca, G. (2014) *Heterogeneous Catalytic Materials: Solid State Chemistry, Surface Chemistry and Catalytic Behaviour*, Newnes.
114. Gartside, R.J. and Greene, M.I. (2007) Processing C4 olefin streams for the maximum production of propylene. US Patent 7214841 B2, Abb Lummus.
115. Cui, Y., Liu, N., Xia, Y., Lv, J., Zheng, S., Xue, N., Peng, L., Guo, X., and Ding, W. (2014) *J. Mol. Catal. A: Chem.*, **394**, 1–9.
116. Botha, J.M., Justice Mbatha, M.M., Nkosi, B.S., Spamer, A., and Swart, J. (2003) Production of propylene. US Patent 6586649B1, Sasol Tech.
117. Liu, H., Zhang, L., Li, X., Huang, S., Liu, S., Xin, W., Xie, S., and Xu, L. (2009) *J. Nat. Gas Chem.*, **18**, 331–336.
118. Popoff, N., Mazoyer, E., Pelletier, J., Gauvin, R.M., and Taoufik, M. (2013) *Chem. Soc. Rev.*, **42**, 9035–9054.
119. Dwyer, C.L. (2006) in *Metal-Catalysis in Industrial Organic Processes* (eds G.P. Chiusoli and P.M. Maitlis), Royal Society of Chemistry, pp. 201–217.
120. Maksasithorn, S., Debecker, D.P., Praserthdam, P., Panpranot, J., Suriye, K., and Ayudhya, S.K.N. (2014) *Chin. J. Catal.*, **35**, 232–241.
121. Stoyanova, M., Rodemerck, U., Bentrup, U., Dingerdissen, U., Linke, D., Mayer, R.W., Lansink Rotgerink, H.G.J., and Tacke, T. (2008) *Appl. Catal., A: Gen.*, **340**, 242–249.
122. Hua, D., Chen, S.L., Yuan, G., Wang, Y., and Zhang, L. (2011) *Transition Met. Chem.*, **36**, 245–248.
123. Ikeda, K., Kawamura, Y., Yamamoto, T., and Iwamoto, M. (2008) *Catal. Commun.*, **9**, 106–110.
124. Alvarado Perea, L., Wolff, T., Veit, P., Hilfert, L., Edelmann, F.T., Hamel, C., and Seidel-Morgenstern, A. (2013) *J. Catal.*, **305**, 154–168.
125. Phongsawat, W., Netivorruksa, B., Suriye, K., Dokjampa, S., Praserthdam, P., and Panpranot, J. (2012) *J. Nat. Gas Chem.*, **21**, 83–90.
126. Chen, S.L., Wang, Y., Yuan, G., Hua, D., Zheng, M., and Zhang, J. (2013) *Chem. Eng. Technol.*, **36**, 795–800.
127. Chauvin, Y. (2006) *Angew. Chem. Int. Ed.*, **45**, 3741–3747.
128. Li, X., Guan, J., Zheng, A., Zhou, D., Han, X., Zhang, W., and Bao, X. (2010) *J. Mol. Catal. A: Chem.*, **330**, 99–106.
129. Hahn, T., Kondratenko, E.V., and Linke, D. (2014) *Chem. Commun.*, **50**, 9060–9063.
130. Lwin, S. and Wachs, I.E. (2014) *ACS Catal.*, **4**, 2505–2520.
131. Zhang, D., Barri, S.A.I., and Chadwick, D. (2011) *Appl. Catal., A: Gen.*, **403**, 1–11.

2
Aromatics from Biomasses: Technological Options for Chemocatalytic Transformations

Fabrizio Cavani, Stefania Albonetti, and Francesco Basile

2.1
The Synthesis of Bioaromatics

Bio-based benzene, toluene, xylenes(BTX) can be produced starting from biomasses via several different routes, as shown in Figure 2.1 [1]. Main pathways include (i) pyrolysis or catalytic pyrolysis of lignocellulose fractions; (ii) aqueous-phase reforming (APR) of aqueous carbohydrate solutions, followed by catalytic transformations; (iii) Diels–Alder reactions between furanic compounds, the latter having been obtained from monosaccharides, and ethylene; this approach has several variants, with either differently functionalized furanics or dienofile compounds; and (iv) dimerization and dehydrocyclization of bioisobutanol. The latter two pathways are aimed at the synthesis of *p*-xylene, whereas the former two produce mixtures of aromatics, which can be either fractionated to single compounds or used as a blend for fuels. The reaction involving the Diels–Alder reaction is not discussed here, being dealt with in the chapter devoted to reactions for C–C bond formation (see Chapter 16).

2.2
The Synthesis of Bio-*p*-Xylene, a Precursor for Bioterephthalic Acid

Many companies – including Danone, Coca-Cola, Pepsi, WalMart, Nike, Heinz, and others – have initiated sustainability goals such as the introduction of partially or fully bio-based polyethylene terephthalate (PET), driven by the need to reduce their environmental footprint and respond to the growing demand for more sustainable packaging, as well as by the need to find an alternative to crude oil as a feedstock [1]. Coupling bio-based mono ethyleneglycol (MEG) and terephthalic acid (TPA) produced from bio-based *p*-xylene would make it possible for packaging companies to offer 100% renewable (and recyclable) PET bottles, fibers, and films. The various synthetic strategies investigated and implemented up to semi-commercial units are described in the following.

Chemicals and Fuels from Bio-Based Building Blocks, First Edition.
Edited by Fabrizio Cavani, Stefania Albonetti, Francesco Basile, and Alessandro Gandini.

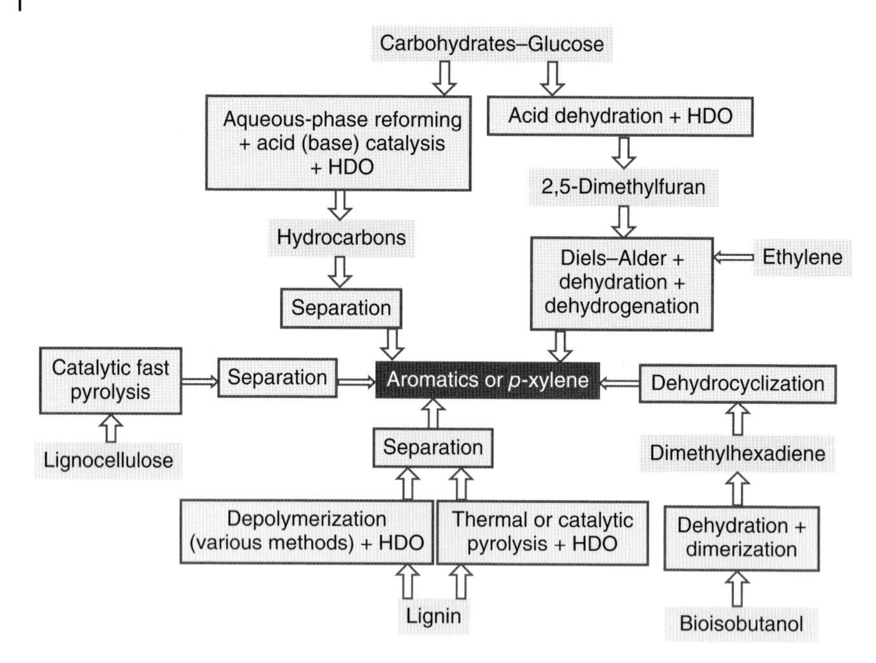

Figure 2.1 Main alternative routes for the production of bioaromatics.

2.2.1 Aromatic Hydrocarbons from Sugars

2.2.1.1 The Virent Technology

Virent is producing bio-*p*-xylene, branded BioFormPX, through its patented technology. Virent's process, trademarked BioForming®, is based on a combination of APR technology and downstream conventional catalytic processing technologies. The process converts aqueous carbohydrate solutions into a mix of hydrocarbons using heterogeneous catalysts at a temperature from 180 to 300 °C and pressure from 10 to 90 bar to reduce the oxygen content in the feedstock.

Various reactions occur during this process: reforming (with generation of hydrogen), dehydrogenation of alcohols, hydrogenation of ketones and aldehydes, deoxygenation, hydrogenolysis, and cyclization. Alcohols, ketones, and aldehydes formed during the APR can be converted to nonoxygenated hydrocarbons in a continuous process using conventional cyclization and hydrotreating techniques (Figure 2.2). For example, the acid-catalyzed condensation step uses a modified ZSM-5 catalyst, which produces a hydrocarbon stream similar to a reformate, containing alkanes, aromatics, and alkenes branded BioFormate™. This stream can be either blended into a gasoline pool or processed to obtain chemical intermediates: for example, purified *p*-xylene can be obtained by crystallization. In alternative approaches, base-catalyzed condensation followed by hydrodeoxygenation can yield kerosene blend, alcohol dehydration produces

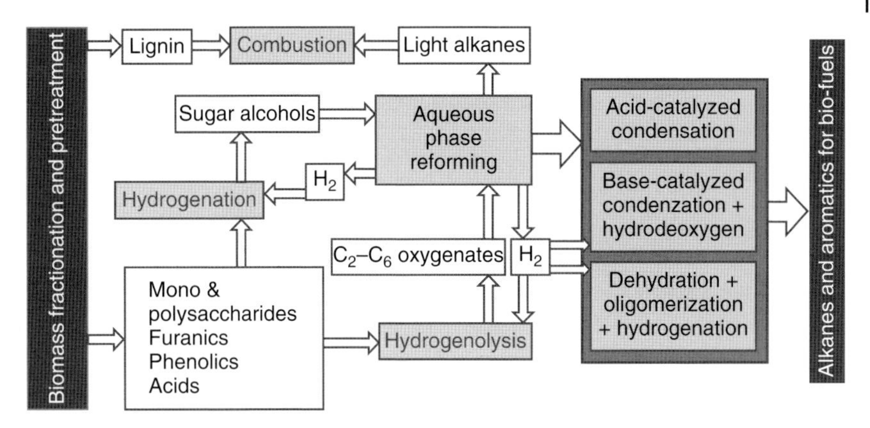

Figure 2.2 The BioForming concept.

olefins, and oligomerization and hydrogenation may provide alkanes for diesel engines.

An example of the possible synthetic pathway for obtaining aromatics is illustrated in [2]. A hydrogenation catalyst based on C-supported Ru and a combined APR/deoxygenation catalyst based on Pt/Re supported over ZrO_2 were used for the conversion of sucrose to a mixture of oxygenates. The two catalysts were loaded in a combined catalytic bed in a flow reactor, with the hydrogenation catalyst after the APR catalyst. The temperature in the hydrogenation bed was kept at approximately 110–150 °C and that of the APR/deoxygenation bed at 150–265 °C; hydrogen was combined with the sucrose feed. The organic phase yield reported was between 20% and 27%, and the reactor outlet contained approximately 30% total mono-oxygenates (11% alcohols, 9% ketones, 10% THF), 13% alkanes, 20% CO_2, and 25% other aqueous species (Figure 2.3). The vapor-phase condensation of oxygenates over a Ga/ZSM-5 catalyst led to aromatics as the main product; as a model molecule, acetone could be transformed into aromatics (79 wt% yield) and alkanes, at a temperature of 375 °C, and 625 psig

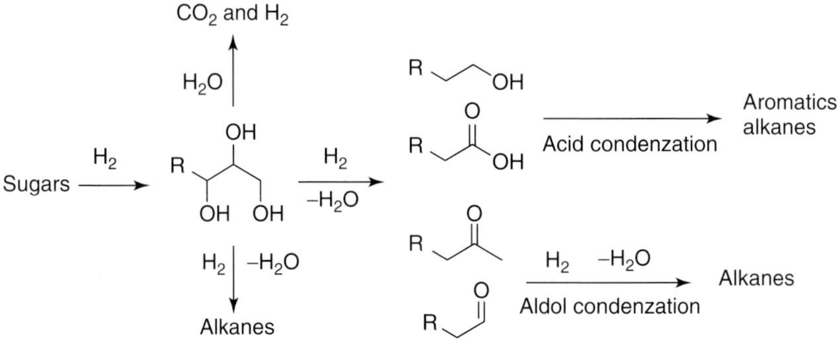

Figure 2.3 From sugars to aromatics and alkanes according to the technology developed by Virent.

pressure. By combining a APR/deoxygenation catalyst with tungstated zirconia and a condensation catalyst, it was possible to convert a glycerol feed to aromatics with a 48 wt% yield (~21% yield to other hydrocarbons), while a sucrose/xylose feed was converted to aromatics with a 25% yield.

2.2.2 Aromatic Hydrocarbons from Lignocellulose or Other Biomass

2.2.2.1 The Anellotech Technology

The process developed by Anellotech, a spin-off funded by Prof. G.W. Huber and D. Sudolsky, is based on catalytic fast pyrolysis (CFP) [3a]. This process is aimed at directly converting the solid biomass into liquid fuels (mainly aromatic compounds) in a single step at high temperature and with a short residence time (<2 min); on the small scale this is an economic technology. The solid biomass is heated rapidly (heating rate > $500° min^{-1}$), up to 400–600 °C, followed by a rapid cooling. During fast pyrolysis, a bio-oil containing more than 300 compounds is produced directly. These oils are usually thermally unstable, have a low heating value, and also degrade with time; therefore, they must be catalytically upgraded in order to be used as fuels. The introduction of zeolite catalysts into the pyrolysis process was shown to convert the oxygenated compounds produced during pyrolysis into aromatic and aliphatic hydrocarbons, which are suitable for blending with gasoline. Zeolites, such as ZSM-5, had already been added to pyrolysis reactors [3b–d], and the formation of aromatic compounds had already been previously highlighted. However, in Huber and collaborators' work, the formation of aromatic compounds was clearly highlighted, while the focus was on the optimization of the yield to these compounds.

During the CFP process, the biomass decomposes thermally into vapors, and the cellulose and hemicellulose form anhydrosugars via pyrolysis; the latter undergo dehydration to form furanic compounds, which enter the zeolite pores and undergo a series of dehydration, decarbonylation, decarboxylation, isomerization, oligomerization, and dehydrogenation reactions that lead to aromatics, CO, CO_2, and water; the lignin fraction forms coke [3e] (Figure 2.4).

In the case of glucose, CFP conducted in the presence of ZSM-5 led to approximately 31% C yield (13.3 wt%) to aromatics, around 12% CO_2, and approximately 16% CO, the remaining C being in the form of coke and unidentified species. Among aromatics, the prevailing compound was naphthalene, followed by toluene, ethylbenzene and xylenes, benzene, and others [3f,g]. The model molecule furan [3h] undergoes either decarbonylation to form allene (C_3H_4) or Diels–Alder condensation to form benzofuran (C_8H_6O) and water. The allene may either oligomerize to form olefins or alkylate the aromatics to form heavier aromatics and ethylene; olefins may react with furan to form aromatics and water. Using a fluidized bed configuration, the CFB of pinewood sawdust with a ZSM-5 catalyst, a 14% C yield of aromatics (mainly the more valuable monoaromatics) was obtained at 600 °C together with a low biomass weight hourly space velocity

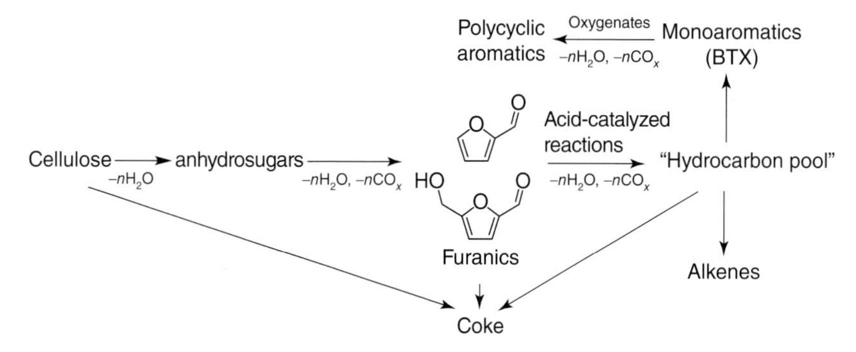

Figure 2.4 General reaction scheme of the transformation of the cellulosic and hemicellulosic fraction of lignocellulosic biomass into aromatics and aliphatic hydrocarbons by means of catalytic fast pyrolysis.

(WHSV); the recycle of the olefins produced could increase the C yield of aromatics up to 20%. The zeolite had to undergo reaction–regeneration cycles [3i]. The highest aromatic yield and the smallest amount of coke were produced with medium-pore zeolites with moderate internal pore space and steric hindrance, such as ZSM-5 and ZSM-11 [3j]. For ZSM-5, the optimal Silica-to-Alumina ratio (SAR) ratio was 30 [3k]. Yield of aromatics and olefins was increased by cofeeding the pinewood sawdust with alcohols (e.g., methanol) at 450–500 °C; the yield was affected by the H/C ratio of the feed. Cofeeding olefins also increased the selectivity to aromatics, for example, propylene may react with the intermediate dimethylfuran via Diels–Alder and dehydration to produce *p*-xylene [3l].

Later, two major improvements were reported, which were achieved through modification of the ZSM-5 catalyst:

1) The addition of Ga to the ZSM-5, with the generation of a bifunctional catalyst, enhanced the rate of aromatics formation; Ga fostered decarbonylation and olefin aromatization. The best C yield to aromatics (23.2%) was obtained at 550 °C (while the unpromoted ZSM-5 showed only a 15.4% C yield), together with an overall aromatics + olefins C yield of 42.7% [3m].
2) A reduced pore opening of the ZSM-5 was achieved by the chemical liquid deposition of silicon alkoxide on the ZSM-5 zeolite. Cofeeding dimethylfuran and propylene over the silylated zeolite had a remarkable effect on the nature of the aromatic compounds produced, with a strongly enhanced yield to *p*-xylene, because of space confinement on the products. For example, the overall selectivity for *p*-xylene with ZSM-5 was 5%, but it was as high as 15% with the silylated zeolite; the para-selectivity increased from 32% to 96%; the silylated Ga-ZSM-5 also showed a 96% para-selectivity. The same positive effects induced by silylation were also observed with furan, although a lower conversion was achieved compared to the untreated zeolite [3n].

In a recent paper, the CFP of wood was tested in a process development unit consisting of a bubbling fluidized bed reactor with the continuous addition and removal of spray-dried ZSM-5 catalyst particles. Reaction parameters

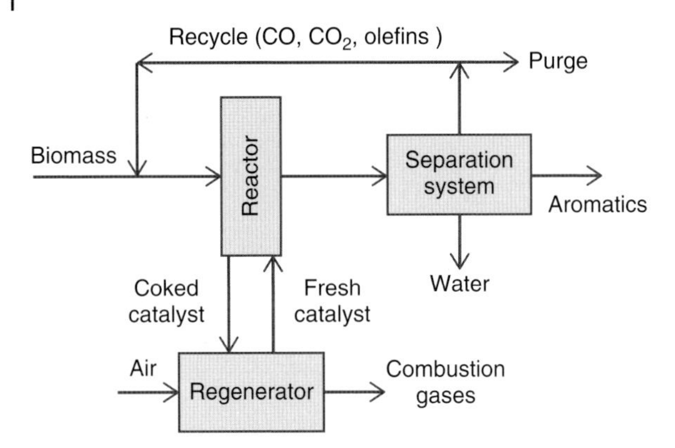

Figure 2.5 Simplified block flow diagram for aromatics production by means of CFP.

(temperature, biomass WHSV, catalyst-to-biomass ratio) were optimized and the results obtained were similar to those achieved with smaller units, thus demonstrating the feasibility of a continuous process with catalyst addition [3o]. A C yield to aromatics as high as 39.5% was obtained from cellulose when all reaction parameters were optimized [3p]. The fluidization can be achieved by using the gas produced during the process.

The simplified scheme of the process is shown in Figure 2.5: the lignocellulosic biomass is first dried and ground, then injected into the fluidized bed reactor in the presence of the zeolite catalyst. The desired aromatic and olefinic hydrocarbons are separated from the gas fraction containing CO and CO_2, which is recycled at the bottom of the reactor; olefins may also be recycled to improve the aromatic yield. The hydrocarbon mixture may be used as such as a transportation fuel or, as an alternative, single compounds are fungible commodities for the petrochemical industry.

A step forward was achieved by combining hydroprocessing with the acid-catalyzed transformation induced by the zeolite catalyst; this treatment increased the H content of pyrolysis oil, producing polyols and alcohols, and the zeolite catalyst converted these compounds into light olefins and aromatics, with a much higher yield than that achieved with pyrolysis alone. For example, when using the water-soluble fraction of a pinewood bio-oil (WSBO) as the feed, with the zeolite alone the percentage of C yield to aromatics was 8.2%, whereas using a Ru-based catalyst for hydrodeoxygenation prior to the zeolite-catalyzed step produced a yield of 21.6%, with 30.2% C yield to olefins. Among aromatics, the highest selectivity was to toluene and xylenes [3q].

In Anellotech patents, the main focus is on aspects such as: (i) the recycling to the pyrolysis reactor of a portion of the products, for example, olefins, for further conversion into useful products [4a,b]; (ii) the use of a gas jet to carry biomass into the fluidized bed reactor [4c]; and (iii) a method to regenerate the zeolite catalyst, which includes the oxidation of carbonaceous deposits and washing the zeolite to remove the minerals deposited [4d].

2.2.3 p-Xylene from Bioalcohols

2.2.3.1 The Gevo Technology

The technology developed by GEVO is based on bioisobutanol as the key reactant. Isobutanol is transformed into an unsaturated, branched C_8 aliphatic compound, which is then dehydrocyclized to *p*-xylene. In May 2014, Gevo announced that it was selling *p*-xylene derived from isobutanol to Toray, one of the world's leading producers of fibers, plastics, and films. Toray also provided funding assistance for the construction of Gevo's demoplant at its biorefinery in Silsbee, Texas, in collaboration with South Hampton Resources Inc. Other partners of Gevo include The Coca-Cola Company, Total SA, Sasol Chemical Industries, and LANXESS Inc.

One key step of the process is the transformation of isobutanol into the C_8 hydrocarbon, such as 2,5-dimethyl-2,4-hexadiene [5a]. Figure 2.6 shows the general reaction scheme, illustrating the different products which can be obtained depending on reaction conditions and the catalyst used. The main reaction pathway includes the dehydration of isobutanol to isobutene and its dehydrogenation into isobutyraldehyde. Then, alkene and aldehyde are made to react via a Prins-like mechanism, with acid catalysis, in order to form the desired diene. In the mentioned patent, isobutanol is dehydrogenated with a Cu

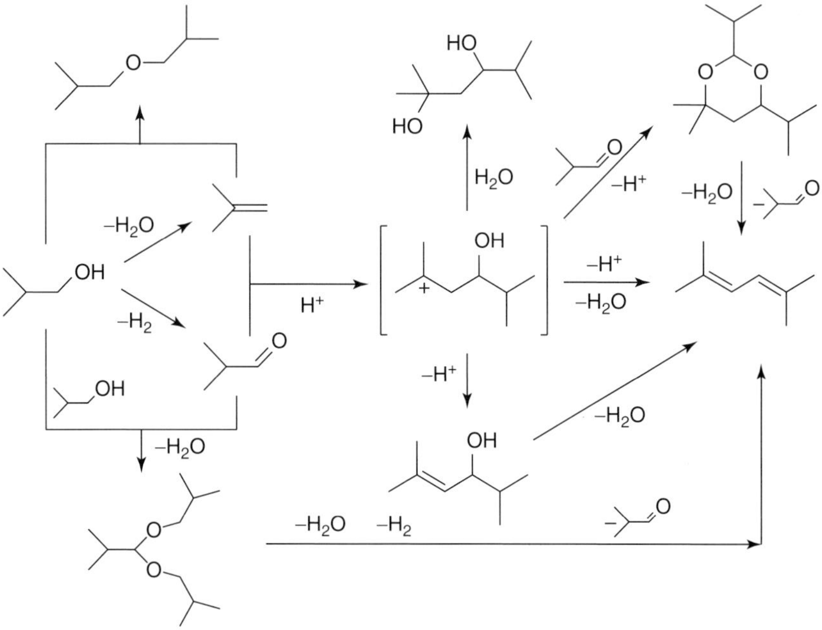

Figure 2.6 The general reaction scheme of isobutanol transformation into 2,5-dimethyl-2,4-hexadiene, with the various products formed based on the reaction conditions and catalyst type.

chromite catalyst at 320 °C to obtain a mixture of aldehyde and unconverted isobutanol; by-products of this step are propane, CO_2, and isobutyric acid. Afterward, the mixture is reacted over a dehydrating catalyst (e.g., alumina, at 300 °C) in order to convert the remaining isobutanol to isobutene; lastly, the isobutene/isobutyraldehyde mixture is reacted over an acid catalyst to produce 2,5-dimethyl-2,4-hexadiene.

The reaction can be carried out in the liquid phase, at 225 °C under pressure with a niobic acid catalyst, to produce diene with a per-pass yield of 35%; by-products are higher oligomers. In an alternative approach, isobutanol and isobutyraldehyde are made to react in the proper molar ratio to produce acetal (1,1-diisobutoxy-3-methylbutane) and enol ether (isobutyl isobutenyl ether). These two compounds are then reacted in the liquid phase, again with a niobic acid catalyst to produce the desired C_8 diene. If the reaction between isobutene and isobutyraldehyde is carried out in the presence of water and with an excess of the aldehyde, high yields of 2,4-diisopropyl-6,6-dimethyl-1,3-dioxane are obtained; however, even this cyclic acetal can be converted to the desired diene with an 80% yield by using a Ca phosphate catalyst at 400 °C, with a minor formation of 2-methyl-2,4-heptadiene. A Nb/P/O catalyst was also used to directly transform the isobutanol/isobutyraldehyde mixture (via intermediate isobutene) into the desired diene, with a 25% yield.

The final step of the process is the dehydrocyclization of diene to produce aromatic compounds; 2,5-dimethyl-2,4-hexadiene is the precursor of *p*-xylene, whereas dimethylheptatriene is the precursor of *o*-xylene. The dehydrogenation of diene with a chromia/alumina catalyst is carried out at 450 °C, with 98% conversion of diene, producing *p*-xylene with an 82% yield, 2% *m*-xylene, less than 1% *o*-xylene, and 13% of various saturated C_8 hydrocarbons.

In other patents, Gevo claimed alternative routes for the production of *p*-xylene precursor. For example, in [5b,c], isobutanol is first oxidized to isobutyraldehyde, by means of either transition-metal oxidants (Cr-, Fe-, Mn-based reagents) or by employing sulfur-type oxidants (Swern-type reagents) or hypervalent iodine reagents. The aldehyde is then homocoupled by means of Wittig or other aldehyde condensation (with subsequent dehydration and/or dehydrogenation) to obtain 2,5-dimethyl-3-hexene. The latter is then purified and dehydrocyclized to *p*-xylene using conventional catalysts and technologies. In the same patents, an alternative pathway includes the homometathesis of either 3-methyl-1-butene (the latter having been obtained by the dehydration of 3-methyl-1-butanol) or a mixture of 3-methyl-1-butene and 3-methyl-2-butene to produce ethylene and 2,5-dimethyl-3-hexene. Lastly, homometathesis can be carried out with isoprene to produce ethylene and a mixture of trienes (2,5-dimethyl-1,3,5-hexatriene, 3,4-dimethyl-1,3,5-hexatriene, and 2,4-dimethyl-1,3,5-hexatriene); the latter may undergo dehydrocyclization to xylene isomers.

In [5d,e], the C_8 unsaturated compound is produced by isobutene dimerization. Isobutene is produced by the vapor-phase dehydration of isobutanol with a heterogeneous γ-alumina catalyst [5f]. Dimerization can be carried out at 450–600 °C under oxidizing conditions with Bi oxide or other oxide catalysts to produce 2,5-dimethyl-1,5-hexadiene with a 50% yield; indeed, it is also claimed that dimerization and dehydrocyclization may be carried out simultaneously by using the Bi oxide catalyst. In alternative, dimerization may be carried out at mild temperatures (50–150 °C), using Pd or Rh chloride catalysts, or $Co(acac)_2$ and triethylaluminum, to obtain a 2,5-dimethyl-1-hexene with a yield of 70–90%, by-products being other C_8 isomers. According to the patent claims, another desirable dimer is 2,4,4-trimethylpentene, and preferred dimerization catalysts are based on zeolites or solid phosphoric acid. The same patents describe in detail the procedure to obtain bioisobutanol from glucose, with a modified *Escherichia Coli* [5g].

The final step, dehydrocyclization, is carried out with either a chromia/alumina catalyst or a Pt/Sn-zeolite. Starting from 100 kg of isobutanol, the yields obtained per pass were 18.7 wt% to *p*-xylene, 1.1% to other C_8 and heavier aromatics, 1.5% to lighter aromatics, 19.1% to isooctene (to be recycled), 26.3% to isobutene (to be recycled), 1.4% to hydrogen and fuel gas, 7.6% to heavier aliphatic hydrocarbons, and 24.3% to water. Figure 2.7 shows a simplified flowchart of the overall process, where the xylenes produced comprise at least 75% *p*-xylene.

2.2.3.2 *p*-Xylene from Bioethanol

An original multistep route for the synthesis of *p*-xylene from ethylene – which can be obtained from bioethanol, among others – has been recently proposed by Brookhart *et al.* [6]. Ethylene is first trimerized to 1-hexene, using a well-known

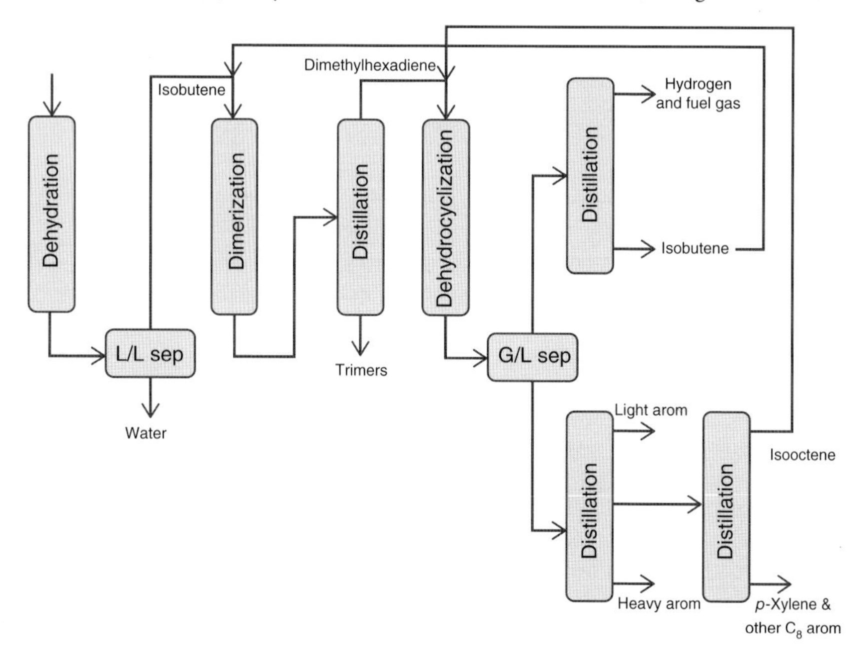

Figure 2.7 Simplified flowchart of the Gevo process for the production of *p*-xylene from isobutanol.

process with the reaction catalyzed by chromium. Hexene is then disproportionated via transfer hydrogenation to hexadienes and *n*-hexane, with catalysts based on Ir pincer complexes, and the formation of the equilibrium mixture. The next step is the uncatalyzed Diels–Alder reaction of both 2*E*,4*E*-hexadiene and 1,3-hexadiene with ethylene, carried out at 250 °C and 600 psi, with a high yield to the corresponding alkylcyclohexene adduct. The final step is the dehydrogenation of adducts *cis*-3,6-dimethylcyclohexene and 3-ethylcyclohexene to *p*-xylene and ethylbenzene, respectively, at 400 °C with a Pt/Al_2O_3 catalyst and high yield. A one-pot approach, combining H-transfer onto hexene and cycloaddition in the presence of ethylene (with the latter also acting as H-acceptor), and even a thermal dehydrogenation to aromatics, was also shown. Overall, the process requires only ethylene (or a source for it) as the reactant for the synthesis of *p*-xylene, uncontaminated by the ortho and meta isomers.

2.2.4 Aromatic Hydrocarbons from Lignin

The depolymerization and deoxygenation of lignin lead to a mixture of aromatics, such as BTX (mainly the meta isomer), and phenolics. The separation of xylenes and the isomerization of *m*-xylene can lead to *p*-xylene in a relatively high yield. The various technologies proposed differ in the way lignin is depolymerized [7a,b,w,x].

Once the biomass is treated and lignin has been isolated, it must be dissolved or transformed into a slurry. Lignin is insoluble in most common solvents, but shows good solubility in dimethyl sulfoxide/SO_2, CO_2-expanded solvents, and ionic liquids. However, the various treatments described in the following are typically carried out after dispersing the lignin or lignin–oil in water or in water/alcohol mixtures.

The depolymerization of lignin can be carried out by catalytic hydrogenation or hydrogenolysis [7c–g], oxidation [7h], hydrolysis [7i–k], or reforming [7l–n]. With an alternative approach, lignin can first be converted by a thermochemical (catalytic) pyrolysis process to produce a bio-oil [7o–r]. The mixture obtained by the deoxygenation of lignin can eventually be fed to a cracker, where a composition mixture close to that of naphtha is produced.

The stream obtained contains oxygenated aromatics, in either the monomeric or oligomeric form; therefore further deoxygenation is needed in order to yield phenolics or BTX [7s,t]. Different types of catalysts have been proposed for the hydrodeoxygenation of lignin, such as Ni, Co, or Mo-based systems. Isolation of single components in the mixture – which is achieved, for example, by hydrolysis – can be carried out by means of nanofiltration or pervaporation.

The main problems in the conversion of lignin into aromatics are [7a]: (i) the formation of char and condensation products, both very common at high temperatures, so moderate temperatures should be used; (ii) the diverse chemical structure of lignin, which can vary significantly depending on the biomass source; and (iii) the difficulty in removing side chains on the aromatic ring.

Recently, Stahl and coworkers reported on the depolymerization of an oxidized lignin in aqueous formic acid at mild conditions to give high yields of aromatics, mainly phenolics and other oxygenated monoaromatic compounds. The strategy adopted by Stahl consists of a two-step process, in which the secondary alcohol of the β-O-4 linkage is selectively oxidized, while in the second step the bond is cleaved with formic acid/sodium formate at 110 °C. Remarkably, the cleavage does not occur via H-transfer, and so no formic acid is consumed [7u,v]. The authors claimed that this procedure made it possible to obtain the highest yields of oxygenated monoaromatics reported to date.

Initiatives in this area are being carried out, among others, by the Dutch/Flemish Consortium Biorizon (joining efforts of TNO, VITO, and Green Chemistry Campus), and the Energy Research Center of the Netherlands (ECN). ECN has developed a process which consists of three main steps: (i) the MILENA technology converts the biomass into gasses at moderate temperatures (700–900 °C, lower than the 1400–1600 °C used, for instance, in coal gasification); this technology can convert all kinds of biomass into a syngas containing large amounts of hydrocarbons using air instead of oxygen; (ii) polyaromatic hydrocarbons are separated from the gas in order to avoid the plugging or fouling of the system (OLGA technology); and (iii) BTX components are separated from the gas; >97% separation efficiency of BTX from the gas is claimed. The typical ratio of B/T/X produced is 90/9/1 at 850 °C without using a catalyst.

2.2.4.1 The Biochemtex MOGHI Process

The key steps in the process developed by Biochemtex for the transformation of lignocellulosic biomass into chemicals are [8a–d]:

1) Pretreating the biomass feedstock with water by means of soaking and steam explosion; starting from *Arundo donax* lignocellulosic biomass, the soaking can be carried out at, for example, 155 °C for 155 min; the soaked liquid (containing sugars) is separated by soaked solids by means of a press; the latter fraction is then subjected to steam explosion (e.g., at 195 °C for 4 min).
2) Mixing the steam-exploded solid with water to obtain a slurry (e.g., containing 7.5 wt% dry matter) to be placed into an enzymatic hydrolysis reactor; feeding the hydrolysate into a bioreactor with a fermenting yeast; separating the fermentation broth into a liquid fraction (containing ethanol and other fractions) from the solid-rich fraction (45–60 wt% lignin, 10–15 wt% ash, and 25–35 wt% residual sugars) by means of a rotary drum filter; this fraction is the pretreated lignin used for further transformations.
3) Converting the lignin using H_2 under catalytic conditions (MOGHI technology).

The integration of these steps is shown in Figure 2.8.

With regard to the MOGHI technology, the key steps are:

1) The under-vacuum preparation of a slurry comprised of lignin, which is needed to facilitate the dispersion of lignin particles in water, by applying

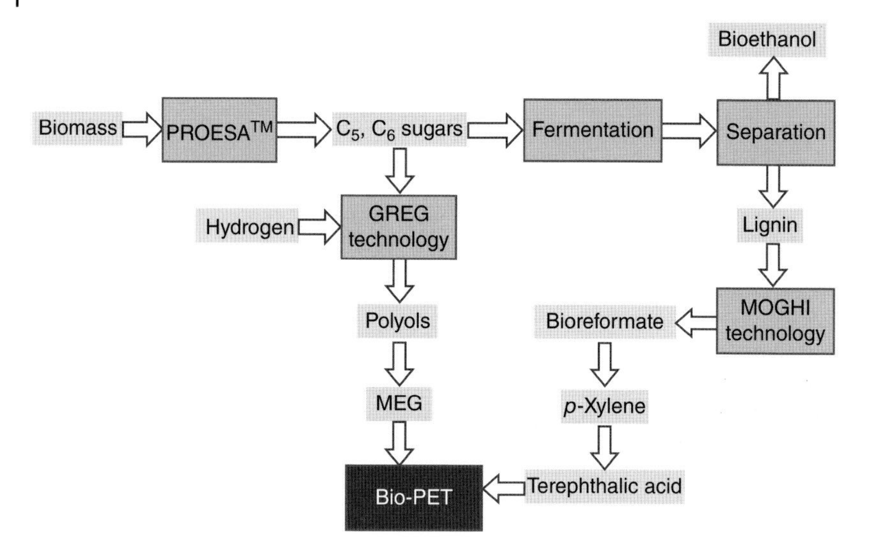

Figure 2.8 The synthesis of bio-PET from lignocellulosic biomass according to the Biochemtex concept: process integration.

high shear forces and adding a slurry liquid (e.g., water) to the lignin-rich material.

2) The charging of the slurry comprised of lignin into a continuous three-phase bubble reactor, where the catalytic hydrodeoxygenation of lignin is carried out.

For example, a reaction carried out under the following conditions: T 325 °C, P 166.5 bar, with lignin-rich material having a 5% concentration, residence time 42 min, and catalyst/lignin-rich wt ratio 0.25, leads to high lignin conversion (a theoretical amount of just 0.2% unreacted lignin). The distribution of products, as determined by GC analysis, gave 2-methoxyphenol, 2,6-dimethoxyphenol, and 4-ethyl-2-methoxyphenol as the major oxygenated product (~14, 10, and 10 GC% area, respectively), with an overall yield to phenolics based on a total fed lignin of 35–40% (yield ~10% if referring to the starting biomass). Many other phenolic compounds were also formed, and the nature of the prevailing phenolic compounds obtained was a function of the reaction conditions. The catalyst used is a sponge nickel or other similar Ni Raney catalysts. Either hydrogen or other hydrogen donors (e.g., ethylene glycol generated during the hydrogenolysis of carbohydrates) are reported to be used for hydrodeoxygenation in the lignin conversion step.

Figure 2.9 shows a simplified scheme of the MOGHI technology, as inferred from the patent literature.

The mixture of the phenolic compounds generated can then be hydrodeoxygenated further to produce aromatic hydrocarbons. In an alternative approach [8e,f], when the hydrodeoxygenation step is carried out in the presence of a zeolite CBV600 catalyst (from Zeolyst Co.), at 350 °C, under H_2 pressure between 146 and

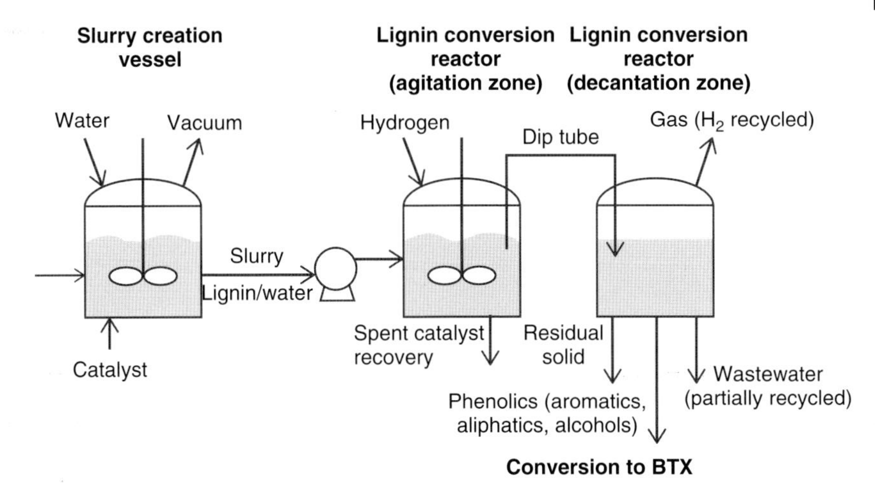

Figure 2.9 The section of the Biochemtex process for the transformation of lignin into aromatics (MOGHI process).

201 bar, with reaction time 1.5 h, remarkable amounts of aromatic and aliphatic hydrocarbons were formed: approximately 15% (as determined by the area percentage of the GC analysis) toluene, 10% ethylbenzene, 3% of *p*-xylene, and others in smaller amounts. In this case, the bioreformate produced by the MOGHI technology can be used as a drop-in mixture which is fully compatible with the classical reformate obtained from naphtha, since it may contain, for example, 10–15 vol% of naphthenes, 75–85% aromatics, and <5% alkanes. At even harsher conditions (H_2 P 233–237 bar, reaction time 3 h) and 305 °C temperature, naphthenic hydrocarbons and alcohols (ethylcyclohexane, C>15 alkanes, ethylcyclohexanol isomers, etc.) were the prevailing products. The use of a higher pressure but lower temperature (e.g., 255 °C) led to the prevalence of alcohols (alkylcyclohexanols).

Notably, prior to being processed in the lignin conversion reactor, the liquid fraction was dark brown because of some soluble contaminants from the water of preceding treatment steps. After separating the water from the organic products downstream from the lignin conversion reactor, the water was no longer dark because of the considerable Chemical Oxygen Demand (COD) reduction; this water can be recycled to the carbohydrate conversion step as soaking water or water for the steam explosion. These results have some positive implications for the wastewater treatment step, which is a significant component of the operating costs for the process.

Another issue concerns avoiding the formation of char; this makes it necessary to shut down the continuous process to permit removing the char from the reactor; the conditions for limiting the formation of char, in terms of temperature and pressure, are identified in Biochemtex patents. For example, pressure above a minimum value must be applied, as it is based on the liquid-phase composition; temperature must be above the boiling point of the slurry liquid and below the critical temperature.

A demo plant for the production of 1000 t per year of bioreformate is going to be built, with raw materials being delivered by the PROESA^{TM} plant in Crescentino.

2.2.5 Other Initiatives

Cool Planet investigated the agricultural waste pyrolysis with coproduction of "bioreformate," which resembles the refinery stream from which aromatics are extracted, and "carbon-negative" biochar, which is a valuable soil amendment for enhancing agricultural productions [9a]. In the process described by Cool Planet, catalytic processes are used to transform volatile gas streams from biomass decomposition. For example, by combining a hydrotreating step with a catalyzed dehydration and a downstream aromatization step, it is possible to produce an aromatic-rich stream, which can be employed for either fuel blending or xylene separation. The patent covers a programmable number of processing stations for subjecting the biomass to programmed reaction conditions and transformation steps.

Stora Enso Oyj patented the transformation of tall oil-based raw material by hydrodeoxygenation using a NiMo catalyst; in addition, a cracking catalyst such an acidic zeolite was also used. The aromatic fraction produced contained xylenes and *p*-cymene [9b]. The resin acid fraction of tall oil consists of acids such as abietic acid, dehydroabietic acid, isopimaric acid, neoabietic acid, palustric acid, pimaric acid, and sandaracopimaric acid which, through deoxygenation and degradation, form monoaromatic compounds. Compounds suitable for the manufacture of TPA can be separated from the liquid phase.

2.3 The Synthesis of Bioterephthalic Acid without the Intermediate Formation of *p*-Xylene

A technology involving muconic acid as the key intermediate was developed by Michigan State University and Draths Corporation; the latter was acquired by Amyris in 2011. Carbohydrates can be converted to a mixture of *cis, trans*- and *cis, cis*-muconic acid isomers using microbial synthesis; these isomers are then converted to the *trans, trans* isomer via reaction in a water/tetrahydrofuran mixture at reflux temperature and in the presence of iodine. The *trans, trans*-muconic acid precipitates upon cooling to room temperature, and is then made to react with ethylene at 150 °C, to yield cyclohex-2-ene-1,4-dicarboxylic acid via a Diels–Alder reaction, carried out in the presence of a Lewis acid. The olefin is then dehydrogenated catalytically to TPA [10a–c]. Figure 2.10 summarizes the main steps of the process.

In an alternative but similar approach, muconic acid is reacted with acetylene to yield the corresponding cyclohexadiene adduct, which is then dehydrogenated to TPA [10d].

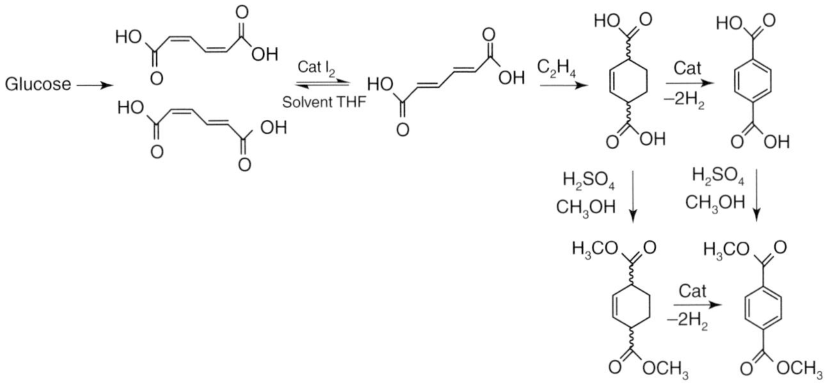

Figure 2.10 The Amyris (former Draths) technology for terephthalic acid production from muconic acid.

The use of terpenes or terpenoids, such as α- and β-pinene, and limonene to make TPA is reported by SABIC [11]. Limonene can be extracted from citrus peels and transformed into *p*-cymene with a 99% yield by means of isomerization and dehydrogenation, using a mixture of ethylenediamine, $FeCl_3$, and metallic Na, at 50 °C; *p*-cymene was then oxidized in two steps into TPA, by first reacting the aromatic compound in H_2O/HNO_3 at reflux, and then completing the oxidation by separating the crude product obtained from the oxidation of *p*-cymene and reacting it with $KMnO_4$, in H_2O/NaOH; the overall yield to TPA was 85%.

Patents by BP describe the Diels–Alder reaction of 2,5-furandicarboxylic acid (FDCA) with ethylene to form the bicyclic ether intermediate 7-oxa-bicyclo[2.2.1]-hept-2-ene-1,4-dicarboxylic acid; the latter is then dehydrated to produce TPA [12].

2.4 Technoeconomic and Environmental Assessment of Bio-*p*-Xylene Production

A few recent papers compared the technical, economic, and environmental assessments of the various processes reported in scientific and patent literature for the synthesis of aromatic compounds starting from different renewable resources [13]. Briefly, the main outcomes of these studies are the following:

1) In the case of the production of *p*-xylene from dimethylfuran via Diels–Alder reaction with ethylene (this process is described in the chapter dealing with C–C formation, see Chapter 16), the main contribution to the *p*-xylene cost comes from the 5-hydroxymethylfurfural production, which is the precursor for the synthesis of dimethylfuran [13a,c]. This is an emblematic example of how biomass treatment and chemical transformation to produce so-called "platform molecules" may be an economic burden for the development of new technologies focusing on biochemical production.

2) Regarding the environmental impact of the various technologies, as inferred from a Life-Cycle-Assessment (LCA) perspective, lignocellulose-based *p*-xylene was shown to have an impact similar to that of petroleum-based *p*-xylene, while starch-based *p*-xylene was less environmentally friendly [13b]. This is mainly due to the cultivation and processing of maize starch. The use of nonrenewable chemicals (e.g., solvents) heavily contributed to the overall impact of lignocellulose-based *p*-xylene. High selectivity made a greater contribution than high biomass conversion to improving process sustainability.

References

1. (a) Christensen, C.H., Rass-Hansen, J., Marsden, C., Taarning, E., and Egeblad, K. (2008) *ChemSusChem*, **1**, 283–289; (b) Dapsens, P.Y., Mondelli, C., and Pérez-Ramìrez, J. (2012) *ACS Catal.*, **2**, 1487–1499; (c) Lanzafame, P., Centi, G., and Perathoner, L. (2014) *Chem. Soc. Rev.*, **43**, 7562–7580; (d) Thomas, J.M. (2014) *ChemSusChem*, **7**, 1801–1832; (e) Harmsen, P.F.H., Hackmann, M.M., and Bos, H.L. (2014) *Biofuels, Bioprod. Biorefin.*, **8**, 306–324; (f) Serrano-Ruiz, J.C., Luque, R., and Sepulveda-Escribano, A. (2011) *Chem. Soc. Rev.*, **40**, 5266–5281; (g) de Jong, E., Higson, A., Walsh, P., and Wellisch, M. (2012) *Biofuels, Bioprod. Biorefin.*, **6**, 606–624; (h) Climent, M.J., Corma, A., and Iborra, S. (2014) *Green Chem.*, **16**, 516–547; (i) Sheldon, R.A. (2014) *Green Chem.*, **16**, 950–963; (j) Collias, D.I., Harris, A.M., Nagpal, V., Cottrell, I.W., and Schultheis, M.W. (2014) *Ind. Biotechnol.*, **10**, 91–105.
2. Cortright, R.C. and Blommel, P.G. (2013) Synthesis of Liquid Fuels and Chemicals From Oxygenated Hydrocarbons. US Patent 2013/0185992 A1, assigned to Virent Inc.
3. (a) Huber, G.W., Gaffney, A.M., Jae, J., and Cheng, Y.-T. (2012) Systems and Processes for Catalytic Pyrolysis of Biomass and Hydrocarbonaceous Materials for Production of Aromatics with Optional Olefin Recycle, and Catalysts Having Selected Particle Size for Catalytic Pyrolysis. US Patent 2012/0203042 A1, assigned to Anellotech Inc., and University of Massachusetts; (b) Olazar, M., Aguado, R., and Bilbao, J. (2000) *AIChE J.*, **46**, 1025–1033; (c) Pattiya, A., Titiloye, J.O., and Bridgwater, A.V. (2008) *J. Anal. Appl. Pyrolysis*, **81**, 72–79; (d) Lappas, A.A., Samolada, M.C., Iatridis, D.K., Voutetakis, S.S., and Vasalos, I.A. (2002) *Fuel*, **81**, 2087–2095; (e) Lin, Y.C., Cho, J., Tompsett, G.A., Westmoreland, P.R., and Huber, G.W. (2009) *J. Phys. Chem. C*, **113**, 20097–20107; (f) Carlson, T.R., Vispute, T.P., and Huber, G.W. (2008) *ChemSusChem*, **1**, 397–400; (g) Carlson, T.R., Tompsett, G.A., Conner, W.C., and Huber, G.W. (2009) *Top. Catal.*, **52**, 241–252; (h) Cheng, Y.-T. and Huber, G.W. (2011) *ACS Catal.*, **1**, 611–628; (i) Carlson, T.R., Cheng, Y.-T., Jae, J., and Huber, G.W. (2011) *Energy Environ. Sci.*, **4**, 145–161; (j) Jae, J., Tompsett, G.A., Foster, A.J., Hammond, K.D., Auerbach, S.M., Lobo, R.F., and Huber, G.W. (2011) *J. Catal.*, **279**, 257–268; (k) Foster, A.J., Jae, J., Cheng, Y.-T., Huber, G.W., and Lobo, R.F. (2012) *Appl. Catal., A*, **423–424**, 154–161; (l) Zhang, H., Carlson, T.R., Xiao, R., and Huber, G.W. (2012) *Green Chem.*, **14**, 98–110; (m) Cheng, Y.-T., Jae, J., Shi, J., Fan, W., and Huber, G.W. (2012) *Angew. Chem. Int. Ed.*, **51**, 1387–1390; (n) Cheng, Y.-T., Wang, Z., Gilbert, C.J., Fan, W., and Huber, G.W. (2012) *Angew. Chem. Int. Ed.*, **51**, 11097–11100; (o) Jae, J., Coolman, R., Mountziaris, T.J., and Huber, G.W. (2014) *Chem. Eng. Sci.*, **108**, 33–46; (p) Karanjkar, P.U., Coolman, R.J., Huber, G.W., Blatnik, M.T., Almalkie, S., de Bruyn Kops, S.M., Mountziaris, T.J., and Conner, W.C.

(2014) *AIChE J.*, **60**, 1320–1335; (q) Vispute, T.P., Zhang, H., Sanna, A., Xiao, R., and Huber, G.W. (2010) *Science*, **330**, 1222–1227.

4. (a) Mazanec, T. and Whiting, J. (2014) Fast Catalytic Pyrolysis with Recycle of Side Products. US Patent 2014/0027265 A1, assigned to Anellotech Inc.; (b) Mazanec, T., Whiting, J., Pesa, F., and Norenberg, G. (2014) Olefin Conditioning in a Fast Catalytic Pyrolysis Recycle Process. US Patent 2014/0031583 A1, assigned to Anellotech Inc.; (c) Mazanec, T., Whiting, J., Song, R., Goodman, Z.W., and Schmidt, C. (2014) Gas Jet Injector Reactor for Catalytic Fast Pyrolysis Process. US Patent 2014/0206913 A1, assigned to Anellotech Inc.; (d) Mazanec, T., Whiting, J., Pesa, F., Chen, Y.-T., and Song, R. (2014) Regeneration of Catalytic Fast Pyrolysis Catalyst. US Patent 2014/0303414 A1, assigned to Anellotech Inc.
5. (a) Taylor, T.J., Taylor, J.D., Peters, M.W., and Henton, D.E. (2012) Variations on Prins-Like Chemistry to Produce 2,5-Dimethylhexadiene from Isobutanol. US Patent 2012/0271082 A1, assigned to Gevo, Inc.; (b) Peters, M.W., Henton, D.E., Taylor, J.D., Taylor, T.J., and Manzer, L.E. (2012) Renewable Xylenes Produced from Biological C4 and C5 Molecules. WO Patent 2012/061372, assigned to Gevo Inc.; (c) Peters, M.W., Taylor, J.D., Taylor, T.J., and Manzer, L.E. (2012) Renewable Xylenes Produced from Bological C4 and C5 Molecules. US Patent 2012/0171741, assigned to Gevo Inc.; (d) Peters, M.W., Taylor, J.D., Jenni, M., Manzer, L.E., and Henton, D.E. (2011) Integrated Process to Selectively Convert Renewable Isobutanol to P-Xylene. WO Patent 2011/044243 A1, assigned to Gevo Inc.; (e) Peters, M.W., Taylor, J.D., Jenni, M., Manzer, L.E., and Henton, D.E. (2011) Integrated Process to Selectively Convert Renewable Isobutanol to P-Xylene. US Patent 2011/0087000 A1, assigned to Gevo Inc.; (f) Taylor, J.D., Jenni, M.D., and Peters, M.W. (2010) *Top. Catal.*, **53**, 1224–1230; (g) Atsumi, S., Wu, T.-Y., Eckl, E.-M., Hawkins, S.D., Buelter, T., and Liao, J.C. (2010) *Appl. Microbiol. Biotechnol.*, **85**, 651–657.
6. Lyons, T.W., Guironnet, D., Findlater, M., and Brookhart, M. (2012) *J. Am. Chem. Soc.*, **134**, 15708–15711.
7. (a) Zakzeski, J., Bruijinincx, P.C.A., Jongerius, A.L., and Weckhuysen, B.M. (2010) *Chem. Rev.*, **110**, 3552–3599; (b) Deuss, P.J., Barta, K., and de Vries, J. (2014) *Catal. Sci. Technol.*, **4**, 1174–1196; (c) Thring, R.W. and Breau, J. (1996) *Fuel*, **75**, 795–800; (d) Gosselink, R.J.A., Teunissen, W., van Dam, J.E.G., de Jong, E., Gellerstedt, G., Schott, E.L., and Sanders, J.P.M. (2012) *Bioresour. Technol.*, **106**, 173–177; (e) He, J.Y., Zhao, C., and Lercher, J.A. (2012) *J. Am. Chem. Soc.*, **134**, 20768–20775; (f) Xu, W., Miller, S.J., Agrawal, P.K., and Jones, C.W. (2012) *ChemSusChem*, **5**, 667–675; (g) Li, C., Zheng, M., Wang, A., and Zhang, T. (2012) *Energy Environ. Sci.*, **5**, 6383–6390; (h) Stark, K., Taccardi, N., Bosmann, A., and Wasserscheid, P. (2010) *ChemSusChem*, **3**, 719–723; (i) Roberts, V.M., Stein, V., Reiner, T., Lemonidou, A., Li, X., and Lercher, J.A. (2011) *Chem. Eur. J.*, **17**, 5939–5948; (j) Lavoie, J.M., Bar, W., and Bilodeau, M. (2011) *Bioresour. Technol.*, **102**, 4917–4920; (k) Toledano, A., Serrano, L., and Labidi, J. (2012) *J. Chem. Technol. Biotechnol.*, **87**, 1593–1599; (l) Zakzeski, J.J., Jongerius, A.L., Bruijnincx, P.C.A., and Weckhuysen, B.M. (2012) *ChemSusChem*, **5**, 1602–1609; (m) Zakzeski, J.J. and Weckhuysen, B.M. (2011) *ChemSusChem*, **4**, 369–378; (n) Jongerius, A.L., Bruijnincx, P.C.A., and Weckhuysen, B.M. (2013) *Green Chem.*, **15**, 3049–3056; (o) Misson, M., Haron, R., Kamaroddin, M.F.A., and Amin, N.A.S. (2009) *Bioresour. Technol.*, **100**, 2867–2873; (p) Patwardhan, P.R., Brown, R.C., and Shanks, B.H. (2011) *ChemSusChem*, **4**, 1629–1636; (q) Dorrestijn, E., Laarhoven, L.J.J., Arends, I.W.C.E., and Mulder, P. (2000) *J. Anal. Appl. Pyrolysis*, **54**, 153–192; (r) Ma, Z., Troussard, E., and van Bokhoven, J.A. (2012) *Appl. Catal., A*, **423–424**, 130; (s) Jongerius, A.L., Gosselink, R.W., Dijkstra, J., Bitter, J.H., Bruijnincx,

P.C.A., and Weckhuysen, B.M. (2013) *ChemCatChem*, **5**, 2964; (t) Jongerius, A.L., Jastrzebski, R., Bruijnincx, P.C.A., and Weckhuysen, B.M. (2012) *J. Catal.*, **285**, 315–323; (u) Rahimi, A., Ulbrich, A., Coon, J.J., and Stahl, S.A. (2014) *Nature*, **515**, 249–252; (v) Rahimi, A., Azarpira, A., Kim, H., Ralph, J., and Stahl, S.A. (2013) *J. Am. Chem. Soc.*, **135**, 6415–6418; (w) Pandey, M.P. and Kim, C.S. (2011) *Chem. Eng. Technol.*, **34**, 29–41; (x) Wang, H., Tucker, M., and Ji, Y. (2013) *J. Appl. Chem.*, **2013**, 1–9.

8. (a) Ryba, S., Murray, A., Elliott, G.A., and Gastaldo, D. (2013) Continuous Process for Conversion of Lignin to Useful Compounds. US Patent 2013/0225856 A1, assigned to Chemtex Italia SpA; (b) Ryba, S., Murray, A., Elliott, G.A., and Gastaldo, D. (2013) Continuous Process for Conversion of Lignin to Useful Compounds. US Patent 2013/0225872 A1, assigned to Chemtex Italia SpA; (c) Ryba, S., Murray, A., Elliott, G.A., and Gastaldo, D. (2014) Continuous Process for Conversion of Lignin to Useful Compounds. US Patent 2014/0096830 A1, assigned to BioChemtex Italia SpA; (d) Murray, A. and Ryba, S. (2014) Lignin Conversion Process. US Patent 2014/0135470 A1, assigned to BioChemtex Italia SpA; (e) Murray, A. and Ryba, S. (2012) WO Patent 2012/174429 A1, assigned to Chemtex Italia SpA; (f) Murray, A. and Ryba, S. (2014) Lignin Conversion Process. US Patent 2014/0135470 A1, assigned to BioChemtex Italia SpA.
9. (a) Cheiky, M.C. and Malyala, R. (2014) System for Making Renewable Fuels. WO Patent 2014/117136 A1, assigned to Cool Planet Energy Systems Inc.; (b) Harlin, A., Räsänen, J., and Penttinen, T. (2013) Hydrogen Treatment of Impure Tall Oil for The Production of Aromatic Monomers. US Patent 2013/0178650, assigned to Stora Enso Oyi.
10. (a) Frost, J.W., Miermont, A., Schweitzer, D., Bul, V., and Wicks, D.A. (2013) Novel Terephthalic and Trimellitic Based Acids and Carboxylate Derivatives Thereof. US Patent 8,367,858 B2, assigned to Amyris; (b) Frost, J.W., Miermont, A., Schweitzer, D., Bul, V., Paschke, E., and Wicks, D.A. (2011) Biobased Polyesters. US Patent 2011/0288263 A1, assigned to Draths Co.; (c) Frost, J.W., Miermont, A., Schweitzer, D., Bul, V., and Wicks, D.A. (2011) Novel Terephthalic and Trimellitic Based Acids and Carboxylate Derivatives Thereof. US Patent 2011/0288311 A1, assigned to Draths Co.; (d) Burk, M.J., Osterhout, R.E., and Sun, J. (2011) Semi-Synthetic Terephthalic Acid via Microorganisms that Produce Muconic Acid. US Patent 2011/0124911.
11. (a) Berti, C., Binassi, E., Colonna, M., Fiorini, M., Kannan, G., Karanam, S., Mazzacurati, M., and Odeh, I. (2014) Bio-Based Terephthalate Polyesters. US Patent 2014/0163195 A1, assigned to SABIC Innovative Plastics IP B.V.; (b) Berti, C., Binassi, E., Colonna, M., Fiorini, M., Kannan, G., Karanam, S., Mazzacurati, M., and Odeh, I. (2015) Bio-Based Terephthalate Polyesters. US Patent 2015/0112100, assigned to SABIC Innovative Plastics IP B.V.; (c) Colonna, M., Berti, C., Fiorini, M., Binassi, E., Mazzacurati, M., Vannini, M., and Karanam, S. (2011) *Green Chem.*, **13**, 2543–2548.
12. Gong, W.H. (2012) Terephthalic Acid Composition and Process for the Production Thereof. US Patent 8,299,278 B2, assigned to BP Corporation North America Inc.
13. (a) Lin, Z., Ierapetritou, M., and Nikolakis, V. (2013) *AIChE J.*, **59**, 2079–2087; (b) Lin, Z., Nikolakis, V., and Ierapetritou, M. (2015) *Ind. Eng. Chem. Res.*, **54**, 2366–2378; (c) Lin, Z., Nikolakis, V., and Ierapetritou, M. (2014) *Ind. Eng. Chem. Res.*, **53**, 10688–10699; (d) Akanuma, Y., Selke, S.E.M., and Auras, R. (2014) *Int. J. Life Cycle Assess.*, **19**, 1238–1246.

3
Isostearic Acid: A Unique Fatty Acid with Great Potential

Sophie C.C. Wiedemann, Pieter C.A. Bruijnincx, and Bert M. Weckhuysen

3.1
Introduction

The total world production of the major oils and fats in 2014 is estimated at around 207 million metric tons [1, 2], of which palm, soybean, and rapeseed together account for two-thirds (Figure 3.1). Animal fats (tallow, lard, butter, and fish) contribute around approximately 12.5% of the total, and specialty vegetable oils, such as corn, linseed, and castor oil, constitute 2.6%. The annual growth rate in production of the major vegetable oils, over the period 2010–2014 is around 4.2%.

In 2011, it was estimated that about 70–75% of the global production of oils and fats was consumed in human nutrition, with the remainder going into animal feed, biofuels, and oleochemicals [3]. In recent years, fuel uses, biodiesel production in particular, and cogeneration in power stations have increased the nonfood demand for vegetable oils. Their large-scale use for fuel and energy is however generally uneconomic in the absence of financial incentives. The ongoing debate on the use of food-grade vegetable oils in both industrial applications and for fuel production has an impact on the assessment of the net environmental and sustainability benefits of (bio-based) raw material choices in these sectors. The focal point of these discussions has been biodiesel and power generation since, compared to chemical and industrial uses, these have the potential to consume vast amounts of natural resources. A dramatic illustration of the potential problems was seen in 2006–2007, when riots in Mexico over a growing shortage of the food staple corn tortillas, were blamed on the rapidly rising demand for corn ethanol created by the United States biofuel industry [4].

Recognition that food-grade vegetable oils alone cannot economically or practically meet a significant part of the demand for vehicle fuels has led to research into second- and third-generation biofuels based on abundant alternative feedstocks, such as lignocellulosic biomass, and municipal and agricultural waste streams. Biodiesel producers have also adapted their processes to handle tallow, used cooking oil, yellow greases, and other lipid streams not suitable for

Chemicals and Fuels from Bio-Based Building Blocks, First Edition.
Edited by Fabrizio Cavani, Stefania Albonetti, Francesco Basile, and Alessandro Gandini.

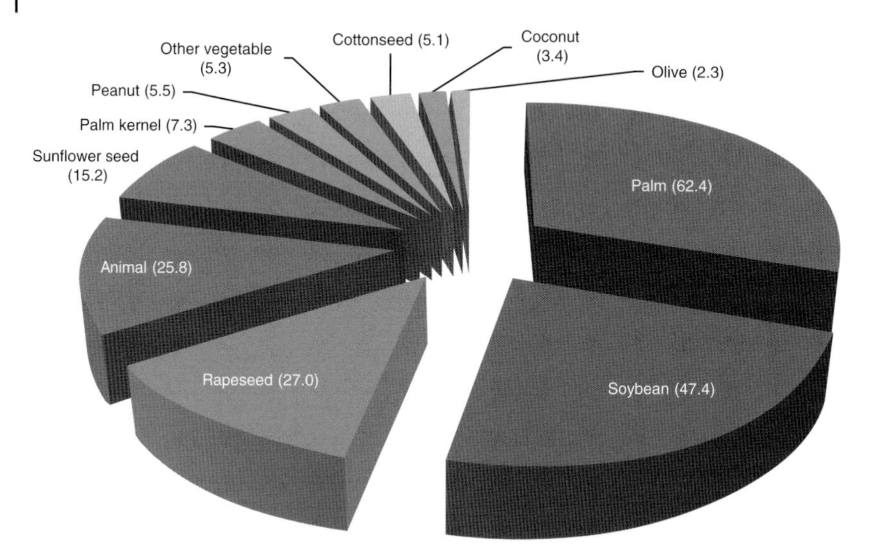

Figure 3.1 World production of major vegetable oils and fats in 2014 (in million metric tons) [1, 2].

human consumption. Alternative lipid sources, such as algae, are also under intense investigation [5], both for fuel and chemicals production.

3.2 Biorefinery and Related Concepts

The use of natural and agricultural feedstocks for the production of chemicals has grown rapidly in recent years and has led to the development of integrated "biorefinery" concepts [6] as a way to achieve the scale and economics to compete with current mineral and petrochemical routes. A parallel strategy has been followed to adapt existing mineral oil refinery infrastructure, so that the bio-based feedstocks can be used directly, or with limited modification, in the downstream refinery operations. The development of supporting technologies, and the relative prices of mineral sources and agricultural commodities, will determine the rate at which this transformation occurs. A perspective on opportunities for bio-based chemical production in the Port of Rotterdam, for instance, estimated that 10–15% of the fossil oil-based bulk chemicals could be substituted by bio-based equivalents within 10 years [7].

Sugars, starches, and other carbohydrates can be converted to small molecules with high functionality, making them good starting points for production of bio-based monomers and chemicals. Some commercial examples of bio-based chemical production include (poly)ethylene from sugarcane ethanol (Braskem [8]), (poly)lactic acid from corn sugar (NatureWorks [9]), 1,3-propanediol from fermentation of corn syrup (DuPont [10]), succinic acid from fermentation of various sugars (Bioamber [11]), and methanol from glycerol and biogas

(BioMCN [12]). Economically feasible alternative feedstocks are available for glycols, isopropanol, acetone, and methyl ethyl ketone, and production processes already exist or are in advanced stages of development [7].

By contrast, oils and fats have long carbon chains and limited functionality, making them more suited to production of hydrocarbon-rich chemicals and products. The already well-established oleochemical industry deals with such conversions and can be considered as biorefineries, in which crude vegetable oils and animal fats are used as raw materials for a diverse range of chemicals [13]; indeed, the origins of the industry can be traced back to the French scientist Michel Eugene Chevreul (1786–1889), following the publication of his pioneering research "Recherches Chimiques sur les Corps Gras d'Origine Animale" in 1823 [14]. The sector has continued to develop, however, and current lipid research in Europe focuses mainly on functionalization of vegetable oils and their derivatives, especially of the double bond, for the production of specialty chemicals and monomers. Enzymatic and fermentation processes are also developing alongside more traditional chemical routes [15–22]. Furthermore, developments in crop breeding technology have a growing importance in industrial oleochemistry, opening up the possibility for new and improved oilseed varieties with more tailored compositions in an economically viable way [23]; previously, oleochemical feedstocks were intrinsically linked to food production and were rarely developed solely for the production of chemicals and industrial products.

The renaissance in oleochemistry is not confined to academia, and important industrial developments have been seen in the last decade: in the United States, Elevance Renewable Sciences [24] has pioneered the use of metathesis to produce a range of specialty chemicals, ingredients, and feedstocks based on vegetable oils, many of which are novel. In Europe, Novamont [25] is combining both oleochemistry and carbohydrate chemistry in the development of biodegradable polymers for packaging materials, within a broad and highly integrated biorefinery concept; in this example, the unsaturated vegetable oils are subjected to a catalyzed oxidative cleavage reaction in the presence of hydrogen peroxide, forming mono- and difunctional acids, which can be considered basic (lipid-derived) platform chemicals.

In summary, the future of oleochemistry looks bright; in addition to the growth of the traditional oleochemical products, driven in part by improving economics compared to mineral oil-based chemicals, developments in raw materials, process technologies, and in the combination with other plant-based materials, promise to further expand the utilization of oils and fats in industrial products.

3.3 Sustainability of Oils and Fats for Industrial Applications

In 2005, production of palm oil overtook soybean oil to gain pole position as the major vegetable oil source. In 2014, palm accounted for 62 million metric tons versus 47 million tons for soybean oil. The rapid rise of palm oil production can be attributed to both economic and agronomic factors; palm oil yields (including the palm kernel oil) are around 5 metric tons ha^{-1} yr^{-1} (predicted to rise in the short term to more than 6 metric tons ha^{-1} yr^{-1}), compared to 0.7 for rapeseed and 0.4 for soybean [26]. In this comparison, it is important to add that soybeans contain only 20% of oil and are grown primarily for their protein-rich meal. Palm oil is therefore the most efficient of the major vegetable oil crops in terms of land use.

This rapid growth has come with an environmental price tag; unchecked deforestation, destruction of natural habitats (e.g., of the orang-utan), and air pollution from burning of rain forests are major issues connected to the palm oil industry. However, due to the work of the certification body Roundtable on Sustainable Palm Oil [27] (RSPO) and the intervention of activist groups such as Greenpeace, there is a strong momentum in the European oil and fat industry toward the development of sustainable palm oil supply chains. Major multinational food producers, such as Unilever, Nestlé, and Procter & Gamble (P&G), are committed to sourcing palm oil from sustainable production sites, allowing the oil to be traced back to the individual mills. With this development and well-publicized commitments from the multinationals, the primacy of palm in the oils and fats industry looks assured for years to come.

Several Malaysian and Indonesian palm producers have purchased European oleochemical companies in recent years, and others have set up distribution networks to supply the European market. As a result, imported palm streams have, to a large extent, displaced animal fats as the key oleochemical raw material. These vertically integrated producers are now moving further down the supply chain into production of fatty alcohols, specialty chemicals, and surfactants [28]. This has parallels with the growth of the petrochemical industry during the postwar years, in which oil producers moved downstream to convert the crude oil fractions into value-added chemicals and products, mitigating the volatility in oil price and helping to stabilize profits.

Producers of other major oil crops are developing their own criteria for sustainability relevant for the conditions and geographies where they are grown: schemes such as the German REDCert certification scheme [29] and International Sustainability and Carbon Certification (ISCC) [30] have been set up to support the development and monitoring of sustainable supply chains. Imported crude vegetable oils, such as soybean and canola, have additional supply chain complexities to meet the consumer demand in Europe for traceable non-genetically-modified (GM) oils. Although current European Union (EU) rules allow GM crops to be grown anywhere within the EU after approval by the European Food Safety Authority (EFSA) and the oils of GM crops can be imported for industrial and food use, consumer-related markets may demand a non-GM option.

Both fuel and industrial producers target by-products from vegetable oil refining as raw materials: acid oils, fatty acid distillates, and similar. However, these streams have a limited availability. They are typically about 5% of the total vegetable oil refinery output, and there are established applications in oleochemicals, animal feed, and others with which any new use must compete. The reality for vegetable oil refining is that very little of the output is a genuine "waste stream," when that is defined as a stream with zero or negative value – indeed some by-products have a higher value than the incoming crude oil, and thus positively contribute to the economics of the refinery. By-products are generally indexed in value to the crude vegetable oil, which they may substitute when the difference in price is sufficient to overcome the extra costs to purify or process them further. Tall oil provides an interesting example of such a lipid by-product and is obtained from the Kraft or sulfate process for paper pulp production from pinewood. It is used exclusively for industrial purposes and consists of a complex mixture of about 50% fatty acids, 40% rosin acids, and 10% unsaponifiable fraction, which are separately purified and sold for use in a wide range of applications [31].

The use of oils and fats for industrial applications has thus been growing steadily over recent years, at a faster rate than for human nutrition. Furthermore, vegetable oils experience a growing share of the industrial and chemical sector as a sustainable source of hydrocarbon-rich feedstock, through substitution of traditionally mineral oil-based chemicals and products as a sustainable source of hydrocarbon-rich feedstock.

The development of green chemistry and biorefinery concepts also gives a new momentum to the field of oleochemistry. In this chapter, we give a short introduction to the modern oleochemical industry and illustrate the renewed interest in oil and fat chemistry with recent research in the catalytic isomerization of linear fatty acids to branched acids – "isostearic acid (ISAC)." The history of ISAC production and current routes to branched fatty acids (natural and fossil derived) are discussed. The special properties of ISAC are illustrated through its use in applications and processes inaccessible to conventional fatty acids. Finally, the latest developments in zeolite catalysis for its selective manufacture are presented.

3.4 Fatty Acids

Whereas food and food production uses predominantly refined oils and fats, fatty acids are the major raw materials for the production of chemicals and industrial products (excluding direct fuel and biodiesel uses). World consumption and production of fatty acids in 2011 were estimated at 6.0 million tons (excluding production of fatty alcohols), expected to rise to 7.0 million tons by 2016. Western Europe consumes 1.5 million tons (including 170 000 t of tall oil fatty acids (TOFAs)) but produces only 0.9 million tons, the balance being mainly palm-based import streams from Southeast Asia [32]. Soaps, chemicals, surfactants, lubricants, and coatings are some of the important applications of fatty acids; in general, their growth follows Gross Domestic Product (GDP) growth in the region of consumption, reflecting the broad range of end-use industries.

Several processes are described for the hydrolysis ("splitting") of fats and oils into fatty acids and glycerol [33], including enzymatic splitting using lipases, but the dominant commercial process is high-pressure continuous splitting, known as the *Colgate–Emery process*; water is contacted in countercurrent flow with the oil or fat at around 250 °C and 50 bar pressure, in a mass ratio of approximately

2 : 1 fat/water, for residence times of 1–3 h. The resulting hydrolysis of the oil or fat to fatty acids and glycerol is almost complete, with yields of up to 99% of the theoretical value. The dilute glycerol stream (10–15 wt%) can be purified and concentrated to 85–99 wt% by evaporation and distillation and is used in many industrial, consumer, and food products.

The crude fatty acids can then be subjected to a variety of downstream physical processes, such as:

- *Distillation.* The crude fatty acids are purified by distillation to remove odor, color, impurities, and volatile components.
- *Crystallization.* Saturated acids can be separated from unsaturated acids by crystallization, using processes based on solvents, water/surfactants, or simply cooling the fatty acids and pressing out the liquid fraction ("dry fractionation"). Commercial-grade oleic acid is made by all of these methods, starting from fatty acids derived from tallow fat or palm oil.
- *Fractionation.* The term "fractionation" in the oleochemical industry refers to a distillation process in which fatty acids are separated based on the boiling point of the different chain lengths. For example, coconut fatty acids, which consist of carbon chains from C_8 to C_{18}, are fractionated into a C_8–C_{10} fatty acid cut, which has a high value as a raw material for lubricants, cosmetics, and food supplements, and a C_{12}–C_{18} fatty acid cut, which is suitable for soaps and detergents, for example.

In addition to physical processes, chemical modifications, such as hydrogenation, esterification, amidation, ethoxylation, and polymerization, are all currently performed on industrial scale (Figure 3.2).

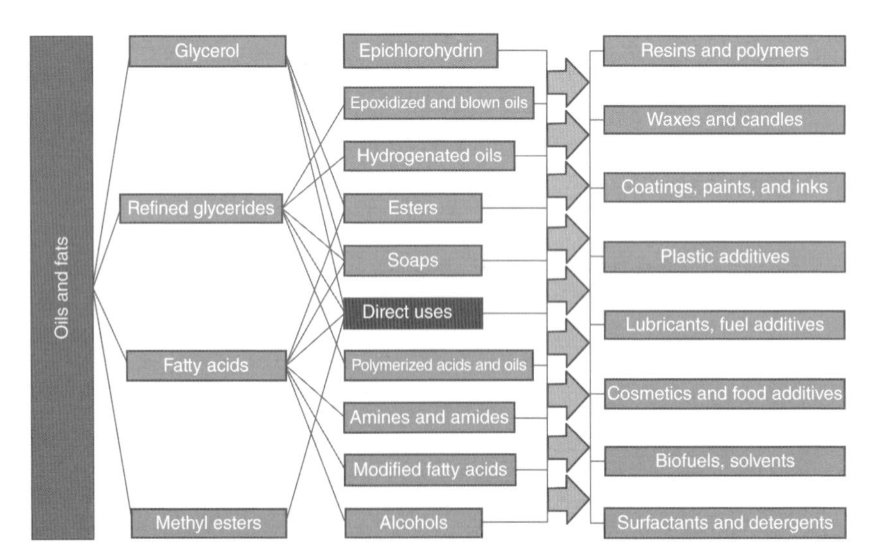

Figure 3.2 Schematic overview of the different conversion routes of oils and fats and the applications of the oleochemicals derived from them. The "modified fatty acids" group includes conjugated, branched, hydroxylated, and hydrogenated fatty acids.

3.5 Polymerization of Fatty Acids

3.5.1 Thermal Polymerization

Thermal polymerization of vegetable oils to produce bodied (thickened) oils was practiced from the beginning of the twentieth century, but in the 1940s attention turned to polymerization of unsaturated fatty acids and their monoesters for use in the manufacture of polymers [34]. The investigations of Bradley and Johnston ([35] and subsequent publications) showed that the main products from thermal polymerization of fatty methyl esters were dimeric and that the structures formed were determined mainly by the degree of unsaturation of the starting material. Johnston hydrogenated the resulting oligomers to polyhydric fatty alcohols using a copper chromite catalyst [36].

The uncatalyzed thermal polymerization of unsaturated fatty acids was performed industrially by Emery Industries in the United States in 1948. TOFA, a by-product of paper production, was the preferred feedstock, being highly unsaturated and therefore sufficiently reactive. The dimer acids were used in the production of polyamide resins for inks and adhesives. However, a process for polymerization of unsaturated fatty acids, and their purification and use in making synthetic resins with improved properties, was described as early as 1935 [37].

3.5.2 Clay-Catalyzed Polymerization

A step change in fatty acid polymerization was achieved in 1957 when Barrett, Goebel, and Myers developed a clay-catalyzed process based on swelling montmorillonite-rich smectite clays (Scheme 3.1) [38, 39].

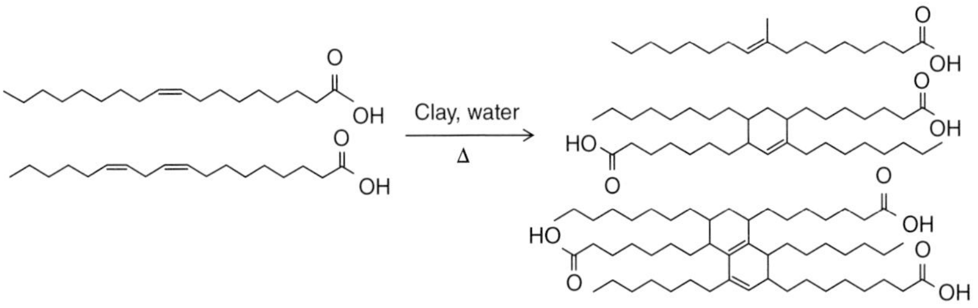

Scheme 3.1 Clay-catalyzed polymerization of commercial oleic acid, which contains mainly linoleic acid in addition to oleic acid. The reaction product is a complex mixture of branched (monomeric) oleic acid, C_{36} dimeric acids, and C_{54} trimeric acids (typical structures are shown for each fraction; note that all fractions contain many isomers).

There are several key advantages of this process over thermal polymerization, namely:

- Reaction times are shorter, and polymerization temperatures lower.
- A wider range of feedstocks can be used; the clay-catalyzed process can polymerize both polyunsaturated and predominantly monounsaturated fatty acids, such as tallow oleic acid.
- Better control is achieved over the ratio of dimeric to trimeric and higher fatty acids, with less dependence on the feedstock composition, due to the constrained reaction sites within the clay interlayers.
- Reaction products with lighter colors are achieved.

A further advantage of the catalyzed process over the thermal one was the quality of the "monomer acids," that is, the non-polymerized fraction of the feed. In the thermal process, it consisted mainly of inert and unreacted fatty acids. However, a different monomer composition was observed in the clay-catalyzed process; the monomeric fraction was not characterized in detail until the 1970s, when McMahon identified it to be rich in methyl-branched saturated and unsaturated acids [40]. Haase, in the late 1980s, further characterized the hydrogenated branched fraction, dividing it into mono- and multiple methyl-branched fatty acids [41]. These methyl-branched isomers, once purified and hydrogenated, became the commercial product now better known as *isostearic acid*.

In later improvements to the clay-catalyzed process, certain alkali and alkaline-earth metal salts were shown to modify the clay interlayer distances and therefore provide a means to fine-tune the process toward higher polymerized fatty acid yields [42]. The modern process includes downstream modifications to the polymerized acids, such as hydrogenation, and further purification of the polymerized acids into dimeric (>98%) and trimeric (>80%) fractions. Derivatives, such as the diol and the diamine, are also commercially available.

3.6 ISAC

During the first decades of the clay-catalyzed process, monomer acids were distilled off from the more valuable polymerized fatty acids and sold as cost-effective fatty acids for a range of applications. Later, the branched acids were purified by solvent separation from the saturated and straight-chain acids and hydrogenated to form what became known as *isostearic acid*, commonly abbreviated as ISAC (Figure 3.3). The first commercial ISAC was produced by Emery Industries, most likely based on the patented process described by Peters in 1957 [43].

It took time for the unique properties of ISAC to be fully appreciated, and markets and applications had to be developed; the relatively high production costs meant that performance benefits had to be demonstrated versus cheaper liquid fatty acids such as tallow oleic acid, rapeseed, and TOFAs (themselves feedstocks for the isomerization/polymerization process). Furthermore, the process could be

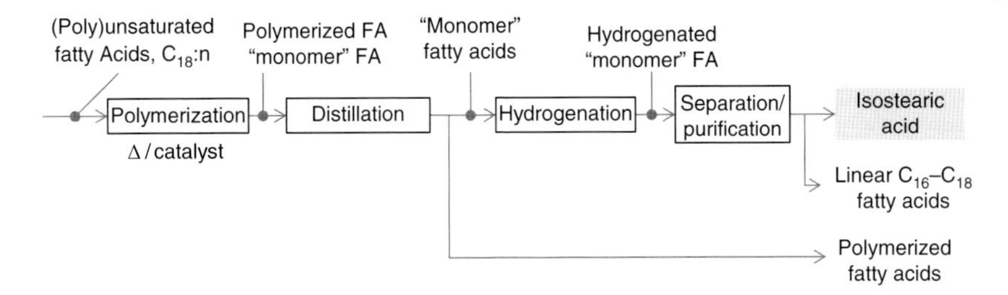

Figure 3.3 Current process for production of isostearic acid and polymerized fatty acids.

steered toward ISAC only to a limited extent, effectively linking ISAC production to polymerized fatty acid demand.

Despite these practical hurdles, the use of ISAC has grown steadily to the point where it can no longer be considered as a "coproduct," but as a high-value functional oleochemical in its own right. The properties of ISAC are unique within the class of vegetable-derived fatty acids, allowing it to be used in applications where the technical performance of commodity fatty acids is insufficient, thus expanding the use of renewable resources in industrial and consumer applications.

3.7 Other Branched Chain Fatty Acids

3.7.1 Natural

C_{20} chain fatty acids with monomethyl branching at the iso- (ω-1) and anteiso- (ω-2) positions are the major constituents of fatty acids bonded to the surface of human hair and mammalian fur [44]. Similar branched structures are found in the milk fat of cows (and other mammals), constituting about 2% of the total fatty acids [45]. Iso- and anteiso-fatty acids with carbon chain lengths of $C_4 - C_{15}$, in the form of wax esters, are major components of lanolin (purified sheep wool grease). Lanolin is a complex mix of lipid components, comprising esters and alcohols as the major fractions [46].

Branched chain fatty acids with methyl groups present on all positions along the alkyl chain have furthermore been isolated from bacteria, although only in minor amounts [47]. Dembitsky presented a comprehensive survey of naturally occurring neoacids (in this definition, "neo" refers to an alkyl chain terminating with a *tert*-butyl group, which differs from the definition for synthetic branched acids covered in Section 3.7.2). The sources identified include plants, fungi, algae, marine invertebrates, and microorganisms [48].

Although interesting, and in many cases serving important biological functions, most of these sources of natural branched lipids are not sufficiently abundant, accessible, or concentrated to provide a viable feedstock source for industrial-scale production. Lanolin is a notable exception; it is purified, separated into different fractions and further functionalized, and represents an important specialty lipid with applications in cosmetics, pharmaceuticals, and industrial products.

3.7.2 Petrochemical

Most of the branched alcohols, acids, and other branched derivatives are currently provided by the petrochemical industry, using ethylene, propylene, paraffins, and (linear and branched) olefins as hydrocarbon sources. Coal and natural gas, current feedstocks for Fischer–Tropsch synthesis processes, can also be considered as raw materials.

Branched acids from petrochemical sources generally have shorter carbon chains and more and/or longer chain branching compared to ISAC. The main commercial route to synthetic branched acids is by carboxylation of olefins, using a modified oxo (carbonylation) process (Scheme 3.2). The olefins themselves are derived from a variety of petrochemical routes and can be branched or linear, internal, or terminal (α olefins). Alternatively, though less common, oxo alcohols can be oxidized under basic conditions to the corresponding acids.

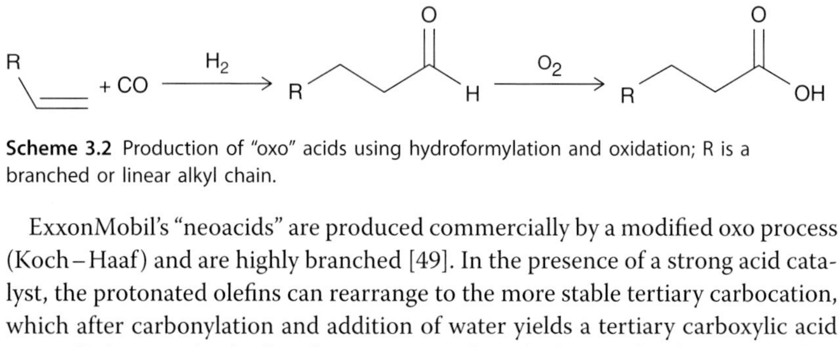

R + CO $\xrightarrow{H_2}$ R–CH₂CH₂–CHO $\xrightarrow{O_2}$ R–CH₂CH₂–COOH

Scheme 3.2 Production of "oxo" acids using hydroformylation and oxidation; R is a branched or linear alkyl chain.

ExxonMobil's "neoacids" are produced commercially by a modified oxo process (Koch–Haaf) and are highly branched [49]. In the presence of a strong acid catalyst, the protonated olefins can rearrange to the more stable tertiary carbocation, which after carbonylation and addition of water yields a tertiary carboxylic acid group (Scheme 3.3). This "neo" structure results in high steric hindrance, which in turn imparts excellent thermal and hydrolytic stability to derivatives. Momentive Specialty Chemicals produce the Versatic™ Acid 10 (neodecanoic acid) by a similar route. The most important oxo-derived acids have total carbon C_9-C_{11} chains and are used as intermediates in the production of, for example, PVC stabilizers, organic peroxide initiators, and metal salt-based catalysts.

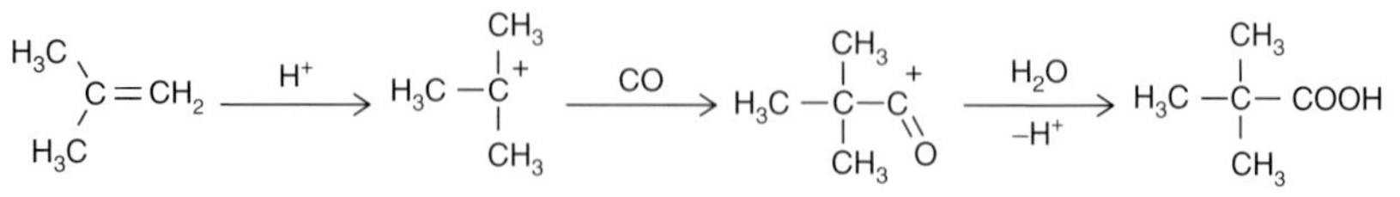

$$(H_3C)_2C{=}CH_2 \xrightarrow{H^+} H_3C-\overset{+}{C}(CH_3)_2 \xrightarrow{CO} H_3C-C(CH_3)_2-\overset{+}{C}{=}O \xrightarrow[-H^+]{H_2O} H_3C-C(CH_3)_2-COOH$$

Scheme 3.3 Production of neopentanoic acid from isobutene via the two-stage Koch reaction.

Closer in structure to ISAC are the $C_{16}-C_{18}$ "isostearic" acids of Nissan Chemical (Fineoxocol® trade name), also derived from an oxo process via oxidation of the corresponding alcohol – a highly branched and a less branched alternative are offered. Due to the absence of linear fatty acids and relatively highly branched structures, the pour points of these synthetic acids are very low (below −30 °C). The published applications of the Fineoxocol® acids are similar to those of ISAC.

Paraffin waxes (linear and branched) can be oxidized with air directly to fatty acids at about 110–130 °C in the presence of cobalt or manganese salts. This route was particularly prevalent in Russia and former Eastern Europe, where an ample supply of wax was available from the petroleum refineries. However, it involves a free radical oxidation process which produces a complex mixture of products including linear and branched acids, esters, aldehydes, and ketones, and isolation of the branched fatty acids requires an expensive work-up process [50]. Production of fatty acids by this route is now of relatively minor importance.

Some specific branched chain alcohols use similar synthetic routes to the acids and can be considered alternatives to isostearyl alcohol, the hydrogenated derivative of ISAC. Isostearyl alcohol is an important personal care ingredient. Shell's Neodol® 67 process, developed in collaboration with P&G for the detergent market, produces lightly branched C_{16}–C_{17} fatty alcohols, structurally close to isostearyl alcohol (Figure 3.4). Notably, the process uses a medium-pore zeolite with unidimensional channels of approximately 0.4–0.7 nm (hydrogen-form ferrierites being preferred) to isomerize linear internal olefins to predominantly (mono-) methyl-branched olefins [51], which are then selectively hydroformylated to the primary alcohol (Figure 3.4). The alcohols, when ethoxylated and/or sulfated, form detergent actives with excellent properties; advantages cited include excellent cold water solubility (despite the longer alkyl chain length compared to conventional detergent surfactants), good tolerance to calcium salts ("hard water"), acceptable biodegradability, and superior surface activity [52].

It is anticipated that the possible growth in Fischer–Tropsch synthesis processes will yield alternative sources of lightly branched olefins from wax cracking as potential raw material sources for the Neodol® 67-type products.

Guerbet alcohols are produced from linear alcohols at high temperature and in the presence of a base and a hydrogen transfer catalyst (copper or Raney nickel). The four-step synthesis route involves oxidation to the aldehyde, aldol condensation, dehydration, and finally hydrogenation to form the branched alcohol (Scheme 3.4). The branching is exclusively on the beta position and of

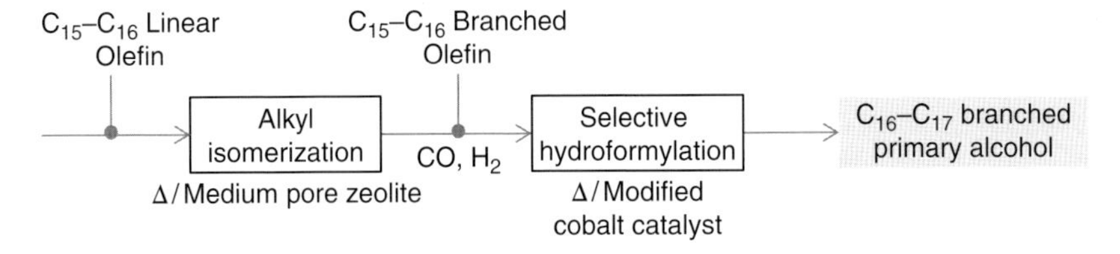

Figure 3.4 Production of C_{16}–C_{17} branched primary alcohols (Neodol® 67) from C_{15} to C_{16} linear olefins.

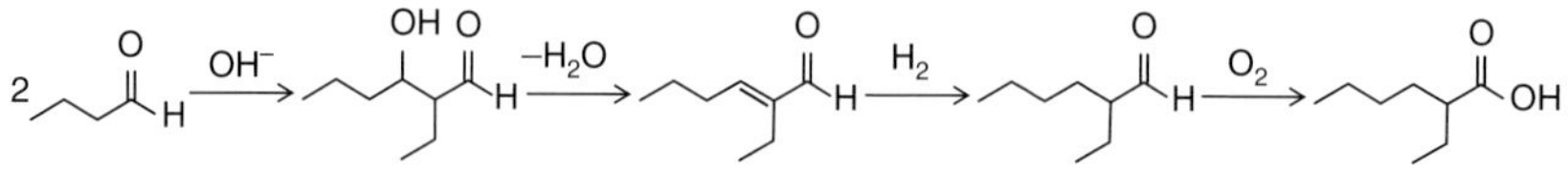

Scheme 3.4 Production of 2-ethylhexanoic acid via the aldol condensation of butanal.

chain length smaller by two carbons than the parent alcohol. This structure gives a significant depression in melting point, making them useful fatty materials for cosmetics and homecare products, a.o. [53]. The corresponding acids can be formed by oxidation of the aldehyde; one of the most important commercial synthetic acids, 2-ethylhexanoic acid, is produced in this way starting from 1-butanal.

Biermann *et al.* have reviewed a range of (mainly) preparative synthesis routes to branched acids, starting from natural (linear) unsaturated fatty acids and methyl esters [54]. Radical additions, Friedel–Crafts alkylations, transition metal-catalyzed additions, and zeolite-induced skeletal isomerization are covered in detail.

3.8 Properties of ISAC

Kinsman has collated and reviewed the physical properties of ISAC and other branched chain acids [55]. This work is further expanded in a book chapter dedicated to branched fatty acids in [56]. More recently, Ngo has measured a number of important physical properties of ISAC (and some of its esters), produced by a zeolite-catalyzed route [57].

In most cases, ISAC exhibits properties which combine the "best" features of oleic acid (*cis*-9-octadecenoic acid, OA) and stearic acid (octadecanoic acid). Some of the key attributes for industrial and consumer applications are discussed later in the following sections.

3.8.1 Thermal and Oxidative Stability

ISAC is comparable in thermal and oxidative stability to stearic acid: this makes it well suited to applications where long shelf life and/or elevated usage temperatures are required. Data from the Kinsman work [55] is illustrative; in an oxygen absorption test performed at 60 °C, ISAC was stable up to 100 days, while commercial-grade oleic acid only reached 1–7 days and technical-grade stearic acid (with a residual unsaturation given by the iodine value of 5 g I_2/100 g) 25 days. Similarly, Ngo compared the oxidative stability of the methyl ester of purified ISAC with soybean oil methyl esters at 110 °C; the ISAC fatty acid methyl esters were 18 times more stable relative to the soybean fatty acid methyl esters according to test [57].

3.8.2
Low-Temperature Liquidity

Liquidity at low temperature is a required property for fatty acid-based ingredients used in outdoor and refrigerated environments, especially in markets such as lubricants. More generally, a liquid fatty acid is easier to store, transport, process, and formulate.

When purified to remove most of the linear saturated fatty acids, ISAC has good cold flow properties, comparable to oleic acid; this is carried over into derivatives such as esters. The branching pattern of oleic acid isomerized using a zeolite shows a skewed distribution around the midpoint [58]. This midpoint methyl branching gives the maximum melting point depression, as confirmed by the study of Cason on C_{19} chain monomethyl-branched fatty acids in which the branching position was systematically varied [59]; it is, however, the complex isomeric mix which accounts for the liquidity in commercial ISAC, since the study of Cason indicates that the pure isomers are all solid at room temperature.

An additional feature in commercial ISAC is the presence of multiple (methyl-) branched species, which when purified by solvent crystallization reach melting points as low as −54 °C [41]; by comparison, a eutectic mixture of (unsaturated) oleic and linoleic fatty acids has a melting point of about −10 °C [60].

3.8.3
Solubility

Derivatives of ISAC show improved solubility across a range of solvent types and polarities compared to the analogous stearic acid derivatives, which makes them easier to formulate and apply in industrial and consumer products. For example, the isostearate salts of zinc and lead were tested in twelve solvents of broadly different polarities and were shown to be compatible with more solvents than either the oleate or stearate analogs [61]. In aqueous solution, the sodium soap of ISAC has a solubility of approximately 15 wt% in water at room temperature, comparable to sodium oleate, whereas sodium stearate is only soluble to approximately 1 wt% [61].

3.8.4
Biodegradability

In general, the inclusion of an alkyl branch in the hydrocarbon chain of a fatty molecule hinders its biodegradability; hence it is important to establish the impact when ISAC and its derivatives are to be developed for use in environmentally sensitive applications.

Commercial ISAC has been tested for biodegradability according to OECD Guideline 301 B (Ready Biodegradability: CO_2 Evolution Test), and the results published on the website of the European Chemicals Agency [62] in support of

REACH registration. ISAC is confirmed to be "readily biodegradable" in this test, meeting the criteria of at least 60% conversion to CO_2 in a 28-day period.

Ngo has studied the biodegradation of ISAC produced by the ferrierite-catalyzed isomerization of high-purity oleic acid [63]. Three strains of *Pseudomonas* bacteria were selected, and all were able to metabolize the ISAC to a good extent, while new linear and/or branched chain fatty acids were formed as products of partial degradation. The results are not directly comparable to the OECD 301 B test, but confirm that ISAC can be readily utilized by bacteria as a carbon source and is thus suitable for use in environmentally sensitive applications where losses and spills could occur.

For a more comprehensive review of the physical properties of natural and synthetic branched acids, the reader is directed to the review of Kinsman [55].

3.9 Applications of ISAC

Due to the (currently) relatively high costs of production and purification, ISAC is used in more demanding applications where the properties listed earlier in section 3.8 are particularly valued. Lubricants, cosmetics/personal care ingredients, and a wide range of industrial applications are important markets for ISAC, and these are discussed in more detail in the following sections (Table 3.1).

3.9.1 Lubricants

Currently, biolubricants constitute only a minor part of the total lubricant market, estimated as approximately 3% in the EU in 2012 [64], although with a moderately

Table 3.1 Overview of the major markets and applications of ISAC and its derivatives, based on patent and published literature.

Markets	Applications	Key properties	References
Lubricant	Metalworking, gear oils, 2-/4-stroke engines, friction modifiers	Oxidative and thermal stability, low toxicity, biodegradability, dispersing lubricity and liquidity	[66–74]
Cosmetic and personal care	Skin and hair care, color cosmetic, sun cream, toiletries	Liquidity, odor and color stability, skin feel, substantivity, water resistance, dispersing and emulsifying	[75–82]
Industrials	Textile softening	Rewettability	[83]
	Metal paste	Dispersing	[84]
	Paper sizing	Water resistance, lubricity	[89]

positive legislative climate this share is expected to double by 2020. Wagner has stated that vegetable-derived lubricants have the technical potential to substitute over 90% of the mineral oil-based lubricants currently in use [65]. However, the considerable cost gap between mineral and vegetable oils must reduce further to encourage switching, and some inherent functional weaknesses need to be overcome. ISAC has great potential here, addressing key concerns regarding oxidative/thermal stability and low-temperature properties, which are commonly cited as stumbling blocks to further development of vegetable-based lubricants. Esters of ISAC are used as components of base oils for the production of synthetic lubricants and greases. In addition to their inherent low toxicity and good biodegradability [66], they exhibit high flashpoints, low pour points, and good application properties. Important base oils include polyol ISAC esters based on pentaerythritol [67], trimethylolpropane (TMP) [68], and neopentyl glycol (NPG). In some cases, the purified ISAC can be substituted by the monomer acids when the application is tolerant of the slightly inferior low-temperature properties and stability.

Important applications within the broad classification of lubricants include metalworking fluids, industrial gear oils [69], two- and four-stroke automotive engines [70], hydraulic fluids [71], greases, and textile lubricants. ISAC-derived amines and amides are also used as friction modifiers in transmission fluids [72, 73]. For a recent survey of synthetic lubricant esters, see [74].

3.9.2 Cosmetics and Personal Care

ISAC and derivatives are important raw materials in the production of a.o. emollients and emulsifiers for cosmetic and skin care products. ISAC esters have similar physical properties to those of unsaturated fatty acids like oleic acid, but with the important benefit of superior odor and color stability (and sometimes additional desirable sensory properties). ISAC can form permeable lipid films on the skin, increasing moisture transport and supporting the penetration of other ingredients through the stratum corneum, making them useful carriers for active ingredients [75].

ISAC esters of light alcohols give a light emolliency, with no oiliness or stickiness. The capryl ester is reported to give similar sensory properties to cyclomethicones (cyclic polydimethylsiloxane polymers currently used in skin care and hair care products as a carrier and delivery system for active ingredients [76]). ISAC can be partially esterified to produce nonionic emulsifiers (e.g., polyglycerol, sorbitan, or polyethylene glycol esters). When fully esterified, they are emollient oils; depending on the structure and functionality of the alcohol/polyol used, a wide range of physical and sensory properties can be achieved, from light oils (e.g., isopropyl isostearate) to viscous, highly skin substantive, and water-repellent liquid waxes (e.g., pentaerythrityl tetraisostearate).

ISAC and derivatives are used to disperse metal oxides in sunscreen products [77, 78] and inorganic pigments in colour cosmetics [79]. High molecular weight isostearate esters based on, for example, oligomers of glycerol and pentaerythritol, are highly substantive, hydrolysis-resistant, and function to enhance gloss in, for example, lipsticks [80].

ISAC and derivatives are also used in hair conditioners, shampoos [81] and colorants, hand soaps, shaving products, deodorants, and nail polish removers. Thickening agents based on ISAC have also been reported (e.g., polyoxyalkylene ethers of glycerin or 1,2-propanediol, esterified with ISAC [82]).

3.9.3
Other Industrials

ISAC and derivatives are used in a very broad range of other markets and applications, including (production of) textiles, paper, plastic additives, surfactants, and detergents. Some examples are given in the following as an illustration of their scope of application.

Cationic surfactants, in particular quaternary nitrogen-containing ones, such as Evonik's Rewoquat® W 325 PG [bis-(isostearic acid amidoethyl)-*N*-polyethoxy-*N*-methylammonium methosulfate] are used in industrial textile softening [83]. The isostearic hydrophobe gives superior rewetting of certain fabrics; that is, it conditions and softens the fabric without excessively waterproofing it. This property has a clear analogy with the use of isostearic derivatives in topical skin care products, which permit the formation of protective yet water-permeable films.

ISAC provides a functional coating for metal and mineral particles used as fillers or pigments in polymer-based products such as plastics and paints; the carboxylic acid group binds to the particle surface, while the fatty chain allows solubilization in the polymer matrix, for example, alumina (surface modification of alumina hydrate with liquid fatty acids [84]). In the form of organo-titanate esters, this compatibilizing coating has broad application (inorganic–organic composites and methods of reacting the same with organotitanium compounds [85]), improving dispersion of the inorganic particles leading to lower melt viscosity, higher filler loadings, and improved physical/mechanical properties in polymer systems. ISAC also provides a dedusting coating for aluminum powder for use a.o. in aqueous slurry explosives [86].

In laundry detergents, formulations containing cationic ISAC derivatives impart antistatic properties to fabrics (ISAC, acrylamide cationic polymers [87], N-alkylisostearamides as antistatic agents [88]). Alkyl ketene dimers based on ISAC-rich fatty acids are used in the paper industry as sizing agents, with improved properties for ink-jet printing [89]. In PVC production, isostearate metal salts (such as zinc, calcium, and tin) are components of liquid stabilizer packages, while esters of ISAC can be used as plasticizers. Polyamide ink resins using ISAC as chain stoppers are claimed to show reduced blooming [90].

3.10
Selective Routes for the Production of ISAC

As shown earlier, ISAC has unique properties within the general class of fatty acids, due to its lightly branched structure. These properties are exploited in a variety of more demanding markets and applications, where commodity liquid vegetable fatty acids are unsuitable or inferior. Its current commercial production is linked to that of polymerized fatty acids (Scheme 3.1). This has several disadvantages, which hinder its further development, namely:

- ISAC coproduction with polymerized fatty acids requires a constant balancing of supply and demand; this limits in any case the supply growth rate to that of the polymer acids.
- Purification of ISAC requires multistep separation processes, which makes the end product costly to produce.
- The clay-catalyzed process offers limited scope to increase the yield of ISAC, and modifications to the process in this direction usually impact the quality of the polymerized acid fraction.
- Synthetic routes to branched fatty acids exist, but they are based on nonsustainable petrochemical sources and/or produce acids or alcohols with substantially different properties.

Therefore, there is a need for a more direct and cost-effective synthesis route to production of ISAC, which would decouple supply from polymerized fatty acid production and allow the unique properties to be fully exploited.

3.10.1
Optimization of the clay-catalyzed process

As mentioned in section 3.5.2, the current commercial clay-catalyzed process for production of ISAC uses a swelling clay in which, skeletal isomerization of the unsaturated fatty acids competes with oligomerization. For both products, the main reaction mechanism is believed to go via the same carbocation intermediate [91]. As a result, complete decoupling of these reactions cannot be achieved by optimization of the process conditions alone, and the catalyst system must be modified.

Oligomerization has been found to be favored by swelling of the clay using water and/or cations (e.g., Li^+), with an increase in oligomer yields from 35% to 55 wt% [92]. Although Brønsted acid sites (tetrahedral sites with Al^{3+} substitution of Si^{4+}) were confirmed as active sites, high yields require the simultaneous presence of octahedral sites (e.g., substitution of Si^{4+} by Mg^{2+}); binding of the carboxylic groups to the latter is believed to prevent poisoning of the tetrahedral sites [93]. Conversely, addition of a small amount of Brønsted acid (e.g., phosphoric acid) as cocatalyst significantly increases yields of the isomerized monomeric fatty acids (from 30% to 50 wt%) [92].

3.10.2
Zeolite-catalyzed branching in the petroleum industry

The replacement of the clay catalyst by a zeolite holds considerable promise for the decoupling of the two competing pathways. Their shape-selective properties are well known in the refining and petrochemical industry [94, 95] and have also been exploited for biomass conversion [96, 97]. The limited space of their intracrystalline volume inhibits the transition state formation from bimolecular reactions, suppressing di- and oligomeric fatty acids. Another advantage of zeolites for the OA skeletal isomerization is the ability to tailor their acidity, in terms of type (Brønsted/Lewis), density, and strength [98].

Shape selectivity effects have been widely studied for the skeletal isomerization of butene to isobutene catalyzed by zeolites and molecular sieves [95, 99–101]. This reaction was heavily studied up to the 1990s in order to meet the expected increasing demand in isobutene for the production of methyl *tert*-butyl ether. Since the US ban of this additive, the intermediate needs are restricted to the methacrylate, polyisobutene, and butyl rubber markets. High yields in isobutene require narrow-pore molecular sieves with 10-membered ring (MR) channels, such as MeAPO-11 [102], SAPO-11 [103, 104], Theta-1 [105], and Zeolite Socony-Mobil (ZSM)-22 (both TON, one-dimensional) [106], ZSM-23 (MTT, one-dimensional) [107], and ZSM-35 (FER) [108]. The three-dimensional 10-MR ZSM-5 (MFI) shows higher butene conversion but a much lower selectivity to isobutene; this is attributed to its larger void volume (from somewhat larger channels and their intersection) [107, 109, 110]. On the other hand, the 10-MR channels (3.1×7.5 Å) of heulandite (HEU) are believed to be too narrow for isobutene diffusion [100]. The best overall performance (activity, selectivity, and stability considerations) is achieved with ferrierite; this zeolite has a pore structure, made from the perpendicular crossing of one-dimensional 10-MR channels (4.2×5.4 Å) with one-dimensional 8-MR channels (3.4×4.7 Å); its moderately strong acidity is also thought to contribute to the low level of oligomerization and cracking reactions [99, 101]. Therefore, since its discovery by Shell, ferrierite has been widely adopted by industry for the isomerization of small olefins (e.g., ISOMPLUS® technology from Lyondell/CDtech) [111, 112] and is commercially available (e.g., from Tosoh Corporation and Zeolyst International).

The Pt/zeolite-based hydroisomerization of long-chain *n*-alkanes ($>C_7$) for the upgrade of petroleum fractions to high-grade fuels and lubricants is another reaction of relevance, known for its specific shape selectivity [95, 112]. It has been the research focus of a number of groups, leading to a detailed understanding of the molecular mechanism [113–115]. High selectivity to branching requires an "ideal" bifunctional catalyst with an optimized ratio between Pt sites for dehydrogenation/hydrogenation reactions and acid sites for the isomerization [114, 116]. The formation of multiple-branched alkanes, precursors of cracking, is again limited by the use of one-dimensional narrow molecular sieves with 10-MR channels, including SAPO-11, ZSM-22, and ZSM-23. Conversely, significant cracking has been observed with zeolites with larger pores (beta and Ultra Stable Y (USY)) [117]. Besides the intermediate state selectivity, the use of the one-dimensional medium-pore zeolites also results in a specific methyl branching pattern that varies as a function of the *n*-alkane chain length. Product diffusion shape selectivity has been invoked to explain this observation [118–120]. Alternatively, the specific branching pattern has also been attributed to pore-mouth and key–lock catalysis, involving van der Waals interactions between the external surface and the long-chain alkanes [121–123]. Similar results, although less marked, have been described for the hydroisomerization of *n*-paraffins in the liquid phase [124]. Latest developments in zeolite engineering to enhance the external surface (and the number of pore mouths) while preserving its acidity have resulted in significant increase of the desired branched alkane yields: for example, hierarchical zeolites obtained by desilication in alkaline medium followed by acid washing [125], as well as by demetallation [126], have recently been reported.

3.10.3
Zeolite-catalyzed branching of fatty acids

The examples listed earlier suggest that zeolites could also represent a promising route to selective production of branched unsaturated fatty acids (BUFAs) in high yields, and have inspired several research groups to study the alkyl isomerization of OA. Focus has initially been on molecular sieves with large-sized channels, a choice motivated possibly because of the bulkiness of the BUFA compared to the isobutene and branched alkanes. Most of the information available is contained in patents, and BUFA yields are often difficult to compare due to the limited analytical information provided on product composition. For example, in some cases the chemical analysis data presented suggest significant purification. Table 3.2 summarizes the key developments.

The first example of zeolite catalysis for OA conversion to BUFA dates from 1997 [127]. The patent discloses the use of molecular sieves, such as protonated mordenite, with large-sized channels (12 MR, 6.5 × 7.0 Å) and linear pore structures. Reactions were carried out for 6 h at 280–300 °C with 4–8% catalyst loading. Addition of a small amount of water was shown to promote the conversion to BUFA, postulated to be due to the conversion of Lewis acid sites into Brønsted ones; this conclusion was drawn based on experiments using zeolite with Si/Al = 7.5 and Si/Al = 9.5. On the contrary, calcination of the catalyst for 2 h at 600 °C led to a significant decrease in BUFA yields, confirming the key role of Brønsted acidity. A direct comparison with the clay-catalyzed process demonstrated the superior selectivity of the protonated mordenite to monomeric acids, although the temperature used for the clay-catalyzed reaction was lower (240 °C) and the solidification point of the zeolite-derived product differed significantly from commercial isostearic. Zeolite USY (Si/Al = 3), and to a minor extent ZSM-5 (Si/Al = 7), gave lower BUFA yields under similar conditions, confirming the importance of the microstructure and/or its acidity.

Table 3.2 Literature overview for the selective OA alkyl isomerization in presence of zeolites and mesoporous materials.

Year	Type	Largest channels	Si/Al	Cat. (wt%)[a)]	Reaction parameters	Catalytic testing	References
1997	Mordenite	12 MR (one-dimensional)	7.5 9.5	8	Water 280–300 °C 6 h	BUFA yields >60%; better than with clay, USY, and ZSM-5	[127]
1997	Mordenite, omega, L-type	12 MR (one-dimensional)	10	2.5	No water 265 °C 4 h	BUFA yields >50% with flat crystals	[128, 129]
2004	Beta, beta/Pd or Pt	12 MR (three-dimensional)	12.5	10	No water 250 °C 5 h	OA conversion up to 50%; range of feedstocks; large fraction of multiple branching	[132, 133]
2007	Beta	12 MR (three-dimensional)	13.5[b)]	3	Water 250 °C 4 h	Conversion >65%	[134]
2004	MAS-5	Mesoporous	25	10	No water 250 °C 5 h	Conversion >60%; more efficient filtration; slower deactivation	[135, 136]
2007	Ferrierite	10 MR (two-dimensional)	8.5 27.5	2.5	Water 250 °C 6 h	Conversion >95%; BUFA yields >70%	[59, 137]
2012	Ferrierite/TPP	10 MR (two-dimensional)	8.5	5	Water 280 °C 6 h	Conversion >95%; BUFA yields >75%; lower oligomer yields	[137]
2014	Ferrierite	10 MR (two-dimensional)	8.5	5	Water 260 °C 4 h	10 recycles with: conversion: 90–99%; BUFA yields: 75–85%	[139]

a) Relative to the feedstock.
b) Calcination at 450 °C instead of 550 °C.

The same year, the potential of one-dimensional zeolites with linear large-sized channels (mordenite, omega, and L-type) was confirmed by Roberts *et al.* [128, 129]. Reactions were carried out for 4 h with 1–5% catalyst loading without addition of water. Optimum temperature around 265 °C and minimum Si/Al ratios of 10 were disclosed. Further improvements were obtained by crystal morphology manipulation, such as flat-plate crystals with the pores along the shortest dimension ($L/D > 10$, where L is the crystallite diameter and D is the crystallite thickness); for both mordenite (hexagonal disks) and L-type zeolites (circular plates), BUFA yields above 50 wt% could be reached with 2.5% loading.

In 2004, large-pore zeolites (>6 Å) were claimed to promote OA conversion to BUFA [130, 131]. Reactions were carried out for 5 h at 250 °C with a high catalyst loading (10%). OA conversions of (up to) 50% were disclosed using beta zeolite (three-dimensional, BEA) with Si/Al = 12.5. The oligomeric fraction is not reported in the product composition, and the comparative example is based on mordenite with very high acidity (Si/Al below 10), so the claim cannot be deduced based on the data provided. The patent discloses the application of H-beta zeolite for a wide range of unsaturated fatty acid sources (a.o. tallow oleic acid, erucic acid), while up to this point most studies used only high-purity oleic acid. Zhang *et al.* also described a "one-pot" process of isomerization followed by hydrogenation, catalyzed by Pt or Pd supported on Beta. In addition, intermediate state shape selectivity was used to explain the observed branching pattern [132, 133]: the large pore size results in both ethyl- and methyl-branched fatty acids, while medium pore size limits the branching to methyl. Compared to the clay-catalyzed process, the large-pore zeolites increase the fraction of multiple-branched fatty acids; quaternary carbon atoms, indicative of geminal, doubly alkyl-substituted ISAC, were not detected in the analysis by ^{13}C NMR, however.

OA conversion to BUFA in the presence of beta zeolite could again be improved by the addition of a small amount of water (as earlier with the mordenite) and by optimizing the zeolite calcination temperature [134]. A compromise between template removal and preservation of the strong Brønsted sites was found at 450 °C instead of 550 °C (conversion of 67% vs. 46% after 5 h at 250 °C for Si/Al = 13.5 and a loading of 3%).

Zhang *et al.* also disclosed the use of mesoporous sieves (Mesoporous Aluminosilicate (MAS)-5, Si/Al = 25) for OA conversion to BUFA [135, 136]. After 5 h reaction at 250 °C with 10% catalyst loading, a conversion of up to 60% was achieved. Such structures are believed to further enhance acid site accessibility to the large fatty acid molecules. Again, no details of the oligomeric fraction are reported in the product composition, and its potential increase due to the larger external surface cannot be ruled out. On the other hand, filtration of such mesoporous sieves from the reaction mixture left little residue in the spent catalyst, and deactivation was less marked (OA conversion only decreased from 60% to 44% upon reuse after filtration, acetone washing, and drying).

3.10.4
Ferrierite – a breakthrough in fatty acid isomerization

In 2007, Ngo *et al.* reported a step change in both OA conversion (>95%) and selectivity toward BUFA (>70%) with protonated zeolites from the ferrierite group [58]. Two commercial ferrierite catalysts ex Tosoh (Si/Al = 8.5, H-Fer-K) and Zeolyst (Si/Al = 27.5, H-Fer-NH_4) were used, the former activated by HCl exchange and the latter by calcination at 500 °C. A number of cocatalysts were tested, of which water was found to be the most effective. Reactions were carried out for 6 h at 250 °C with a catalyst loading of 2.5–5%. Comparative examples with clay (2.5 h at 250 °C with a 4.3% loading) and protonated mordenite (Si/Al ~ 9, 6 h at 250 °C with a 2.5% loading) clearly confirmed the superiority of ferrierite. This zeolite was however found to suffer from deactivation and could not be reused.

Later, the ferrierite selectivity to the monomeric fraction was further improved by the addition of a small amount of a bulky Lewis base as promoter, for example, 2.5% (relative to the catalyst) of triphenylphosphine (TPP) [137]. It is believed that such a base interacts with the non-shape-selective acidic sites on the external zeolite surface. However, this method did not allow the complete suppression of residual oligomers and significantly slowed down the reaction; the temperature was increased to 280 °C to compensate for this effect. TPP also promoted reaction selectivity with mordenite.

Picolinyl esters of the hydrogenated BUFA were analyzed by GC × GC-TOF-MS. Ngo *et al.* reported that zeolites ferrierite and mordenite gave different branching fingerprints (both type and number of isomers differed) [137]. These results are in line with earlier findings of Zhang *et al.* [132, 133]. Surprisingly, when analyzing the picolinyl esters of the BUFA, only 30 isomers were counted in the distillate from ferrierite catalysis [138]; this is hardly higher than the 28 species found after hydrogenation [137], suggesting a preferred geometric configuration for the remaining double bonds.

Ngo *et al.* proposed a series of protocols to regenerate the spent ferrierite, allowing for 10–20 times reuse. After an initial solvent washing, the first method involves a laborious acid treatment that consumes a large amount of water [137]. The second procedure alternates a number of heat (110–260 °C) and acid treatments (after five to six cycles) [139]. The focus of this study was on regeneration, and no details were reported on the deactivation mechanisms themselves. Recently, [140], spent ferrierite catalyst samples obtained from after OA isomerization under similar conditions to Ngo *et al.* were analyzed by a combination of spectroscopic and characterization techniques to get detailed insight into the mechanism of deactivation. The active sites were shown to be poisoned by (poly)enylic species thought to be formed from hydrogen transfer reactions on the OA. In addition, zeolite pore blockage was detected at very early stages of the reaction, due to the formation of mainly alkyl aromatic species. Importantly, this latter observation suggests that only the pore mouth is actively employed in catalysis.

The relationship between ferrierite structure and activity has also been studied in depth for the OA isomerization reaction [141]; in this work, subtle differences in the morphology and acidity (including type, strength and accessibility of the active site) were shown to lead to significant variations in activity and selectivity to (mono)branching. More specifically, a highly active zeolite for OA isomerization requires low external acidity (to inhibit non-selective reactions such as oligomerization of the OA) and an optimal density of Brønsted acid sites in the 10-MR channels.

The examples described earlier clearly show that zeolites hold considerable promise for the selective manufacture of ISAC. The patent literature bears witness to the research effort of the oleochemical industry in searching for the ideal microstructure. Inspired from the development of zeolites in the refining and petrochemical industry in the 1990s, catalytic testing initially targeted the microstructures with medium to large channels. Ferrierite, extensively researched for its superior activity and stability in the isomerization of butene to isobutene, has only recently been tested in the skeletal isomerization of OA. In this reaction, it has been shown to give complete conversion, and high selectivity to the BUFA, representing a major improvement over previously tested zeolites. The most recent research has provided key insights into the mechanism of the reaction and the origins of deactivation, and highlights how certain elements in the catalyst structure play in a key role in determining activity and selectivity. This new knowledge paves the way to even more effective catalyst designs for alkyl chain branching of lipids, which remains an underexploited reaction in the oleochemical toolkit.

3.11 Summary and Conclusions

Vegetable oils and animal fats are important and established raw materials for the production of chemicals and ingredients. From its origins in the production of stearine candles more than 150 years ago, and shaped by major world events in the twentieth century, the oleochemical industry has grown in scale and diversity to become an important branch of the chemical industry. With increasing emphasis on renewable and sustainable industrial practices, oleochemistry has gained a new momentum as a means to develop alternatives to fossil-derived chemicals and products. In the last decades, research in oil and fats for industrial use has extended beyond the traditional boundaries into areas previously occupied by the food, chemical, and other industries. Three directions in particular can be identified:

- *New lipid sources.* Through crop breeding and agronomics, development of alternative lipid sources from plants, yeasts and algae, and exploitation of existing non-food plant oil sources.

- *Catalysis.* Novel catalysts, including enzymes, adapted for oleochemical feedstocks from other fields of development, to achieve new functional derivatives and expand the range of application of oils and fat-derived chemicals.
- *Biorefinery concepts.* Integration of processes and streams from a wide range of plant-derived feedstocks to achieve the economics and scale of production. This is strongly linked to the above developments but goes further in combining traditionally distinct sources (e.g., wood/cellulose, starches, sugars, plant proteins, etc.) and stimulating new developments in process technology, biotechnology, and catalysis required to achieve a true biorefinery concept.

In this chapter, we reviewed the role of catalysis and catalyst development in the production of ISAC, both from a historical perspective and through recent research into more selective zeolite-catalyzed routes. ISAC holds a unique position in the oleochemical portfolio, due to its special properties, and this has been illustrated through a discussion of the markets and applications where its functionality is particularly valued. As an example of the direction of oleochemical research, it illustrates the potential of proper catalyst design to open up significant new markets and applications for oleochemicals.

Acknowledgments

The authors thank Croda for providing financial support. Hans Ridderikhoff, Hans Vreeswijk, Dr Bas Wels (Croda), and Peter Tollington (Cargill) are acknowledged for their helpful comments.

References

1. U.S. Department of Agriculture, Economic Research Service, Oil Crops Yearbook (March 2013) Table 47, at: http://usda.mannlib.cornell.edu/MannUsda/viewStaticPage.do.
2. Oil World Monthly, ISTA Mielke GmbH (March 27 2015).
3. Malveda, M.P., Blagoev, M., and Funada, C. (April 2012) CEH Industry Overview. IHS Report 220.5000, www.ihs.com/chemical (accessed 26 September 2015).
4. National Geographical News (October 28 2010) Mexico's Poor Seek Relief from Tortilla Shortage.
5. Ahmad, A.L., Mat Yasin, N.H., Derek, C.J.C., and Lim, J.K. (2011) *Renewable Sustainable Energy Rev.*, **15**, 584.
6. Kamm, B., Gruber, P.R., and Kamm, M. (eds) (2005) *Biorefineries – Industrial Processes and Products*, Wiley-VCH Verlag GmbH, Weinheim.
7. van Haveren, J., Scott, E.L., and Sanders, J. (2008) *Biofuels, Bioprod. Biorefin.*, **2**, 41.
8. Braskem www.braskem.com.br (accessed 24 September 2015).
9. NatureWorks www.natureworksllc.com (accessed 24 September 2015).
10. DuPont Tate & Lyle Bio Products Company www.duponttateandlyle.com/zemea (accessed 24 September 2015).
11. Bioamber www.bio-amber.com (accessed 24 September 2015).
12. BioMCN www.biomcn.eu (accessed 24 September 2015).
13. Johnson, R.W. and Fritz, E. (1989) *Fatty Acids in Industry: Processes, Properties, Derivatives, Applications*, Marcel Dekker Inc., New York.

14. Cheuvreul, M.E. (1823) *Recherches chimiques sur les corps gras d'origine animale*, Chez Levrault, Paris.
15. Meier, M.A.R., Metzger, J.O., and Schubert, U.S. (2007) *Chem. Soc. Rev.*, **36** (11), 1788.
16. Metzger, J.O. and Meier, M.A.R. (2008) *Eur. J. Lipid Sci. Technol.*, **110**, 787.
17. Metzger, J.O. (2009) *Eur. J. Lipid Sci. Technol.*, **111**, 865.
18. Arno, B. and Pérez Gomes, J. (2010) *Eur. J. Lipid Sci. Technol.*, **112**, 31.
19. Metzger, J.O. and Meier, M.A.R. (2011) *Eur. J. Lipid Sci. Technol.*, **113**, 1.
20. Biermann, U., Bornscheuer, U., Meier, M.A.R., Metzger, J.O., and Schäfer, H.J. (2011) *Angew. Chem. Int. Ed.*, **50**, 3854.
21. Metzger, J.O. and Meier, M.A.R. (2012) *Eur. J. Lipid Sci. Technol.*, **114**, 1.
22. Metzger, J.O. and Meier, M.A.R. (2014) *Eur. J. Lipid Sci. Technol.*, **116**, 1.
23. TCI www.techcrops.com (accessed 24 September 2015).
24. Elevance www.elevance.com (accessed 24 September 2015).
25. Novamont www.novamont.com (accessed 24 September 2015).
26. European Palm Oil Alliance www.palmoilandfood.eu/en/palm-oil-production (accessed 24 September 2015).
27. RSPO www.rspo.org (accessed 24 September 2015).
28. Peng, K.K. (2014) *Chem. Eng.*, **880**, 40.
29. RedCert www.redcert.org (accessed 24 September 2015).
30. ISCC www.iscc-system.org (accessed 24 September 2015).
31. Formo, M.W. (1982) in *Bailey's Industrial Oil and Fat Products*, 4th edn, vol. 2, Chapter 6 (ed. D. Swern), John Wiley & Sons, Inc., p. 384.
32. Malveda, M.P., Blagoev, M., and Funada, C. (July 2012) CEH Industry Overview. IHS Report 657.5000, his.com/chemical.
33. Johnson, R.W. and Fritz, E. (1989) *Fatty Acids in Industry: Processes, Properties, Derivatives, Applications*, Marcel Dekker Inc., New York, pp. 23–72.
34. Mackley, K.S. (1947) *Fatty Acids – Their Chemistry and Physical Properties*, Interscience Publishers Inc., New York, pp. 328–332.
35. Bradley, T.F. and Johnston, W.B. (1940) *Ind. Eng. Chem.*, **32**, 802.
36. Johnston, W.B. (1944) High molecular weight polyhydric alcohol. US Patent 2,347,562, to American Cyanamid Co.
37. Hill, A. and Walker, E.E. (1935) Improved drying oils and their application. GB Patent 428,864, to ICI Ltd.
38. Barrett, F.O., Goebel, C.G., and Peters, R.M. (1957) Process of dimerizing monounsaturated fatty acids. US Patent 2,793,219, to Emery Industries, Inc.
39. Barrett, F.O., Goebel, C.G., and Peters, R.M. (1957) Method of making polymeric acids. US Patent 2,793,220, to Emery Industries, Inc.
40. McMahon, D.H. and Crowell, E.P. (1974) *J. Am. Oil Chem. Soc.*, **51**, 522.
41. Haase, K.D., Heynen, A.J., and Laane, N.L.M. (1989) *Lipid/Fett*, **91**, 350.
42. Barrett, F.O., Goebel, C.G., and Myers, L.D. (1960) Polymerization of unsaturated fatty acids. US Patent 2,955,121, to Emery Industries, Inc.
43. Peters, R.M. (1957) Hydrogenation of structurally modified acids and products produced thereby. US Patent 2,812,342, to Emery Industries, Inc.
44. Liljeblad, J.F.D., Tyrode, E., Thormann, E., Dublanchet, A., Luengo, G., Johnson, C.M., and Rutland, M.W. (2014) *Phys. Chem. Chem. Phys.*, **16**, 17869.
45. Ran-Ressler, R.R., Bae, S.E., Lawrence, P., Wang, D.H., and Brenna, J.T. (2014) *Br. J. Nutr.*, 1.
46. Gunstone, F.D. (1993) *Chem. Phys. Lipids*, **65**, 155.
47. Polgar, N. (1971) in *Topics in Lipid Chemistry*, vol. 2 (ed. F.D. Gunstone), Logos, London, p. 207.
48. Dembitsky, V.M. (2006) *Lipids*, **41** (4), 309.
49. Fefer, M. (1978) *J. Am. Oil Chem. Soc.*, **55**, A342.
50. Stade, H.W. (ed.) (1995) *Anionic Surfactants: Organic Chemistry*, Marcel Dekker Inc., New York, pp. 29–30.
51. Kravetz, L., Murray, B.D., and Singleton, D.M. (1998) Highly branched primary alcohol compositions, and

biodegradable detergents made therefrom. US Patent 5,849,960, to Shell Oil Company.
52. Zoller, U. and Sosis, P. (eds) (2009) *Handbook of Detergents, Part F: Production*, CRC Press, Boca Raton, FL, pp. 129–130.
53. Olenick, A.J. and Bilbo, R.E. (1987) *Soap Cosmet. Chem. Spec.*, **63**, 52.
54. Biermann, U. and Metzger, J.O. (2008) *Eur. J. Lipid Sci. Technol.*, **110**, 805.
55. Kinsman, D.V. (1979) *J. Am. Oil Chem. Soc.*, **56**, 823A.
56. Johnson, R.W. and Fritz, E. (1989) *Fatty Acids in Industry: Processes, Properties, Derivatives, Applications*, Marcel Dekker Inc., New York, pp. 237–255.
57. Ngo, H.L., Dunn, R.O., Sharma, B., and Foglia, T.A. (2011) *Eur. J. Lipid Sci. Technol.*, **113**, 180.
58. Ngo, H.L., Nunez, A., Lin, W., and Foglia, T.A. (2007) *Eur. J. Lipid Sci. Technol.*, **108**, 214.
59. Cason, J. and Winans, W.R. (1950) *J. Org. Chem.*, **15**, 139.
60. Stewart, H.W. and Wheeler, D.H. (1941) *Oil Soap*, **18**, 69.
61. Johnson, R.W. and Fritz, E. (1989) *Fatty Acids in Industry: Processes, Properties, Derivatives, Applications*, Marcel Dekker Inc., New York, pp. 245–247.
62. European Chemicals Agency www.echa.europa.eu (accessed 24 September 2015).
63. Ngo, H.L., Ashby, R.D., and Nuñez, A. (2012) *J. Am. Oil Chem. Soc.*, **89**, 1885.
64. On, D. (2012) *Formule Verte*, 10.
65. Wagner, H., Luther, R., and Mang, T. (2001) *Appl. Catal., A: Gen.*, **221**, 429.
66. Hartley, R.J., Duncan, C.B., and Tiffany, G.M. (1999) Synthetic biodegradable lubricants and functional fluids. US Patent 5880075, to Exxon Chemical Patents Inc.
67. Beimesch, B.J., Schnur, N.E., and Hughes, C.J. (1984) Di- and tripentaerythritol esters of isostearic acid. US Patent 4477383, to National Distillers and Chemical Corporation.
68. Leleu, G., Bedague, P., and Sillion, B. (1977) Trimethylolpropane esters useful as base lubricants for motor oils. US Patent 4061581, to Institut Francais Du Petrole, Rhone Progil.
69. Okada, T. and Hara, S. (2009) Gear oil composition. European Patent 2133405A1, to Idemitsu Kosan Co., Ltd.
70. Svarcas, L.R., Brenner, M.S., and Nau, T.A. (2006) Lubricant additive composition suitable for lubricating, preventing deposit formation, or clean-up of two-stroke engines. WO Patent 2006004806, to Lubrizol Corporation.
71. Bunemann, T.F., Kardol, A.D., and de Mooij, A.C. (2004) Hydraulic fluids. US Patent 20040075079, to Unichema Chemie B.V.
72. Adams, P.E., Bush, J.H., Lahiri, S., and Tipton, C.D. (2005) Friction modifiers for improved anti-shudder performance and high static friction in transmission fluids. European Patent 1534805, to The Lubrizol Corporation.
73. Buitrago, J.A., Shiga, M., Shiroi, T., and Takayama, H. (2009) A lubricating oil composition for automatic transmissions. European Patent 1803796B1, to Chevron Oronite Company LLC, Chevron Texaco Japan Ltd.
74. Boyde, S. and Randles, S.J. (2013) in *Esters. Synthetics, Mineral Oils, and Bio-Based Lubricants: Chemistry and Technology*, Chapter 3 (ed L.R. Rudnick), CRC Press, p. 51.
75. Vetter, W. and Wegner, I. (2009) *Chromatographia*, **70**, 157.
76. Fogel, A.W. (2000) Emollient esters based upon capryl alcohol and isostearic acid. US Patent 6126951, to Bernel Chemical Company, Inc.
77. Yagi, K. and Kanda, N. (2012) Oily dispersion and cosmetic material incorporating this oily dispersion. European Patent 2455061A1, to Tayca Corporation.
78. Strobridge, J.R. (1991) Oil in water emulsion sunscreen composition. US Patent 5041281, to Amway Corporation.
79. Ehara, T. and Yamaguchi, K. (2014) Trehalose fatty acid ester composition. European Patent 1958612B1, to The Nisshin OilliO Group, Ltd.
80. Hutchison, R.B. and Mores, L.R. (1975) Polyglycerol partial esters and their use in lipstick formulations. US Patent 3,890,358, to Emery Industries Inc.

81. Hutcheson, K., Flemington, R., and Gomolka, M.A. (1971) Shampoo composition. US Patent 3590122, to Colgate Palmolive Company.
82. Huettinger, R. and Holtschmidt, U. (1986) Polyoxyalkylene ethers of glycerin or 1,2-propanediol, esterified with fatty acid and/or isostearic acid, their synthesis and use as thickening or solubilizing agents. US Patent 4614622, to Th. Goldschmidt AG.
83. Evonik www.evonik.com (accessed 24 September 2015).
84. Bonsignore, P.V. (1981) Surface modification of alumina hydrate with liquid fatty acids. US Patent 4283316, to Aluminum Company of America.
85. Monte, S.J. and Bruins, P.F. (1978) Inorganic-organic composites and methods of reacting the same with organo-titanium compounds. US Patent 4098758, to Kenrich Petrochemicals, Inc.
86. Kondis, T.J. and Rolles, R. (1973) Isostearic acid coated, non-dusting aluminum particles. US Patent 3,781,177, to Aluminum Company of America.
87. Bauman, R.A. and Pierce, R.C. (1983) Detergent composition providing antistatic properties. US Patent 4418011, to Colgate-Palmolive Company.
88. Bauman, R.A. (1985) N-Alkylisostearamides as antistatic agents. US Patent 4497715, Colgate-Palmolive Company.
89. Malmstrom, O *et al.* (2001) Method of making sized paper, a sized paper grade, and a paper size. US Patent 20010054491, to Raisio Chemicals Ltd.
90. Van Beek, D.A. (1992) Polyamide of dimer acids, alkylene diamine and polyalkylene polyamine. US Patent 5,138,027, to Henkel Corporation.
91. Brütting, R. and Spiteller, G. (1994) *Fat Sci. Technol.*, **96**, 361.
92. Nakano, Y., Foglia, T.A., Kohashi, H., Perlstein, T., and Serota, S. (1985) *J. Am. Oil Chem. Soc.*, **62**, 888.
93. Koster, R.M., Bogert, M., de Leeuw, B., Poels, E.K., and Bliek, A. (1998) *J. Mol. Catal. A: Chem.*, **134**, 159.
94. Ribeiro, F.R., Alvarez, F., Henriques, C., Lemos, F., Lopes, J.M., and Ribeiro, M.F. (1995) *J. Mol. Catal. A: Chem.*, **96**, 245.
95. Degnan, T.F. (2003) *J. Catal.*, **216**, 32.
96. Perot, G. and Guisnet, M. (1990) *J. Mol. Catal.*, **61**, 173.
97. Jae, J., Tompsett, G.A., Foster, A.J., Hammond, K.D., Auerbach, S.M., Lobo, R.F., and Huber, G.W. (2011) *J. Catal.*, **279**, 257.
98. Hunger, M. (2010) in *Zeolites and Catalysis, Synthesis, Reactions and Applications*, Chapter 17 (eds J. Čejka, A. Corma, and S. Zones), Wiley-VCH Verlag GmbH, Weinheim, pp. 493–546.
99. Butler, A.C. and Nicolaides, C.P. (1993) *Catal. Today*, **18**, 443.
100. Houžvička, J. and Ponec, V. (1997) *Catal. Rev. Sci. Eng.*, **39**, 319.
101. van Donk, S., Bitter, J.H., and de Jong, K.P. (2001) *Appl. Catal.*, **212**, 97.
102. Gielgens, L.H., Veenstra, I.H.E., Ponec, V., Haanepen, M.J., and van Hooff, J.H.C. (1995) *Catal. Lett.*, **32**, 195.
103. Houžvička, J. and Ponec, V. (1996) *Appl. Catal.*, **145**, 95.
104. Gajda, G.J. (1992) Butene isomerization process. US Patent 5,132,484, to UOP.
105. Barri, S.A.I., Walker, D.W., and Tahir, R. (1987) N-olefin isomerization process. European Patent 247 802, to BP Company.
106. Byggningsbacka, R., Kumar, N., and Lindfors, L.E. (1998) *J. Catal.*, **178**, 611.
107. O'Young, C., Pellet, R.J., Casey, D.G., Ugolini, J.R., and Sawicki, R.A. (1995) *J. Catal.*, **151**, 467.
108. Granvallet, P., de Jong, K.P., Kortbeek, A.G.T.G., Kraushaar-Czarnetzki, B., and Mooiweer, H.H. (1992) Process for the conversion of a feedstock comprising linear olefins. European Patent 0501577A1, to Shell.
109. Seo, R.G., Jeong, H.S., Hong, S.B., and Uh, H.S. (1996) *Catal. Lett.*, **36**, 249.
110. Houžvička, J., Klik, R., Kubelkova, L., and Ponec, V. (1997) *Appl. Catal.*, **150**, 101.
111. Degnan, T.F. (2000) *Top. Catal.*, **13**, 349.
112. Vermeiren, W. and Gilson, J.P. (2009) *Top. Catal.*, **52**, 1131.
113. Weitkamp, J., Jacobs, P.A., and Martens, J.A. (1983) *Appl. Catal.*, **8**, 123.

114. Ernst, S., Kumar, R., Neuber, M., and Weitkamp, J. (1987) *Stud. Surf. Sci. Catal.*, **39**, 531.
115. Huybrechts, W., Vanbutsele, G., Houthoofd, K.J., Bertinchamps, F., Narasimhan, C.S.L., Giagneaux, E.M., Thybaut, J.W., Marin, G.B., Denayer, J.F.M., Baron, G.V., Jacobs, P.A., and Martens, J.A. (2005) *Catal. Lett.*, **100**, 235.
116. Guisnet, M. (2013) *Catal. Today*, **218**, 123.
117. Deldari, H. (2005) *Appl. Catal., A: Gen.*, **293**, 1.
118. Maesen, T.L.M., Schenk, M., Vlugt, T.J.H., de Jonge, J.P., and Smit, B. (1999) *J. Catal.*, **188**, 403.
119. Sastre, G., Chica, A., and Corma, A. (2000) *J. Catal.*, **195**, 227.
120. Maesen, T.L.M., Krishna, R., van Baten, J.M., Smit, B., Calero, S., and Castillo Sanchez, J.M. (2008) *J. Catal.*, **256**, 95.
121. Claude, M.C. and Martens, J.A. (2000) *J. Catal.*, **190**, 39.
122. Claude, M.C., Vanbutsele, G., and Martens, J.A. (2001) *J. Catal.*, **203**, 213.
123. Martens, J.A., Vanbutsele, G., Jacobs, P.A., Denayer, J., Ocackoglu, R., Baron, G., Arroyo, J.A.M., Thybaut, J., and Marin, G.B. (2001) *Catal. Today*, **65**, 111.
124. Munoz Arroyo, J.A., Martens, G.G., Froment, G.F., Marin, G.B., Jacobs, P.A., and Martens, J.A. (2000) *Appl. Catal., A: Gen.*, **192**, 9.
125. Martens, J.A., Verboekend, D., Thomas, K., Vanbutsele, G., Gilson, J.P., and Pérez-Ramírez, J. (2013) *ChemSusChem*, **6**, 421.
126. Martens, J.A., Verboekend, D., Thomas, K., Vanbutsele, G., Pérez-Ramírez, J., and Gilson, J.P. (2013) *Catal. Today*, **218**, 135.
127. Abe, H., Matsumura, Y., Sakuma, Y., and Tomifuji, T. (1997) Process for the preparation of branched chain fatty acids and alkyl esters thereof. US Patent 5,677,473, to Kao Corporation.
128. Hodgson, W.R., Lok, C.M., Roberts, G., and Koetsier, W.T. (1997) Fatty acid isomerization. European Patent 0774451 B1, to Unichema Chemie B.V.
129. Hodgson, W.R., Koetsier, W.T., Lok, C.M., and Roberts, G. (1999) Fatty acid isomerization. US Patent 5,856,539, to Unichema Chemie B.V.
130. Zhang, S., Zhang, Z., and Steichen, D. (2004) Skeletal isomerization of fatty acids. US Patent 6,831,184 B2, to Akzo Nobel N.V.
131. Zhang, Z. and Zhang, S. (2006) Metal ion exchanged solid materials as catalysts for the skeletal isomerization of fatty acids and alkyl esters thereof. US Patent 7,098,353 B2, to Akzo Nobel N.V.
132. Zhang, C., Dery, M., Zhang, S., and Steichen, D. (2004) *J. Surf. Deterg.*, **7**, 211.
133. Zhang, S. and Zhang, Z. (2004) Fatty acid isomerization with mesoporous zeolites. US Patent 6,723,862 B2, to Akzo Nobel N.V.
134. Zhang, S. and Zhang, Z.C. (2007) *Catal. Lett.*, **115**, 3.
135. Ha, L., Mao, J., Zhou, J., Zhang, Z.C., and Zhang, S. (2009) *Appl. Catal., A: Gen.*, **356**, 52.
136. Zhang, S. and Zhang, Z. (2004) Fatty acid isomerization with mesoporous zeolites. European Patent 1490171 B1, to Akzo Nobel N.V.
137. Ngo, H.L., Hoh, E., and Foglia, T.A. (2012) *Eur. J. Lipid Sci. Technol.*, **114**, 213.
138. Ngo, H.L., Dunn, R.O., and Hoh, E. (2013) *Eur. J. Lipid Sci. Technol.*, **115**, 676.
139. Ngo, H.L. (2014) *Eur. J. Lipid Sci. Technol.*, **116**, 645.
140. Wiedemann, S.C.C., Stewart, J.A., Soulimani, F., van Bergen-Brenkman, T., Langelaar, S., Wels, B., de Peinder, P., Bruijnincx, P.C.A., and Weckhuysen, B.M. (2014) *J. Catal.*, **316**, 24.
141. Wiedemann, S.C.C., Muñoz-Murillo, A., Oord, R., van Bergen-Brenkman, T., Wels, B., Bruijnincx, P.C.A., and Weckhuysen, B.M. (2015) *J. Catal.* **329**, 195.

4
Biosyngas and Derived Products from Gasification and Aqueous Phase Reforming

Francesco Basile, Stefania Albonetti, Fabrizio Cavani, Erica Lombardi, and Rodolfo Mafessanti

4.1
Introduction

The production of second-generation fuels from biomass is entering a new stage focused on large projects and industrial initiatives. However, the uncertain scenario in terms of climate change commitment and burden share, fossil fuel price, and renewable legislation limits does not provide the best environment for a large-scale spread of new technologies; furthermore, this innovation still requires a real breakthrough in order to lead to a clearly enhanced efficiency and sustainability. There are different approaches driving the conversion of biomass bulk or its fraction into fuel: (i) the fermentation process to bioethanol and biomethane, (ii) the production of oxygenated liquid compounds by several treatment (pyrolysis, liquefaction, triglyceride extraction, etc.) followed by hydrotreating (HDT) to upgrade the bio-oil produced, and (iii) gasification or aqueous phase reforming (APR), which produces syngas and hydrocarbons or alcohols. Gasification uses an oxidizing agent (air, steam, oxygen, or a combination of these) for the high-temperature transformation of a carbonaceous feedstock into a gaseous energy carrier consisting of permanent, non condensable gases, mainly syngas. Subsequently, in a separate stage, the syngas obtained is transformed by Fischer–Tropsch (FT) reaction into diesel or other syngas fuel such as methanol dimethyl ether (DME) or C_{2+} alcohols.

The APR reaction is an alternative pathway to gasification for the transformation of biomass and its main components (sugars, polyols, and even proteins) into several products such as hydrogen, hydrocarbon, and oxygenated molecules, with the main products depending on the reaction conditions, process configuration, and catalysts. The APR treatment is applied to a hydrolyzed biomass in water solutions at 225–300 °C in subcritical conditions.

Chemicals and Fuels from Bio-Based Building Blocks, First Edition.
Edited by Fabrizio Cavani, Stefania Albonetti, Francesco Basile, and Alessandro Gandini.

4.2
Biomass Gasification

Gasification technology was first known three centuries ago, when hot gases from coal and coke furnaces were used in boilers [1]. There are several examples of the gasification of biomass for the production of fuel at low pressure, and some of them even work at moderate pressure, while there is no experience in high-pressure systems using biomass in a full-chain biomass to liquid (BTL) process. Therefore, coal gasification is well established, while biomass gasification is still being developed: these two technologies, however, cannot be compared due to feedstock differences (e.g., char reactivity, composition, ash composition, moisture content, density, and energy content).

The thermodynamics of biomass gasification leads to the formation of CO, CO_2, and H_2 and H_2O in the gas phase, and their specific composition depends on the amount of oxygen, which also determines the overall autothermal temperature in the process. When using an equivalent ratio (ER) of 0.275, the equilibrium leads to

$$CH_{1.4}O_{0.59} + 0.29\,O_2 \rightarrow 0.63\,H_2 + 0.07\,H_2O + 0.90\,CO + 0.10\,CO \qquad (4.1)$$

In practice, CH_4 and light hydrocarbons, benzene toluene and xylene (BTX), and tars are still present in the gas stream. The first stage of gasification is the pyrolysis of the biomass leading to a vapor, a solid, and a gas phase. The solid and vapor are then gasified by using steam and oxygen (Figure 4.1).

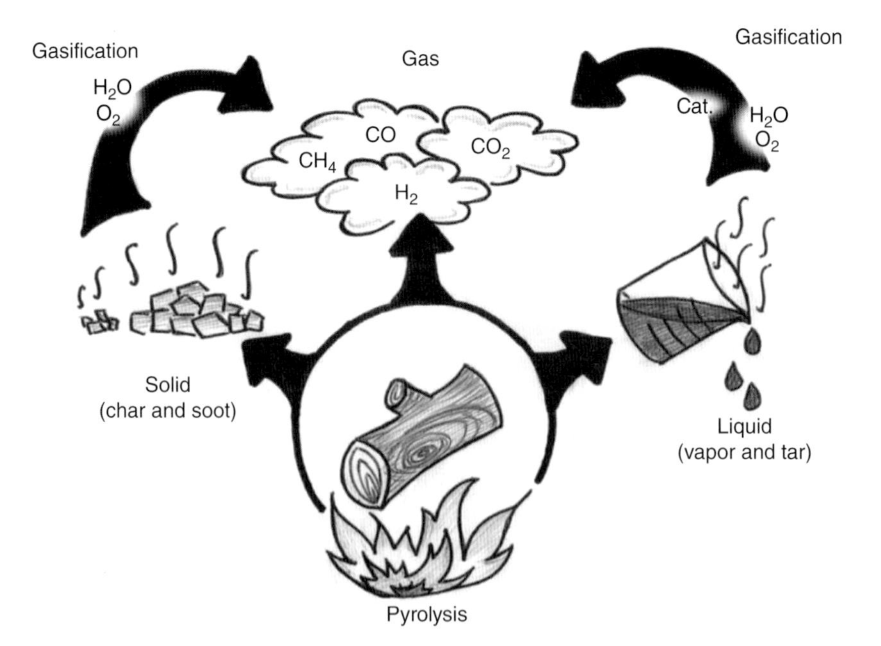

Figure 4.1 Representation of the biomass transformation during the gasification process.

Table 4.1 Composition at the exit of gasifier from Varnamo and Gussing plants [4].

Component	Direct (vol% dry gas)	Indirect (vol% dry gas)
CO	19–21	22–25
CO_2	38–42	20–25
H_2	18–20	35–45
CH_4	10–11	10–12
C_2–HC's	2.5	3–4
BTX	0.2 g m^{-3}	0.3 g m^{-3}
Naphthene	0.1	0.15 g m^{-3}
Tars	20–30 mg m^{-3}	30–40 mg m^{-3}
NH_3	500–1000 ppm	500–1000 ppm
H_2S	100–300 ppm	200–300 ppm
P	20 bar	6 bar
H_2O content	35–40% of the overall gas	>50 of the overall gas

Various gasifiers have been designed, but only those with direct and indirect (or double) fluidized bed and entrained flow configurations are being considered in applications that generate over 10 MW_{th} [2, 3]. Fluidized bed gasifiers are produced by a number of manufacturers in the range from 1 to 150 MW_{th} for operations at atmospheric or high pressures, using air or oxygen as the gasifying agent. This process produces a gas phase that contains several contaminants such as H_2S, NH_3 particulate, tars, alkali metals, fuel-bound nitrogen compounds, and an ash residue containing some char. The gas composition and contamination level vary with the feedstock, reactor type, and operating parameters (Table 4.1).

The main advantage of the gasification process is the transformation of the biomass as such, or after drying and mechanical pretreatments, into a gas mixture having a cold gas efficiency (CGE), that is, the high heating value (HHV) share of the biomass input that is contained in the cold gas up to 80% (Figure 4.2). The energy of the process is stored in the chemical bond, that is, H_2, CO, CH_4, light hydrocarbon, and tars – so the main disadvantage is the partial loss of the energy in the form of heat due to side reactions forming total oxidation species. What is more, the biomass gasification, if compared with other sources of fossil fuels (coke, heavy vacuum distillation residues, etc.), reaches the maximum level of HHV in the gas phase with an ER (ER with respect to the stoichiometry of total combustion or λ) of 0.25–0.33. This is due to the presence of oxygen in the biomass feedstock, which leads to an autothermal temperature below 950 °C in oxygen/steam gasification [5, 6]. At this temperature the process is far from the equilibrium. Some reactions, such as water–gas shift (WGS), are considered to be closer to the equilibrium, while reactions such as hydrocarbon reforming and cracking or char gasification are far from the equilibrium, therefore methane light hydrocarbons and tars are still present. Moreover, the amount of tars present is rich in energy and needs to be subsequently transformed into syngas to obtain a high fuel yield [7].

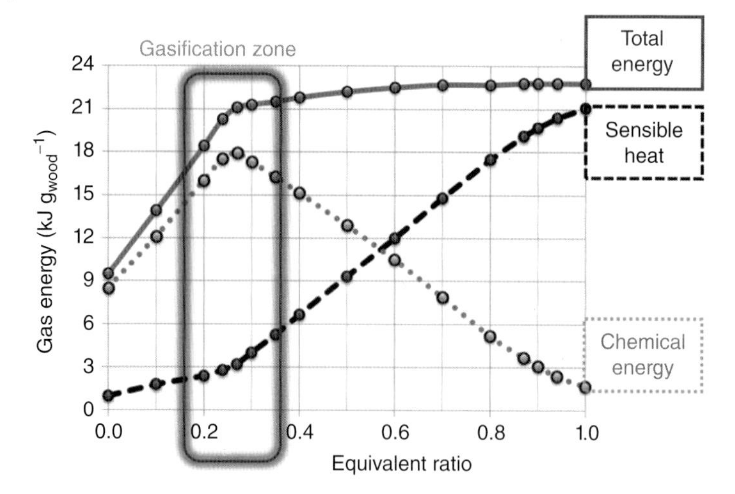

Figure 4.2 Total energy share of the gas products between sensible heat and chemical energy as function of the oxygen addition (ER) [4].

4.2.1
Gasification Process

From a scientific standpoint, the gasification process toward syngas can be run in oxygen/steam at about 850–950 °C. The gas is then cleaned at low to medium temperature using a metallic candle filter at 400 °C and desulfurized using ZnO and then either upgraded by reforming methane, light hydrocarbons and tars or just separated from tar before fuel production. Nevertheless, if the reforming is carried out after cooling, the process is not efficient in terms of energy required since, after cooling, it is necessary to raise the temperature up to 900 °C and higher for the reforming process. Furthermore, it is not economically efficient, because the oxygen production plant either increases the cost or requires a larger plant size (200 MW_{th}), with a consequent larger biomass land extension necessary to feed the plant.

There are several different strategies for increasing the CGE and decreasing the cost. In particular, two main points must be taken into account for efficient and sustainable biomass transformation, and different strategies can be used to implement them in gasification processes [3, 8, 9]. The first point to be optimized is the CGE of the process for the conversion into fuel of the chemical energy contained in the biomass, while the second point is the optimum plant capacity in consideration of both the availability of biomass and the economy of the oxygen plant.

4.2.1.1 Densification and High-Temperature Gasification

An increase in the gasification temperature generally obtained through an increase in the feedstock energy density, which makes it possible to increase both

Table 4.2 Gasification reaction occurring after pyrolysis in the gasifier [4].

Main gasification reactions	ΔH^0_{298} (kJ mol^{-1})
$VOC \leftrightarrows CH_4 + C$	<0
$C_nH_{2m} + nH_2O \leftrightarrows nCO + (m+n)H_2$	<0
$C + 1/2O_2 \rightarrow CO$	−111
$CO + 1/2O_2 \rightarrow CO_2$	−254
$H_2 + 1/2O_2 \rightarrow H_2O$	−242
$C + H_2O \rightarrow CO + H_2$	+131
$C + CO_2 \rightarrow 2CO$	+172
$C + 2H_2 \leftrightarrows CH_4$	−75
$CO + 3H_2 \leftrightarrows CH_4 + H_2O$	−206
$CO + H_2O \leftrightarrows CO_2 + H_2$	−41
$CO_2 + 4H_2 \leftrightarrows CH_4 + 2H_2O$	−165

the heat produced during gasification and the temperature of the process inside the gasifier, consequently increases the kinetics of the reactions listed in Table 4.2 and reduces hydrocarbons to a low level (below 1–2% as carbon content).

The densification process requires a thermochemical pretreatment carried out using different configurations: the most common is torrefaction with the aim of eliminating water by hydroxide condensation; a second option is a low-temperature pyrolysis or gasification which makes it possible to use either the solid and liquid phase as slurry or the solid and gas in high-temperature gasification. The use of the solid and gas in a slurry offers the possibility to carry out the pyrolysis process in a delocalized plant, thus reducing transportation costs and, especially, energy consumption by at least 3/4 while increasing the useful biomass crop distance from 30–40 to 150–200 km (Figure 4.3) [10, 11, 4]. This concept was developed in the bioliq® pilot plant at Karlsruhe, where the pyrolysis step would be carried out in several 50 MW_{th} pyrolysis plants. The liquid and solid products as slurry would be then transported as slurry and transformed into a centralized entrained flow gasification of 2000 MW_{th} at 50 bar and 1200–1400 °C with a very high conversion rate (99%) and low methane slip (0.1%). Obviously, the problems connected with the complexity of running multisite plants are reducing the impact of this solution. A further problem concerns the possibility to produce a stable and transportable slurry.

The use of the gas and solid coming from a low-temperature gasification pretreatment is the system employed by Choren, which has developed the Carbo-V® process. The latter is based on a low-temperature gasifier from which the gas is partially oxidized in a burner (using a higher oxygen ratio than direct gasification) at high temperatures, and the solid is added and gasified downstream of the partial oxidation (PO) by endothermic reaction. This reaction is run in the same reactor as the PO and works as chemical quenching by producing a vitrified slag and a gas with low unconverted levels of hydrocarbon and methane [12]. In the full-plant hypothesis, the first stage will be carried out for a 45 MW_{th} capacity and will be tested full scale in the planned plant in Freiberg (beta plant), while the alpha plant

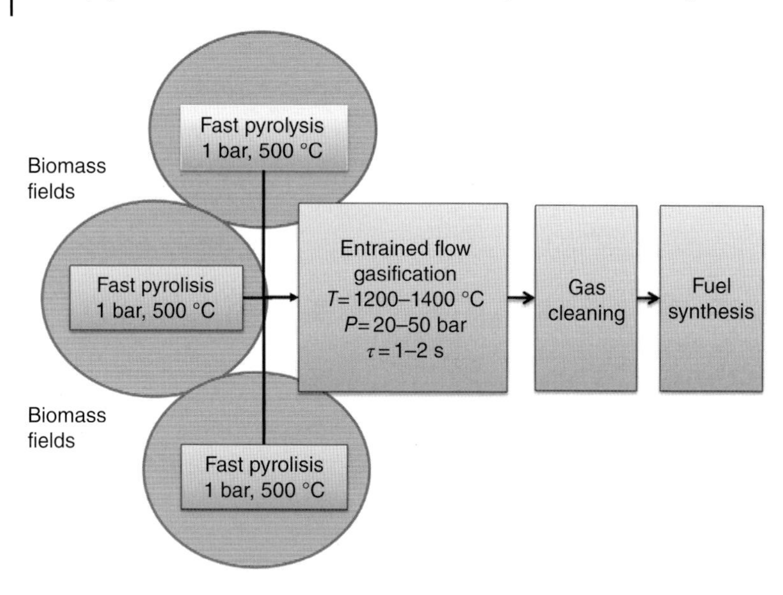

Figure 4.3 Concept at the base of the bioliq pilot plant in Karlsruhe [4].

is a 1 MW_{th} pilot. The beta plant choice to prove the low-temperature gasification at full scale indicates the importance of defining the operability of this part of the plant, which is characterized by mechanical stirring and difficulty in scaling up due to the partial transformation of the biomass. Unfortunately, the project construction has slowed down due to financial problems, and Choren was sold to Linde Engineering in 2012 that has finalized the detailed engineering and optimized the units claiming to be ready for long-term operation. The full-scale plant was claimed to produce 18 Ml of BTL. The second gasification reactor is planned for a capacity of 300–400 MW and will work at moderate pressure (5 bar), producing a gas containing 0.5% of methane, with tar under the detection limit, and vitrified slag. The full-scale plant was planned to produce 18 Ml of BTL and requires several low-temperature stages close to the second gasifier to run continuously. In conclusion, compared to previous processes, it loses the advantage of reducing transportation costs, even though it avoids product stability problems [13, 14].

A very common approach is that of coupling the gasification process with a torrefaction pretreatment that has been proposed in several projects [15]. Uhde has optimized a proprietary entrained flow capable of working at 1200–1400 °C with different feedstocks, including biomass after torrefaction, while Air Liquide is developing several techniques to convert the pretreated biomass into syngas in an oxy burner at 1300–1400 °C; a pilot plant to test this concept at 1 $t\,h^{-1}$ is in the development stage.

British Airways is planning to use 600 000 t of waste to produce over 50 000 tons of biojet fuel and 50 000 t of biodiesel per year in Thurrock, Essex. The plant will use Solena's Plasma Gasification (SPG) technology, which can process 20–50% more waste than conventional gasification technologies, as well as commercial Velocys FT technology for the production of jet fuels.

4.2.1.2 Direct Gasification

The direct gasification process includes a fluidized gasifier, gas cleaning, and upgrading sections, and fuel synthesis (Figure 4.4). Fluidized combustion and gasification for heat and power production are mainly based on fluidized bed reactors. Nevertheless, gasification is not sufficient for obtaining an efficient process for the conversion of biomass into synthesis gas for chemical or fuel production, due to the conversion limits for carbon feedstock in synthesis gas. In steam/oxygen circulating fluidized bed (CFB) gasification, the overall tars and BTX products account for almost the 15% of the heating value of gasification products, and, therefore, they must be transformed into syngas in order to increase the fuel yield. Furthermore, the need to use oxygen in the fuel production entails an increase in the plant size in order to balance the investment cost of the oxygen purification section; as a consequence, in region without forest residues, it is also difficult to ensure the availability of biomass necessary to feed the 200 MW thermal fluidized bed reactor hypothesized as the scale-up size for various pilot plants.

Another solution is Enea's Unique plant project, which carries out the reforming and separation using ceramic candles directly in the fluidized bed. Nevertheless, the system has not been extensively tested for longtime operation.

Since the cost of the air separation unit (ASU) contributes to limiting the possibility to reduce the size of the plant, an alternative solution comes from the use of the allothermal fluidized bed process. The latter can be run in a double fluidized bed reactor with the solid circulating between the two reactors. In this configuration, the second reactor is an endothermic gasifier fed with biomass and steam, and the first step is used to oxidize the char produced with air. The thermal cycle is guaranteed by the circulating material, which is heated to 1200 °C during oxidation in the first reactor and then cooled down to 850 °C during the endothermic gasification process. The unconverted hydrocarbon in the gasification step is

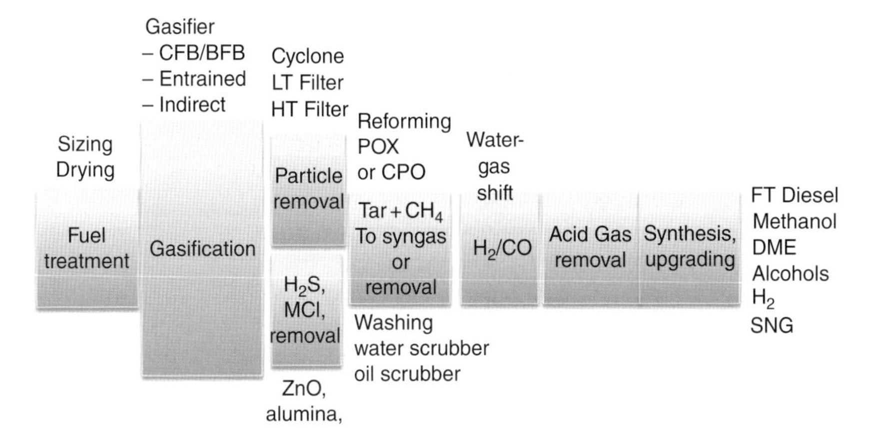

Figure 4.4 Gasification, cleaning, and upgrading unit using direct gasification with a fluidized bed.

higher than that obtained in the gasification with oxygen due to the outlet temperature below 900 °C (i.e., methane concentration within 8–11% instead of 5–7%). This principle was applied with some technical success in Gussing, with the aim of running a combined heat power (CHP) cogeneration system in which the energy of methane and C_2-C_3 hydrocarbons is used to produce heat and power, while tars are absorbed using the Olga process.

The use of the double fluidized bed gasifier is analogous to the double-bed dehydrogenation process based on an endothermic dehydrogenation reactor in which the heat is delivered by the catalyst circulating between the first reactor and a regeneration reactor which oxidizes the carbon deposited on the catalyst surface.

In the case of fuel production, there are two alternative ways to use the Gussing plant: for transforming hydrocarbons into synthesis gas by a reforming process or transforming synthesis gas into biosynthetic natural gas (BSNG)) by methanation. The former requires high energy input to run a classical reforming (because the autothermal reforming (ATR) requires oxygen, which is not present in the plant), while the latter seems more suitable for this type of plant. The Gussing plant and most of the projects described are still running at low or moderate pressure when compared with classical syngas production processes that run at 20–30 bar to facilitate the fuel production and are optimized for high-pressure plants, whether the production is carried out via methanol or via FT reaction.

Extended demonstration activities are required to better identify the advantages and drawbacks of each technology, as well as the improvements in pretreatment, gasification, and gas cleaning steps. In any case, it can be stated that the winning technology has not yet emerged and will certainly be site and biomass dependent, in terms of biomass availability, distribution, and quality [16, 17]. Moreover, the biomass choice has been recently extended to wet biomasses claiming efficient technology [18].

4.2.2 Catalytic Gasification

An increasing trend in using active fluidized bed materials or catalysts for gas cleaning and upgrading, and also in using entrained flow gasifiers, can be seen in research literature and demonstration activities [19].

Primary catalysts can be added directly to the biomass by dry mixing with the bed material or additives. These catalysts operate under the same conditions as the gasifier and are usually made of cheap material. The main purpose is to reduce tar content, but the method has little effect on the conversion of methane and C_{2-3} hydrocarbons into the product gas. Dolomite, with the general formula $MgCO_3 \cdot CaCO_3$, is a catalyst active in tar reduction inside a fluidized bed biomass gasification. It makes it possible to increase gas yields at the expense of condensable and heavy products. The main advantage is that a very high tar conversion can be obtained in suitable conditions, even though it deactivates rapidly, due to carbon deposition and attrition; on the other hand, however, it is cheap and easily replaced.

In another configuration, the catalyst can be located downstream of the gasifier in a fluidized bed reactor working at temperatures over 800 °C, and it is especially active if previously calcined. Dolomite has no acid sites, and, therefore, the interaction with the alkali present in the biomass is very poor, thus avoiding sintering phenomena. Moreover, dolomite is not active for reforming the methane present in the product gas, and its main function is the removal of heavy hydrocarbons prior to the reforming of the light hydrocarbons carried out to maximize syngas yield. Furthermore, when using dolomite- or CaO-containing materials, special attention must be paid to the equilibrium reaction in $CaCO_3$ and CaO. In fact, the latter concentration increases with T and decreases by increasing CO_2 concentration in the feed (e.g., using oxygen as the gasifying agent instead of air).

An alternative material is olivine, a mineral containing magnesium iron silicate. Olivine has the advantage of having higher attrition resistance than dolomite. Pretreated olivine has a good performance in tar reduction, and its activity is comparable to calcined dolomite [20]. Natural olivine can be used as a biomass gasification catalyst in a fluidized bed reactor for tar reduction but also as a nickel support [21] for converting light hydrocarbons and methane also. The iron present in the structure stabilizes the nickel in reducing conditions. Moreover, some nickel seems to be present in the olivine structure as an oxide, and this inclusion leads to an increase in free iron oxide on the surface, which is claimed to induce a reverse WGS reaction. Ni-based catalysts are very effective not only for hydrocarbon reforming but also for decreasing the amount of nitrogenous compounds such as ammonia. Since, in terms of catalytic systems, Ni-modified olivine seems to meet the activity and mechanical stability requirements, its use is suggested downstream of the gasifier because the presence of nickel is not compatible with the recycling of ashes, which is usually prescribed in the biomass gasification sustainability assessment.

Alkaly catalysts are added directly to the biomass by either wet impregnation or dry mixing. They reduce tar content and – to a lesser extent – affect the methane content in the product gas. However, the increase of alkali concentration increase the tendency toward particle agglomeration of bed materials. Alkali and their salts are present in the ashes of several biomass types, so ashes are considered effective catalysts for tar removal. Analysing potassium carbonate supported on alumina it appears to be more resistant to carbon deposition than Ni, although not as active, since it has a much lower hydrocarbon conversion.

In addition to dolomite and olivine, MgO has also been used in several gasifiers as bed material with enhanced properties. Unlike silica-containing materials, its basicity increases the system stability even when using a biomass with a high amount of potassium and sodium contained in the ash. This is because MgO does not interact with Na and K and does not exhibit any sintering or agglomeration phenomena. The peculiar characteristic of MgO is the increase in hydrogen production during gasification, leading to a H_2/CO ratio of 2 [22, 23]. This is believed to occur due to the increase in the WGS reaction, probably affected by the presence of iron oxide in extracted magnesite ore (3–5%). This explanation, however, is not in agreement with the constant trend observed when changing the steam-to-biomass ratio. A further explanation is the dehydrogenation activity of MgO at high temperature by deprotonation mechanism: in this case the increase in H_2 concentration in the early stage of the process will then affect other reactions such as hydrocarbon hydrocracking and phenol hydrodeoxygenation (HDO). It is difficult to arrive at a definitive clarification of this topic, since the system does not lend itself to standardization due to the impurities characterizing natural magnesite.

Another attempt at increasing tar cracking inside the reactor deals with the possibility of recycling the ashes produced after gasification, in the field of additives and fertilizers for biomass crops. Since the bed material cannot be separated from biomass residue after gasification, the elements present as tar cracking catalysts are limited to the harmless ones. Therefore, Fe- and Cu-based materials, together with natural alkali oxides and salts, seem to be the best candidates for producing an active bed material, even though tar cracking and reforming activities must be improved [24].

The combination of proper primary measures with downstream methods is seen to be very effective in reducing tar, while they may also contribute to increasing methane reforming and final CGE. Conversely, secondary methods are conventionally used to treat the hot product gas from the gasifier [25]. Two different methods can be described: low-temperature cleaning and "dry" high-temperature cleaning. The conventional "wet" low-temperature syngas cleaning has some efficiency drawbacks and requires additional wastewater treatment, but it is well established, while hot gas cleaning is based on high-temperature filtration. Metal candle filters are widely used after gasification for heat and power purposes. The metal filter works below 600 °C and is not sufficient for use in the gasification coupled with secondary reforming, which runs at a temperature higher than 800 °C. As an alternative, ceramic candles have been proposed in many plants, and some significant tests have been carried out in Chrisgas and Unique projects and developed by Pall Schumacher, obtaining stable performances between 800 and 850 °C. Furthermore, hot gas filters with catalytic cracking activity have also been proposed [26]. It should be noted that during filtration above 800 °C most of the alkali, especially in the form of chloride, are present in the gas phase and will be present in downstream high-temperature processes (tar cracking and hydrocarbon reforming), while the remaining ashes can be efficiently separated. For this reason, at temperatures of 650–750 °C the filtration becomes problematic due to alkali condensation.

4.2.3
Gas Upgrading by Reforming

The exit gas from the gasifier containing methane and light hydrocarbons must be upgraded to a clean syngas in order to produce fuels or other products. Today, the predominant commercial technology for syngas generation is steam reforming (SR), in which hydrocarbons and steam are catalytically converted into hydrogen and carbon monoxide [27] in an endothermic reaction. An alternative approach is

PO, the exothermic, noncatalytic reaction of hydrocarbon and oxygen to produce syngas mixtures, which have appreciably different compositions. In particular, SR produces a syngas having a much higher H_2/CO ratio. This, of course, represents a distinct advantage for SR in hydrogen production applications. Nevertheless, high-temperature SR is difficult to implement since it requires a source of external heat obtained by burning gaseous or liquid fuels, and this either decreases the overall efficiency or increases CO_2 emissions (if a fossil fuel is used). ATR cofeeds oxygen and steam, and the oxidation reaction occurs in the burner at the entrance of the reactor and produces the heat required for the reforming, which takes place downstream in the same reactor. ATR properly refers to a single-step process for feedstock conversion to syngas; however, the same basic idea can be applied to reactors fed by partially reformed gases from a primary reformer or a gasifier. Due to differences in feed composition, in particular the lower HHV, ATR reactors, and secondary reformers have different thermal behavior and constraints in terms of soot formation and require some changes in the burner and reactor design [28, 29]. ATR technology has considerable potential for further optimization, especially in combination with the gas-heated reforming (GHR) principle. The prereformed feedstock of the secondary ATR may be assimilated to the stream coming from a gasifier. In GHR, part of the heat in the ATR effluent is used for the SR reaction, while the feed is preheated in a heat exchange-type reactor in [30]. The catalytic partial oxidation (CPO) process, in which oxidation and reforming reactions occur on the catalytic bed, may be a possible solution if oxygen is available on-site. The CPO makes it possible to work at low residence time (a few milliseconds), with the advantages of small reactor dimension and high productivity. This process is particularly interesting for small- to medium-sized applications and is being developed by numerous companies, in particular by ENI at a demonstration stage. Catalysts containing Ni, Ni/Rh, or just Rh dispersed on high-temperature-resistant supports (stable above 1000 °C) are suitable for this reaction.

In reforming reactions, commercial plants commonly use supported nickel catalysts [28, 29]. The catalyst contains 15–25 wt% nickel oxide on a mineral carrier (α-Al_2O_3, alumina silicates, calcium aluminate, magnesia). Before starting the process, nickel oxide must be reduced to metallic nickel. This is preferably done with hydrogen but is also possible with natural gas or even with the feed gas itself at a high temperature (above 600 °C, depending on the reducing stream). The required properties of catalyst carriers are relatively high specific surface area, low pressure drop, and high mechanical resistance at temperatures up to 1000 °C. It must be stressed that the main catalyst poison in the SR process is sulfur. Concentrations as low as 50 ppm give rise to the deactivation of the catalyst surface [29]. To some extent, activity loss may be offset by increasing the temperature promoting the inert nickel sulfide transformation into Ni or NiO. Moreover, in the specific case of biomass gasification, alkaline salts and metal oxide particles may act as additional poisons. A further cause of activity loss is carbon deposition, which may be avoided by increasing the steam(S)/C ratio, for which economic evaluations indicate an optimum below 2.5–3 v/v.

Table 4.3 Comparison of partial oxidation and ATR upgrading downstream of the gasifier [4].

Component	Exit gasifier (vol%)	ATR 1000 °C (vol%)	POX 1300 °C (vol%)
Inlet O_2	—	7	10
Inlet T C	—	800	800
C_{2-}	1.6	—	—
CH_4	8.2	—	—
CO	11.9	23.8	24.3
CO_2	27.9	19.8	19.2
H_2	11.8	23.0	16.1
H_2O	37.7	33.4	39.7
NH_3	0.3	0.01	0.01
H_2S	0.01	0.01	0.01
Tars + BTX	0.3	—	—
LHV (MJ kg^{-1})	6.6	5.6	4.8
LHV (MJ Nm^{-3})	7.3	5.4	4.8

In fluidized gasifiers, two different cases may be present: in processes using oxygen as the gasification agent, ATR or PO processes may be used [31] to convert the hydrocarbons present in the gasification (Table 4.3), in particular (i) thermal PO at 1300 °C with a calculated loss of low heating value (LHV)/kg of 27% compared to inlet gases and (ii) ATR at 1000 °C (outlet temperature) with a LHV/kg loss of 15% compared to the inlet gases, but requiring 30% of the oxygen used in the first case. This calculation is based on a filter temperature of 800 °C and a final syngas concentration at the equilibrium (70% on dry and nitrogen-free bases).

Although the processes for the production of H_2 and/or syngas are well established [32], their feasibility applied to a gasification-generated gas depends on the activity and stability of the catalysts. The main problems of the reforming process are the deactivation and, especially, sulfur poisoning and coke formation, while some long-term effects of the alkali, passing through the hot gas filter, cannot be ruled out, especially on Ni sintering [33]. In several research activities, the effects of catalytic performances have been studied using model compounds or simulated atmospheres in tar cracking and tar reforming using Ni-based catalysts, having a perovskite structure or obtained from hydrotalcite precursor [34–37]. In some cases, real gas has been used with a Ni-based catalyst placed after hot gas filtration downstream a gasifier fed with three types of biomass ranging from clean wood to miscanthus [4, 23].

The use of clean wood leads to high tar conversion even at a temperature of 550 °C, while high methane conversion requires 750 °C. On the other hand, the use of woody residues (demolition wood) with a higher sulfur content (H_2S concentration of 100 ppm in the inlet gas) requires higher temperatures to reach a similar conversion increasing syngas production of 40% ($CO + H_2 > 70\%$ dry and N_2-free bases) at 1000 °C. This demonstrates the possibility to increase the efficiency and yield of fuel production compared to the PO option (Figure 4.5).

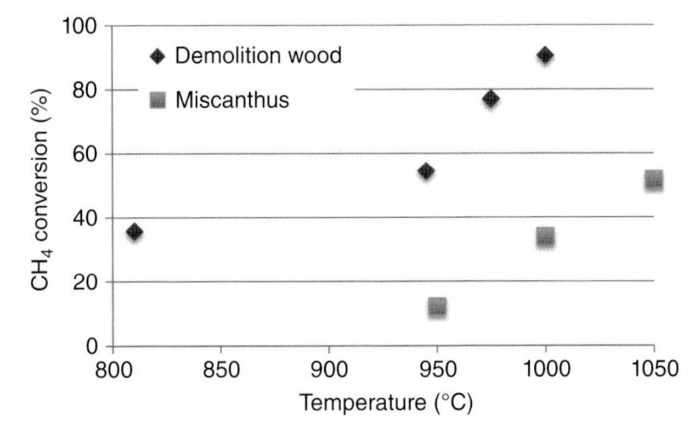

Figure 4.5 Ni/Mg/Al catalyst derived from hydrotalcite used in reforming after gasification of different feedstocks.

Using miscanthus and producing a H_2S content of 200–250 ppm in the inlet gas after gasification, a 50% methane conversion can be reached only at a temperature of 1050 °C (with a $CO + H_2$ concentration of 63%). In these conditions there is still an advantage over the thermal process which must be evaluated in the long term while taking into consideration the deactivation risk due to the presence of alkali at high temperatures, which can induce Ni sintering [38, 39]. When a Ni/Rh catalyst was used with demolition wood, methane conversion was higher than with the commercial Ni catalyst and reached 75% at 945 °C T(out). Methane in the outlet of the reforming was below the detection limit using a T(out) of 1000 °C.

All the tests were stabilized for 1 h and repeated after 2 h to detect any deactivation, showing rather stable results. The Ni/Rh-containing catalyst is therefore more resistant to sulfur deactivation and makes it possible to carry out reactions with woody and herbaceous biomasses at a temperature of 50 °C or more lower than that of Ni-containing catalysts.

The second option is provided by the dual-bed gasifier using air in the combustion zone and steam in the endothermic gasification: in this case, neither ATR nor PO is suitable, since both require oxygen, which is not present in the plant. Therefore, middle-temperature reforming (800 °C) is necessary and sulfur-resistant catalysts must be used. Rh proves resistant to sulfur and can be used at temperatures just above 800 °C, but its cost is very high; furthermore the need to deliver the heat for the reforming requires the use of fuel that probably is of fossil origin. In conclusion, using indirect gasification without any further catalyst development seems more suitable to convert – after tar hydrocracking and WGS reaction – the syngas into methane, producing BSNG.

4.2.4
Downstream of the Reformer

Most of the applications of syngas such as methanol, DME, and FT diesel fuel require an increased H_2/CO ratio after gasification, which usually produces a $H_2/CO = 1$. The presence of several poisons such as HCl and H_2S makes it advisable to avoid CuZn catalysts, which are usually active in the low-temperature WGS with clean stream, and instead use the FeCr system, more resistant to poisons. A FeCr catalyst can be used to shift the synthesis gas without deactivation at the expected H_2S levels (0–200 v/v ppm). The presence of other poisons in the gas phase, such as NH_3, must be considered since they strongly impact the catalyst activity. Even though NH_3 should be mainly transformed into N_2 and H_2 during the reforming step, residual contents of NH_3 could affect the activity of WGS catalysts. The FeCr system requires an inlet temperature of at least 350 °C [40]; alternatively a Cu/Zn catalyst can be used after the ZnO removal of sulfur and HCl. On the other hand, WGS of gases containing high amounts of sulfur and tar requires catalysts mainly consisting of Co and Mo oxides. Moreover, their activity is high only with Co and Mo as sulfides and requires levels of sulfur above 500 ppm.

Synfuel production will not be described and dealt with specifically since it has not yet been specifically developed for biomass. Nevertheless, there are some interesting studies related to BTL production processes focusing on a reduction of the scale factor and pressure of the FT plant. In this framework, due to the exothermic reaction (4.2), the possibility of increasing the heat transfer and the temperature control for small-scale FT reactors appears promising

$$nCO + (2n + 1)H_2 \rightarrow C_nH_{2n+2} + nH_2O \tag{4.2}$$

when using a wall reactor approach or when producing modular reactors that can be increased in number to reach the optimum size. Further optimization and integration have been carried out on catalysts for the FT of biosyngas [41], leading to an increased flexibility of the FT process in terms of H_2/CO ratio by working with Fe-based catalysts.

When focusing on methanol production, the processes aiming for the direct production or coproduction of DME are very interesting because this molecule has good diesel, physical, and chemical properties similar to those of liquid petroleum gas (LPG). The production of DME involves three reactions:

$$CO_2 + 3H_2 \leftrightarrows CH_3OH + H_2O \tag{4.3}$$

$$CO + H_2O \leftrightarrows CO_2 + H_2 \tag{4.4}$$

$$2CH_3OH \leftrightarrows 2CH_3OCH_3 + H_2O \tag{4.5}$$

This synthesis is especially interesting because it makes it possible to minimize the equilibrium constraints inherent in the methanol synthesis by transforming methanol into DME. Moreover, the water formed in reaction (4.3) is to some extent driving reaction (4.2) to produce more hydrogen, which in turn will drive reaction (4.1) to produce more methanol. Thus the combination of these reactions results in a strong synergetic effect, which dramatically increases the synthesis gas conversion potential.

4.2.5
Future Process Breakthrough

Indirect gasification techniques were introduced to lower the cost of the process for small- to medium-sized gasification units (<100 MW). By skipping the Air separation unit (ASU), this process offers great advantages in BSNG production, while syngas and synfuel still require an optimization of the process and the possibility to run it at high pressure in order to facilitate downstream fuel production. The use of other gasification processes requires a larger scale plant to be competitive, even though the present picture is still based on the state-of-the-art system, which does not use catalysts [9]; a 20% increase in syngas yield and a decrease of oxygen consumption can lead to at least a 20% of cost reduction (Figure 4.6).

A possible breakthrough is the integration of the biomass gasification process with water electrolysis, producing H_2 and O_2 from renewable sources. The two systems may be efficiently coupled because oxygen can be used as the gasification agent in a steam/oxygen process skipping the ASU, whereas hydrogen can be added downstream to increase the H_2/CO ratio, which is usually close to 1. This makes it possible to use all the CO produced in the gasification for the production of fuel (i.e., FT) while avoiding the WGS reaction, which consumes CO to form CO_2 (one-third of the CO is used to adjust the H_2/CO ratio).

The process can be integrated using low-temperature electrolyzers but seems much more interesting if a high-temperature electrolyzer (Figure 4.7), where the heat of the reaction can be recovered, is used. In fact, the $\Delta G°$ at 100 °C is 225 kJ mol^{-1}, while $\Delta G°$ at 800 °C is 190 kJ mol^{-1} from a thermodynamic point of view requiring less electric power; furthermore the efficiency of the high-temperature process is 15% higher than that at low temperature and can be even increased using an external heat source as the hot gas at the exit of the gasifier [42].

The direct gasification process using fluidized bed technology is that which can benefit most from the integration at midscale (50 MW$_{th}$). It uses a lower

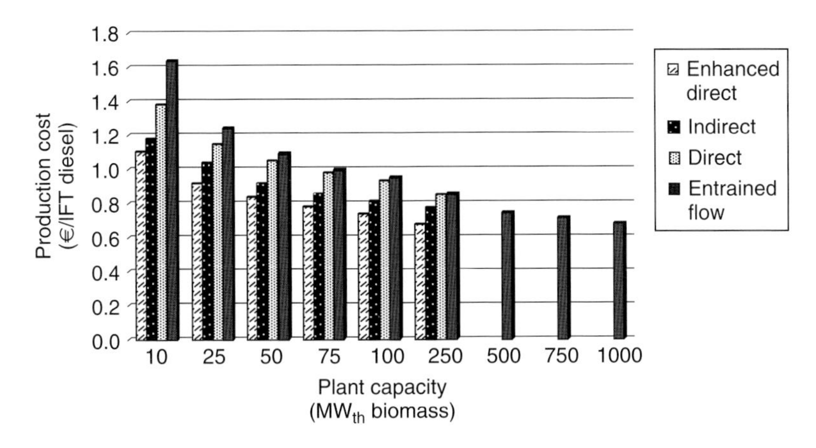

Figure 4.6 Fuel production costs of different gasification technologies. (Adapted from [9].)

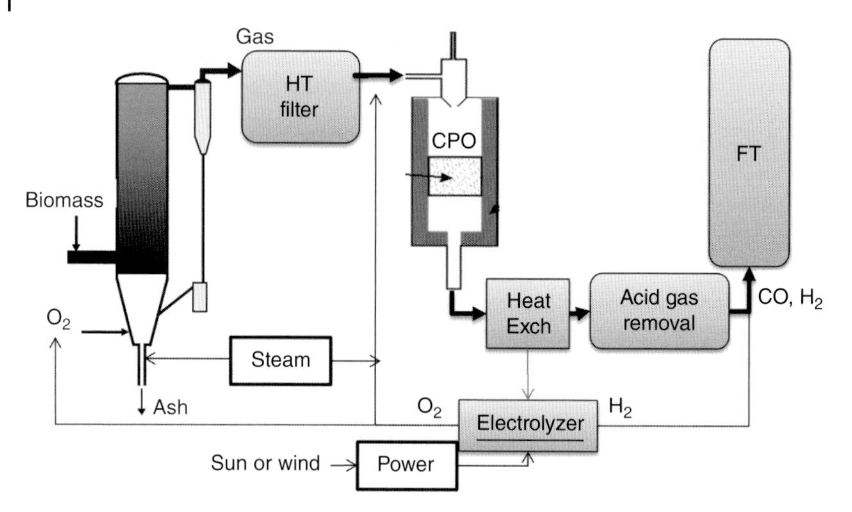

Figure 4.7 Process scheme of integrated electrolysis and gasification for fuel production.

temperature with respect to the entrained flow and ensures a higher heating value on the gasification stream, which can then be efficiently converted into fuel. The increase of 30% of the fuel production can be a further imrpovment of the figure enhanced direct gasification presented in Figure 4.6.

4.3
Aqueous Phase Reforming

APR has recently become of great interest in the production of green fuels and valuable bioplatform molecules, because it offers several advantages over other biomass transformation processes. In fact, sugars and polyols can react with water in the gas phase to generate reforming products, but significant energy is required for the vaporization of the feed, and a high temperature is required to obtain the complete conversion of the hydrocarbons formed during polyol decomposition. Thus one of the main advantages of the APR process in the production of hydrogen and fuels is that the energy requirement is low, because the vaporization of the reaction feed is avoided, while the second advantage is related to the lower temperature, which minimizes undesirable reactions such as thermal degradation and the formation of coke precursors. Moreover, the use of a water solution enhances the safety of the process and increases the economy thanks to the possibility to use hydrolyzed or wet biomass-derived products.

Further specific advantages are related to each process configuration and the desired products; the process producing hydrogen leads to a mixture containing low amounts of CO, due to the possibility of coupling with WGS reaction in a single batch. Hydrogen is produced in a pressurized vessel reaching a partial pressure above 5 atm and can be handled and purified with dense or porous membrane technologies. Furthermore, the overall reaction occurs in a single step, compared to the multistep reforming and WGS processes [43]. By acting on the reaction conditions and the type of catalyst, the process can be driven toward gas-phase products for the production of fuel or liquid-phase molecules such as acids, alcohols, glycols, and aldehydes for the polymer industry, cosmetics, and chemistry building blocks.

4.3.1 Thermodynamic and Kinetic Considerations

The SR reaction of alkanes and carbohydrates are shown in Eqs. (4.6) and (4.7) [43]:

$$C_nH_{2n+2} + nH_2O \leftrightarrows nCO + (2n+1)H_2 \tag{4.6}$$

$$C_nH_{2y}O_n \leftrightarrows nCO + yH_2 \tag{4.7}$$

During the reforming process, hydrogen yield can also be increased by WGS reaction, as indicated in Eq. (4.4).

Figure 4.8 shows the $\Delta G°$ values both for SR reaction of low molecular weight alkanes (methane, ethane, propane, and hexane) and for APR of carbohydrates with a C:O ratio of 1 : 1 (methanol, ethylene glycol, glycerol, and sorbitol).

As indicated, the reforming of alkanes becomes thermodynamically favorable only over 700 K, in particular over 900 K for methane, whereas for the reforming of carbohydrates the reaction is possible at much lower temperatures. Furthermore,

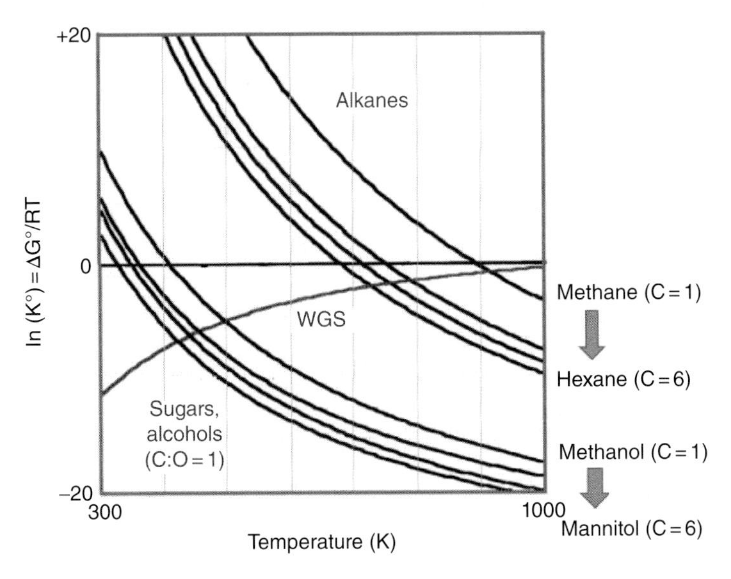

Figure 4.8 Gibbs free energy change with temperature for reforming reactions of alkanes and carbohydrates and WGS reaction. (Adapted from [43].)

the WGS reaction has no important role during the SR of alkanes, due to the high reaction temperature, but it increases hydrogen production significantly in low-temperature reaction processes.

In the case of carbohydrates reforming temperatures, the reagent has low vapor pressure, especially in the case of sugars; thus, unlike in the SR of alkanes, the reaction can take place mainly in the liquid phase and be carried out using liquid water (APR). As shown in Figure 4.8, the reforming reaction of sugars and polyols is thermodynamically feasible, but it is likely that the production of CO and H_2 is also the result of intermediate liquid-phase dehydrogenation and decarbonylation reactions. Independently from the kinetic pathway followed, the overall reaction remains less energy demanding than SR for the production of hydrogen in a single reactor.

At low temperatures, the formation of alkanes via methanation (Eq. (8)) and even the FT (Eq. (2)) reaction is energetically more advantageous than reforming [44]. This means that the C–O bond breaking, followed by hydrogenation, is preferable to the C–C cleavage and WGS. To enhance the reforming reaction and the production of hydrogen, an appropriate catalyst can be used; furthermore, the catalyst serves to promote the WGS reaction to remove CO from the metal surface and avoid parallel reactions onto oxygenated molecules in water (Figure 4.9) [45]:

$$CO + 3H_2 \leftrightarrows CH_4 + H_2O \quad CO_2 + 4H_2 \rightleftarrows CH_4 + 2H_2O \tag{4.8}$$

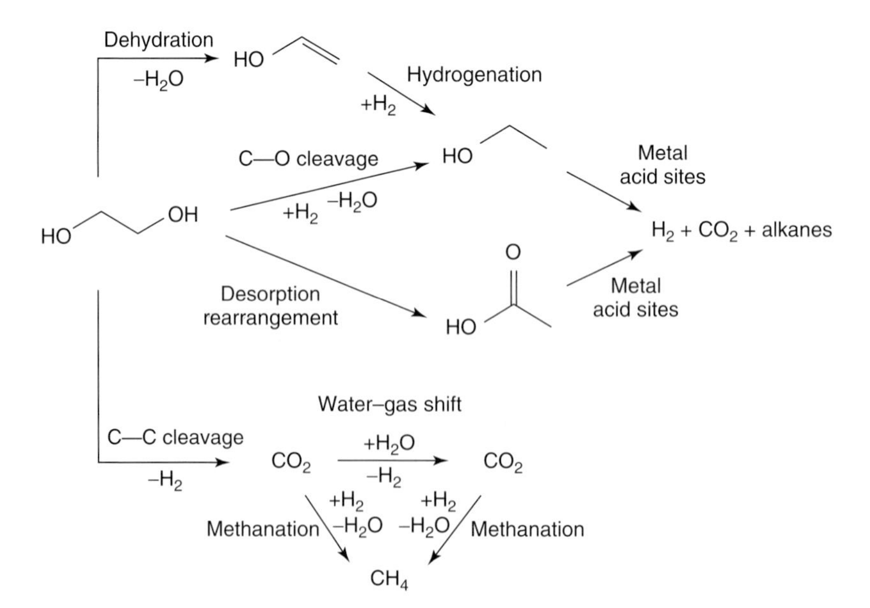

Figure 4.9 Parallel reactions in water for ethylene glycol. (Adapted from [45].)

4.3.2
Catalysts for APR Reaction

Several metal-supported catalysts have been studied for the APR reaction, because the synergic effect between metal and acid/basic features of the support may promote different reactions. An analysis of the active phases to be used in the APR reaction may be conducted by comparing their activity toward C–C cleavage, WGS, and the production of alkanes. Primarily, the study of the hydrogenolysis reaction (C–C cleavage) has been evaluated on ethane using different metals [46]. The report scale of activity gives some indications on APR, even though the scale cannot be directly extended to the C–C cleavage of polyalcohols:

$$Ru \approx Ni > Ir \approx Rh > Fe \approx Co > Pt \approx Pd > Cu$$

In addition, the metal should be active not only for C–C cleavage but also for WGS reaction in order to allow the removal of CO adsorbed on the metal. In this regard, Grenoble *et al.* [47] reported the activity scale of metals over alumina support, which, nevertheless, is depending on the oxidation state of the metal:

$$Cu > Co \approx Ru > Ni > Pt > Fe \approx Pd > Rh$$

Lastly, the selected metal must not present high activity toward undesirable reactions such as methanation and FT, especially when hydrogen is the desired product of the reaction. A work by Vannice [48] indicates the rate of methanation by different metals supported on silica:

$$Co \approx Fe > Ru > Ni > Rh > Ir \approx Pt > Pd$$

Thus the most desired metals are the ones having the lowest methanation activity (i.e., Pt and Pd) but a good activity toward C–C cleavage and WGS reaction. Based on the same consideration, an interesting metal is Ni due also to its lower price, and its use in combination to Pt as bimetallic catalyst.

Several supports have also been tested for this type of reaction. A study by Menezes *et al.* [49] reported an analysis on the basicity of the support at the same Pt loading. Within the oxides studied, MgO and ZrO_2 show the highest hydrogen production and low alkane yield (CH_4) due to the higher electron-donating nature of the metal and the basicity of the MgO support probably involved in the reaction mechanism. As reported by Dumesic *et al.* [50], the highest H_2 production rate from ethylene glycol was reached at 498 K for Pt supported on TiO_2 and Pt black, followed by carbon, Al_2O_3, ZrO, and $SiO_2-Al_2O_3$. On the other hand, the production of alkanes and alkane precursors (acetaldehyde, ethanol, and acetic acid) is significant for Pt black and Pt supported on carbon, so these supports are not suitable for APR reaction if hydrogen is the selected reaction product. Thus Pt/Al_2O_3 and, to a lesser extent, Pt/TiO_2 and Pt/ZrO are the most selective catalysts for this kind of reaction. Interesting results, comparable to Pt/Al_2O_3 catalyst, have also been reached with a Sn-modified Ni Raney catalyst [51] at high Ni loading. This low-cost catalyst has a good activity in the C–C cleavage of small polyols while inhibiting the methanation reaction and producing higher yields compared to an unpromoted Ni catalyst. The catalyst analysis demonstrate the sensitvness of the reaction with respect to metal support and bimetallic systems which need further investigation and development.

4.3.3 Reaction Conditions and Feed

APR reaction is carried out in batch systems and in fixed-bed reactors. Closed-vessel reactors can be both batch and semibatch systems, while fixed-bed reactors can be fed in cocurrent flow and in countercurrent flow. The differences highlighted in the apparatus arise from the need to increase selectivity toward gas- or liquid-phase products. This is because maintaining the H_2 produced in the reaction in a close system can lead to secondary hydrogenation reactions in the liquid phase, thus leading to lower gas yields and a higher number of hydrogenation and HDO products in the liquid phase. Instead, if a high H_2 yield is desired, there is a need for a sweep gas to permit the H_2 removal from the system, both in semibatch reactors and in fixed-bed ones. Most of the work experiments in literature have been conducted on fixed-bed reactors because of their higher flexibility in studying reaction conditions.

Generally speaking, the reaction is carried out in a temperature range of 180–260 °C because at lower temperatures the reaction is not promoted, while at higher temperatures the decomposition of sugars and polyols is dominating and the catalyst will be quickly poisoned by carbon formation. Pressure conditions are generally set in the 25–60 bar range depending on the temperature of reaction, so as to remain close to the equilibrium pressure of water. In some cases, H_2 is added to provide the system with a reducing atmosphere to drive the reaction toward the formation of alkanes. Batch reactors, conversely, work in autogenous pressure, which is the equilibrium pressure reached by water at a given temperature. The pressure at the end of the reaction is clearly incremented by the partial pressure of the gas generated in the reaction, unless a backpressure regulator is present in the plant.

The reagents used for the APR reaction are mainly sugars and polyols because they must originate from biomass for the main purpose of the process, but they must also be soluble in water. The most studied sugar is glucose, but xylitol and sorbitol have also been investigated. As for alcohols, methanol and ethanol are generally tested to analyze the effect of catalysts on primary alcoholic groups; a greater attention, however, is focused on ethylene glycol, propylene glycol, and glycerol. A study on H_2 production and alkane production has been performed by Dumesic *et al.* [45] to compare different feeds in reaction at 225 and 265 °C over 3 wt% Pt/Al_2O_3 catalyst (Figure 4.10). As expected, when increasing the feed complexity, the selectivity toward hydrogen decreases, while alkane production increases, due to the presence of competitive reactions in the liquid phase when reagents with a higher functionality are used. Moreover, in the case of glucose, hydrogen yield decreases when increasing the feed concentration because, unlike with other polyols, in this case H_2 yield is not insensitive to the concentration

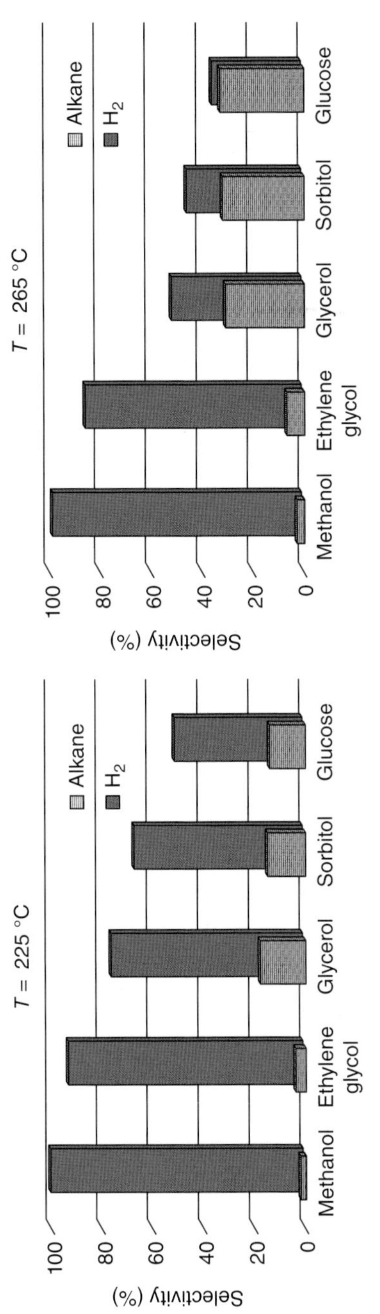

Figure 4.10 H_2 selectivity and alkane selectivity of diluted oxygenated hydrocarbon solutions (1 wt%) at 225 and 265 °C [45].

of feed [52]. Independently from the feed, the sugar and polyol concentration in water is generally in the 1–10 wt% range.

Independently from reactivity issues, glucose and glycerol are certainly the most studied in the APR reaction. The first can be generated by hydrolysis and enzymatic processes from polysaccharides deriving from second-generation biomass. Glycerol, on the other hand, is the main by-product of the esterification process for producing biodiesel fuel, leading to a high availability of this feedstock on the market. As for glucose, the possibility of upgrading this molecule toward APR reaction is of high interest. Attempts have also been made to use crude glycerol in APR reactions [53], but the use of nonpurified feed plays an important role in reducing hydrogen yields.

4.3.4 Mechanism of Reaction

Since APR reaction can be applied to different feeds, the effect of product distribution can be significant. In all cases, gas-phase products are H_2, CO (in low amounts), CO_2, methane, and light alkanes (C_2–C_3), while liquid-phase products may be alcohols, diols, aldehydes, acids, and, in general, oxygenated compounds, and their distribution strongly depends on the type of feed used. In the liquid phase, C–C cleavage, C–O cleavage, dehydrogenation, hydrogenation, hydrogenolysis – depending on the catalyst functionalities – and hydration and dehydrogenation are possible, due to the presence of water. Furthermore, the presence of highly reactive compounds such as aldehydes and alcohols may generate tautomeric equilibria, aldol condensation reactions, and, in some cases, polymerization and cyclization also.

The complexity of the aforementioned reactions, therefore, can lead to different mechanisms with changes in the feed but also to significant changes in selectivity by acting on both the acid/basic sites of the support and the type of active phase. An example is shown in Figure 4.11, on the mechanism of glycerol APR. Solid arrows represent reactions performed with Pt/Al_2O_3 1–5 wt% at 225 °C with glycerol feed concentration 10–30 wt% in water [54], while dashed arrows indicate the work of another group [55] dealing with Pt–Re/C 3 wt% catalyst at 225 °C with a 10 wt% glycerol feed. Bold gray arrows represent the products that are generally accepted from glycerol, and it can be clearly seen how changing the catalyst in the same experimental conditions can affect the product composition. This can be explained by the effect of Re introduction in the active phase in changing the structure of the metal and on the role of the acid site (i.e., alumina) [56].

Another example of the mechanism is shown in Figure 4.12, but, differently from the work reported for glycerol, the reaction focuses on the production of alkanes over Pt/HZSM-5 and Ni/HZSM-5 1–5wt% metal loading from xylitol [57]. Reactions were performed under H_2 atmosphere (2–4 MPa) to inhibit the dehydrogenation pathway and in the 180–250 °C range. Together with hydrogenation reactions on metal sites, C–O cleavage by dehydration on acid sites, and C–C cleavage by decarbonylation on metal sites, C–C coupling reactions on acid sites

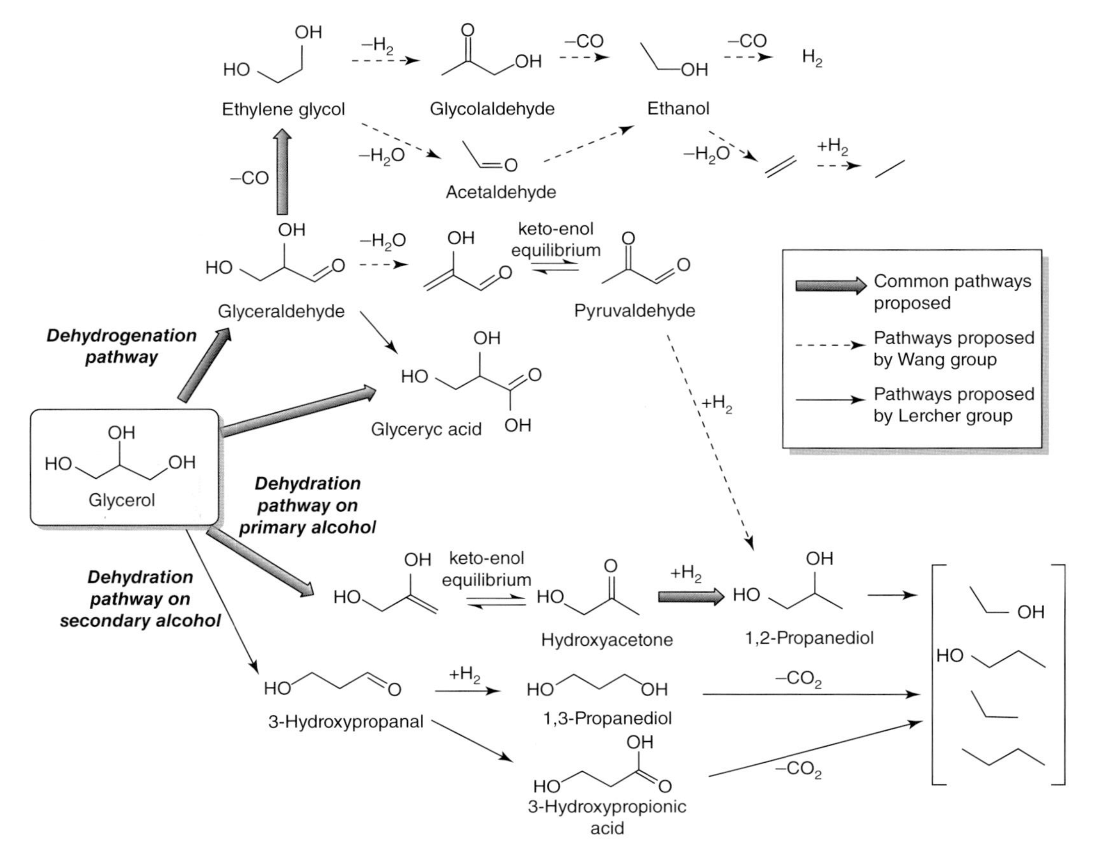

Figure 4.11 Proposed mechanism for the APR reaction of glycerol. Solid arrows pathways proposed by Lercher and coworkers [54] and dashed arrows from Wang and coworkers [55]. Bold gray arrows represent commonly accepted reaction pathways from glycerol. (Adapted from [56].)

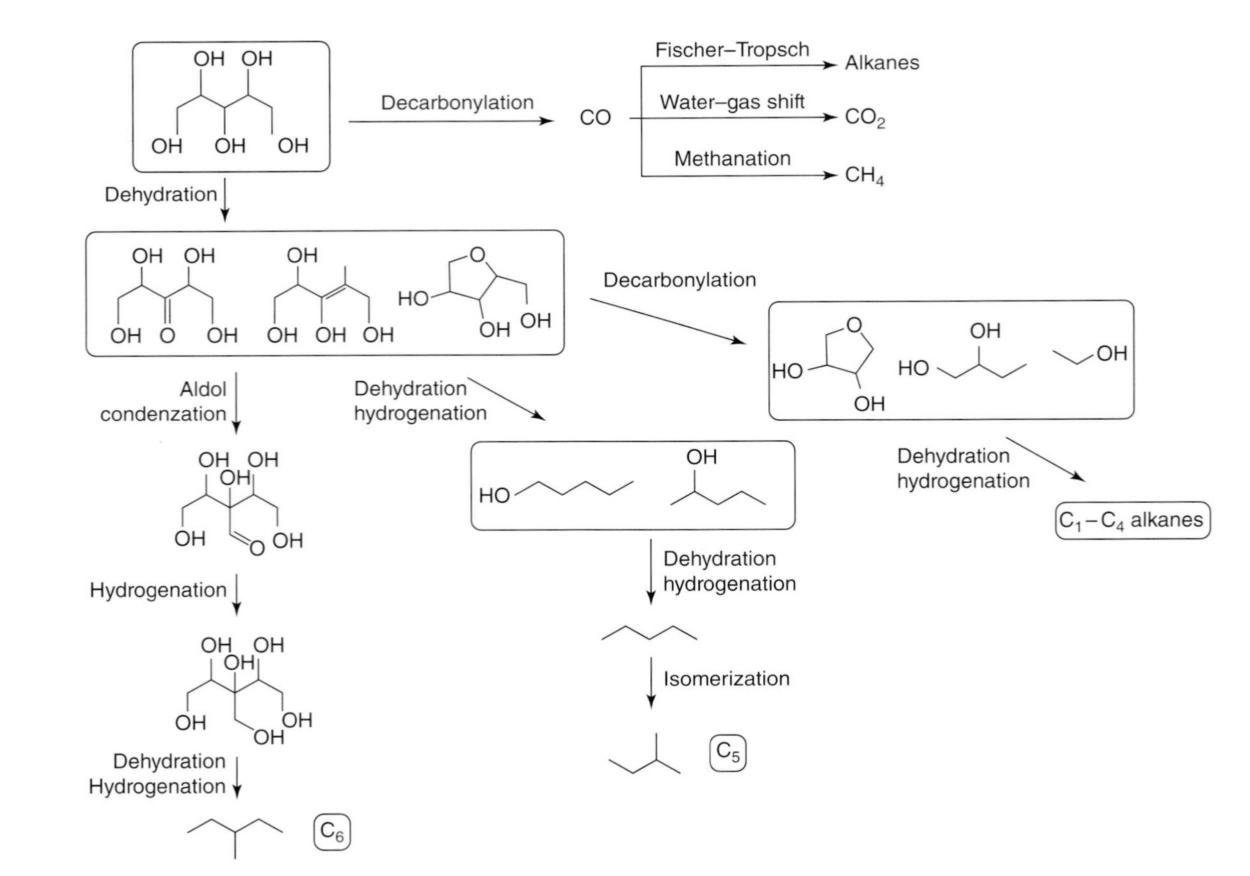

Figure 4.12 Proposed mechanism for the production of alkanes from xylitol over bifunctional metal–acid catalyst. (Adapted from [57].)

and isomerization and condensation processes due to the presence of highly functionalized molecules are also present.

4.3.5
APR on Biomass Fractions

In addition to the study on polyols and monomeric sugars, the use of lignocellulosic fractions is also progressing for APR reaction. Erbatur *et al.* [58] report a study comparing the three fractions obtainable from different biomass (cellulose, hemicellulose, and lignin) after pretreatment in subcritical conditions before APR reaction. The comparison of the three hydrolysate feeds over commercial Pt/C shows that the lignin fraction does not produce hydrogen, but only CO_2 with small amounts of CH_4, while hemicellulose shows 27% H_2 and 10% CH_4 in the total gas produced. Conversely, the cellulose fraction shows the highest quantity of gases produced (48% more than hemicellulose over wheat straw and 28% for kenaf), but the relative percentage for hydrogen over the total gases is lower than with hemicellulose. Further comparison analyses were performed without a catalyst and showed no presence of hydrogen; CO_2 was mainly produced, with only a small amount of CO. In this case, the differences in reactivity within the fractions of the type of biomass analyzed are explained by the crystallinity degree of the two supports. The hydrolytic pretreatment leads to a lower crystal index (calculated as a ratio within the XRD crystal peak area and the amorphous one), compared to the starting material, so the fractions are more sensitive to degradation, producing a higher amount of gaseous products.

The effect of the crystalline fraction over APR reaction was analyzed more in depth by Tian *et al.* [59]. By studying filter paper, degreased cotton, and microcrystalline cellulose, they correlated the decreasing crystalline index of those feeds with the decreasing H_2 selectivity in the same order, while the selectivity value was not correlated to the polymerization degree of the feed. The principal explanation for this behavior can be found in the absence – during APR reaction – of a strong acid catalyst that makes it possible to accelerate the hydrolysis step of the reaction, which is highly dependent on the crystallinity of the feed. Once produced by hydrolysis, the suggested intermediates, that is, glucose, are then readily transformed into hydrogen over a Pt/C catalyst. To avoid the previously discussed limiting step, Lin *et al.* [60] proposed a combined hydrothermal-catalytic treatment to improve the hydrolysis step of the biomass by adding a mineral acid to the traditional APR catalyst. The study reports that the amount of reducing sugar (glucose and other monomers) obtained by this combined treatment is higher than the amount of α-cellulose contained in each cellulosic feed tested, with the simultaneous production of hydrogen in a one-step process. This indicates that substrates such as cotton, sawdust, and pulp can be simultaneously hydrolyzed and reformed with good results as compared to α-cellulose.

In addition to the possibility of producing H_2 by APR of lignocellulosic fraction, one of the main drawbacks in the use of those feeds is the percentage

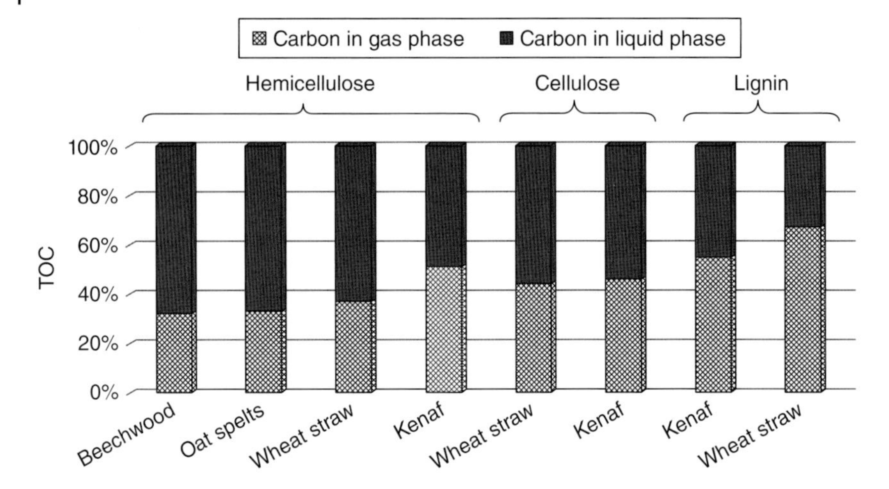

Figure 4.13 Comparison of total organic carbon (TOC) percentage of gaseous products and liquid products after APR reaction over different feeds [58].

of reagent transformed into gaseous products, compared with the amount of solid residues, that is, water-insoluble organics and char. Erbatur *et al.* [58] report (Figure 4.13) the percentage of carbon-containing gaseous products (CO_2, CO, C_2H_4, CH_4) compared with the insoluble residue after reaction. As for hemicellulose, the gaseous carbon is dependent on the feed analyzed and may vary between 30% and 50%, but with a higher amount of ungasified products compared to other fractions. As for cellulose, gaseous carbon value is around 45%, while for lignin the carbon that goes to gas products is higher compared to other fractions, thus indicating that the degradation toward CO_2 of this fraction is higher while no H_2 is produced. Another study by Zakzeski and Weckhuysen [61] studied more in depth the lignin fraction, since it is more difficult to treat than other fractions due to solubility problems and sulfur content. They analyzed the possibility of solubilizing the feed in APR reaction conditions in order to enhance the possibility of obtaining gaseous products by combining the use of a Pt/Al_2O_3 catalyst with H_2SO_4 to enhance the hydrolysis step. H_2 yields up to 8% were obtained over different feeds with ethylene and methane as by-products. As for the liquid phase, several aromatic monomers were identified. A more in-depth study was performed on the effect of this type of reaction on different model molecules, such as guaiacol and polyaromatics containing β-O-4 and 5-5′ bonds, that are found to be susceptible of disruption, thus generating monomers and free methoxy groups susceptible to hydrolysis and consequently to be reformed into H_2. Unfortunately, recondensation reactions can always lead to high molecular weight products causing a low gas yield of reaction.

In conclusion, the use of more complex feeds has not yet been studied in depth, mainly due to the complexity of the reaction mechanism, combined with the low reactivity of the feed; however, some interesting results have been presented that indicate the possibility of H_2 production by APR reaction.

4.3.6 Pilot Plants and Patents

The first pilot plant developed for APR reaction belongs to Virent Energy Systems Inc, a partner of the University of Wisconsin, in Madison [62]. The project started in 2005 with the aim of optimizing reactor, streams, and efficiency of the reaction to produce a 10 kg H_2 per day demonstration system within the Department of Energy USA (DOE) Hydrogen Program goals. The selected feedstock was glucose, hydrogenated to sorbitol in situ to produce a hydrogen-rich stream purified with pressure swing adsorption, as shown in Figure 4.14. In 2009, the catalyst was optimized to maintain hydrogen productivity, and targets were achieved. In particular, the economic analysis showed that the sensitivity of hydrogen costs (including the cost of feedstock, processing, and storage equipment, utilities, and maintenance) is strongly dependent on the cost of feedstock (Figure 4.15), followed by other raw material costs, capital costs, operation, and maintenance. The APR reaction has subsequently been developed by Virent in the BioForming® process combining this technology with a modified conventional catalytic process (i.e., petrochemical process). This process can fit a wide range of water-soluble C_5 and C_6 sugars, as well as further products deriving from the degradation of biomass (polysaccharides, organic acids, furfural), and produces a mixture of chemical intermediates including alcohols, ketones, acids, furans, paraffins, and other oxygenated hydrocarbons. This mixture can further react over a ZSM-5 modified catalyst either via a traditional process to produce a high-octane gasoline blendstock named BioFormate™ or via HDT [63].

Starting in 2005, another demonstration unit was developed by United Technologies Research Center (UTRC) [64] to study biomass slurry hydrolysis and

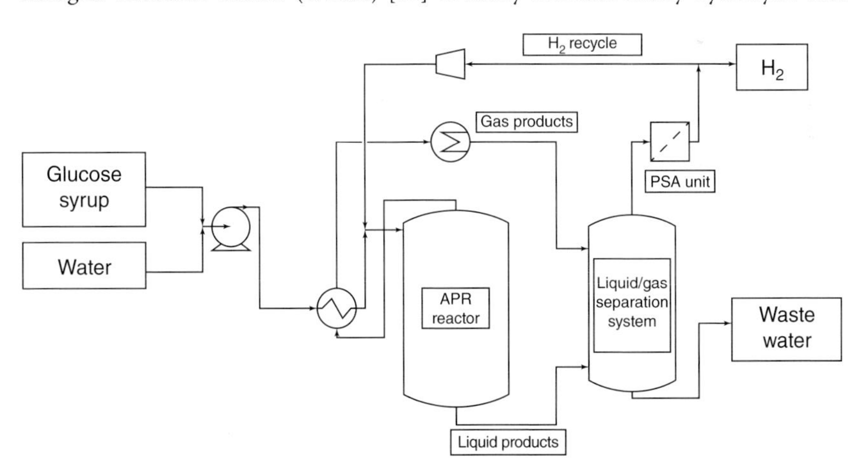

Figure 4.14 Schematic Virent APR pilot plant [58].

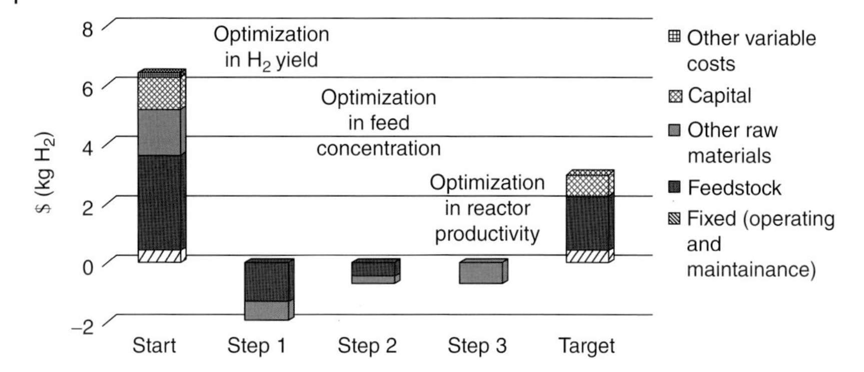

Figure 4.15 Cost breakdown for the Virent APR process [58].

reforming process for hydrogen production from woody biomass. One of the main differences from the Virent process is the use of a slurry of ground biomass, directly hydrolyzed in the reactor with base (K_2CO_3) at high but subcritical pressure and the subsequent application of liquid reforming with a Pt or Ni catalyst. The aim is to deconstruct cellulose fibers through the use of a base and separate H_2 continuously. Unfortunately, before the end of the project in 2009, only a fixed-bed Ni Raney demonstration plant had been developed, with a lower feed rate and a minor biomass concentration than the project objectives because of technical challenges.

There are not many patent submissions regarding the APR reaction, probably because the process is still being developed and so far high hydrogen yields have not been reached. In particular, patent literature can be subdivided between works concerning the reforming process alone and more complex projects that treat rough biomass through several steps, which also include an APR unit.

Monnier *et al.* [65] have developed a single liquid-phase APR unit for the production of H_2 from oxygenated hydrocarbons (i.e., C_5 and C_6 sugars) in a stirred tank reactor with small metal particles dispersed on a solid support. The reaction is performed at low temperature (255–285 °C), but high pressure, in order to maintain the condensed phase in the system. A sweep gas makes it possible to remove H_2 and CO_2 by a membrane separation system. Another single unit process has been patented by Cortright and Dumesic [66]: it can be implemented for both liquid- and gas-phase reforming units, with the use of an alkali and a heterogeneous supported metal catalyst at low temperatures (below 400 °C). The same author presented a work [67] on the reforming of biomass to produce polyols (i.e., propylene glycol, ethylene glycol) and other oxygenated compounds (ketones, aldehydes, carboxylic acids, and alcohols) with O:C ratio from 0.5 : 1 to 1.5 : 1. Differently from what is reported earlier, the reaction is developed in a two-step process: the first one generating H_2 and oxygenated hydrocarbons through APR and the second one to react H_2 with oxygenated hydrocarbons. For both these steps, the bimetallic catalyst was supported on porous material, and the process can be performed separately or in a single fixed-bed reactor loaded with a double catalytic bed. H_2 selectivity can be enhanced by the addition of an alkali solution to increase the pH to the 4–10 range for the first step and an acid solution for the second step in the 1–4 pH range.

As for the integration of the APR process in a biomass treatment plant, Chheda and Powell [68] incorporated the reforming step in a more complex system to make it possible to pretreat rough biomass. The first step is a mild temperature treatment in water to induce the solubilization of carbohydrates; after a separation unit, the solid fraction is digested again thanks to a digestive solvent to obtain a pulp containing cellulose, hemicellulose, and a fraction of soluble lignin. After the removal of nitrogen and sulfur compounds – which are poisonous for APR catalysts – the liquid stream and the pulp are reformed in a liquid phase at 200–280 °C on a metal-supported catalyst. Lastly, the APR outflow (mainly made of oxygenated molecules) can be processed to fuel or, in part, redirected to the plant as digestive solvent. Ma *et al.* [69] proposed the idea of prehydrolyzing the lignocellulose biomass with water and maleic acid and subsequently treating it at first in a low-temperature reforming unit and then in a high-temperature reforming unit. The first unit works at 160–210 °C and 4–6.5 MPa of H_2 with $Ni/Al_2O_3-SiO_2$ catalyst and the second unit at 210–270 °C 2–5.6 MPa of H_2 with Ni/HZSM-5. The advantage of the overall process is the production of an outlet stream with characteristics similar to those of biogasoline, with a cheap catalyst and the use of H_2 to avoid polymerization and coke formation on the catalyst. Furthermore, the use of maleic acid, instead of more traditional mineral acids, permits to avoid the separation step before the two-stage liquid reforming reaction.

4.3.7 Integration of the APR Process in a Biorefinery

Today, aqueous phase reactions of polyols are studied in industries to produce fuels and chemicals, thanks to the possibility of tuning this reaction to different products, as shown in Figure 4.16 [70].

Several pilot and demonstration units are currently in operation, for example, for the production of chemicals such as isosorbide, glycols, and aromatics, or applications such as combined aqueous phase dehydration/hydrogenation systems for the production of liquid fuels [71]. Furthermore, APR processes may be included in a biorefinery plant to autogenerate the necessary hydrogen supply (for HDT, HDO, and pyrolysis reactions), which currently comes from fossil resources. On the other hand, the economic drawbacks of the application of APR sections into existing plants are primarily due to the difficulty of obtaining a high hydrogen yield with the use of crude feeds and the need for a large number of steps involved in the biomass pretreatment. Moreover, the product stream necessitates costly purification steps before its application in subsequent units.

Nevertheless, the good results obtained and the possibility of applying the process to different feeds without significant changes in reactivity suggest that the APR process is a suitable alternative for the production of hydrogen and building blocks from renewable biomass.

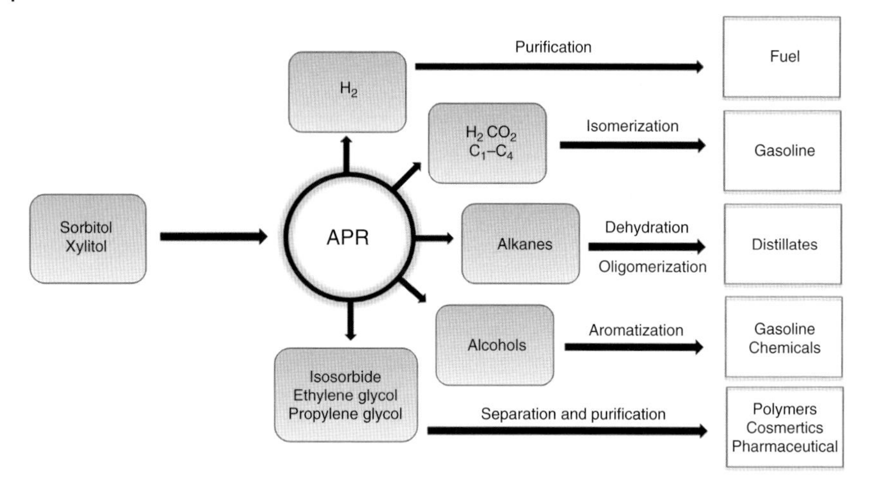

Figure 4.16 Industrial applications of aqueous phase processes.

References

1. Buekens, A.G., Maniatis, K., and Bridgwater, A.V. (1990) in *Commercial and Marketing Aspects of Gasifiers* (eds A.G. Buekens, A.V. Bridgwater, G.L. Ferrero, and K. Maniatis), Commission of the European Communities, Brussels, p. 8.
2. Higman, C. and Van der Burgt, M. (2008) *Gasification Processes, Gasification*, 2nd edn, Elsevier, Amsterdam, pp. 91–191.
3. Beenackers, A.A.C.M. (1999) *Renewable Energy*, **16**, 1180–1186.
4. Basile, F. and Trifirò, F. (2011) in *Renewable Resources and Renewable Energy A Global Challenge*, 2nd edn (eds P. Fornasiero and M. Graziani), CRC Press, pp. 213–242.
5. Kitzler, H., Pfeifer, C., and Hofbauer, H. (2011) *Fuel Process. Technol.*, **92**, 908–914.
6. Ståhl, K. and Neergaard, M. (1998) *Biomass Bioenergy*, **15**, 205–211.
7. Albertazzi, S., Basile, F., Brandin, J., Einvall, J., Hulteberg, C., Fornasari, G., Rosetti, V., Sanati, M., Trifirò, F., and Vaccari, A. (2005) *Catal. Today*, **106**, 297–300.
8. Kirkels, A.F. and Verbong, G.P.J. (2011) *Renewable Sustainable Energy Rev.*, **15**, 471–481.
9. Zhang, W. (2010) *Fuel Process. Technol.*, **91**, 866–876.
10. Iliuta, I., Leclerc, A., and Larachi, F. (2010) *Bioresour. Technol.*, **101**, 3194–3208.
11. Trippe, F., Fröhling, M., Schultmann, F., Stahl, R., and Henrich, E. (2011) *Fuel Process. Technol.*, **92**, 2169–2184.
12. Seiler, J., Hohwiller, C., Imbach, J., and Luciani, J.F. (2010) *Energy*, **35**, 3587–3592.
13. Couhert, C., Salvador, S., and Commandré, J.-M. (2009) *Fuel*, **88**, 2286–2290.
14. Svoboda, K., Pohořelý, M., Hartman, M., and Martinec, J. (2009) *Fuel Process. Technol.*, **90**, 629–635.
15. van der Stelta, M.J.C., Gerhauserb, H., Kielb, J.H.A., and Ptasinski, K.J. (2011) *Biomass Bioenergy*, **35** (9), 3748–3762.
16. Ptasinski, K.J., Prins, M.J., and Pierik, A. (2007) *Energy*, **32**, 568–574.
17. Wang, L., Weller, C.L., Jones, D.D., and Hanna, M.A. (2008) *Biomass Bioenergy*, **32**, 573–581.
18. Van Rossum, G., Potic, B., Kersten, S.R.A., and Van Swaaij, W.P.M. (2009) *Catal. Today*, **145**, 10–18.
19. Leibold, H., Hornung, A., and Seifert, H. (2008) *Powder Technol.*, **180**, 265–270.

20. Devi, L., Ptasinski, K.J., and Janssen, F.J.J.G. (2005) *Fuel Process. Technol.*, **86**, 707–730.
21. Courson, C., Udron, L., Świerczyński, D., Petit, C., and Kiennemann, A. (2002) *Catal. Today*, **76**, 75–86.
22. Siedlecki, M. and de Jong, W. (2015) *Biomass Bioenergy*, **35**, S40–S62.
23. Siedlecki, M., Nieuwstraten, R., Simeone, E., de Jong, W., and Verkooijen, A.H.M. (2009) *Energy Fuels*, **23** (11), 5643–5654.
24. Wang, L., Chen, J., Watanabe, H., Xud, Y., Tamura, M., Nakagawa, Y., and Tomishige, K. (2014) *Appl. Catal. Environ.*, **160–161**, 701–715.
25. Tijmensen, M.J.A., Faaij, A.P.C., Hamelinck, C.N., and Van Hardeveld, M.R.M. (2002) *Biomass Bioenergy*, **23**, 129–152.
26. Nacken, M., Ma, L., Heidenreich, S., and Baron, G.V. (2009) *Appl. Catal., B: Environ.*, **88**, 292–298.
27. Wilhelm, D.J., Simbeck, D.R., Karp, A.D., and Dickenson, R.L. (2001) *Fuel Process. Technol.*, **71**, 139.
28. Aasberg-Petersen, K., Dybkjær*, I., Ovesen, C.V., Schjødt, N.C. Sehested, J., and Thomsen, S.G. (2011) *Journal of Natural Gas Science and Engineering*, **3**, 423–459.
29. Twigg, M. (ed) (1989) *Catalyst Handbook*, Wolfe Publishing LTD, London, 2nd edn.
30. Bakkerud, P.K., Gol, J.N., and Aasberg-Petersen, K. (2004) *Preferred Synthesis Gas Production Routes for GTL, Natural Gas Conversion VII*, Elsevier.
31. Albertazzi, S., Basile, F., Benito, P., Fornasari, G., Trifirò, F., and Vaccari, A. (2009) in *Syngas: Production Methods, Post Treatment and Economics* (eds A. Kurucz and I. Bencik), Nova Publishers, pp. 409–416.
32. Holladay, J.D., Hu, J., King, D.L., and Wang, Y. (2009) *Catal. Today*, **139**, 244–260.
33. Albertazzi, S., Basile, F., Fornasari, G., Trifirò, F., and Vaccari, A. (2007) in *Gasification of Biomass to Produce Hydrogen in Gasification Renewable Resources and Renewable Energy* (eds M. Graziani and P. Fornasiero), Taylor & Francis, pp. 197–213.
34. Li, D., Tamura, M., Nakagawa, Y., and Tomishige, K. (2015) *Bioresour. Technol.*, **178**, 53–64.
35. Albertazzi, S., Basile, F., Brandin, J., Einvall, J., Hulteberg, C., Fornasari, G., Sanati, M., Trifirò, F., and Vaccari, A. (2008) *Biomass Bioenergy*, **32**, 345–353.
36. Sekine, Y., Mukai, D., Murai, Y., Tochiya, S., Izutsu, Y., Sekiguchi, K., Hosomura, N., Arai, H., Kikuchi, E., and Sugiura, Y. (2013) *Appl. Catal., A: Gen.*, **451**, 160–167.
37. Li, D., Koike, M., Wang, L., Nakagawa, Y., Xu, Y., and Tomishige, K. (2014) *ChemSusChem*, **7**, 510–522.
38. Basile, F., Albertazzi, S., Barbera, D., Benito, P., Einvall, J., Brandin, J., Fornasari, G., Trifirò, F., and Vaccari, A. (2011) *Biomass Bioenergy*, **35**, S116–S122.
39. Moud, P.H., Andersson, K.J., Lanza, R., Pettersson, J.B.C., and Engvall, K. (2015) *Fuel*, **154**, 95–106.
40. Einvall, J., Parslanda, C., Benito, P., Basile, F., and Brandin, J. (2011) *Biomass Bioenergy*, **35**, S123–S131.
41. Luque, R., de la Osa, A.R., Campelo, J.M., Romero, A.A., Valverde, J.L., and Sanchez, P. (2012) *Energy Environ. Sci.*, **5**, 5186–5202.
42. Bercero, M.A. (2012) *J. Power Sources*, **203**, 4–16.
43. Davda, R.R., Shabaker, J.W., Huber, G.W., Cortright, R.D., and Dumesic, J.A. (2005) *Appl. Catal., B: Environ.*, **56**, 171–186.
44. Davda, R.R., Shabaker, J.W., Huber, G.W., Cortright, R.D., and Dumesic, J.A. (2003) *Appl. Catal., B: Environ.*, **43**, 13–26.
45. (a) Cortright, R.D., Davda, R.R., and Dumesic, J.A. (2002) *Nature*, **418**, 964–967; (b) Gurbuz, E.I. and Dumesic, J.A. (2013) in *Catalysis for the Conversion of Biomass and Its Derivatives*, Proceedings 2, Chapter 10 (eds M. Behrens and K. Datye), Max Planck Research Library for the History and Development of Knowledge, Berlin.
46. Sinfelt, J.H. and Yates, D.J.C. (1967) *J. Catal.*, **8**, 82.
47. Grenoble, D.C., Estadt, M.M., and Ollis, D.F. (1981) *J. Catal.*, **67**, 90–102.

48. Vannice, M.A. (1977) *J. Catal.*, **50**, 228–236.
49. Menezes, A.O., Rodrigyes, M.T., Zimmaro, A., Borges, L.E.P., and Fraga, M.A. (2011) *Renewable Energy*, **36**, 595–599.
50. Shabaker, J.W., Huber, G.W., Davda, R.R., Cortright, R.D., and Dumesic, J.A. (2003) *Catal. Lett.*, **88**, 1–8.
51. Shabaker, J.W. and Dumesic, J.A. (2004) *Ind. Eng. Chem. Res.*, **43**, 3105–3112.
52. (a) Eggleston, G. and Vercellotti, J.R. (2000) *Carbohydr. J. Chem.*, **19**, 1305–1318; (b) Sapronov, A.R. (1969) *Khlebopek. Konditer. Prom.*, **13**, 12.
53. Lenhert, K. and Claus, P. (2008) *Catal. Commun.*, **9**, 2543–2546.
54. Wawrzetz, A., Peng, B., Hrabar, A., Jentys, A., Lemonidou, A.A., and Lercher, J. (2010) *J. Catal.*, **269**, 411–420.
55. King, D.L., Zhang, L., Xia, G., Karim, A.M., Heldebrant, D.J., and Wang, X. (2010) *Appl. Catal., B*, **13**, 99.
56. Lin, Y.C. (2013) *Int. J. Hydrogen Energy*, **38**, 2678.
57. Jiang, T., Wang, T., Ma, L., Li, Y., Zhang, Q., and Zhang, X. (2012) *Appl. Energy*, **90**, 51.
58. Irmak, S., Kurtulus, M., Hasanoglu, A., and Erbatur, O. (2013) *Biomass Bioenergy*, **49**, 102–108.
59. Wen, G., Xu, Y., Xu, Z., and Tian, Z. (2010) *Catal. Commun.*, **11**, 522–526.
60. Chang, A.C., Lee, Y.S., and Lin, K.H. (2012) *Int. J. Hydrogen Energy*, **37**, 15691.
61. Zakzeski, J. and Weckhuysen, B.M. (2011) *ChemSusChem*, **4**, 369–378.
62. Cortright, D.R.(2005) Hydrogen Generation from Biomass-Derived Carbohydrates vie the Aqueous Phase Reforming (APR) Process, award number: DE-FG36-05G15046 (2005-2008), DOE Hydrogen Program FY 2005 Progress Report.
63. Virent (2015) http://www.virent.com/technology/bioforming (accessed June 2015).
64. Emerson, S.C. (2011) A Novel Slurry-Based Biomass Reforming Process. award number: DE-FG36-05GO15042 (2005-2011), FY 2011 Annual Progress Report DOE Hydrogen and Fuel Cells Program.
65. (a)Monnier, J., Zang, Y., and MacFarlain, A. (2009) Production of hydrogen from oxygenated hydrocarbons. WO Patent 2009/129622, assigned to the minister of Natural Resources, Canada; (b)Monnier, J., Zang, Y., and MacFarlain, A. (2014) Production of hydrogen from oxygenated hydrocarbons. US Patent 2014/8673262 (B2), assigned to the minister of Natural Resources, Canada.
66. Cortright, R.D. and Dumesic, G.A. (2004) Low-temperature hydrogen production from oxygenated hydrocarbons. US Patent 2004/6699457 (B2), assigned to Wisconsin Alumni Research Foundation (Madison, WI).
67. (a)Cortright, R.D. (2014) Methods and systems for generating polyols. US Patent 8754263 B2, assigned to Virent Energy System Inc. (b)Cortright, R.D. (2008) Methods and systems for generating polyols. WO Patent 2008/069830, assigned to Virent Energy System Inc.
68. Chheda, J.N. and Powell, J.B. (2012) Process to produce biofuels from biomass. WO Patent 2012/088078 (A2), Shell Oil Company.
69. Ma, L., Wang, T., Yuan, Z., Zhang, X., Zhuang, X., Wu, C., Jiang, T., and Zhang, Q. (2012) Process for hydrolysed reforming of liquous cellulose biomass to produce bio-gasoline. US Patent 2012/0216451 (A1), assigned to Guangzhou Institute Of Energy Conversion, Chinese Academy Of Sciences.
70. Vilcocq, L., Cabiac, A., Especel, C., Guillon, E., and Duprez, D. (2013) *Oil Gas Sci. Technol.*, **68**, 5.
71. Huber, G.W., Chheda, J.N., Barrett, C.J., and Dumesic, J.A. (2005) *Science*, **308** (5727), 1446–1450.

5
The Hydrogenation of Vegetable Oil to Jet and Diesel Fuels in a Complex Refining Scenario

Giuseppe Bellussi, Vincenzo Calemma, Paolo Pollesel, and Giacomo Rispoli

5.1
Introduction

Since 2008, after the global economic crisis, the refinery margins in Europe have been very low [1]. The most influent factors have been the high oil price, which remained close to $\$100\,\text{bbl}^{-1}$ until mid-2014, and the refining overcapacity. Because of the limited population growth in Europe and of the efficiency increase of passenger car engines, the fuel demand has declined. Fossil fuel demand loss in the period 2005–2050 is considered to be 166 Mt (average $6.6\,\text{Mt}\,\text{yr}^{-1}$), equivalent to the combined capacity of the nine biggest (or the 40 smallest) out of the 90 current active EU refineries [2].

At the same time, there have been some modest reductions in refining capacity, but investments in plant improvements have also continued, and the net reduction in refining capacity has fallen well short of the reduction in demand for fuel. As a result, the loading factor of Europe's refineries has fallen dramatically since 2008 and is likely to remain low for many years.

Another significant factor is the high energy cost. Compared to Middle East, where there is availability of low -cost natural gas, European industry is penalized by the higher cost of energy. From 2006, European and US energy costs also started to diverge. The production of low-cost American shale gas provides cheap natural gas to US refineries and steam crackers, resulting in an industrial energy cost about five times lower than in Europe.

Older and less complex European assets compared to North America and the lower scale of European refineries, particularly with respect to Middle East and Asia, are a further penalization for the competitiveness of the European refining system.

A relevant impact to the refining scenario is due to the environmental constraints: environmental regulation is not uniform all over the world. Europe is historically more ready to update the environmental rules and usually adopts more stringent rules. Europe, for example, has been the first area to decrease sulfur limit in gasoline and diesel below 10 ppm.

Chemicals and Fuels from Bio-Based Building Blocks, First Edition.
Edited by Fabrizio Cavani, Stefania Albonetti, Francesco Basile, and Alessandro Gandini.

At present, European regulations are those that push more the use of biocomponents into the transportation fuel pool. Target share of renewable energy in road fuel is now set at 10% in 2020, with future targets still under debate. According to this framework, share of biofuels in Europe would increase to reach 20Mt by 2020; this substitution of fossil fuels will put more pressure on the refining industry.

EU policies are affecting the biofuel market on different fronts: environmental policy for energy savings, market rates and shares, and general taxation. In particular, the Renewable Energy Sources directive (RESs 2009/28/CE, also known as *20-20-20*), at present under revision, prescribes reducing CO_2 emissions from EU countries by 20%, requiring a 20% decrease as conservation efforts in energy consumption and increasing energy production from renewable sources up to 20%, which includes 10% from renewable and sustainable biocomponents for transportation fuels. The Fuel Quality Directive, approved in 2012, confirmed the 10% target of energy from biofuels and the double counting for energy content of biofuels derived from second-generation feedstocks (used cooking oils, animal fats, tallow oil, etc.); moreover it also introduced quadruple counting for biofuels coming from third-generation feedstocks, that is, biomass from urban waste, algae, raw glycerin, palm oil mill waste effluents, and so on. The proposed revision, which is at present under examination by the European Parliament, also considers the relevance of the indirect land use change (ILUC) effects, which measure CO_2 emission increase due to virgin ground deforestation for agricultural purposes.

Bioethanol (for gasoline blending) and biodiesel (BD) fatty acid methyl esters (FAMEs) are the most common biofuels that can be blended with gasoline and diesel; however the quality of these biofuels compared to the conventional fuels and the new regulations set limitations to their use. Focusing on diesel fuel, which is by far the most important automotive fuel in Europe, FAME is the main biocomponents of reference. Vegetable oil cannot be directly fed to a diesel engine because of some characteristics, like the viscosity that is 10–20 times higher than that of mineral diesel oil. The oil has to be transformed, via transesterification with methanol, into methyl esters of the triglycerides. The transformation process is chemically simple and the production costs are relatively low. However, there are severe drawbacks related to the quality of the obtained product. Main disadvantages with FAME are as follows:

- Quality can vary with the species of the parent crops.
- Tendency to polymerization (low stability).
- Biofouling (filter plugging).
- Poor cold flow properties.
- Low energy content per volume.

For these reasons, there is a "blending treshold" for FAME, with a limitation to 7% maximum blending in conventional diesel fuel.

It appears clear by what was previously reported that the situation of the European refinery system is critical. The mandate for blending 10% on energy base of biofuels to the conventional transportation fuels has, in principle, increased the difficulties. New ideas were necessary to mitigate these inconvenient and eventually turn a drawback into an opportunity. The needs for refinery capacity reduction could be coupled with that of building new facilities for the production of biofuels. In fact, it was found that the existing refinery HDS units could be properly converted for the hydrogenation of vegetable oils to high-quality BD. This solution has the merit of saving existing plants from the closure and reducing the capital cost required for the production of BD. The first example of this kind of conversion occurred at the refinery of Venice, and in the following the main aspects of this venture will be reported.

5.2 The Feedstock

Triglycerides are by all means one of the most suitable and available sources of renewable material for the production of transportation fuels. The world production of oils and fats in 2013 was 187×10^6 t [3, 4] which represents about 5% of the current world production of petroleum [5]. Vegetable oils are about 80% of the overall production; mainly palm and soybean oil (63%), followed in lower quantities by rapeseed (13%), sunflower (9%), cottonseed, coconut, canola, and so on. The remaining 20% is made up of oils and fats from animal of which the major producers are the United States and Europe. Of worldwide production of oils and fats, 81% is intended for human nutrition and 5% for animal nutrition while the remaining 14%, is used by the oleochemical industry to produce mainly FAMEs (Fatty Acid Methyl Esthers), fatty acids, fatty alcohols, and obviously glycerine. Nearly 90% of the production is currently consumed as food or animal feed; therefore the growth of fuel production from renewable source is constrained by their availability and the negative impact on supply/price. It is necessary to develop the use of feedstock produced through cultivation or process not competing with the food chain. In this regard plants cultivated on marginal lands such as *Jatropha* and camelina or the production at industrial scale of oils from algae could represent a possible answer to the problem [6, 7]. At any rate, it is also very important to develop suitable processes, which can convert the vegetable oils in the most efficient way. The global production for 2012 of the main vegetable oils is given in Table 5.1. Palm oil, soybean oil, and rapeseed oil are those produced in larger quantity, while sunflower and other oils have a significantly lower production. The production and availability of oilseeds is based on regional conditions. The largest producing countries of soybean oil are Brazil, Argentina, and the United States. As for rapeseed oil EU27, China, Canada, and India account for nearly 90% of the world production. Palm oil is a tropical oilseed with major share of its production in Malaysia and Indonesia whose production is over 80% of overall worldwide production [8]. The high cost of edible oil and the need of not competing with the food chain have prompted the development of alternative sources of oleaginous feedstock. In this regard, the production of lipid feedstock from microalgae and *Jatropha* crop has been a subject of numerous efforts but till now without

Table 5.1 World production 2011/2012 of the main vegetable oil production.

Oil source[a)]	10^6 t yr^{-1}
Palm	53.33
Soybean	43.42
Rapeseed	23.50
Sunflower seed	13.62
Palm kernel	6.25
Cottonseed	5.25
Peanut	5.18
Coconut	3.52
Olive	2.89

a) Other important oils produced in lower quantities include corn oil, linseed oil, safflower oil, sesame oil, and grape seed oil.
Adapted from [4].

success [4]. In 2009, Exxon-Mobil in collaboration with Synthetic Genomics founded a $600 million research program aimed at studing the technology for the production of transportation fuel from algae. In May 2013, after spending $100 million ExxonMobil announced a drastic downsize of the program as a result of the fact that almost 4 years of work failed to produce economically viable results.

Another significant source of oleaginous feedstocks can be obtained from animal through the rendering process. Tallow is a rendered cattle fat, and its world production in 2012 was approximately 6.7×10^6 t. The top three producers are the United States, Brazil, and Australia. Lard is a rendered pig fat. Its world production in 2012 is 5.7×10^6 t, while the top three producers are China, Germany, and Brazil.

5.2.1 Vegetable Oils

Vegetable oils are essentially made up of a mixture of triglycerides. They are composed of long-chain fatty acid esters with glycerol as depicted in Figure 5.1.

The three fatty acid chains forming the triglyceride can be equal or different. Fatty acids are distinguished by the aliphatic chain length, the number of double bonds, and their position along the aliphatic chain. The specific composition of fatty acids present in vegetable oils varies with the source. Example of typical fatty acids of the most common vegetable oils is shown in Table 5.2 [9–11].

The chain length of most abundant fatty acids ranges between 14–16 and 18–20. However, there are exceptions: peanut oil presents significant percentages of long-chain fatty acids C_{20-24}, while coconut and palm kernel oil show a high content of fatty acids C_{8-16}. *Jatropha* oil is characterized by a fatty acid composition where the main component is the oleic acid $C_{18:1}$ (32–45%), followed by linoleic $C_{18:2}$ (15–32%) and palmitic $C_{16:0}$ (13–15%), while the concentration of stearic acid $C_{18:0}$ is modest [12]. Palm and coconut oils are among the few

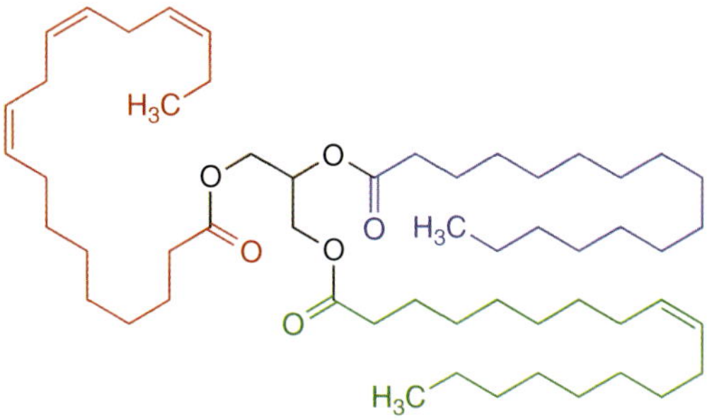

Figure 5.1 Triglyceride molecule containing linolenic (red), palmitic (blue), and oleic (green) acids.

Table 5.2 Fatty acid composition of vegetable oils.

Type	Fatty acids [a]									
	Caprylic $C_{8:0}$	Capric $C_{10:0}$	Lauric $C_{12:0}$	Myristic $C_{14:0}$	Palmitic $C_{16:0}$	Stearic $C_{18:0}$	Oleic $C_{18:1}$	Linoleic $C_{18:2}$	Linolenic $C_{18:3}$	Others
Palm	—	—	—	1	44	4	40	10	Trace	1
Soybean	—	—	—	—	10	4	23	53	8	C_{20-22} 1.5
Rapeseed	—	—	—	—	5	2.5	59	21	9	C_{20-22} 2
Sunflower	—	—	—	—	6	5	18	69	<0.5	C_{20-22} 1.5
Palm kernel	4	3	47	16	8	3	16	2	—	—
Cotton seed	—	—	—	1	25	2	18	53	0.3	$C_{16:1}$ 0.7
Peanut	—	—	—	—	13	3	38	41	Trace	C_{20-24} 5
Coconut	7	7	48	18	9	3	6	2	Trace	—
Olive	—	—	—	—	10	2	78	7	1	2
Lineseed	—	—	—	—	6	3	17	14	60	—
Corn	—	—	—	—	13	3	31	52	1	—
Jatropha	—	—	—	—	20	7	41	32	—	—

a) C*n*:*y*; *n* stands for chain length and *y* stands for the number of double bonds.

oils with a high content of saturated fatty acids. Palm oil is semisolid at room temperature, and as for the fatty acid distribution, the main compound is palmitic $C_{16:0}$ (44%), followed by the stearic $C_{18:0}$ and low concentration of myristic $C_{14:0}$ (1%). From a general standpoint, the number of double bonds along the aliphatic chain can reach up to six, while the cis configuration is predominant. The melting point increases along the chain length, while the presence of double bonds has an opposite effect. Not surprisingly the melting point increases in proportion to the presence of double bonds with trans configuration. Crude oil consists primarily of triglycerides whose concentration ranges between 95% and 97% and other compounds present in lower concentration such as free fatty acids, phospholipids, tocopherols, and carotene. In addition to organic impurities, crude oils contain also inorganic matter such as alkali, alkali earth metals, P, Cl, and transition metals at parts per million level. Upon refining, the concentration of minor components is reduced.

Table 5.3 Characteristics of crude and refined palm oil.

	Crude palm oil	Refined palm oil
S	7.6	4.5
N	9.2	1.4
Cl	3.7	2.9
Fe	7.6	<0.3
Na	4.8	<1
K	6	<0.1
Ca	12	1.4
Mg	3.2	0.35
P	10.5	1.5
Al	1.5	0.1
Zn	0.3	<0.1

Typical concentration of metals and heteroatoms in crude and refined palm oil is reported in Table 5.3.

The effect of contaminants on hydroconversion of vegetable oil over sulfided CoMo/γ-Al_2O_3 catalyst was investigated by Kubička and Horáček [13]. The authors used several rapeseed oils with different refining degrees containing different concentrations of metals, phosphorus, free fatty acids, and water. The main findings were as follows: the presence of alkalis promotes catalyst deactivation due to their deposition on catalyst surface leading to blockage/poisoning of active sites and phosphorus showed a twofold effect. When charge-compensating alkalis were present, corresponding phosphates were deposited above and at the beginning of the catalyst bed leading to gradual buildup of deposits. On the other hand, in the absence of alkalis, decomposition of phospholipids yielded phosphoric acid that catalyzed oligomerization reactions leading to rapid catalyst deactivation by carbonaceous deposits.

5.2.2 Animal Oils and Fats

As an alternative to vegetable oils, sources of triglycerides for the production of green diesel (GD) through hydroconversion process are rendered animal fats, waste fats, oils, and greases. The rendering process consists in boiling under pressure with steam injection, material like meat trimming, packing house by-products, bones and fallen animal, followed by a settling stage. The overall world production is $30 \times 10^6\,t\,yr^{-1}$, and 60% of which is made up of tallow and lard. Another significant source of animal oil/fat is fish oil ($1 \times 10^6\,t\,yr^{-1}$). A distinctive feature of the latter is the high content of unsaturated fatty acids $C_{20:4}$, $C_{20:5}$, $C_{22:5}$, and $C_{22:6}$ which makes it less suitable for GD production owing to the high hydrogen consumption associated with their hydroconversion.

Table 5.4 Fatty acid composition of animal fats.

Fatty acids[a]	Fish	Tallow	Lard	Chicken
$C_{14:0}$	3.5–8.2	1.4–6.3	0.5–2.5	1
$C_{16:0}$	13.6–21.9	20–37	20–32	25
$C_{16:1}$	4.2–7.9	0.7–8.8	1.7–5	8
$C_{18:0}$	2.4–6.2	6.0–40	5–24	6
$C_{18:1}$	9.8–15.8	26–50	35–62	41
$C_{18:2}$	1.3–3.3	0.5–5	3–16	18
$C_{18:3}$	1–2.6	—	—	—
$C_{18:4}$	1.3–7	—	—	—
$C_{20:1}$	0.7–14.5	—	—	—
$C_{20:4}$	1.3–2.4	—	—	—
$C_{20:5}$	5.1–13.8	—	—	—
$C_{22:1}$	0.4–1.6	—	—	—
$C_{22:5}$	0.9–3.3	—	—	—
$C_{22:6}$	10.5–27.7	—	—	—
$C_{24:1}$	0.7–1.2	—	—	—
Other FA	2.0–11.4	—	—	—

a) $Cn:y$; n stands for chain length and y stands for the number of double bonds.

A typical composition of triglycerides of different origins is given in Table 5.4 [14, 15].

Excluding fish oil, the data show that triglycerides from animals are particularly rich in saturated fatty acids C_{16} (25–38%) and C_{18} (15–28%). Although animal fats do not show significant differences in terms of chain length distribution of fatty acids in comparison with vegetable oil, the high content of saturated fraction gives the product a high melting point. Such a characteristic is a problem for the production of biofuel through transesterification because the product obtained presents poor cold flow properties. Besides triglycerides, animal fats contain in varying degrees other compounds like phospholipids, nitrogen, and sulfur-containing compounds. Likewise the vegetable oil, animal fats may contain significant amount of inorganic elements such as P, Na, Ca, and Mg.

5.2.3
Triglycerides from Algae

The production of biofuels from renewable sources is a strongly debated matter because feedstock used in the existing commercial processes is largely derived from the production intended for human food. Such a situation may cause negative effects on the prices of food and land used for the production of food. These considerations have generated a growing interest in the development of processes for the production of biofuels from feedstocks not in competition with

Table 5.5 Fatty acid compositions of triglycerides from algae.

Fatty acids	*Chaetoceros calcitrans*	*Skeletonema costatum*	*Phaeodactylum tricornutum*	*Chlamydomonas reinhardtii*	*Chlorella vulgaris*	*Dunadiella salina*	*Dunadiella tertiolecta*	*Scenedesmus obliquus*	*Neochloris oleoabundans*
10:0	—	—	24.6	—	0.5	—	0.4	1	—
12:0	—	—	—	—	2.7	1.5	1.9	0.5	—
13:0	—	—	—	—	0.7	—	1.8	0.2	—
14:0	18.6	14.6	7.5	2.3	2.5	0.6	1	0.9	1.9
14:1	—	0.4	25.5	—	0.9	0.4	0.7	21.7	0.4
15:0	—	2.7	8	—	—	1.9	9.4	2.3	1.4
15:1	—	—	—	—	2.4	—	2.4	6.2	—
16:0	26.3	12.4	15.4	32.4	17.4	19.4	13.2	11.6	36.3
16:1	27.5	22.8	22.6	1.7	3.1	1.7	5.5	5.6	2.5
16:2	—	4.1	4.4	1.6	8.1	1.5	2.7	3.2	2.1
16:3	—	10.2	8.3	2.1	2.6	7.2	2.8	0.7	1
17:0	—	0.3	0.1	—	3.9	—	1.4	20.4	11.7
17:1	—	—	—	—	31.6	—	4.1	1.2	1
18:0	2.6	1.8	1.8	—	1	1.5	2.6	10	5
18:1	4.5	2.9	5.5	17.7	9.2	5.3	5.7	9.5	23.4
18:2	08	1.4	1.6	10.8	20.7	6.2	14.2	17.5	10.2
18:3	—	0.9	0.9	21.6	14.3	38.7	35	1.9	10
18:4	—	2.4	0.5	—	—	0.7	1.3	0.2	2.1
20:0	—	—	1.3	—	1.5	—	—	2	2.1
20:1	—	—	0.5	—	0.9	—	—	—	2.5
20:2	—	0.2	0.5	—	—	0.1	—	0.4	—
20:3	—	—	0.1	—	0.8		—	—	—
20:4	—	1.4	2.2	—	0.4	—	0.3	—	—
20:5	6.7	14.2	26.1	—	0.5	0.1	0.4	—	—
22:0	—	—	—	—	—	—	—	—	—
22:1	—	—	—	—	—	—	—	—	—
22:2	—	0.2	—	—	—	—	—	—	—
22:3	—	—	—	—	—	—	—	—	—
22:4	—	1.4	—	—	—	—	—	—	—
22:5	—	14.2	—	—	—	—	—	—	—
22:6	0.6	2.1	0.9	—	0.5	—	—	—	—

Adapted from Ref. [10].

food production. In this context, the production of triglycerides from microalgae cultivation appears to have excellent potential in this area [16]. Fatty acid composition of triglycerides extracted from different species of algae is reported in Table 5.5 [10].

The comparison with the composition of vegetable oils points out some differences:

- Composition of triglycerides from algae presents a higher variance.
- Although most of the algae species show high concentration of fatty acids between C_{16} and C_{18}, their average concentration is lower.
- Several algae species present a wider fatty acid distribution than vegetable oils with a significant presence of lighter C_{12-14} and heavier C_{20-22} fractions.

5.3 Hydroconversion Processes of Vegetable Oils and Animal Fats

From a general standpoint, the technology of catalytic hydroconversion of fats and oil to produce fuels for automotive transport largely is derived from the hydroconversion processes used in the refining industry. Presently, existing technology for producing gas oil from renewable sources is largely based on the production of FAME [5, 17]. World production capacity of BD grew at a rate of 32% from 2000 to 2005 to 7 million tons per year [18] to further increase up to 18 million tons per year in 2012. While FAME has some desirable qualities such as high cetane, there are other issues associated with its use such as poor stability, high solvency, and low heating value. In addition, the transestherification reaction involved in the production of FAME leads to formation of glycerol which is a low-value product. Another approach to produce gas oil from renewable source requiring low capital investment is co-feeding [19]. Co-feeding consists in feeding vegetable oils together with diesel fuel of mineral origin to the HDS unit. In this case, the deoxygenation of the vegetable oil takes place simultaneously to the process of HDS of the mineral matrix in the same reaction conditions. However, the presence of significant concentrations of oxygenated compounds requires more drastic reaction conditions (higher temperature) to achieve the same level of sulfur removal. Several companies such as BP, Cepsa, ConocoPhillips, and Petrobras have used this approach for the production of biofuels. Although the co-feeding seems to be at first glance, the most simple, and straightforward, method to convert vegetable oils into transportation fuel, it suffers from a number of limitations related to the process and product specifications. Regarding the first facet, the concentration of oil fed is limited by the amount of hydrogen available, the control of the temperature of the reactor (the deoxygenation reaction is highly exothermic typically from 5 to 10 times greater than that for typical HDS), the lifetime of the catalyst, and the processing of by-products such as H_2O, CO, and CO_2. As regards the quality of the products, the conversion of vegetable oils leads to the formation of *n*-paraffins, and consequently there is an increase in the cetane number and a deterioration of the cold flow properties [20]. An emerging technology alternative to FAME and co-feeding is catalytic hydroconversion processes, specifically developed for the conversion of oil and natural fats, for the production of biofuels. Although this type of approach requires investment costs significantly higher than the co-feeding, it allows considerable flexibility for the use of different charges and control of the properties of the final product [20].

One of the first examples of industrial process for the hydroconversion of triglycerides is due to Neste Oil with the NExBTL process [21]. Since 2007 the refinery is operating in Porvoo (Finland), a plant with a capacity of $190\,\mathrm{kt\,yr^{-1}}$, revamped in 2009 with by a second unit of the same capacity. In November 2010, Neste Oil opened a production plant in Singapore, based on NExBTL technology, with a capacity of $800\,\mathrm{kt\,yr^{-1}}$, while another plant of the same capacity was put into operation in September 2011 in Rotterdam. The process produces GD from vegetable oils and animal fats that help, according to Neste Oil a reduction in CO_2, emissions by 40–80% depending on the feedstock used. Eni and UOP Honeywell have developed a hydroconversion process named Ecofining^{TM} for the production of gas oil/kerosene from vegetable oils and animal fats [20, 22–24]. The process of UOP/Eni Ecofining is constituted by a first stage for the hydrotreatment of triglycerides leading to the complete removal of oxygen functionality [25], followed by a second stage of hydroisomerization to improve the cold flow properties of diesel fuel. Since August 2013 is running in Norco (Louisiana, USA) at the Valero refinery of St Charles corporation a plant with a capacity of 360 kt yr^{-1} for the production of renewable diesel fuel [26, 27]. The process is based on the Ecofining^{TM} technology and provides the use of recycled animal fat, used cooking oil, and others as feedstocks. In May 2014, entered into production, at Venice refinery, a plant based on Ecofining technology for the production of GD. The core of the project consists in converting two existing hydrodesulfurization units to the new hydrorefining process. The plant has a capacity of 300 kt yr^{-1} of renewable GD using as feedstock vegetable oils, animal fats, and used cooking oils [28]. Furthermore, UOP has developed and commercialized technology that converts nonedible, second-generation natural oils and wastes to Green Jet Fuel™ [29]. Since 2004, Haldor Topsoe has developed a technology called HydroFlex for the production of biofuels from various renewable feedstocks such as tall oil, natural fats, vegetable oils, and pyrolysis oils. A distinctive feature of the technology seems to be the flexibility as for the feedstock [30, 31]. HT recently announced the development of a new hydrotreating technology for the production of GD and green kerosene (GK) from tall oil [32]. In January 2015, a wood-based biorefinery producing renewable diesel began commercial production in Lappeenranta, Finland. The annual production capacity of the new plant is 90/100 t yr^{-1} of renewable diesel [33].

ConocoPhillips began in 2006 the commercial production of renewable diesel fuel at the company's Whitegate refinery in Cork, Ireland. The refinery is producing 1000 barrels (40 000 gal) of renewable diesel fuel per day. The production process developed by ConocoPhillips hydrogenates vegetable oils to produce renewable diesel fuel component that meets EU standards. Soybean is the primary feedstock, although the plant can use other vegetable oils and animal fats [34, 35]. Vegan^{TM} is a hydroprocessing technology developed by Axens that appears ready to be licensed [36, 37]. Likewise, the Eni/UOP and Neste Oil technology conversion of renewable oils and fats into gas oil and kerosene occurs through a two-step process. In the first stage, the vegetable oil is deoxygenated to produce a mixture of *n*-paraffin, while in the second stage, the *n*-paraffins are isomerized to improve their cold flow properties.

5.3.1 Ecofining^{TM} Process

Ecofining^{TM} is a stand-alone two-stage process to convert vegetable oils, animal fats, and used cooking oil to GD [20, 25]. A simplified scheme of the process Eni/UOP Ecofining^{TM} is shown in Figure 5.2.

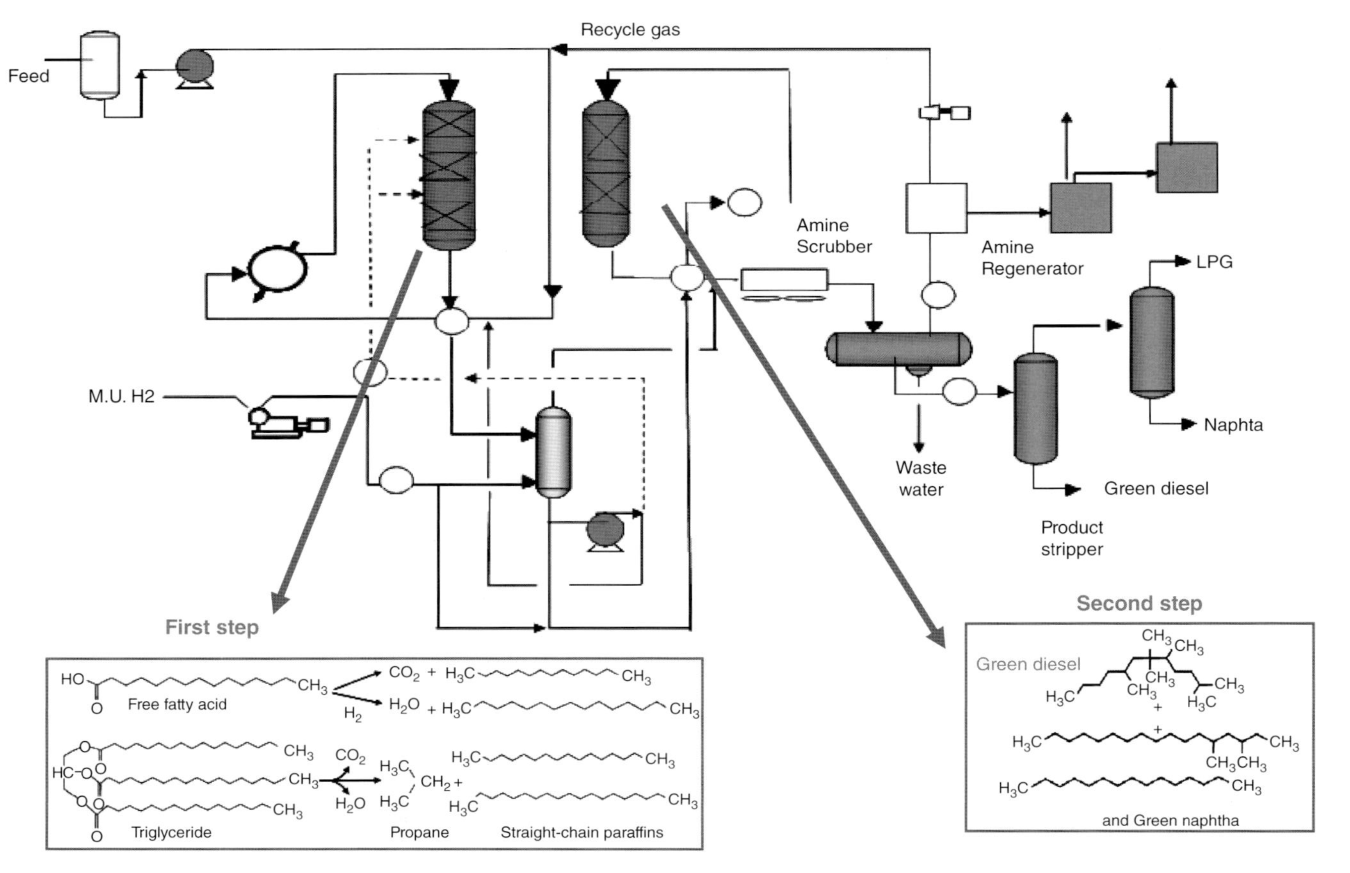

Figure 5.2 Process scheme of Eni/UOP Ecofining™.

The conversion of triglycerides occurs in two stages both under hydrogen pressure; in the first stage double bonds are saturated and triglyceride molecule is cracked, removing the oxygen as CO, CO_2, and H_2O and producing a mixture of normal paraffins. In the second stage, the mixture of normal paraffins is isomerized to improve the cold flow properties. The distribution of normal paraffins leaving the first reactor largely mirrors the distribution of fatty acids in the triglyceride. The most significant difference concerns the presence of *n*-paraffins with odd carbon atoms arising from the removal of oxygen of the triglyceride molecule via decarbonylation (DCO)/decarboxylation (DCO_2), which leads to the formation of *n*-paraffins with a carbon atom less than the fatty acids present in the triglyceride. Differently, when the oxygen is removed as H_2O, the *n*-paraffin produced presents the same number of carbon atoms of the corresponding fatty acid. The relative importance of the two-reaction pathway will depend on catalyst and operating conditions.

Conversion of triglycerides in the first stage is conducted under hydrogen pressure of 3–10 Mpa, at temperatures around 300 °C in the presence of sulfided transition bimetallic catalysts deposited on alumina. The overall reaction is highly exothermic, and depending on the reaction route for oxygen removal and content of double bonds, it may vary from 800 to 550 kJ mol^{-1}. Similarly, the overall chemical hydrogen consumption depends on the content of double bonds and reaction path for oxygen removal, and it may range from 70 to 270 Nl mol^{-1} for DCO_2 and hydrodeoxygenation (HDO), respectively, plus 23 Nl mol^{-1} for each double bond. In order to obtain an efficient management of the reaction heat, a portion of the product from the first reactor is recycled back and mixed with fresh vegetable oil. The products leaving the first reactor are sent to a separator for the removal of H_2O, CO_2, and light hydrocarbons (HC) consisting almost exclusively of propane. Subsequently, the mixture of liquid HC is mixed with hydrogen and sent to the isomerization reactor where the *n*-paraffins are converted to iso-paraffins. In the case of the production of GD, operating conditions of the second stage are selected to maximize the yield in GD and at the same time obtain a product with the required cold flow properties. The isomerization reaction does not imply a hydrogen consumption. Therefore, hydrogen consumption in the second stage is very low and is associated with the hydrocracking, which to a more or less extent, depending on the catalyst characteristics and operating condition, is always present in hydroisomerization reaction. The products leaving the second reactor are sent to a separator where the hydrogen and HC with low molecular weight are separated from the rest of the mixture. Hydrogen after purification is sent to the reactor, while the liquid fraction is sent to the fractionation section. The operating conditions of the second stage can be adjusted in order to produce a fraction of biojet fuel. Typical yields of Ecofining™ process are given in Table 5.6 [20, 25].

5.3.2
Product Characteristics and Fuel Specification

The product leaving the first stage resulting from the hydrotreatment of triglycerides is a mixture made up almost completely of *n*-paraffins with few percent

Table 5.6 Typical yields of Ecofining™ process in gas oil mode.

Feed	
Vegetable oil (wt%)	96.5–97
Hydrogen (wt%)	3–3.5
Products	
Diesel (wt%)	75–85
Naphtha (wt%)	1–8
Propane (wt%)	4–5
$CO + CO_2$ (wt%)	3–4
Water (wt%)	6–8

Table 5.7 Chemical physical characteristics of normal paraffins.

Paraffin	Cetane number	Melting point (°C)	Density ($g\ ml^{-1}$)	Boiling point (°C)
n-C_{12}	80	−10	749	216
n-C_{13}	88	−5	756	234
n-C_{14}	93	6	760	250
n-C_{15}	95	10	769	267
n-C_{16}	100	18	770	281
n-C_{17}	105	22	778	302
n-C_{18}	110	28	778	316

of *iso*-paraffin with a distribution very similar to that of the fatty acids present in the feed. These normal paraffins as shown in Table 5.7 are characterized by a high cetane number (the highest among the various classes of HC), low density, and poor cold flow properties as witnessed by the melting point [38, 39]. While cetane number is well beyond the specifications, depending on season and climate, the specifications of EN590 for diesel provide a cold filter plugging point (CFPP) ranging between 0 and −45 °C. Consequently, the mixture of *n*-paraffins can be blended with diesel fuel only in small percentages.

In order to improve the cold flow properties, the products from the first stage are sent to an isomerization stage converting *n*-paraffins to *iso*-paraffins. The latter are molecules with the same molecular weight and composition of the reacting molecules but have side branches along the paraffin chain. The isomerization process is frequently used in the refining industry for several purposes: producing high-octane components from paraffins C_5–C_6 [40], improving the cold properties of middle distillates [41], and productions of lubricant oils [42]. The process is conducted in the presence of hydrogen under pressure and in the presence of a suitable noble metal-based bifunctional catalyst. If the objective is to maximize the production of gas oil, cut is important that the catalyst has a high selectivity to isomerization and simultaneously a low activity toward the cracking reaction. Characteristics of the catalyst, which maximize the selectivity toward isomerization, are discussed in the paragraph concerning the isomerization

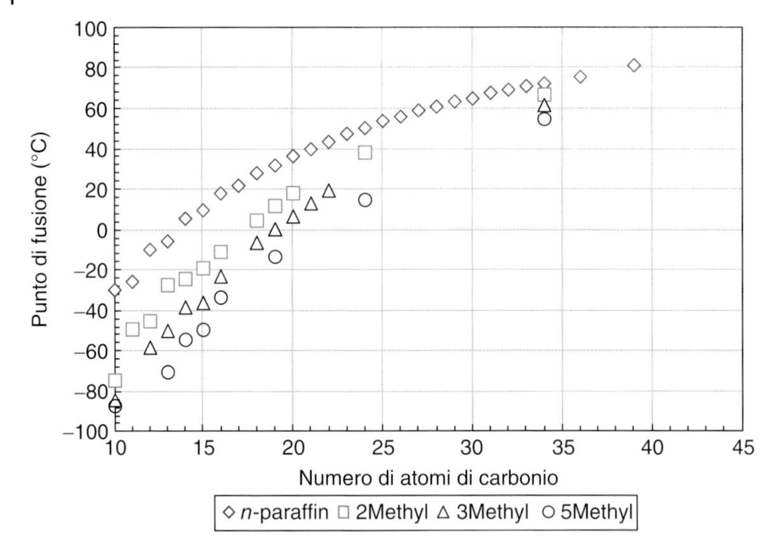

Figure 5.3 Melting points of normal paraffins and iso-paraffins with methyl branching.

chemistry. Otherwise, if the objective is the production of jet fuel which has a distribution shifted toward molecular weights lower than the mixture leaving the first stage, in addition to isomerization, another important aspect of the products closely connected with the cracking reaction is to minimize the formation of products lighter than C_9. Given the absence of compounds containing heteroatoms and particularly sulfur, for this type of reaction, catalysts may be used based on noble metal (Pt, Pd), which generally have significantly higher performance compared to the sulfided transition metals. The data reported in Figure 5.3 show that the melting point of n-paraffins increases with increasing molecular weight. The isomerization significantly lowers the melting point. Notably, it is observed that the decrease of the melting point [43–46]:

- Increases when the branching is shifted toward the middle of the paraffinic chain
- Increases with the length of the branching
- Increases with the branching degree that is the number of branching present on the paraffinic chain

The correlation between GD cloud point and *iso*-paraffin concentration is shown in Figure 5.4 [20].

Notably the increase of iso-paraffin content results from increasing the severity of the process which leads to a product distribution slightly shifted toward lower molecular weights and higher naphtha yields as shown in Table 5.8 [Eni data].

In addition to the change in the melting point, the isomerization influences also the cetane number. The data reported in the literature in this respect are less numerous; any way from a qualitative point of view is observed roughly the same effect observed for the melting point. The cetane number decreases [38, 47]:

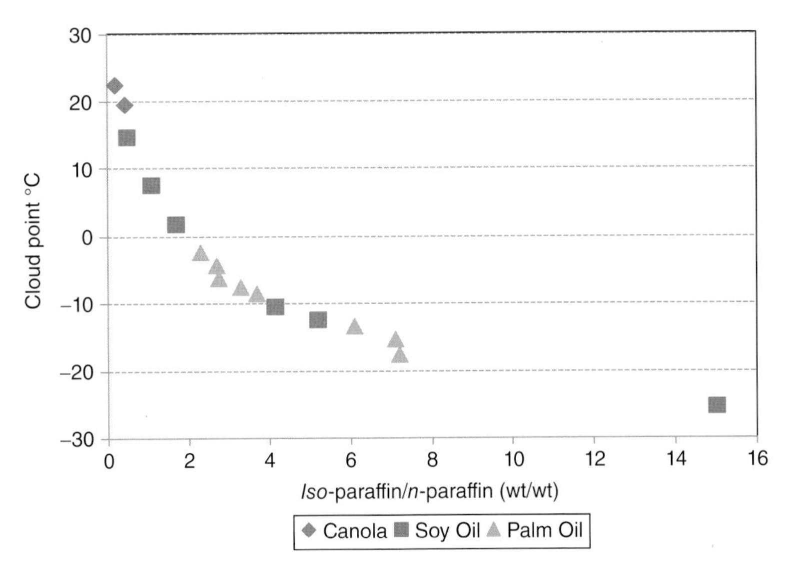

Figure 5.4 Green diesel cloud point as a function of *iso*-/*n*-paraffin ratio.

Table 5.8 Product yields and product characteristics.

Feed	Palm oil	Palm oil	Soybean oil	Soybean oil
	Cloud Point 0 °C	Cloud Point −8 °C	Cloud Point 0 °C	Cloud Point −8 °C
CO_2	5.4	5.4	5.4	5.4
H_2O	8.3	8.3	8.3	8.3
C_1-C_4	5.9	6.9	6.6	7.0
Naphtha	2.9	9.5	7.7	9.6
Diesel	80.3	75.5	75.5	73.0

- With the increase of the branching degree
- With the increase of the branching chain length
- For branching shifted toward the middle of the paraffinic chain

However, the type of branching formed during hydroisomerization of normal paraffins over bifunctional catalysts leads to a consistent improvement of cold flow properties, while the cetane number still remains very high [48].

As for the quality of the products, the conversion of vegetable oils through a hydrotreating process offers several advantages over the process of transesterification, exemplified in Table 5.9:

- The GD has a high cetane number, by far superior to current specifications. In blending with conventional diesel source component, GD can act actually as a cetane improver.

Table 5.9 Comparison of physical properties.

	Ultralow Sulfur diesel	Biodiesel FAME	Green diesel Ecofining™
Oxygen (%)	0	11	0
Specific weight (kg l^{-1})	0.84	0.88	0.78
Sulfur content (ppm)	<10	<1	<1
Lower heating value (MJ kg^{-1})	43	38	44
Cloud point (°C)	−5	−5 to +15	−20 to +10
Boiling point range (°C)	200–350	340–355	200–320
Polyaromatic (wt%)	11	0	0
Emission hydrocarbons	Baseline	nd	−60%
Emission CO	Baseline	nd	−40%
Cetane number	51	50–65	70–90
Oxidation stability	Baseline	Poor	Excellent

nd = not determined

- The heating value of the GD is 15% higher than FAME. Since the calculation of the quantity of fuels from renewable sources is carried out on an energy basis, the amount of GD required is less than FAME. Furthermore, if methanol used for transesterification does not come from renewable sources, only 93% of FAME should be considered from the renewable source.
- Being the GD exclusively composed of a mixture of normal and *iso*-paraffins, its use leads to a reduction in CO and PHA emissions [25].
- In the case of GD, it is possible, by tuning the operating conditions of isomerization stage, to change cold flow properties over a wide range.
- The presence of double bonds makes FAME susceptible to oxidation. Differently, the GD being composed of saturated structures presents an excellent stability toward oxidation.

5.4 Chemistry of Triglycerides Hydroconversion

5.4.1 Deoxygenation over Sulfided Catalysts

It is known for a long time that catalysts based on transition metal sulfide are active in the reaction of HDO, and one of the first reviews on this subject dates back to 1983 [49]. More recent reviews on HDO were written by Elliott [50], Furimsky [51], Jensen and colleagues [52], and Furimsky [53]. The latter has the merit of covering many of the aspects associated with hydroprocessing of biomass for biofuel production. Until about 10 years ago, the HDO reaction has been studied mainly in relation to the processes of hydrotreatment of petroleum cuts, and in this context, its importance compared to the reactions of the removal of sulfur and nitrogen was secondary. Consequently, almost all of the studies concerned the deoxygenation reaction of compounds such as phenols, furans, and ethers, which are the most representative of the oxygenated compounds, which can be found in petroleum-derived cuts. Although the situation is rapidly changing, relatively little has been published on the hydroconversion of triglycerides or compounds containing oxygen functionality related to the conversion of triglycerides such as carboxylic acid esters and alcohols. Generally, the catalysts used for the HDO reaction are as almost all of the hydrotreating catalysts used in refinery processes, based on a pair of sulfides of transition metals consisting in mainly a metal of group VIB (molybdenum, tungsten) and a metal of group VIIIB (cobalt, nickel) which acts as a promoter. The weight concentration of the metals is commonly 1–4% for Co and Ni, 8–16% for Mo, and 12–25% for W. Typical supports are alumina, silica-alumina, zeolite, kieselguhr, and magnesia. The choice of a particular combination between the metal phase and support obviously depends on the type of application and desired activity [54–56]. Several studies conducted with sulfided transition metal catalysts supported on alumina have shown that it is possible to obtain a complete conversion of the triglycerides with the formation of a mixture of *n*-paraffins with a distribution similar to that of the fatty acids present in the triglyceride molecule [19, 57, 58]. Hydroconversion of triglycerides consists of a treatment under hydrogen pressure (typically 2–7 MPa) in the presence of a suitable catalyst of the type mentioned previously in the range of temperatures between 270 and 350 °C. The reaction mechanism of HDO of triglycerides in the presence of sulfided transition metals is relatively complex, and so far only the general reaction pathway is sufficiently known, while information concerning the mechanism at molecular level are relatively scarce and several facets remain to be elucidated. Based on the studies reported in the literature, in addition to the hydrogenation of the double bonds present on the fatty acid chain, the conversion of triglycerides occurs via two main reaction pathways [19, 59, 60]:

- *HDO*: In this case, the oxygen present in the triglyceride molecule is removed forming to water with the simultaneous formation of a linear paraffin chain having the same number of the carbon atoms of the fatty acid present in the triglyceride starting.
- *Decarboxylation (DCO_2)/decarbonylation (DCO)*: The removal of oxygen through the formation of CO and CO_2 leads to the formation of linear paraffins with a carbon atom less in the starting fatty acid.

In each case the glyceric part is converted to propane. The process described is presented Figure 5.5 in the case of a triglyceride containing palmitic acid. We note that according to the reaction path, there are significant differences for the consumption of hydrogen and carbon yield, distribution of product gas composition, and heat of reaction. Without considering the hydrogen involved in the double bond saturation if the triglycerides are converted by HDO route, the hydrogen consumed is 12 mol, whereas conversion of triglycerides via DCO_2 route requires 3 mol of H_2. Of course, according to the reaction pathway, also the yield will be affected with a lowering of roughly 5–6% in the case of DCO_2 route. Another

Hydrodeoxygenation (HDO)
$$C_{51}H_{98-2x}O_6 + (12+X)H_2 \xrightarrow{\text{catalyst}} 3\,C_{16}H_{34} + C_3H_8 + 6\,H_2O \qquad \Delta H^{25}: -718\ \text{kJ mol}^{-1}$$

Decarbonylation (DCO)
$$C_{51}H_{98-2x}O_6 + (9+X)H_2 \xrightarrow{\text{catalyst}} 3\,C_{15}H_{32} + C_3H_8 + 3\,CO + 3\,H_2O \qquad \Delta H^{25}: -110\ \text{kJ mol}^{-1}$$

Decarboxylation (DCO_2)
$$C_{51}H_{98-2x}O_6 + (3+X)\,H_2 \xrightarrow{\text{catalyst}} 3\,C_{15}H_{32} + C_3H_8 + 3\,CO_2 \qquad \Delta H^{25}: -102\ \text{kJ mol}^{-1}$$

Figure 5.5 Triglyceride conversion: stoichiometry of hydrodeoxygenation, decarbonylation, and decarboxylation reactions. 2x stands for the decrease of hydrogen atoms due to the presence of double bonds. Enthalpy of reaction refers to the model triglyceride molecule triacylglycerol of palmitic acid.

Table 5.10 Theoretical paraffin yields and hydrogen consumption associated with DCO_2 and HDO.

Oil	Rapeseed		Palm		Soybean	
	HDO	DCO_2	HDO	DCO_2	HDO	DCO_2
Consumption H_2 (wt%)	3.6	1.5	3.2	1.2	3.8	1.7
Paraffin yield (wt%)	86.3	81.5	85.3	80.4	86.4	81.6
Yield H_2O (wt%)	12.3	—	12.7	—	12.4	—
Yield propane (wt%)	5.0	5.0	5.2	5.2	5.0	5.0
Yield CO_2 (wt%)	—	15.0	—	15.6	—	15.1

Adapted from Ref. [61].

facet which deserves to be considered is the heat of reaction, significantly lower in the case where the oxygen is removed via the formation of CO_2. Calculated reaction enthalpy at 300 °C for a typical palm oil is ΔH_{HDO}: $-788\ \text{kJ mol}^{-1}$; ΔH_{DCO2}: $-559\ \text{kJ mol}^{-1}$.

Table 5.10 shows for some vegetable oils the relative calculated consumption of hydrogen in the, together with the theoretical yields for the two reaction paths, hydrogenation and decarboxylation.

For example, the conversion of rapeseed oil requires depending on the relative presence of the two reaction paths a hydrogen consumption of 3.6% and 1.5 wt% of the charge corresponding to 403 and 168 Nl kg^{-1} (charge), respectively. Also the paraffin yields vary significantly depending on the relative presence of HDO or DCO_2. Since DCO_2 implies a loss of one carbon atom, the product yields are on average about 5% lower than in the case where the oxygen is removed as H_2O. In summary, the decarboxylation involves a much lower hydrogen consumption and no water is produced. Differently the removal of oxygen by HDO against a much higher hydrogen consumption leads the higher yields and implies a significantly higher reaction enthalpy. In any case, the process leads to the formation of significant yields of propane resulting from hydrogenation of the glycerol part of triglycerides and significant amounts of H_2O and CO_2 as by-products. The relative importance of the HDO and DCO/DCO_2 depends on the type of catalyst and

$CO + H_2O \quad CO_2 + H_2 \quad \Delta G = -23 \div -11\ kJ\ mol^{-1}$

$CO_2 + 4H_2 \quad CH_4 + 2H_2O \quad \Delta G = -74 \div -35\ kJ\ mol^{-1}$

$CO + 3H_2 \quad CH_4 + H_2O \quad \Delta G = -98 \div -50\ kJ\ mol^{-1}$

Figure 5.6 Water gas shift and methanation reactions of CO and CO_2 in the range of temperatures between 500 and 700 K.

the reaction conditions [20, 62, 63]. Since triglycerides contain fatty acids with even number of carbon atoms, the concentration of paraffins with odd and even carbon atoms in the products is indicative of the relative presence of the decarboxylation + DCO in respect to the HDO. The ratio C_{odd}/C_{even} increases with temperature, while the increase of hydrogen pressure leads to an increase of the relative importance of HDO path with formation of water [20]. However, under typical operating conditions of the process, both the water gas shift reaction and methanation reaction of CO and CO_2 are thermodynamically favored (Figure 5.6). In these circumstances, it is therefore difficult to distinguish experimentally if the deoxygenation is via decarboxylation or DCO because the concentration of CO and CO_2 is linked to the thermodynamic equilibrium and the activity of the catalyst considered toward the aforementioned reactions. However, we notice that on the basis of the ΔG values reported, CO and CO_2 would be completely transformed in CH_4, while if we consider the methanation reaction not occurring, the water gas shift reaction would be shifted toward the formation of CO_2. The activity of the catalyst toward the previously reported reactions is of major importance owing to the hydrogen consumption and heat balance. In case all CO or CO_2 is transformed into CH_4, there would be a higher consumption of hydrogen and higher heat of reaction. Data reported by Donnis *et al.* indicate that at 350 °C, about 30% of CO was converted into CH_4 [64]. According to the scheme proposed by Huber *et al.* [19] shown in. Figure 5.7, the hydrogenation of the double bond present in the aliphatic chain of the fatty acids is the faster stage. Subsequently, the hydrogenated triglyceride undergoes a series of reactions leading to the formation of various intermediates such as diglycerides, monoglycerides, and carboxylic acids. These compounds are, in turn, through HDO and DCO, converted to the corresponding HC. More recently, Kubička and Kaluža [60] studied the deoxygenation of rapeseed oil in the presence of three sulfided catalysts consisting in NiMo, Mo, and Ni deposited on alumina. The experimental data collected over the three sulfided catalysts allowed to propose a reaction scheme presented in Figure 5.8. Not surprisingly, hydrogenation of the double bond on the fatty acid chain is the fastest reaction. Subsequently, the saturated triglycerides can either:

- Undergo subsequent reduction stages that lead to the formation of *n*-paraffins with an even number of carbon atoms
- Or decarboxylate thus leading directly to the formation of *n*-paraffins with an odd number of carbon atoms

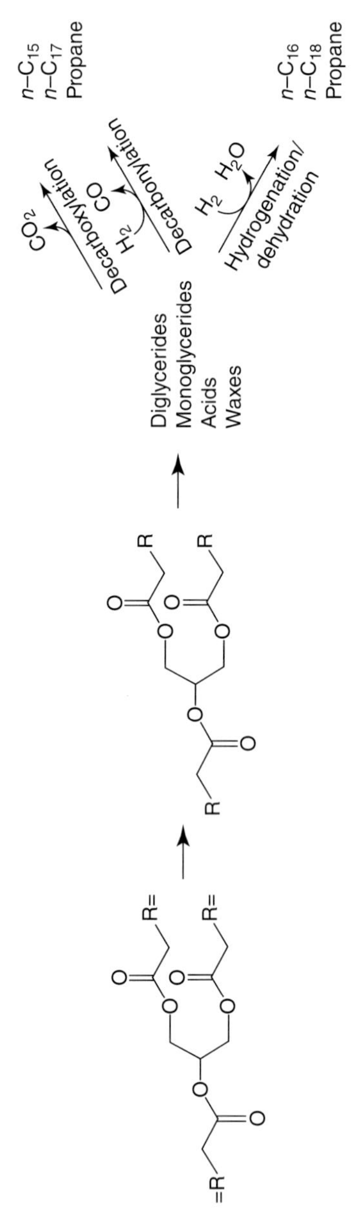

Figure 5.7 Triglyceride conversion scheme proposed by Huber *et al.* (Adapted from Ref [19].)

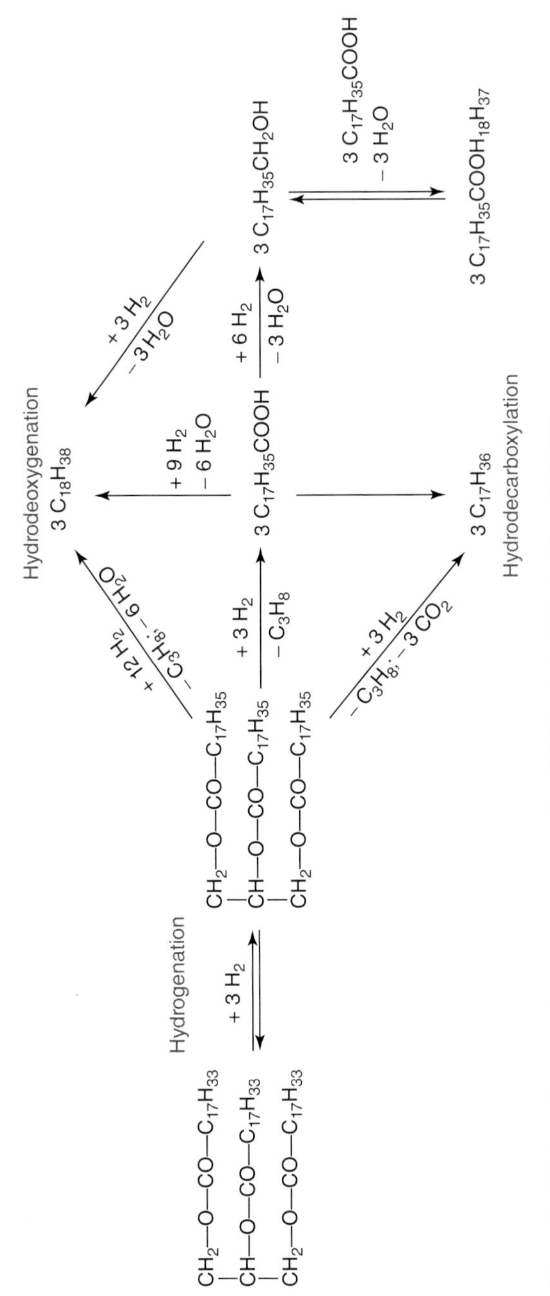

Figure 5.8 Triglyceride conversion scheme proposed by Kubička *et al.* (Adapted from Ref. [60].)

In addition to the indications of the different reaction paths, the work of Kubička and Kaluža showed that the composition of the catalyst significantly influences the relative importance of HDO and DCO/DCO_2 reaction. Particularly it was shown that the catalyst based on NiMo yields a mixture of products resulting from reactions of HDO and DCO. Differently, the catalyst based on NiS_2 promotes the decarboxylation reaction, while the catalyst based on MoS_2 yields to the formation mixture of n-paraffins which is derived almost exclusively from reactions of HDO.

Ruinart de Brimont *et al.* [66] investigated through a combined experimental and DFT study the deoxygenation of ethyl heptanoate in the presence of three bulk sulfided catalysts: MoS_2, Ni_3S_2, and Ni-promoted MoS_2. Besides the two competitive reaction paths HDO and DCO, it has been shown that the presence of Ni in both the mixed phase NiMoS and Ni_3S_2 changes the selectivity between HDO and DCO. Particularly, it was found that in presence of bulk MoS_2, the reaction proceeds in a highly selective way through the HDO path whereas with Ni_3S_2 the oxygen removal occurs preferentially through a DCO/DCO_2 pathway. With Ni-promoted MoS_2, both reaction pathways are present, and the Ni/Mo ratio determines the relative presence of HDO and DCO reaction pathways. Figure 5.9 shows a decrease of HDO contribution at increasing values of Ni/Mo ratio. In a recent paper Kubička *et al.* [63] investigated the effect of support-phase active interaction on activity and selectivity in deoxygenation of triglycerides. Three different NiMo catalysts with the same metal composition were prepared using as support SiO_2, TiO_2, and Al_2O_3. It was found that the dispersion of the active phase

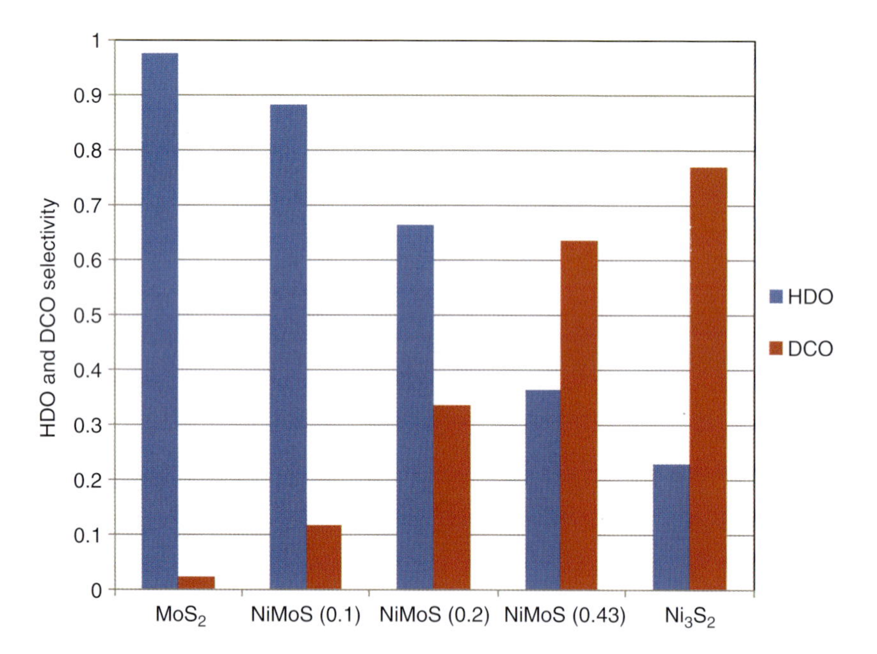

Figure 5.9 Effect of Ni/Mo ratio on HDO and DCO pathway for oxygen removal. (Adapted from [66].)

decreased in the following order: $SiO_2 > Al_2O_3 > TiO_2$. The activity of the catalyst was similar, while a significant change was found as for the relative importance of decarboxylation versus HDO. The SiO_2-supported catalyst showed a larger extent of decarboxylation reaction, while TiO_2-supported catalyst exhibited the opposite behavior. Studies with model molecules such as esters, carboxylic acids, and aldehydes have allowed to better define the chemistry of the reaction from a mechanistic point of view. Ryymin *et al.* [68–70] have shown that similarly to what is observed in the case of triglycerides, deoxygenation of an aliphatic ester (ethyl heptanoate) takes place through two reaction paths, namely, HDO and DCO. Furthermore, the authors have shown the role that the aldehyde group, formed as a result of the reduction of the carboxyl group, has in determining the selectivity HDO/DCO. The results of a more recent study in which methyl laurate was used as model compound mimicking the triglyceride molecule showed a more complex mechanism where the enol form the aldehyde is the active intermediate, which reacts in the subsequent hydrogenation phase [64]. The proposed reaction scheme shown in Figure 5.10 points out that conversion of methyl laurate occurs through two distinct reaction pathways leading to the formation of $n-1$ paraffin via DCO_2 and *n*-paraffin via HDO. In the latter case convincing evidences were presented that *n*-paraffin is not formed via direct carbonyl hydrogenation but through a mechanism where the keto–enol tautomerism plays an important role. This view was supported by the fact that reactivity of ketone with α-hydrogen presented a much higher reactivity than ketone without α-hydrogen. In this case the enolized form of aldehyde offers a plausible reaction pathway for the removal of oxygen as water.

5.4.2 Hydroisomerization

The production of GD or GK via the two-stage processes of Ecofining, NExBTL, or Vegan involves sending the product coming out from the first reactor to a second stage of hydrocracking/hydroisomerization. As the products of the first stage consist essentially of a mixture of normal paraffins in the range C_{14}–C_{20}, the process hydroisomerization/hydrocracking fulfills a dual purpose: to change as required the carbon distribution of *n-iso*-paraffins and increase the degree of branching of the products to improve the cold flow properties. The relative importance of hydroisomerization and hydrocracking reactions depends on the objective of the hydrotreatment, that is, if you want maximize gas oil or kerosene yield. The production of GK whose distribution in terms of chain length is approximately between C_9 and C_{15} requires the simultaneous presence of "selective cracking" and isomerization in order to shift the distribution paraffins from C_{14} to C_{20} toward the kerosene cut and give in the meanwhile to the products the appropriate isomerization degree. In this case it is important to avoid the reaction of "overcracking," that is, the consecutive cracking reaction of the first formed product from the cracking of paraffins present in the feed. Differently, in case the process is aimed at producing GD, it is necessary to maximize the isomerization

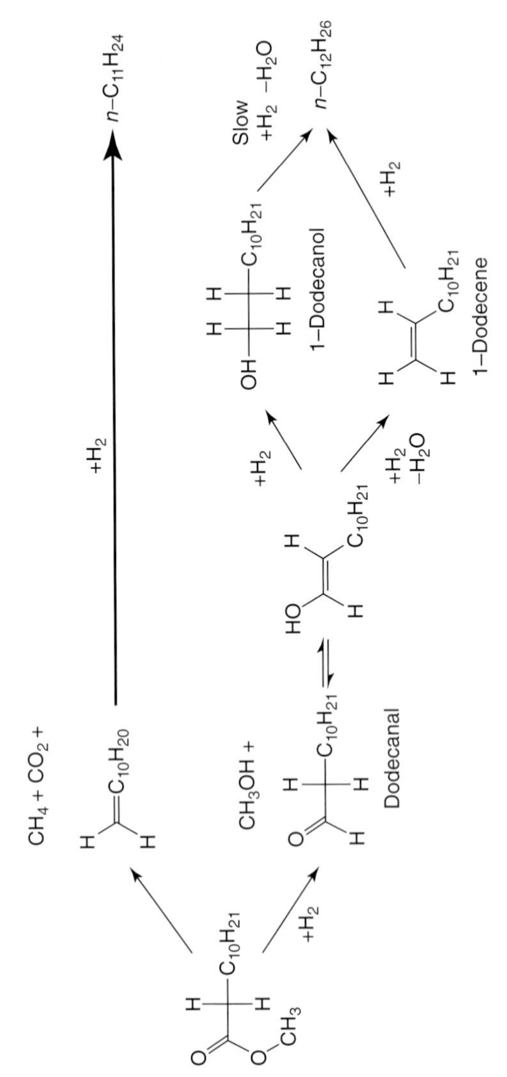

Figure 5.10 Reaction scheme for catalytic hydroconversion of methyl laurate to normal paraffin. (Adapted from Ref. [64].)

reaction and to limit as much as possible the hydrocracking reaction, which inevitably leads to a decrease of the yields in the desired product. From a general point of view, the hydrocracking and hydroisomerization of *n*-paraffins are reactions that can be performed on a broad spectrum of catalysts which according to the needs have been developed for specific applications [65, 67].

Hydroconversion catalysts (meaning by this term the set of both hydrocracking and hydroisomerization) are bifunctional, that is, they are characterized by the presence of two types of active sites:

- Brønsted acid sites which play an isomerizing and cracking function
- Metal sites or transition metal sulfide which play a hydro/dehydrogenating function

Typical acidic substrates are [65, 67, 71–78] amorphous oxides or mixtures of oxides (e.g., HF-treated Al_2O_3, $SiO_2-Al_2O_3$, ZrO_2/SO_4^{2-}); zeolites (Y, beta, mordenite, ZSM-5); and silicoaluminophosphate (SAPO-11, SAPO-31, SAPO-41). The most commonly used metals are Pt, Pd, or bimetallic systems of transition metals (e.g., Ni/Co, Ni/W, Ni/Mo, W/Mo). The latter are active as sulfide and are mainly used in the refining industry for hydrotreatment and hydroconversion of oil charges which are characterized by the presence of high concentrations of sulfur compounds. The hydrocracking and hydroisomerization of the paraffinic chains is an active field of research subject of numerous studies for both applicative and fundamental reasons [72–74, 79–84]. The results of research activity have led to a better understanding of the chemistry and the major factors affecting the catalyst activity, selectivity toward isomerization and distribution of the reaction products. The appropriate balance between the acidity of the support in terms of concentration of acid sites and their strength on the one hand and the activity of hydro/dehydrogenating function metal on the other is of primary importance in determining the activity, the selectivity toward isomerization, the distribution of the cracking products, and the stability of the catalyst [85, 86]. The presence of a metal phase with a high hydro/dehydrogenating activity on a support with Brønsted acid sites with medium–low strength favors the isomerization, while the cracking products have a more favorable distribution for the production of middle distillates. Consequently, catalysts loaded with noble metals (Pt, Pd) show a selectivity to isomerization that is considerably higher and a better distribution of the cracking products than the catalysts where the hydro/dehydrogenating function is constituted by transition metal sulfides [87–89]. The effect of the different natures of the hydro/dehydrogenating function is shown in Figure 5.11. The catalyst with the active phase made up of CoMo sulfide leads to the formation of cracking products shifted toward lighter compounds, while the catalyst loaded with Pt characterized by a higher hydro/dehydrogenating activity leads to a symmetrical distribution of the cracking product indication that the consecutive reaction of the products of the first formation (overcracking) is negligible. In addition to the factors discussed earlier, the nature of the support acid exerts a significant influence on product distribution and selectivity for isomerization. Although the behavior may change significantly depending on the synthesis conditions and/or

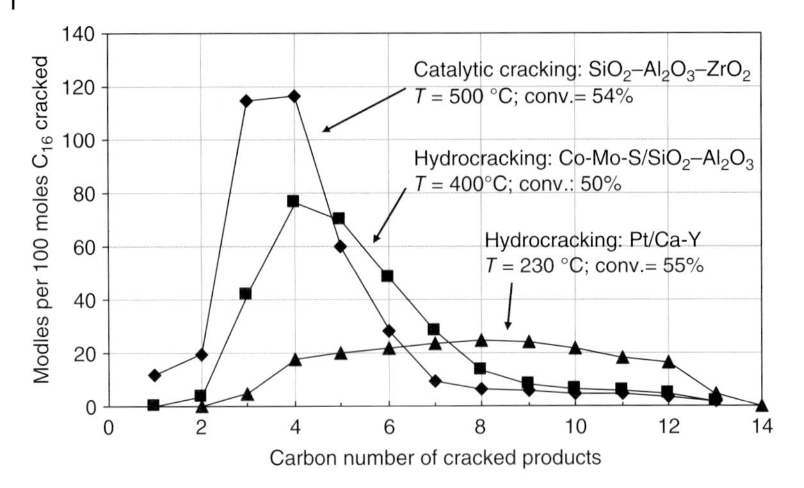

Figure 5.11 Molar carbon number distribution in catalytic cracking and hydrocracking of *n*-hexadecane at 50% conversion. (Adapted from Ref. [89].)

postsynthesis modifications [91], catalysts wherein the acidic support is made up of zeolites are more active than catalysts based on amorphous silico-aluminas. However, the latter give rise to cracking products with a distribution that will maximize the yield in middle distillates, whereas catalytic systems based on zeolite lead to a distribution of the products of cracking shifted toward the formation of gasoline [65]. High selectivity for the isomerization has been reported for amorphous silico-aluminas [92] and amorphous SAPO [93]. Unlike amorphous materials, the zeolitic materials may exhibit the phenomena of shape selectivity leading to high selectivity for isomerization as in the case of ZSM-22 [72] and SAPO-11 [73]. The mechanism of the conversion of bifunctional catalysts began to be actively investigated in the early 1960s. Based on previous works of Mills *et al.* [94] and Weisz [95], a mechanism that proceeded through the formation of an intermediate carbenium, hydro/dehydrogenation stages, and isomerization was proposed [90]. Later works allowed to obtain a clearer picture, with regard to the role of both acid sites and metal and the interaction between the two functions [81, 96]. Currently, the scheme more widely accepted by the scientific community (see Figure 5.12) provides that the reaction on a bifunctional catalyst proceeds through the adsorption of *n*-paraffin on the surface of the catalyst, the subsequent formation of olefin to the metal site, followed by diffusion toward the Brønsted acid site, and formation of secondary carbenium. The latter rearranges to tertiary carbenium, which in turn can desorb to form the *iso*-olefin or undergo β-scission reaction with the production of two fragments: a smaller carbenium and an olefin. The olefin and/or *iso*-olefin are subsequently hydrogenated to the metal site with the formation of the corresponding *n*-paraffin or *iso*-paraffin. In this scheme, in the presence of a catalyst characterized by a high activity of the function hydro/dehydrogenating, the formation of the intermediate olefin is fast enough not to be the rate-determining step, and the acid function determines the kinetics of the system [86, 96]. In these conditions, the rearrangement of the secondary

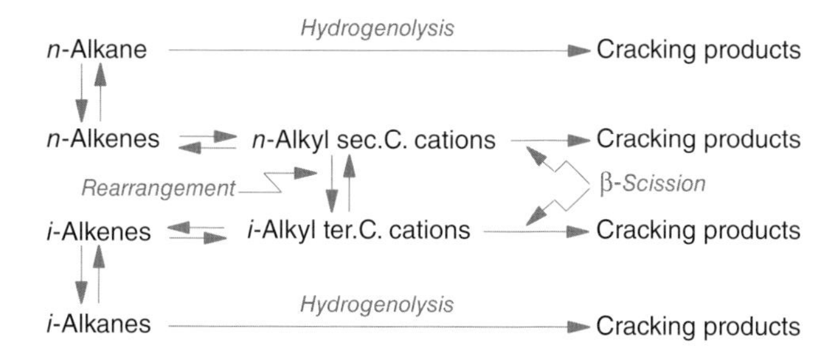

Figure 5.12 Hydroisomerization/hydrocracking reaction scheme of *n*-paraffins on bifunctional catalyst. (Adapted from Ref. [81].)

carbocation to the tertiary carbocation is considered to be the rate-determining step. Both isomerization and hydrocracking occur through the same carbenium intermediate. Consequently, the selectivity toward one of the two reaction paths will depend on the factors that influence the stability of the intermediate of the reaction. In this context, strong acid sites stabilize the intermediate carbenium thus promoting β-scission instead of the isomerization. The rearrangement of secondary carbenium occurs most likely via protonated cyclopropane (PCP) intermediate as proposed originally by Brower and Hogeveen [97] and Condon [98]. In agreement with the postulated mechanism, the activation energies for the isomerization and hydrocracking show very similar values, [99–101], while the reaction rate shows a negative reaction order with respect to hydrogen partial pressure [79, 99, 100, 102, 103]. This evidence together with the similarity of the activation energies reported previously constitutes the strongest evidences in favor the mechanism described earlier. In addition to the mechanism via carbenium, Figure 5.12 considers also the presence of the hydrogenolysis reaction catalyzed by metallic sites whose importance than the path of the main reaction depends on the operating conditions and the type of metal [89]. From a macroscopic point of view with a catalyst characterized by an optimal balance between the acid and metal function, the hydroconversion process of *n*-paraffins can be described as a series of consecutive and parallel reactions (Figure 5.13) where the *n*-paraffin is converted to the monobranched *iso*-paraffin and subsequently through consecutive reactions to di- and multibranched *iso*-paraffins. Simultaneously with the isomerization reactions, the molecules undergo also to a cracking process through a mechanism of β-scission that leads to the formation of fragments of lower molecular weight. In this scheme along the series *n*-paraffin, iso-paraffin monobranched *iso*-paraffin, and multibranched iso-paraffin, the ratio between the isomerization and cracking rate tends to decrease rapidly, and consequently there is a decrease in the selectivity toward the isomerization with the increase of the degree of conversion. As the data reported in Table 5.11 show, the β-scission of type A is by far the fastest reaction and requires the presence of a carbenium with a tribranched structure where branching are in position ααγ on paraffin chain. Scission of type B1 is significantly slower and implies the presence of gem-type structures, and the C–C scission occurs through a process energetically less

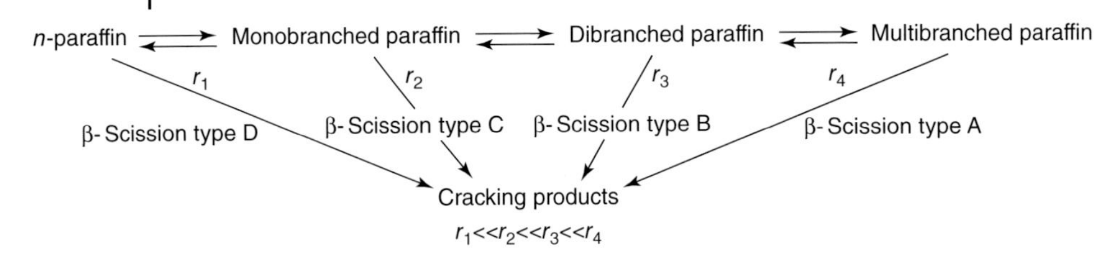

Figure 5.13 Reaction scheme for the formation of isomers and cracking products.

favorable than C–C scission of type A. In descending order of rate, other modes of cleavage of C–C bond are type B2 for *iso*-paraffins with isolated branching, type C for the monobranched paraffins, and type D for normal paraffins.

As for the isomerization, it occurs through two modes: type A and type B. Type A is faster and does not lead to a change of the degree of branching. It corresponds to the reaction where the branching moves along the paraffin chain and occurs through an intermediate corner protonated cyclopropane (CPCP) or π-complex. Differently, the reaction of type B occurs through an intermediate edge protonated cyclopropane (EPCP), leads to a change of the degree of branching, and is significantly slower than the previous. In view of the reaction rate, pertaining to the different modes of β-scission and isomerization, it can be easily explained why during the conversion of *n*-paraffins selectivity tends to decrease strongly with the degree of progress of the reaction and is difficult to obtain high concentrations of tribranched *iso*-paraffins. We notice that once formed, the tribranched *iso*-paraffin may take through consecutive isomerization reactions of type A, a configuration suitable to β-scission. The most abundant branching that is formed during the isomerization reaction is the methyl group followed in descending order by the groups ethyl and propyl [48, 104]. It is widely accepted that the formation of the methyl group occurs through an intermediate CPCP, while the formation of ethyl and propyl groups by analogy with the mechanism PCP can occur, respectively, via the formation of protonated cyclobutane (PCB) and cyclopentane [104, 105]. The lower formation rate of the ethyl group of its higher homologs most likely depends on the formation of a less energetically favorable intermediate. It has been reported that the PCB is less stable than the PCP of 130 kJ mol^{-1} [106].

5.5 Life Cycle Assessment and Emission

The European directive promoting the use of renewable energy ("RED," directive 2009/28/EC) requires that at least 10% by energy used in the transport sector has to be from renewable sources by 2020. In addition, the directive regulating fuel quality and overall greenhouse gas emissions ("FQD," directive 2009/30/EC) demands a 6% reduction of fuels' greenhouse gases by 2020. The purpose of a life cycle assessment (LCA) applied to renewable diesel is to quantify and compare

Table 5.11 β-Scission an isomerization mode occurring during hydroconversion of paraffins. Adapted from Ref. [115].

β-scissione and isomerization mode		Relative rates	Minimum carbon number
β-scission type A	 →	[1050; 170[a]]	n≥8
Isomerization type A	 CPCP[b] →	[56]	n≥6
β-scission type B1	 →	[2,8]	n≥7
β-scission type B2	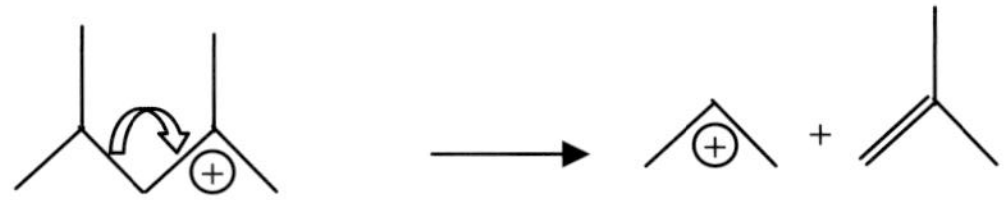 →	[1[c]]	n≥7
Isomerization type B	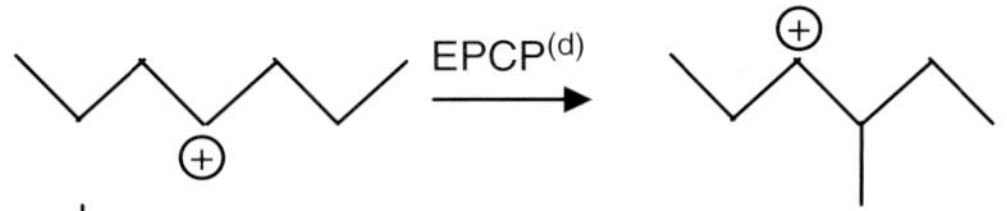 EPCP[d] →	[0,8]	n≥5
β-scission type C	 →	[0,4]	n≥6
β-scission type D	 →	[≈0]	n≥4

a) Reference [114].
b) CPCP: corner protonated cyclopropane.
c) Reference hydrocarbon.
d) EPCP: edge protonated cyclopropane.

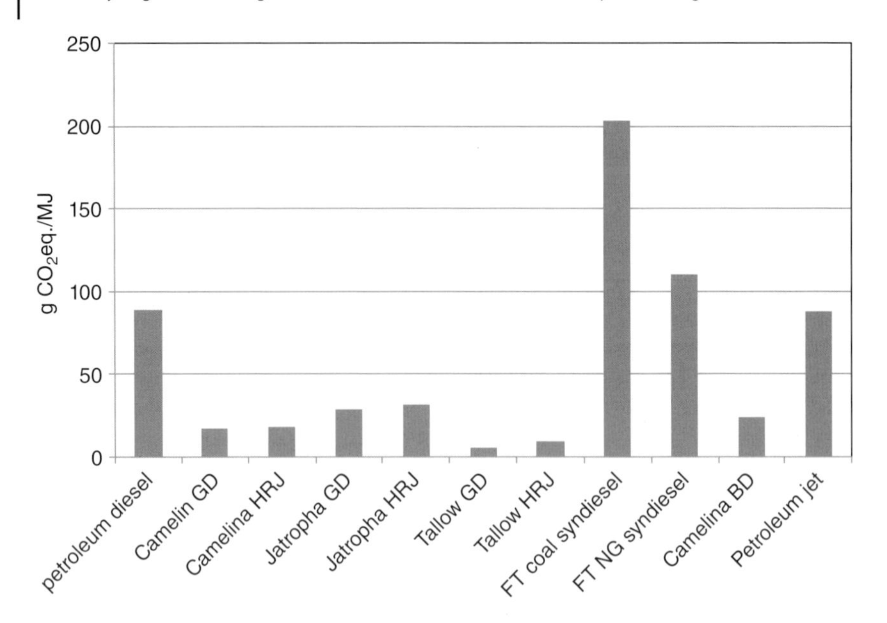

Figure 5.14 GHG emission associated with the production of green diesel syndiesel and petroleum fuels. (Adapted from [25].)

environmental flows of resources and pollutants (to and from the environment) associated with both renewable diesel and petroleum-based diesel, over the entire life cycle of the respective products. The method is based on the analysis of all the steps involved in the production of a product, from supply of raw material to its final use. Many life cycle analyses have been conducted to assess the impact of biofuel production on the environment [107, 108]. However, relatively few dealt with comparison of different feedstocks.

Figure 5.14 shows the results, obtained by Kalnes *et al.* [25], concerning the GHG emission for each final product examined. From a general standpoint, GD, hydrogenated renewable jet (HRJ), and BD show remarkably lower GHG emissions in comparison with the products derived from fossil sources. The highest reductions in GHG emission are realized with GD from tallow. Considering the products obtained from camelina, GHG emissions for GD production are lower than HRJ. We notice that the production of syndiesel via Fischer–Tropsch synthesis implies a remarkably higher GHG emission even higher than petroleum-derived fuels. LCA analysis carried out by Neste Oil for several feedstocks pointed out GHG savings were 52% palm oil, 51% rapeseed oil, and 78% animal fat [109].

5.5.1
Emissions

As previously shown the overall result of hydrotreatment of vegetable oil is the production of a mixture of normal and *iso*-paraffins falling in the distillation range of gas oil. In the last two decades, environmental factors have progressively led to the promulgation of more stringent automotive emission standards. The European Union has established emission standards to set limits for vehicle exhaust gases, in terms of NO_x, CO, HC, and particulate emissions [110], and these are regularly updated. Particularly, the introduction of Euro 5 standards in 2009 reduced the emission of particulate matter (PM) of diesel vehicles from 25 to 5 mg km^{-1}, while the introduction of Euro 6 requires a significant reduction of NO_x and $NO_x + HC$ in respect to Euro 5. The current fuel regulations require a minimum diesel cetane number of 51 and a maximum sulfur content for diesel of 10 ppm [111]. The present scenario calls for the use of cleaner fuels characterized by a cleaner combustion allowing a more efficient posttreatment of the exhaust gas. As previously shown the overall result of hydrotreatment of vegetable oil is the production of a mixture of normal and *iso*-paraffins falling in the distillation range of gas oil. Such a gas oil is characterized by the absence of sulfur and aromatic compounds whose presence in petroleum-derived fuels invariably leads to higher pollutant emissions [112, 113]. Eni performed engine testing aimed at comparing emission characteristics of petroleum-derived diesel with blending containing variable concentration of the so-called GD from the Ecofining process located in Venice refinery. Emission test investigation have been carried out according to a modified New European Driving Cycle using as a cycle test only the urban part. The vehicle used for the tests carried out with gas oil blended with 7% FAME and 10% HVO was Ford Focus Station Wagon Euro 5, displacement 1560 cc, while the vehicle used for the test with gas oil blended with 100% HVO was WW Golf Euro 6, displacement 1968. Table 5.12 shows the properties of neat petroleum-derived gas oil together with the blending with FAME, HVO, and neat HVO.

The data reported in Figure 5.15 indicate that the emissions of gas oil blended with HVO are remarkably lower than neat gas oil. Reduction of HC, CO, and PM is particularly strong. The reduction of emissions of gas oil blended with FAME is lower than those associated with the gas oil blended with HVO. It is also important to point out that a significant decrease of NO_x emission was observed, while the use of FAME typically increases NO_x emissions.

The results obtained with neat HVO (Figure 5.16) even in this case show a strong reduction of emission in comparison with neat petroleum-derived gas oil. Even though in both cases a strong reduction of emissions is observed, it should be noted that the results are not directly comparable because they are obtained with different cars having also different Euro standards. In the last years Neste Oil has carried out a rather extensive activity aimed at assessing the effect of blending petroleum-derived gas oil with HVO on pipe air emission. All in all, as reported by a Neste Oil report [109], exhaust emission tests have been performed totally with over 32 trucks and buses or their engines, several passenger cars in vehicle, and engine test beds. A summary of the results is shown in Figure 5.17 where remarkable reduction of particulate mass, carbon monoxide (CO), and HC emissions can be noticed.

Table 5.12 Ecofining HRD diesel blend used for engine tests [Eni data].

Property	Method	Unit	Reference gas oil (RG)	RG + 10% HVO	RG + 7% FAME	HVO 100%
Density@15 °C	ASTM D4052-96	$kg\,m^{-3}$	840.3	833.8	833	779.0
Cetane number	ASTM D613-36	—	54.0	55.7	—	74.8
Distillation	ASTM D86-01	—	—	—	—	—
I.B.P.	—	°C	203.1	196.7	204.8	158.6
10%	—	°C	236.8	234.1	235.7	222.6
50%	—	°C	284.9	282.9	288.7	281.8
90%	—	°C	345.2	340.9	342.1	295.1
F.B.P.	—	°C	368.9	369.3	365	313.9
Aromatics	EN12916-00	—	—	—	—	—
Total aromatics	—	%wt	24.5	21.7	22.4	0
Monoaromatics	—	%wt	19.7	17.7	18.3	0
Polyaromatics	—	%wt	4.5	4.0	4.1	0
Kinematic viscosity, 40 °C	ASTM D445-03	$mm^2\,s^{-1}$	3.453	3.285	—	2.632
CFPP	—	°C	−6	−6	—	−10
Lubricity (H.F.R.R.)	EN 12156-1	μm	—	422	—	—

I.B.P. = Initial Boiling Point; F.B.P. = Final Boiling Point

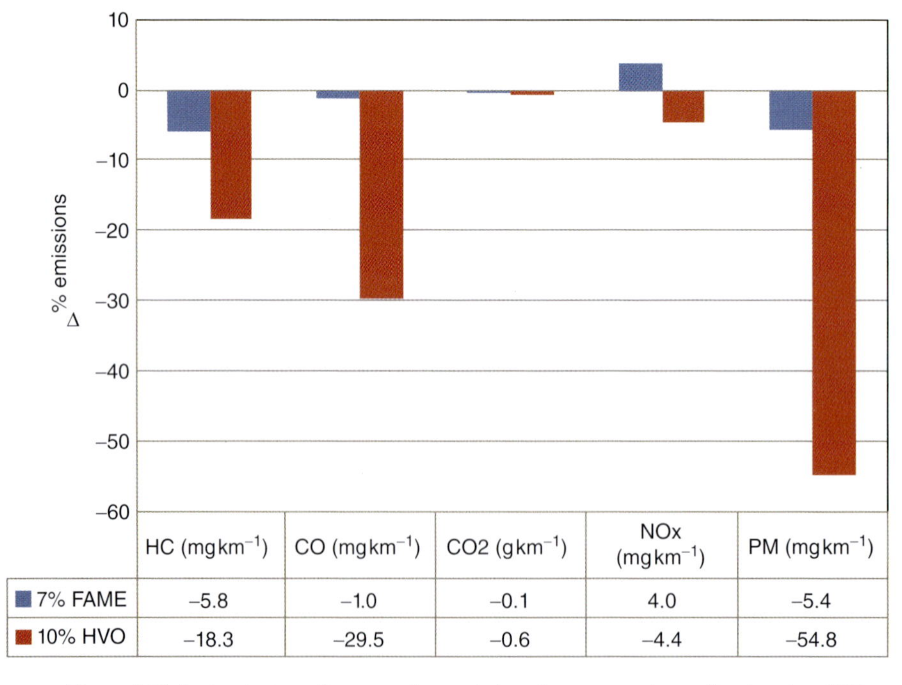

Figure 5.15 Engine test results comparing emissions from a petroleum diesel and an HRD petroleum diesel blend.

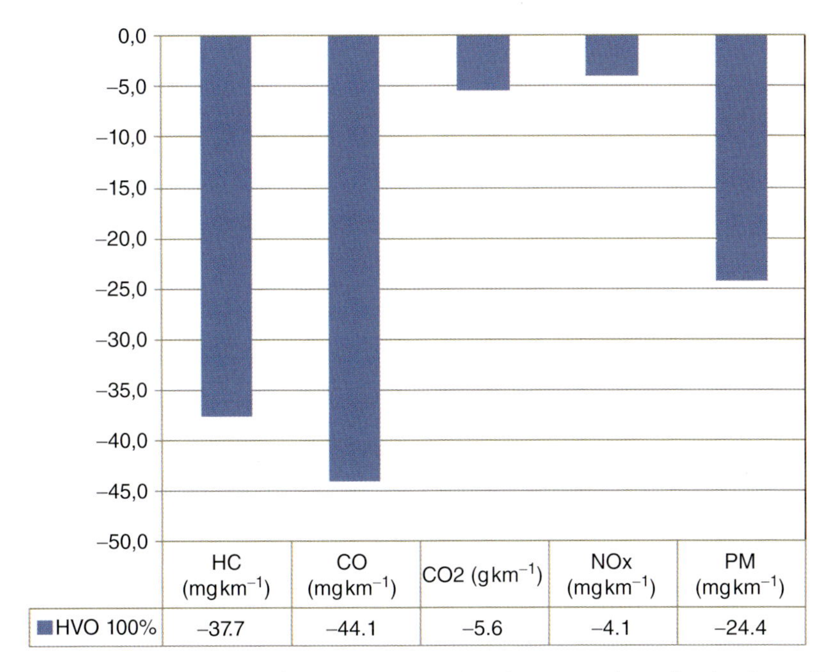

	HC (mg km^{-1})	CO (mg km^{-1})	CO2 (g km^{-1})	NOx (mg km^{-1})	PM (mg km^{-1})
HVO 100%	–37.7	–44.1	–5.6	–4.1	–24.4

Figure 5.16 Engine test results comparing emissions from a petroleum diesel and neat HVO.

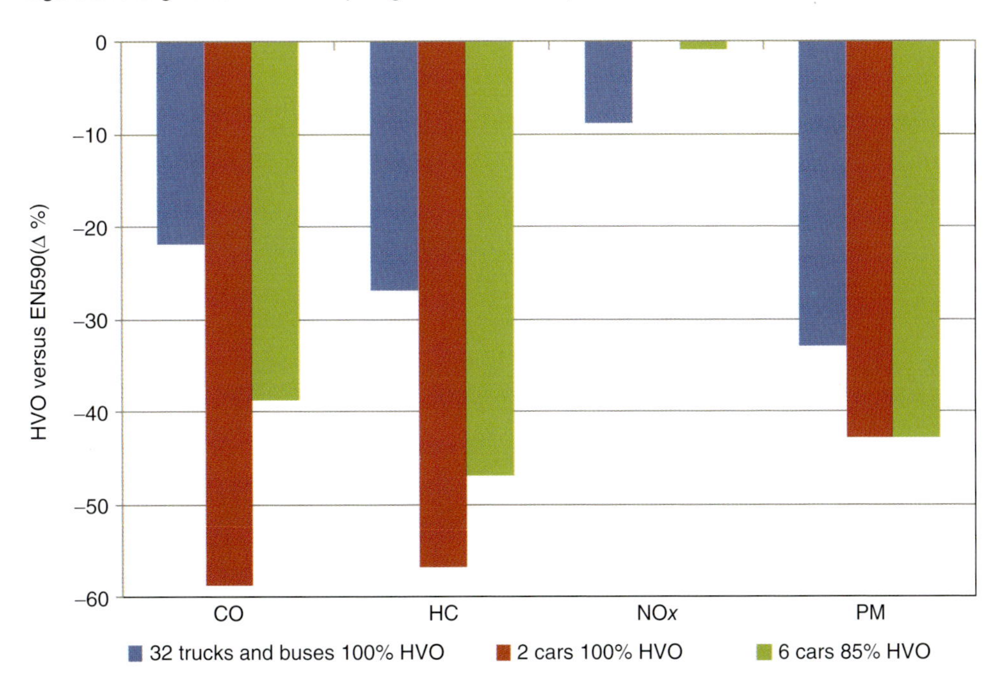

Figure 5.17 Average effect of neat and 85% HVO–15% EN590 blending on tailpipe emissions in Euro II to Euro V vehicles compared to a sulfur-free EN590 diesel fuel.

5.6 The Green Refinery Project

According to present regulations, the target of 10% energy content from renewable sources in transportation fuels, with a 6% reduction of GHG emission, cannot be reached using only FAME. Under these conditions, Eni SpA and Honeywell UOP decided to develop an innovative process, named Ecofining, to produce a next-generation high-quality biofuel. The Ecofining process accomplishes a complete hydrogenation of vegetable oils. The process is based on a two-stage reaction. The main features of the process and the characteristics of the products are described in Chapter 5.3.

The characteristics of the GD have already been discussed, in comparison with conventional ultralow sulfur diesel and FAME (see Table 5.9). GD does not contain sulfur, aromatics, and polynuclear aromatics. The extremely high cetane number (similar to that of Fischer–Tropsch diesel) and the low density allow blending into the final diesel fuels lower-quality fractions, still matching the diesel fuel specifications. The complete removal of the oxygen results in an energy content clearly higher than FAME. One of the main advantages of the process is the possibility to precisely control the cloud point of the final GD and alpine diesel quality can also be obtained. However, the most important feature is probably the extreme flexibility of the process with respect to the feedstock: product properties can be obtained and controlled independently from the type of the biofeedstock used [28]. Some years ago, Eni started a project for the construction of the first Ecofining plant at the Livorno refinery (Italy), but due to unfavorable market and economic scenarios of that period, the project did not get to the construction phase. Under the present European situation with a decreased demand for petroleum products and refining oversupply, as described in Chapter 1, and considering the need for an efficient process to produce high-quality biodiesel, Eni evaluated the possibility to convert an existing refinery in a refinery focused on the production of biofuels. This idea was the basis for the "Green Refinery project" that involved the conversion of Eni's Venice refinery into a biorefinery for the production of high-quality biofuels. The project has a particular strategic meaning for Eni, not only because it is aimed at increasing the profitability of an industrial site and at developing a new business, but also because this project provides new opportunities for the Venice refinery, which was facing economic pressure due to its low-conversion processing scheme. Venice refinery is an old complex in operation since 1926. The refinery had a simple scheme that included a hydroskimming section with a visbreaker/thermal cracker unit. The refining capacity was 80 000 bpd, and the conversion index was only 20%.

On the other side of Eni's biofuels, consumption was approximately 1 million tons per year (year 2012), and the biofuel demand was mostly met with external supplying. The proper conversion of the refinery into a "biorefinery" would have allowed transforming an industrial complex, which was no more profitable in a new production unit, based on a proprietary technology, which could produce new high-quality biofuel according to the market demand.

The core of the Green Refinery project is the conversion of the two existing refinery hydrodesulfurization units into the new hydroconversion process [116]. The hydrogen necessary for the process is produced by the existing naphtha catalytic reforming unit, while in a second phase, a proper steam reforming unit will be installed. The green refinery configuration will be completed by a pretreatment unit, able to reduce the contaminant level in the Ecofining feed, assuring flexibility in feedstock supply. The feedstock in the first phase is vegetable oil, but the plant will also treat "second-generation" feeds like animal fats, used cooking oils, and oils from treated biomass. Several innovative solutions had to be designed in order to cope with some features of the existing plants that did not match with the new configuration. For example, a dedicated study was required for gas separation and recycle hydrogen purification sections, which in the new configuration have to treat a low-sulfur, high-CO_2, and high-water mixture very different from the original design specifications. The project schedule has been very strict: the project was announced in September 2012, the Venice refinery stopped conventional operations in August 2013, and the Ecofining plant operations started in May 2014. In the first phase of the project, the plant feedstock capacity is 350 kt yr^{-1}; after the completion of the second phase, the total capacity will reach 550 kt yr^{-1}. The total investment of the project is about 100 million euro, which represents a very large saving compared to the capital investment required for a new grassroots unit of the same capacity [117].

To summarize the main advantages, the project allows to:

- Reduce four to five times the investment costs compared to a grassroot plant
- Produce high-quality biofuels (primarily diesel plus naphtha, propane/LPG, and kerosene as secondary products)
- Reuse plants, which produced conventional fuels, whose market demand is declining in Europe, for a process that produces biofuels, that is, converting a conventional refinery in a "biorefinery"

The conversion of the refinery into a biorefinery will also bring great benefit in terms of environmental impact in the Venice area, with a significant decrease of the global emissions from the site, regarding NO_x, SO_x, and PM. The proper rearrangement of several unit operations and equipments usually already existing in a conventional refinery allows to transform "traditional" refinery plants in a high-efficiency unit that produces GD and other biofuels from vegetable oils. The project has a particular strategic meaning for Eni, not only because it is aimed at increasing the profitability of an industrial site and at developing a new business, but also because this project provides new opportunities for the Venice refinery, which was facing economic pressure due to its low-conversion processing scheme.

5.7 Conclusions

The possibility to convert an existing refinery to a "green refinery" dedicated to the production of biofuels is an interesting option to face the refinery overcapacity in Europe and to comply with the regulation on the use of renewable fuels. The hydrogenation of vegetable oils and other renewable feedstocks allows producing high-quality "GD" that helps to satisfy the evolving regulations. The investment costs associated to the hydrogenation process are higher than those for the more conventional production of FAME; however the proper revamping of existing refinery units can sharply decrease the capital costs and make economically sustainable the process. The transformation of Venice refinery is a positive experience and can open the way to new promising projects for the production of hydrotreated vegetable oil.

References

1. Bartelloni, A. (2014) Staying ahead of the curve in the changing market Global Refining Summit, Barcelona, March, 2014.
2. Bellussi, G., Rispoli, G., Millini, M., and Pollesel, P. (2014) *Chem. Ing. Tech.*, **86** (12), 2150–2159.
3. REA Holdings Plc http://www.rea.co.uk/rea/en/markets/oilsandfats/worldproduction (accessed 23 September 2015).
4. Nexant (2013) Next Generation Biofeedstocks: Resources for Renewables, Nexant special report, October 2013.
5. ENI (2014) O&G World Oil and Gas Review 2014, http://www.eni.com/it&uscore;IT/azienda/cultura-energia/world-oil-gas-review/world-oil-gas-review-2014.shtml (accessed 23 September 2015).
6. Francis, G., Edinger, R., and Becker, K. (2005) *Nat. Resour. Forum*, **29**, 12–24.
7. Wealth Works http://wealthworks.org/economic-development-resources/success-stories/camelina-biofuel-thought.
8. Fediol http://www.fediol.eu/web/world%20production%20data/1011306087/list1187970075/f1.html (accessed 23 September 2015).
9. Daudin, A., Dupassieux, N., and Chapus, T. (2013) in *Catalysis by Transition Metals* (eds H. Toulhoat and P. Raybaud), Editions TECHNIP, p. 742.
10. Hoekman, S.K., Broch, A., Robbins, C., Ceniceros, E., and Natarajan, M. (2012) *Renewable Sustainable Energy Rev.*, **16**, 143–169.
11. Gunstone, F.D. (ed) (2002) *Vegetable Oils in Food Technology: Composition, Properties and Uses*, Blackwell Publishing.
12. Sarin, R., Shanna, M., Sinharay, S., and Malhotra, R.K. (2007) *Fuel*, **86**, 1365–1371.
13. Kubička, D. and Horáček, J. (2011) *Appl. Catal., A: Gen.*, **394**, 9–17.
14. Frankel, E.N., Sauté-Garcia, T., Meyer, A.S., and German, J.B. (2002) *J. Agric. Food. Chem.*, **50**, 2094–2099.
15. Extension http://www.extension.org/pages/30256/animal-fats-for-biodiesel-production (accessed 23 September 2015).
16. Mata, T.M., Martins, A.A., and Caetano, N.S. (2010) *Renewable Sustainable Energy Rev.*, **14**, 217–232.
17. Badu, V., Borugadda, B., and Goud, V. (2012) *Renewable Sustainable Energy Rev.*, **16**, 4763–4784.
18. Ronald, G., Bray (2007) Advances in Biofuels and Renewable Diesel Production, PEP Report 251A, December 2007, SRI Consulting.
19. Huber, G.W., O'Connor, P., and Corma, A. (2007) *Appl. Catal., A: Gen.*, **329**, 120–129.
20. Baldiraghi, F., Di Stanislao, M., Faraci, G., Perego, C., Marker, T., Gosling, C., Kokayeff, P., Kalnes, T., and Marinangeli, R. (2009) in *Sustainable Industrial Chemistry*, Chapter 8 (eds F. Cavani, G. Centi, S. Perathoner, and F. Trifiró), Wiley-VCH Verlag GmbH & Co.

21. Neste Oil (2015) http://www.nesteoil.com/default.asp?path=1,41,535,547,23110 (accessed 23 September 2015).
22. UOP http://www.uop.com/processing-solutions/renewables/green-diesel/.
23. Holmgren, J., Gosling, C., Marinangeli, R., Marker, T., Faraci, G., and Perego, C. (2007) *Hydrocarbon Process.*, 67–71.
24. Holmgren, J., Gosling, C., Marinangeli, R., Marker, T., Faraci, G., and Perego, C. (2007) A new development in renewable fuels: green diesel. 105th National Petrochemical and Refiners Association (NPRA) Annual Meeting, San Antonio, TX, Technical Paper AM-07-10.
25. Kalnes, T.N., McCall, M.M., and Shonnard, D.R. (2010) in *Thermochemical Conversion of Biomass to Liquid Fuels and Chemicals*, Chapter 18 (ed M. Crocker), RCS Publishing, p. 468.
26. Diamond Green Diesel http://www.diamondgreendiesel.com/Pages/default.aspx (accessed 23 September 2015).
27. Biofuels Digest http://www.biofuelsdigest.com/bdigest/2013/07/01/largest-green-diesel-plant-completed-in-us-diamond-green-diesel-can-it-make-money/ (accessed 23 September 2015).
28. Amoroso, A., Rispoli, G., and Prati, C. (2013) *Hydrocarbon Process.*, **92** (2), 95–100.
29. UOP http://www.uop.com/processing-solutions/renewables/green-jet-fuel/.
30. Haldor Topsoe http://www.topsoe.com/business&uscore;areas/refining/Renewable&uscore;fuels.aspx (accessed 23 September 2015).
31. Haldor Topsoe http://www.digitalrefining.com/data/literature/file/274435681.pdf (accessed 23 September 2015).
32. Egeberg, R.G., Michaelsen, N.H., and Skyum, L. Novel Hydrotreating Technology for Production of Green Diesel, http://www.topsoe.com/sites/default/files/novel&uscore;hydrotreating&uscore;technology&uscore;for&uscore;production&uscore;of&uscore;green&uscore;diesel.ashx&uscore;.pdf (accessed 23 September 2015).
33. Haldor Topsoe http://www.topsoe.com/news/2015/03/wood-based-renewable-diesel-bio-refinery-goes-stream-finland (accessed 23 September 2015).
34. Biodiesel Magazine http://www.biodieselmagazine.com/articles/1481/conocophillips-begins-production-of-renewable-diesel (accessed 23 September 2015).
35. Green Car Congress http://www.greencarcongress.com/2006/12/conocophillips&uscore;.html (accessed 23 September 2015).
36. Scharff, Y., Asteris, D., and Fédou, S. (2013) *OCL*, **20** (5), D502.
37. Axens http://www.axens.net/product/technology-licensing/11008/vegan.html (accessed 23 September 2015).
38. Murphy, M.J., Taylor, J.D., and McCormick, R.L. (2004) *Compendium of Experimental Cetane Number Data*, NREL.
39. NIST http://webbook.nist.gov/chemistry/ (accessed 23 September 2015).
40. Cusher, N.A. (2003) in *Handbook of Petroleum Refining Processes*, Chapter 9 (ed R.A. Meyers), McGraw-Hill.
41. Perego, C., Calemma, V., and Pollesel, P. (2010) in *Zeolites and Catalysis*, vol. 2, Chapter 19 (eds J. Cejka *et al.*), Wiley-VCH Verlag GmbH, p. 585.
42. Sequeira, A. (1994) *Lubricant Base Oil and Wax Processing*, Chapter 8, Marcel-Dekker Inc., p. 194.
43. Chen, N.Y., Garwood, W.E., and Dwyer, F.G. (1989) *Shape Selective Catalysis in Industrial Applications*, Marcel Dekker, New York, p. 176.
44. Burch, K.J. and Whitehead, E.G. (2004) *J. Chem. Eng. Data*, **49** (4), 858.
45. American Petroleum Institute (1940-1966) *Properties of Hydrocarbons of High Molecular Weight Research Project*, vol. 42, American Petroleum Institute, New York.
46. Lynch, T.R. (2008) *Process Chemistry of Lubricant Base Stocks*, Chapter 2, CRC Press, p. 21.
47. Creton, B., Dartiguelongue, C., de Bruin, T., and Toulhoat, H. (2010) *Energy Fuels*, **24**, 5396–5403.
48. Calemma, V., Gambaro, C., Parker, W.O. Jr., Carbone, R., Giardino, R., and

Scorletti, P. (2010) *Catal. Today*, **149**, 40–46.

49. Furimsky, E. (1983) *Catal. Rev. Sci. Eng.*, **25**, 421–458.
50. Elliott, D.C. (2007) *Energy Fuels*, **2**, 1792–1815.
51. Furimsky, E. (2000) *Appl. Catal., A: Gen.*, **199**, 147–190.
52. Mortensen, P.M., Grunwaldt, J.-D., Jensen, P.A., Knudsen, K.G., and Jensen, A.D. (2011) *Appl. Catal., A: Gen.*, **407**, 1–19.
53. Furimsky, E. (2013) *Catal. Today*, **217**, 13–56.
54. Scherzer, J. and Gruia, A.J. (1996) *Hydrocracking Science and Technology*, Marcel Dekker.
55. Toulhoat, H. and Raybaud, P. (eds) (2013) *Catalysis by Transition Metals Sulphides From Molecular Theory to Industrial Application*, Editions Technip, Paris.
56. Topsoe, H., Clausen, B.S., and Massoth, F.E. (1996) *Hydrotreating Catalysis Science and Technology*, Springer-Verlag.
57. Nunes, P.P., Brodzki, D., Bugli, G., and Djéga-Mariadassou, G. (1986) *Oil Gas Sci. Technol. – Rev. IFP*, **41** (3), 421–431.
58. Gusmão, J., Brodzki, D., Djéga-Mariadassou, G., and Frety, R. (1989) *Catal. Today*, **5**, 533–544.
59. Daudin, A. and Chapus, T. (2009) *Prep. Pap. Am. Chem. Soc., Div. Petr. Chem.*, **54** (2), 122–124.
60. Kubička, D. and Kaluža, L. (2010) *Appl. Catal., A: Gen.*, **372**, 199–208.
61. Ballerini, D., Chapus, T., Hillion, G., and Montagne, X. (2012) in *Biofuels*, Chapter 3 (ed D. Ballerini), Editions Technip, Paris.
62. Veriansyah, B., Han, J.H., Kim, S.K., Hong, S.-A., Kim, Y.J., Lim, J.S., Shu, Y.-W., Oh, S.-G., and Kim, J. (2012) *Fuel*, **94**, 578–585.
63. Kubička, D., Horáček, J., Setnička, M., Bulánek, R., Zukal, A., and Kubičková, I. (2014) *Appl. Catal. B: Environ.*, **145**, 101–107.
64. Donnis, B., Egeberg, R.G., Blom, P., and Knudsen, K.G. (2009) *Top. Catal.*, **52**, 229–240.
65. Scherzer, J. and Gruia, A.J. (1996) *Hydrocracking Science and Technology*, Chapters 3 and 7, Marcel Dekker, New York, pp. 13–39 and 96–111.
66. Ruinart de Brimont, M., Dupont, C., Daudin, A., Geantet, C., and Raybaud, P. (2012) *J. Catal.*, **286**, 153–164.
67. Sequeira, A. Jr., (1994) *Lubricant Base Oil and Wax Processing*, Chapters 6 and 8, Marcel Dekker, New York, pp. 119–152 and 194–224.
68. Ryymin, E.-M., Honkela, M.L., Viljava, T.-R., and Krause, A.O.I. (2010) *Appl. Catal., A: Gen.*, **389**, 114–121.
69. Senol, O.I., Viljava, T.R., and Krause, A.O.I. (2005) *Catal. Today*, **100**, 331–335.
70. Senol, O.I., Viljava, T.R., and Krause, A.O.I. (2005) *Catal. Today*, **106**, 186–189.
71. Wen, M.Y., Wender, I., and Tierney, J.W. (1990) *Energy Fuels*, **4** (4), 372–379.
72. Souverijns, W., Martens, J.A., Froment, G.F., and Jacobs, P.A. (1998) *J. Catal.*, **174**, 177–184.
73. Miller, S.J. (1994) *Microporous Mater.*, **2**, 439–449.
74. Taylor, R.J. and Petty, H.R. (1994) *Appl. Catal., A: Gen.*, **119**, 121–138.
75. Garwood, W.E., Le, Q.N., and Wong, S.S. (1990) High Viscosity Index Lubricants. US Patent 4975177 A, assigned to Mobil Oil Corp.
76. Cody, I.A., Dumfries, D.H., Neal, A.H., and Riley, K.L. (1993) High porosity, high surface area isomerization catalyst and its use. US Patent 5182248 A, assigned to Exxon Research and Engineering Company.
77. Mériaudeau, P., Tuan, V.A., Nghiem, V.T., Lai, S.Y., Hung, L.N., and Naccache, C. (1997) *J. Catal.*, **169**, 55–66.
78. Perego, C., Zanibelli, L., Flego, C., Del Bianco, A., and Bellussi, G. (1994) Catalyst for the hydroisomerization of long-chain N-paraffins and process for preparing it. EP Patent 582347, assigned to Eniricerche.
79. Weitkamp, J., Jacobs, P.A., and Martens, J.A. (1983) *Appl. Catal., A: Gen.*, **8**, 123–141.
80. Steijns, M., Froment, G., Jacobs, P., Uytterhoeven, J., and Weitkamp, J.

(1981) *Ind. Eng. Chem. Prod. Res. Dev.*, **20**, 654–660.

81. Weitkamp, J. (1975) in *Hydrocracking and Hydrotreating*, ACS Symposium Series, vol. 20 (eds J.W. Ward and S.A. Quader), American Chemical Society, Washington, DC, pp. 1–27.
82. Martens, J.A., Jacobs, P.A., and Weitkamp, J. (1986) *Appl. Catal.*, **20**, 239–281.
83. Martens, J.A., Jacobs, P.A., and Weitkamp, J. (1986) *Appl. Catal.*, **20**, 283–303.
84. Martens, J.A., Tielen, M., and Jacobs, P.A. (1989) in *Zeolites as Catalysts, Sorbents and Detergent Builders*, vol. 49 (eds H.G. Karge and J. Weitkamp), Elsevier, Amsterdam, pp. 20–25.
85. Alvarez, F., Ribeiro, F.R., Perot, G., Thomazeau, C., and Guisnet, M. (1996) *J. Catal.*, **162**, 179–189.
86. Giannetto, G.E., Perot, G.R., and Guisnet, M. (1986) *Ind. Eng. Chem. Prod. Res. Dev.*, **25** (3), 481–490.
87. Archibald, R.C., Greensfelder, B.S., Holzman, G., and Rowe, D.H. (1960) *Ind. Eng. Chem.*, **52** (9), 745–750.
88. Gibson, J.W., Good, G.M., and Holzman, G. (1960) *Ind. Eng. Chem.*, **52** (2), 113–116.
89. Weitkamp, J. and Ernst, S. (1990) in *Guidelines for Mastering the Properties of Molecular Sieves* (eds D. Bathomeuf, E.G. Derouane, and W. Hölderich), Plenum Press, New York, p. 343.
90. Coonradt, H. and Garwood, W.E. (1964) *Ind. Eng. Chem. Process Des. Dev.*, **3** (1), 38–45.
91. Kühl, G.H. (1999) in *Catalysis and Zeolites Fundamental and Applications*, Chapter 3 (eds J. Weitkamp and L. Puppe), Springer, p. 81.
92. Corma, A., Martinez, A., Pergher, S., Peratello, S., Perego, C., and Bellussi, G. (1997) *Appl. Catal., A: Gen.*, **152**, 107–125.
93. Calemma, V., Flego, C., Carluccio, L.C., Parker, W., Giardino, R., and Faraci, G. (2009) Process for the Preparation of Middle Distillates and Lube Bases Starting from Synthetic Hydrocarbon Feedstocks. US Patent 7534340, assigned to Eni SpA and Institut Français du Pétrole.
94. Mills, G.A., Heinemann, H., Milliken, T.H., and Oblad, A.G. (1953) *Ind. Eng. Chem.*, **45**, 134–137.
95. Weisz, D.B. (1962) *Adv. Catal.*, **13**, 137–190.
96. Guisnet, M., Alvarez, F., Giannetto, G.E., and Perot, G.R. (1987) *Catal. Today*, **1** (4), 415–433.
97. Brower, D.M. and Hogeveen, H. (1972) *Prog. Phys. Org. Chem*, **9**, 179–240.
98. Condon, F.E. (1958) in *Catalysis*, vol. 6 (ed P.H. Emmet), Reinhold, New York, p. 121.
99. Steijns, M. and Froment, G.F. (1981) *Ind. Eng. Chem. Prod. Res. Dev.*, **20**, 660–668.
100. Ribeiro, F., Marcilly, C., and Guisnet, M. (1982) *J. Catal.*, **78**, 267–274.
101. Debrabandere, B. and Froment, G.F. (1997) in *Hydrotreatment and Hydrocracking of Oil Fractions*, Studies in Surface Science and Catalysis, vol. 106 (eds G.F. Froment, B. Delmon, and P. Grange), Elsevier, pp. 379–389.
102. Calemma, V., Peratello, S., and Perego, C. (2000) *Appl. Catal., A: Gen.*, **190**, 207–218.
103. Roussel, M., Norsic, S., Lemberton, J.-L., Guisnet, M., Cseri, T., and Benazzi, E. (2005) *Appl. Catal., A: Gen.*, **279**, 53–58.
104. Weitkamp, J. (1982) *Ind. Eng. Chem. Prod. Res. Dev.*, **21**, 550–558.
105. J.A. Martens and P.A. Jacobs (2008) in *Handbook of Heterogeneous Catalysis*, Part A General Principles, Methods and Reaction Engineering (Eds. G. Ertl, H. Knozinger, J. Weitkamp), Wiley-VCH Verlag GmbH, Vol. 3, pp. 1137-1149.
106. Fiaux, A., Smith, D.L., and Futrel, J.H. (1977) *J. Mass Spectrom. Ion Phys.*, **25**, 281–294.
107. Edwards, R., Larivé, J.-F., and Beziat, J.-C. (2011) Well to Wheels Analysis for Future Automobile Fuels and Powertrains in European Context, Well to Wheels Report version 3c, July 2011, http://iet.jrc.ec.europa.eu/about-jec/sites/iet.jrc.ec.europa.eu.about-jec/files/documents/wtw3&uscore;wtw&uscore;report&uscore;eurformat.pdf (accessed 23 September 2015).

108. Sheehan, J., Camobreco, V., Duffield, J., Graboski, M., and Shapori, H. (1998) A Life Cycle Inventory of Biodiesel and Petroleum Diesel for Use in an Urban Bus. NREL/SR-580-24089, National Energy Laboratory, US Department of Energy.
109. Neste Oil (2014) Hydrotreated Vegetable Oil-Premium Renewable Biofuels for Diesel Engines. NESTE OIL Report, February 2014, http://www.nesteoil.fi/binary.asp?path=35;52;11990;22214;22951;23473 (accessed 23 September 2015).
110. (2007) Regulation (EC) No. 715/2007 of the European Parliament and of the Council of 20 June **2007** on type approval of motor vehicles with respect to emissions from light passenger and commercial vehicles (Euro 5 and Euro 6) and on access to vehicle repair and maintenance information. *Off. J. Eur. Communities*, L171/1.
111. (2009) Directive 2009/30/EC of the European Parliament and of the Council, 23.04.2009, amending Directive 98/70/EC relating to the quality of petrol and diesel fuels. *Off. J. Eur. Communities*, L140/88.
112. Alleman, T.L. and Mc Cormick, R.L. (2003) Fischer–Tropsch Diesel Fuels–Properties and Exhaust Emissions: A Literature Review. SAE Paper 2003-01-0763.
113. Nakakita, K., Ban, H., Takasu, S., Hotta, Y., Inagaki, K., Weissman, W., and Farrell, J.T. (2004) Effect of Hydrocarbon Molecular Structure in Diesel Fuel on In-Cylinder Soot Formation and Exhaust Emissions. SAE Paper 2003-01-1914.
114. Martens, J.A., Tielen, M., and Jacobs, P.A. (1987) *Catal. Today*, **1**, 435.
115. Marcilly, C. (2003) *Catalyse Acido-basique Application au raffinage et à la pétrochimie*, vol. 1, Chapter 4, Editions Technip, p. 217.
116. Rispoli, G., Prati, C., Amoroso, A., and Pollesel, P. (2012) Method for Revamping a Conventional Mineral Oils Refinery to a Biorefinery. EP 2855638.
117. Prati, C., Rispoli, G., and Anumakonda, A. (2012) Case Study & Value Propositions for Implementing 2nd Generation Biofuels Projects in Europe. ERTC 17th Annual Meeting, Vienna, November, 2012.

Part II
Bio-Monomers

Chemicals and Fuels from Bio-Based Building Blocks, First Edition.
Edited by Fabrizio Cavani, Stefania Albonetti, Francesco Basile, and Alessandro Gandini.

6
Synthesis of Adipic Acid Starting from Renewable Raw Materials

Thomas R. Boussie, Gary M. Diamond, Eric Dias, and Vince Murphy

6.1
Introduction

Historically, the majority of large-volume chemical products have been manufactured from crude oil, natural gas, and coal-derived intermediates, including CO, H_2, ethylene, propylene, butadiene, benzene, and *p*-xylene as the primary building blocks for petrochemicals production. Important exceptions include oleochemicals, derived from seed oil triglycerides, and industrial fermentation products such as ethanol, citric acid, and lactic acid. The past decade has seen significant effort both in academia and industry to expand and diversify the scope of raw materials employed in the production of industrial chemicals. These efforts are driven by a combination of high price volatility in global crude oil markets and interest in reducing the environmental footprint of industrial chemicals manufacturing.

One option to diversify the source of industrial chemical feedstocks is to expand the use of renewable raw materials. This can be accomplished by direct conversion of biomass (e.g., through pyrolysis or gasification) or through fractionation of biomass and conversion of its component fractions: lignin and polysaccharides. Sugars derived from biomass processing are of particular interest due to their homogeneous composition and high chemical functionality. Primary sources of sugars at scales suitable for industrial chemicals production include glucose from hydrolysis of starch or cellulose; sucrose from cane, beet, or sorghum processing; and xylose from hemicellulose hydrolysis. Sugars can be converted to chemicals using "biological" processes (enzymatic or fermentation), or through "chemical" processes, typically involving homogeneous or heterogeneous catalysis. Rennovia is focused on developing processes for the conversion of sugars (glucose and xylose) to industrial chemical products using heterogeneous catalysts.

This chapter describes Rennovia's development of process technology for the conversion of glucose to adipic acid. Currently produced from benzene, adipic acid is a commodity chemical with global production capacity exceeding $6 billion per year. Adipic acid is used in the production of nylon-6,6 resins and

Chemicals and Fuels from Bio-Based Building Blocks, First Edition.
Edited by Fabrizio Cavani, Stefania Albonetti, Francesco Basile, and Alessandro Gandini.

fibers, polyesters, polyurethanes, and diester plasticizers. Rennovia's development of a cost-advantaged process for production for adipic acid from a widely available renewable raw material illustrates the potential for sustainable technologies to compete directly with, and potentially to displace, traditional petroleum-based chemical production technologies.

6.2
Challenges for Bio-Based Chemicals Production

There are two major challenges to the commercialization of industrial chemicals from renewable raw materials. The first challenge is economics. Petrochemical feedstocks are relatively inexpensive on a carbon basis, and existing production is operated highly efficiently at enormous scale within a globally integrated manufacturing infrastructure. Importantly, a fundamental feature of the petrochemical value chain is that it begins with hydrocarbon feedstocks (ethylene, propylene, benzene, etc.) and typically *adds* oxidative functionality to produce higher-value products. In so doing, the molecular weight of the product is higher than that of its precursor, or the "pound-to-pound conversion" is positive. For example, propylene (MW 42) is converted to acrylic acid (MW 72), benzene (MW 78) is converted adipic acid (MW 146), and so on. This is known in the industry as *selling air*. For carbohydrates, the opposite situation prevails. Starting with a C:O ratio of 1, conversion to industrial chemicals typically entails selective *removal* of oxidative functionality and commensurate reduction in molecular weight. This negative pound-to-pound conversion has profound influence on determining which chemical products can be made competitively from renewable raw materials.

The second major challenge to renewable chemicals production is technical. In order to compete with petroleum-based raw materials and mature petrochemical process technologies, conversion technologies for bio-based raw materials must operate at high process efficiencies. This entails both high carbon efficiency (high process selectivity to minimize raw material costs), as well as high manufacturing efficiency (high volumetric space velocities, concentrated reactant and product streams, and energy-efficient separations to minimize capital and operating costs). For fermentation processes this translates to engineering high on-pathway metabolic carbon flux to targeted end products and the ability to operate at high titer in concentrated media. Chemical catalytic processes must achieve high process selectivities, with minimal catalyst costs and high reactor space-time yields.

While the petrochemical industry enjoys the fruits of 80+ years of technology development and refinement, these are early days for industrial-scale bio-based raw material conversion. Fundamental conversion technologies must be developed and then scaled to commercial production while competing directly with highly optimized petrochemical processes. In light of these challenges, judicious choice of target products is critical to maximize the potential for success.

6.3
Choice of Adipic Acid as Product Target by Rennovia

There are several metrics by which one can assess the commercial viability of potential renewable chemical products. Ideally, one would develop a full production cost model for each product based on a comprehensive mass and energy balance, with well-developed capital and operational expenditure estimates. This is unrealistic when evaluating a wide range of potential products, particularly if new technologies must be developed and implemented. Because feedstock is the single largest contributor to commodity chemical manufacturing costs [1], a simpler approach is to focus on raw material utilization as the key metric for evaluating and ranking potential product opportunities. One simple figure of merit (FOM) is the difference between the selling price of a chemical and the raw material costs for its production at 100% theoretical yield:

$$\text{FOM} = \text{Product selling price} - \text{Cost of raw materials at 100\% theoretical yield}$$

This analysis assumes similar manufacturing costs and depreciation for various processes, but given that these compose a minority of overall production costs, this is acceptable for a high-level analysis. Calculating this FOM for the production of a range of high-volume industrial chemicals from glucose reveals some interesting trends (Figure 6.1):

1) The general trend that emerges is that products with more oxidative functionality (higher O : C ratios) tend to represent more attractive economic opportunities. This is driven by (i) generally higher market values for more highly oxygenated products and (ii) lower glucose costs for oxygenated products due to reduced pound-to-pound conversion losses.
2) Purely hydrocarbon products are generally the poorest candidates. These products combine relatively low selling prices with high glucose raw material costs. The exception to this class of products is isoprene due to its relatively high selling price.
3) Ethanol, the largest volume product from glucose, appears to be a poor candidate in this simple analysis. In practice, the sale of dry distillers grains (DDGs) as a co-product of corn ethanol produced from the fermentation of corn kernel biomass contributes significantly to the economic return from ethanol production. It is possible that similar coproduct credit could potentially apply to related products such as *n*-butanol, isobutanol, isopropanol, and acetone.
4) Products currently commercially produced from glucose (lactic acid, citric acid, gluconic acid, sorbitol, lysine, succinic acid, 1,3-propanediol) appear as attractive opportunities by this analysis.
5) The two most economically attractive opportunities by this analysis are 1,4-butanediol (1,4-BDO) and adipic acid.

Notably, bio-based 1,4-BDO production is entering commercial production by both Genomatica and Bioamber [2]. Bio-based adipic acid, both by fermentation

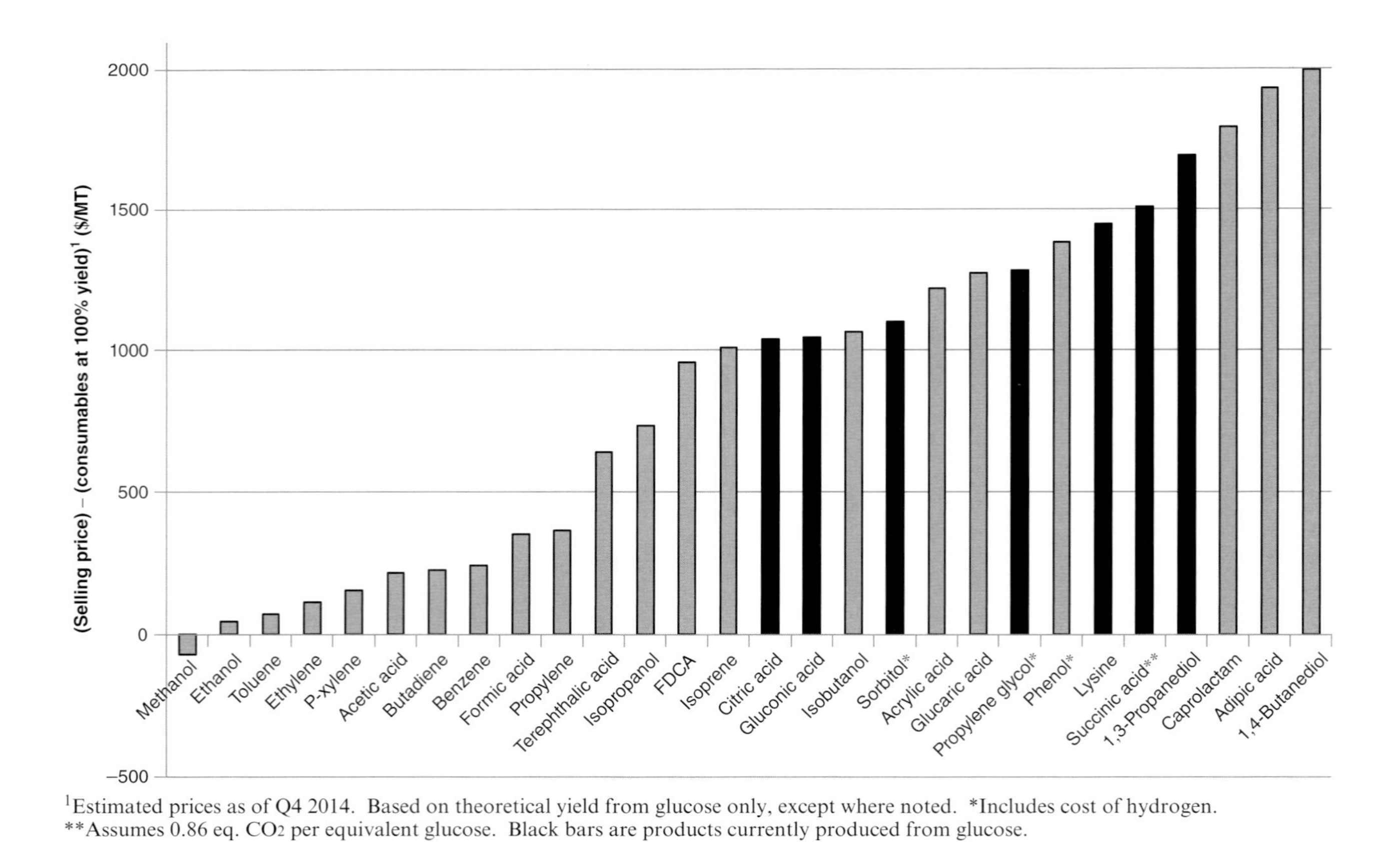

[1]Estimated prices as of Q4 2014. Based on theoretical yield from glucose only, except where noted. *Includes cost of hydrogen.
**Assumes 0.86 eq. CO_2 per equivalent glucose. Black bars are products currently produced from glucose.

Figure 6.1 Economic opportunity analysis for chemical production from glucose.

and chemical catalytic processes, is under development by multiple companies and is the first commercialization target of Rennovia. It is also worth noting that some bio-based products that do not appear by this analysis to be economically attractive from glucose may be viable from less expensive renewable carbon sources – for example, from biomass or municipal waste. However, such opportunities are outside of the scope of this analysis.

6.4 Conventional and Fermentation-Based Adipic Acid Production Technologies

The dominant technology for petroleum-based adipic acid production is via a multistep process from benzene (Scheme 6.1). This process has many drawbacks, including high process complexity, high energy consumption, and the production of nitrous oxide (N_2O) as a by-product. With over 300 times the global warming potential of CO_2, N_2O is a potent greenhouse gas (GHG) subject to stringent abatement requirements. While N_2O emission from adipic acid production is 90–98% mitigated in most advanced manufacturing environments, mitigation technology adds additional cost and complexity to the production process, and the residual N_2O emissions contribute significantly to the GHG footprint of conventional adipic acid production [3].

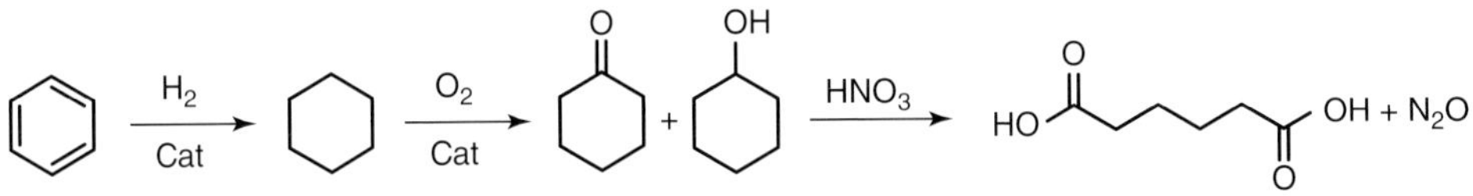

H_2, Cat; O_2, Cat; O + OH; HNO_3; HO, O, O, OH + N_2O

Scheme 6.1 Conventional adipic acid production from benzene.

Several approaches have been taken to produce adipic acid from bio-based raw materials. These approaches have been the subject of a recent comprehensive review by Cavallaro [4]. One of the advantages of a fermentation approach is the potential to utilize a variety of carbon sources including sugars (glucose, sucrose, xylose), as well as glycerol and CO/H_2. Challenges to fermentation include:

1) Maximizing carbon flux through often highly engineered catalyst pathways
2) Minimizing feedstock requirements for basal metabolic processes
3) The need for additional nutrients and media modifiers
4) Restrictions in operational temperature and pH
5) Titer limitations due to product inhibition and/or toxicity of the product to the organism
6) Susceptibility to contamination and the need to maintain a sterile fermentation environment.

The isolation and purification of nonvolatile products such as adipic acid from complex and heterogeneous fermentation broths can also add cost and complexity to the overall process.

In the evaluation of the economic viability of fermentation approaches to adipic acid production, it is of paramount importance to consider the carbon efficiency of each pathway. Adipic acid is net reduced versus glucose. In the absence of an external reductant, some glucose must therefore be consumed to provide the required reductive equivalents. This results in a minimum consumption of 1.08 equiv. of glucose per equivalent adipic acid, which corresponds to a 92% maximum theoretical yield from glucose only:

$$\frac{13}{12}C_6H_{12}O_6 \rightarrow C_6H_{10}O_4 + \frac{1}{2}CO_2 + \frac{3}{2}H_2O$$

Depending on the fermentation pathway, there can be additional carbon losses (typically via CO_2 evolution) that can significantly reduce the theoretical yield below 92%. The earliest fermentative approach to adipic acid, reported by Draths and Frost [5], involved engineering of the ischemic acid fermentation pathway to produce *cis, cis*-muconic acid, which could then be subsequently catalytically hydrogenated to adipic acid. This approach suffers from a modest 85% theoretical maximum molar yield. Additional adipic acid fermentation pathways have been published (as patents or patent applications) by Invista [6], DSM [7], Celexion [8], Verdezyne [9], and Genomatica [10]. Theoretical yields for these pathways vary from 50 to 92%.

6.5 Rennovia's Bio-Based Adipic Acid Production Technology

Rennovia's approach to bio-based adipic acid production involves not fermentation but heterogeneous catalysis. Heterogeneous catalytic processes are ubiquitous in petroleum refining and petrochemicals production. Rennovia seeks to adapt conventional catalysts and process technologies to convert renewable raw materials to chemical products currently produced from petrochemical feedstocks. The development of an economically competitive process for adipic acid production from glucose requires:

1) High-yield conversions to intermediate and final products
2) Minimum product loss on isolation and purification
3) Low consumables costs
4) Acceptable energy (steam, electricity) requirements
5) High space-time efficient unit operations to minimize capital and operating costs.

Rennovia's bio-based adipic acid process involves two catalytic steps: (i) aerobic oxidation of glucose to glucaric acid and (ii) hydrodeoxygenation of glucaric acid to adipic acid (Scheme 6.2).

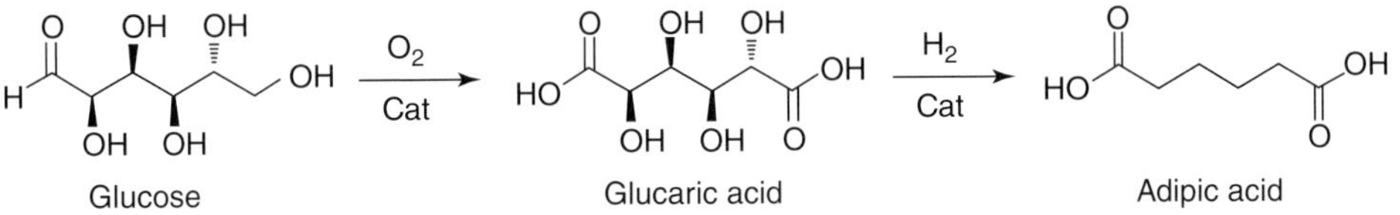

O_2 / Cat → H_2 / Cat →

Glucose — Glucaric acid — Adipic acid

Scheme 6.2 Rennovia's adipic acid production from glucose.

6.6 Step 1: Selective Oxidation of Glucose to Glucaric Acid

Glucaric acid is currently produced at small commercial scale through nitric acid oxidation of glucose [11]. In addition to the cost and difficulty of handling nitric acid, the reaction is only moderately selective with significant formation of lower-chain diacid and internal keto by-products. In addition, glucaric acid is typically isolated from the reaction mixture via salt formation (as the K or Ca salt) and reacidification. This adds significant cost and complexity and limits the scalability of the process. Initially Kiely *et al.* [12], and now Rivertop Renewables [13], have sought to improve the conventional nitric acid oxidation process in two ways: first, by improving reaction selectivity through careful control of reaction temperature, and second by reducing nitric acid costs through recycling and regeneration of NO_x reaction by-products. This process has been piloted and is entering small-scale commercial production [14].

Glucose can also be converted to glucaric acid with high selectivity through catalytic oxidation using TEMPO or related nitroxyl radical oxidants [15]. The reported systems operate at high (11 – 12) pH and typically employ hypochlorite as the terminal oxidant. While highly selective, the reagent costs and pH limitations likely limit the scalability of this approach for commercial applications.

Finally, there have been several papers describing the electrocatalytic oxidation of glucose to glucaric acid, some with high reported reaction selectivities [16]. The scalability and commercial viability of such a process is difficult to judge.

In our view the most attractive option for glucaric acid production, with minimal non-glucose consumables costs, ease of operation, and flexibility for product isolation and purification, is through the direct heterogeneous aerobic oxidation of glucose in water at native pH (Scheme 6.3). To implement such a process, several key technical challenges must be overcome. These include:

1) High catalyst selectivity to glucaric acid product
2) High catalyst activity operating at native pH

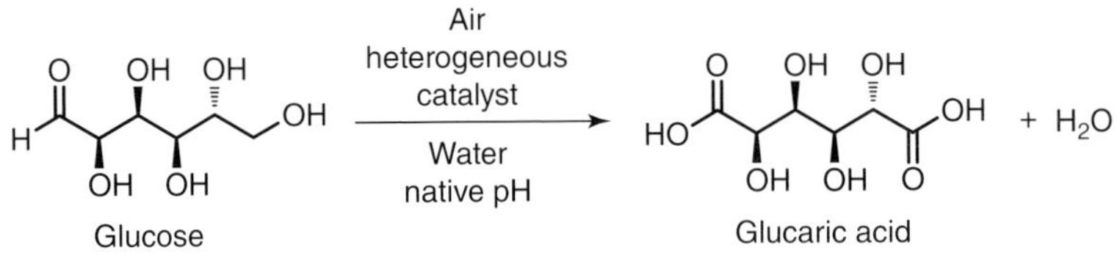

Glucose → (Air, heterogeneous catalyst / Water, native pH) → Glucaric acid + H_2O

Scheme 6.3 Aerobic oxidation of glucose to glucaric acid over heterogeneous catalyst.

3) Long-term catalyst stability in aqueous acidic and oxidative media
4) Reactor design capable of operating at space-time yields and at substrate concentrations required for favorable process economics.

6.6.1 Identification of Selective Catalysts for Aerobic Oxidation of Glucose to Glucaric Acid at Native pH

There are a number of reports in the academic and patent literature describing catalysts for the heterogeneous aerobic oxidation of glucose [17, 18]. The majority of these reports describe the production of gluconic acid, although there are some reports of oxidation of glucose (or gluconic acid) to glucaric acid. Achieving high selectivity for glucose to gluconic acid via selective oxidation of the aldehyde functional group (C1 position) is relatively straightforward, whereas conversion of glucose to glucaric acid poses a much higher selectivity challenge as the catalyst must be capable of oxidizing both the aldehyde and primary alcohol (C6 position) in the presence of four secondary alcohols (C2–C5 positions) (Scheme 6.3). Oxidation of the secondary alcohols to internal keto products is known to give rise to C–C fragmentation reactions with the production of lower-chain diacid products and CO_2. Furthermore, the collective literature (glucose to gluconic and glucose/gluconic to glucaric) almost invariably describes operation at elevated pH through the continuous addition of base [17, 18].

Representative data from screening a variety of heterogeneous catalysts for the conversion of glucose to glucaric acid in water at native pH are plotted in Figure 6.2. These data were generated using Rennovia's high-throughput catalyst synthesis and screening infrastructure, which has been described elsewhere [19,

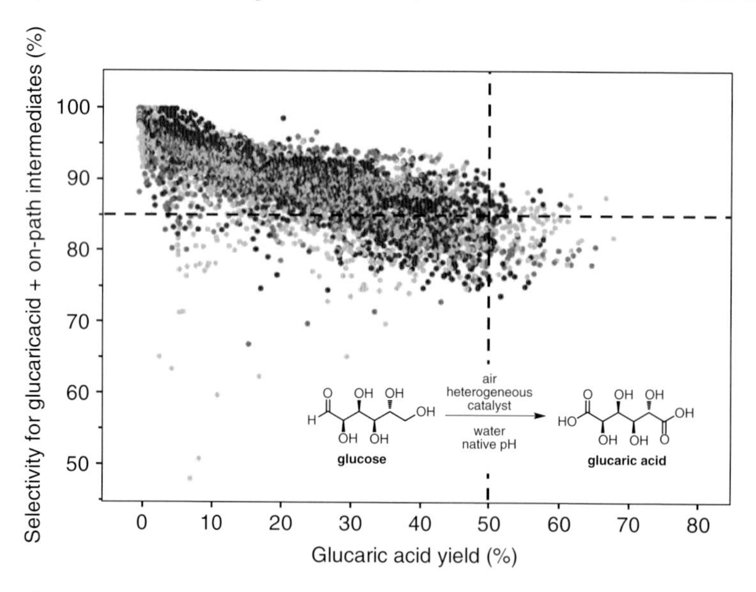

Figure 6.2 Representative data for high-throughput screening of glucose oxidation catalysts.

20]. Reactions were performed in multi-well parallel batch reactors containing about 10 mg powdered catalyst and 200 μl of 10 wt% glucose solution per well. Product analyses were performed post-reaction after selected reaction times. Typical reaction temperatures and O_2 pressures were 80–120 °C and 75 psig, respectively. The data are plotted as glucaric acid yield (X-axis) versus "on-path selectivity" (Y-axis). *On-path selectivity* is defined as the sum of the selectivity to glucaric acid and the selectivities to all intermediates that are on the reaction pathways to glucaric acid (see Scheme 6.4).

The data in Figure 6.2 demonstrate that it is possible to identify catalyst formulations that maintain high selectivity to on-path products (>85%) at relatively high glucaric acid yields (>50%) while operating at native pH (area in upper-right quadrant defined by dashed lines). To achieve high (>90%) glucaric acid process yields with these catalysts, the process is run at approximately 50% single-pass conversion to glucaric acid, followed by separation and recycling reaction intermediates (Figure 6.3).

6.6.2
Demonstration of Long-Term Catalyst Stability for Glucose Oxidation Reaction

A significant challenge to the development of heterogeneous catalysts for the conversion of sugars to chemicals (or fuels) is the long-term stability of catalysts operating in aqueous or other polar reaction media, typically at elevated temperatures. Under these conditions heterogeneous catalysts can degrade over time through a variety of irreversible and potentially reversible mechanisms including:

- Irreversible
  - Support dissolution
  - Changes in support surface area, pore size, pore size distribution, or support crystallinity
  - Physical degradation of the support (production of fines)
  - Loss (dissolution) of supported metal(s)
  - Sintering, redistribution, or other changes in metal morphology.
- Potentially reversible
  - Catalyst fouling (coking)
  - Metal poisoning.

In a commercial setting, a 1-year catalyst lifetime requires >8000 h on stream. Testing for long-term catalyst stability in the laboratory environment requires formulating catalysts on supports of suitable dimensions (i.e., powders or shaped extrudates) and operating under commercially relevant reaction conditions of substrate concentration, operating temperature, and liquid and gas flow velocities for extended time periods. Figure 6.4 shows representative data for the catalytic oxidation of glucose to glucaric acid on shaped catalysts operated at partial conversion in a fixed-bed reactor. The data are plotted as on-path selectivity versus time. Note that these catalyst formulations maintain high on-path selectivity for greater than 1000 h time on stream. Rennovia's most advanced glucose oxidation catalysts have operated for nearly 3000 h on stream without significant degradation or loss of performance.

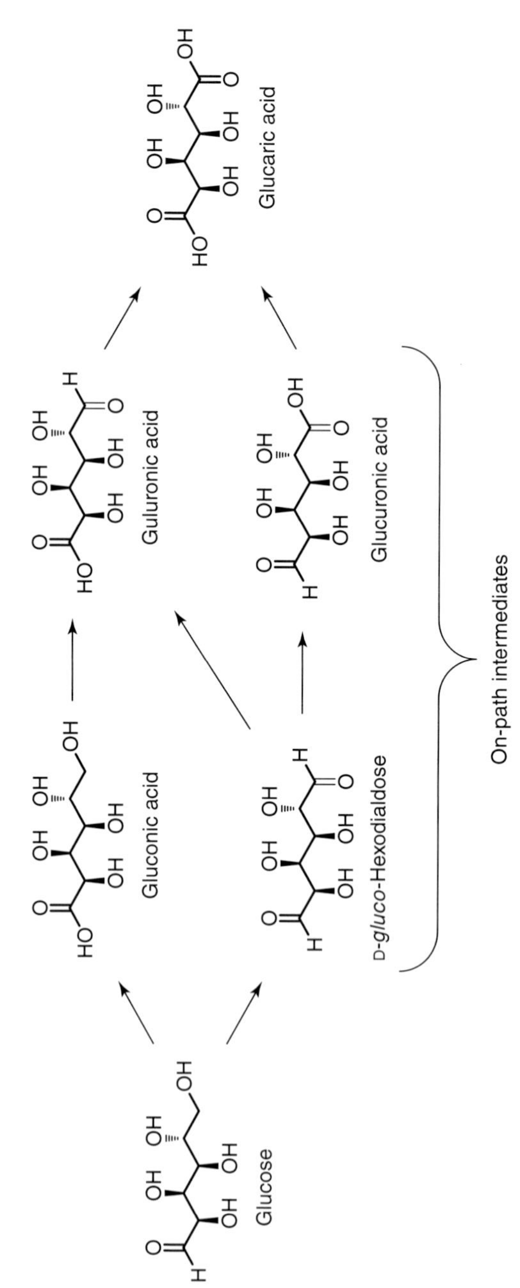

Scheme 6.4 On-path intermediates in the oxidation of glucose to glucaric acid.

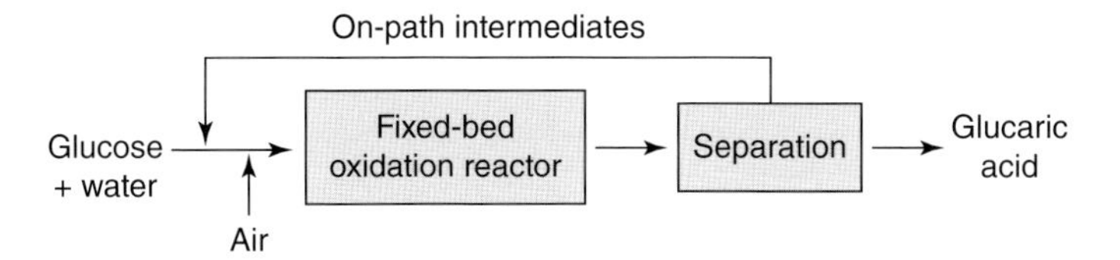

Figure 6.3 Scheme for partial conversion process for glucaric acid production.

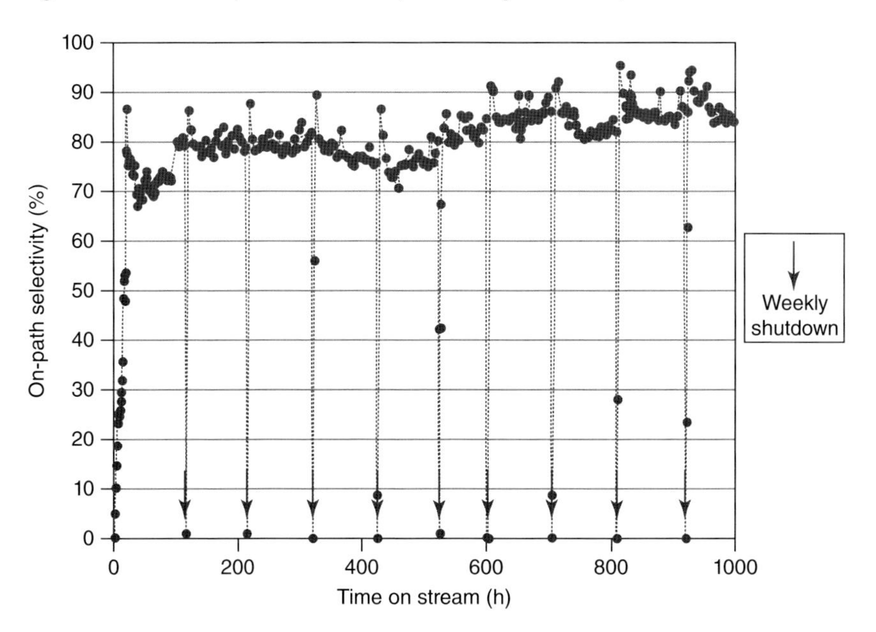

Figure 6.4 Long-term stable operation of glucose oxidation catalyst.

6.7
Step 2: Selective Hydrodeoxygenation of Glucaric Acid to Adipic Acid

The reduction of glucaric acid to adipic acid with hydrogen entails hydrogenolysis of the C2–C5 secondary alcohol groups of glucaric acid without reducing the C1 and C6 carboxylic acids or cleaving C–C bonds (Scheme 6.5). While the direct catalytic reduction of alcohols to alkanes with hydrogen is a relatively uncommon reaction, it has begun to receive more attention recently due to its importance for the conversion of bio-oils or fatty acid derivatives to hydrocarbon fuels. This can be viewed as a natural extension of conventional hydrotreating, a unit operation of petroleum refining for removal of sulfur and nitrogen from crude oil fractions in the process of converting them to transportation fuels. As such, most of the reported systems employ catalysts and reaction conditions that not only reduce carboxylic acids but also often result in fragmentation of the carbon chain. While this is suitable for the production of fuel-like compositions, the lack of selectivity

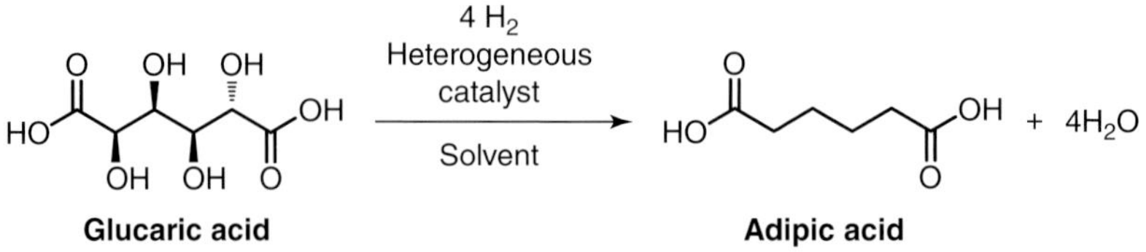

Glucaric acid $\xrightarrow[\text{Solvent}]{4\ H_2,\ \text{Heterogeneous catalyst}}$ Adipic acid + $4H_2O$

Scheme 6.5 Selective hydrodeoxygenation of glucaric acid to adipic acid.

does not lend itself to the selective conversion of glucaric acid to adipic acid. Selective hydrodeoxygenation of C–OH bonds is an excellent example of a reaction that is highly desirable for the conversion of bio-based feedstocks to bulk and specialty chemicals that has no analog in petrochemicals production. Enabling such a reaction requires development of a fundamentally new and scalable catalytic transformation.

One potential approach to hydrodeoxygenation of polyhydroxylated substrates is through dehydration of each alcohol to produce an intermediate olefin followed by catalytic hydrogenation. While this method may be suitable for certain specific substrates and products, it also has several limitations. In particular, the general applicability of this method is complicated by the poor selectivity of the dehydration reaction and the potential formation of cyclic ethers as by-products. For glucaric acid, it is expected that the β-OH groups (C3 and C4) would be particularly susceptible to dehydration, while the α-OH groups (C2 and C5) are expected to be highly resistant. The closest analog to the dehydration of C2 and C5 OH-substituted adipic acid is the dehydration of lactic acid to acrylic acid which requires temperatures in excess of 250 °C (for activated systems) or 350 °C (for unactivated systems). Under these conditions glucaric acid, its reaction intermediates, or other polyhydroxylated substrates are unlikely to be stable.

One approach to reduce the severity of reaction conditions and to increase selectivity for C–OH reduction over carboxylic acid reduction and C–C bond cleavage is to convert the –OH group to a more reactive substituent. Conventional synthetic organic methods such as tosylation are prohibitively expensive and not industrially scalable. An alternative method is to employ a hydrohalic (HX) acid as a catalytic activator of C–OH bonds. By this approach, halodehydroxylation of the alcohol to produce an alkyl halide intermediate is followed by catalytic hydrodehalogenation of the alkyl halide intermediate, which liberates the deoxygenated product and regenerates the HX for continued reaction (Figure 6.5).

6.7.1 Identification of Catalysts and Conditions for the Selective Reduction of Glucaric Acid to Adipic Acid

To test this approach for the conversion of glucaric acid to adipic acid, a broad set of potential hydrogenation catalysts were screened in a variety of solvents with the addition of substoichiometric amounts of HI, HBr, or HCl as co-catalysts. Representative data from these screening experiments are shown in Figure 6.6.

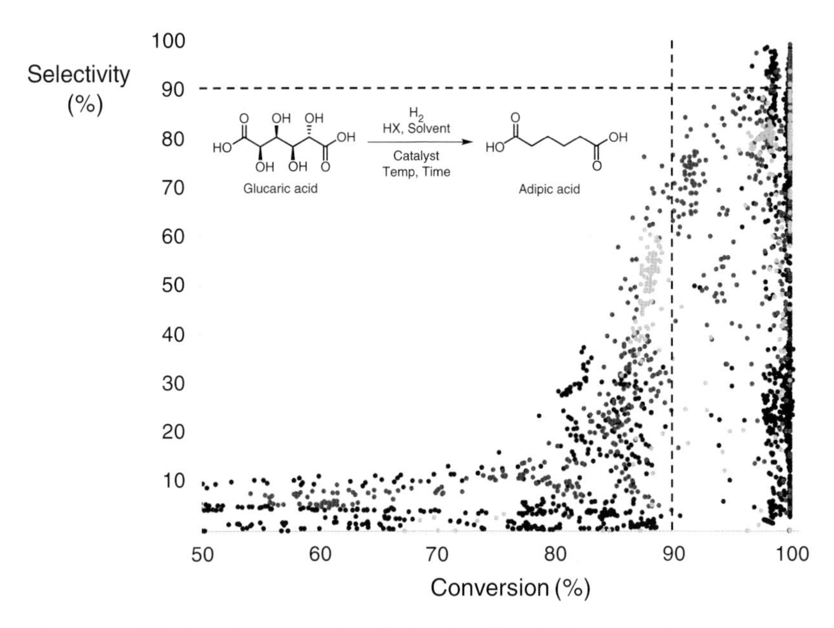

Figure 6.5 High-throughput screening of glucaric acid hydrodeoxygenation.

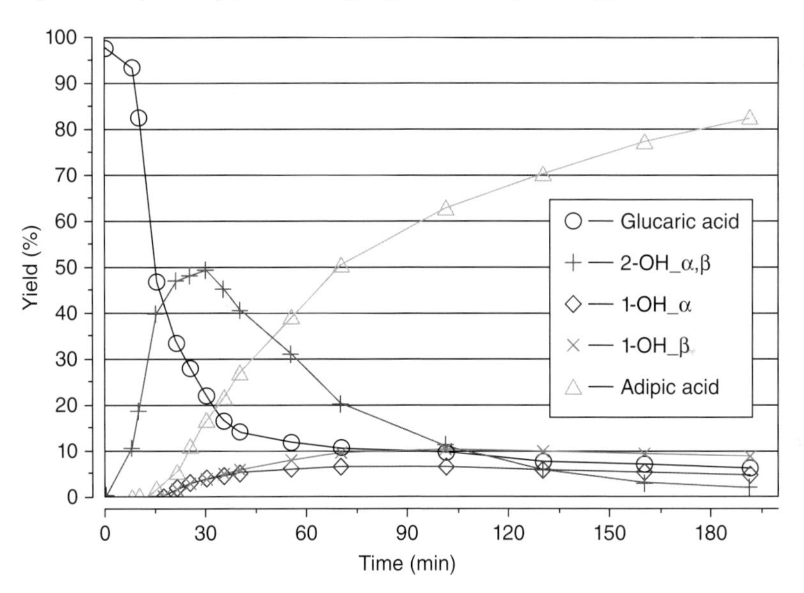

Figure 6.6 Reaction profile for hydrodeoxygenation of glucaric acid.

The data are plotted as adipic acid yield on the Y-axis *versus* individual experiment on the X-axis. While there are many combinations of catalyst, solvent, and HX promoter that give poor or moderate yields of adipic acid, notably there are several examples where adipic acid yields exceed 95%. The highest reaction yields were attained when HBr was the cocatalyst and acetic acid was the solvent (Scheme 6.6).

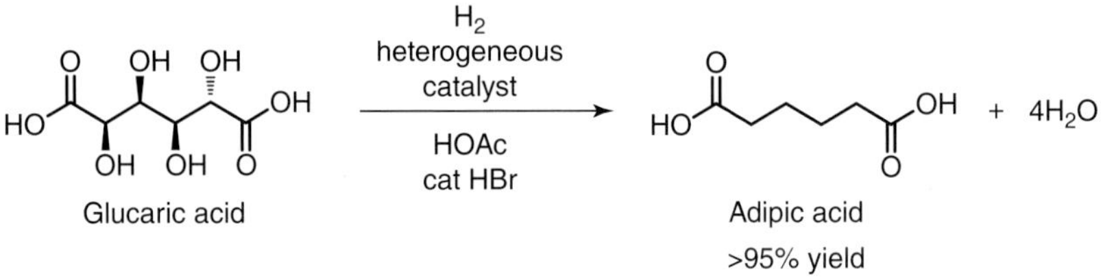

Glucaric acid $\xrightarrow[\text{HOAc, cat HBr}]{H_2\text{, heterogeneous catalyst}}$ Adipic acid (>95% yield) + $4H_2O$

Scheme 6.6 High-yield hydrodeoxygenation of glucaric acid to adipic acid.

6.7.2 Reaction Pathways for the Selective Reduction of Glucaric Acid to Adipic Acid

Initial investigations indicated that both acetate esters (formed via reaction of the –OH groups with the solvent) and lactones (formed via intramolecular esterification) were important reaction intermediates, suggesting that this transformation was more complex than the "simple" pathway shown in Figure 6.6 and providing some insight into the observation that acetic acid was necessary for a high-yielding conversion of glucaric acid to adipic acid. Closer examination of the reaction profile as a function of time (Figure 6.7) provides additional evidence that a different pathway could be operating. Note that for the following discussion the nomenclature "x-OH" refers to a compound that has x number of –OH functional group equivalents. Glucaric acid, lactones of glucaric acid, and acetylated forms

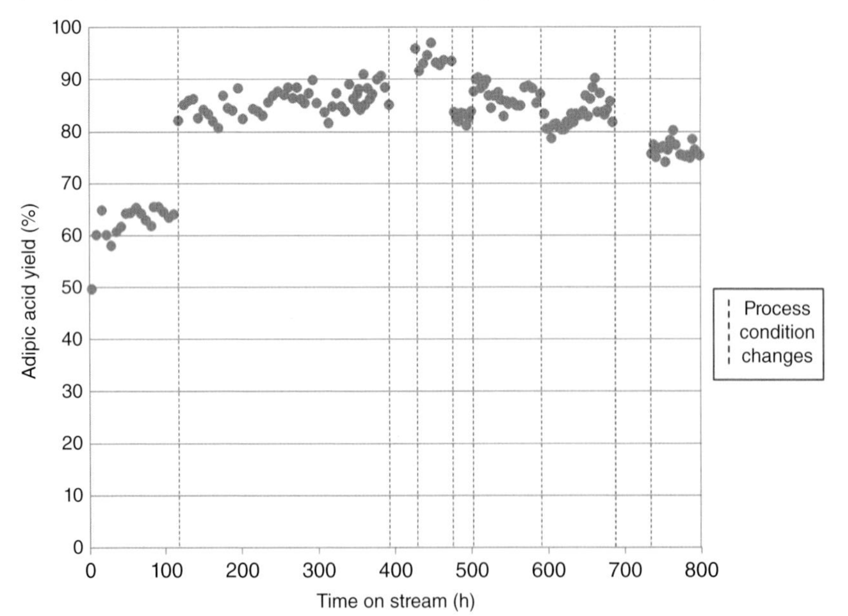

Figure 6.7 Long-term stable operation of glucaric acid hydrodeoxygenation catalyst.

of both glucaric acid and its lactones are designated as “4-OH” species. Likewise “3-OH,” “2-OH,” and “1-OH” are deoxygenated intermediates along the reaction pathway that can be alcohols, lactones, or acetylated version thereof.

The data in Figure 6.7 show disappearance of glucaric acid (or its lactone/acetylated derivatives) to be concomitant with the appearance of “2-OH” species with no observable “3-OH” intermediates. Likewise, adipic acid formation is concomitant with the disappearance of the “2-OH” intermediates. Two explanations for this behavior are that (i) the reaction kinetics are such that removal of one hydroxyl group is followed by a much faster reaction that removes a subsequent hydroxyl group or (ii) there is a pathway by which two hydroxyl groups are removed simultaneously resulting in the direct conversion of glucaric acid → 2-OH → adipic acid via a “coupled hydrodeoxygenation.” Examining the remaining data in Figure 6.7, it is apparent that while the “1-OH” compounds (both α- and β-isomers are observed) are detected, they grow in concurrently with the appearance of adipic acid and then disappear more slowly, indicating that they are not exclusively intermediates between “2-OH” and the adipic acid product.

Reaction of model substrates has provided additional evidence for a “coupled hydrodeoxygenation” pathway. Under conditions similar to those used for the glucaric acid to adipic acid conversion, tartaric acid reacts to produce succinic acid without detection of malic acid (or equivalent “1-OH” intermediates); however, when malic acid is used as the substrate, it reacts much more slowly to produce succinic acid (Scheme 6.7). Taken with the kinetic data in Figure 6.7, it appears that the primary pathway for the conversion of glucaric acid to adipic acid is one in which a “coupled hydrodeoxygenation” takes place to remove two hydroxyl groups at once, and the removal of single hydroxyl groups is a secondary pathway that is much slower.

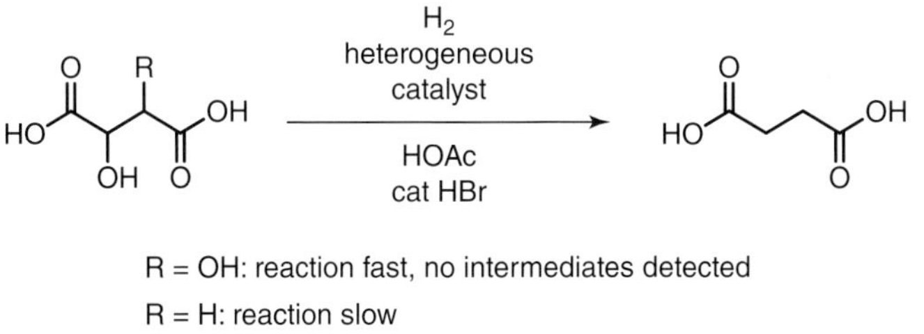

H_2
heterogeneous
catalyst

HOAc
cat HBr

R = OH: reaction fast, no intermediates detected
R = H: reaction slow

Scheme 6.7 Reaction of tartaric acid and malic acid under hydrodeoxygenation conditions.

It has also been determined analytically that the major isomer of the “2-OH” intermediate along the pathway from glucaric to adipic acid is one in which the hydroxyl groups are adjacent to each other in the α,β-positions, suggesting that the “coupled hydrodeoxygenation” is likely specific for vicinal diols. While the mechanism for this transformation is not fully understood, evidence from both the literature and our own experiments has led us to propose the reaction sequence shown in Scheme 6.8. Reaction of vicinal diols, including those in closely related gluconic acid, with HOAc/HBr, can produce 1,2-bromoacetylated products [21] via an acyl onium intermediate. Elimination of acetic acid gives the vinyl bromide intermediate [21], which subsequently undergoes catalytic

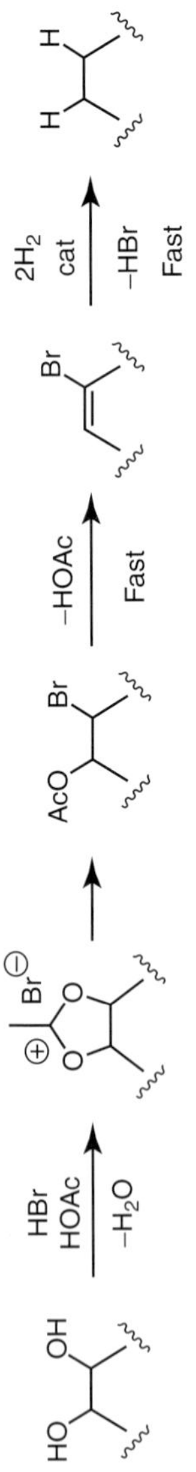

Scheme 6.8 Potential mechanism for coupled hydrodeoxygenation of vicinal diols.

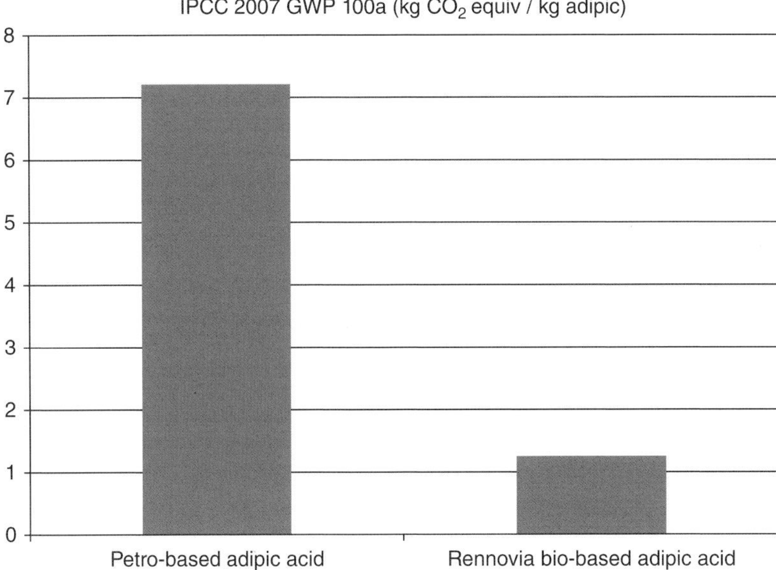

Figure 6.8 Comparison of carbon footprint of petro- and bio-based adipic acid.

hydrogenation and hydrodehalogenation to liberate the saturated product and regenerate HBr. A more thorough examination of this intriguing and useful transformation is underway.

6.7.3
Demonstration of Long-Term Catalyst Stability for Glucaric Acid Hydrodeoxygenation Reaction

As in the oxidation of glucose to glucaric acid, it is a significant challenge to develop heterogeneous catalysts that are stable under the highly polar and acidic reaction conditions (HOAc/H_2O/HBr) involved in the conversion of glucaric acid to adipic acid. Figure 6.8 shows data both for conversion of glucaric acid to adipic acid in a fixed-bed reactor. These data demonstrate that with our catalysts and conditions there is no loss of catalyst selectivity or activity with over 800 h time on stream.

6.8
Current Status of Rennovia's Bio-Based Adipic Acid Process Technology

Rennovia has successfully demonstrated long-term pilot operation of the glucose to glucaric acid and glucaric acid to adipic acid catalytic reactions, as well as pilot-scale operation of all ancillary unit operations (solvent recovery and recycle, adipic acid crystallization, etc.) of the overall glucose to adipic acid process.

The adipic acid process is currently being operated in an integrated miniplant with Johnson Matthey Davy Technologies, including incorporation of all major recycle streams [22]. Operation of the miniplant will provide design data for a commercial-scale production plant.

6.9 Bio- versus Petro-Based Adipic Acid Production Economics

The barrier to entry for new production technologies in the chemical industry is extremely high. New technology is nearly always in direct competition with installed, large scale, often fully depreciated production assets. The financial justification for capital investment in first-of-a-kind technology requires a clear and compelling economic advantage in the marketplace. Rennovia's bio-based adipic acid process, and its competitiveness versus conventional adipic acid, is the subject of recent process economic reports from IHS and Nexant [23]. These independent reports concluded that Rennovia's technology has high potential to be competitive with conventional cyclohexane oxidation. Depending on operating scale, Rennovia's production costs are projected to be 20–25% below those of the conventional cyclohexane-based adipic acid.

6.10 Life Cycle Assessment

The use of bio-based feedstocks offers the potential to dramatically reduce the carbon footprint of chemicals production, both by reducing the energy required for chemical manufacturing and through the utilization of biogenic carbon. Because adipic acid used in the production of durable goods, such as nylon-6,6 and polyurethanes, carbon from atmospheric CO_2 (in the form of glucose) is assumed in an end-of-life analysis to be effectively sequestered in the final product. This has the net effect of reducing atmospheric CO_2 by the amount of carbon contained in the final product. As detailed previously, conventional adipic acid manufacturing also produces N_2O as a by-product, contributing significantly to its GHG emissions profile. Rennovia's bio-based adipic acid, which requires less energy to produce, sequesters biogenic CO_2, and does not emit N_2O, is projected to have a significantly reduced carbon footprint versus current petroleum-based technology (Figure 6.8). In addition to the 80% reduction in GHG emissions, large reductions are also predicted for the life cycle assessment (LCA) metrics of human health, ecosystem quality, and resource depletion [24].

6.11 Conclusions

The use of renewable feedstocks for chemical production has several potential advantages over conventional petroleum-based technologies, including reduced environment footprint, reduced GHG emissions, insulation from petroleum market volatility, and most importantly a lower fundamental production cost structure. The development of Rennovia's bio-based adipic acid technology illustrates the possibility of realizing these advantages, particularly through the use of chemical catalytic processes that enable the selective transformation of glucose in an economically viable process.

References

1. (a) Ereev, S.Y. and Patel, M.K. (2012) *J. Bus. Chem.*, **9** (1), 31–48; (b)Wilting, H. and Hanemaaijer, A. (2014) Share of Raw Materials Costs in Total Production Costs. PBL Netherlands Environmental Assessment Agency, The Hague, Netherlands. Report PBL 1506.
2. (a) Bioamber http://www.bio-amber.com (accessed 30 September 2015).; (b) Genomatica http://www.genomatica.com (accessed 30 September 2015).
3. (a) Musser, M.T. (2005) *Ullmann's Encyclopedia of Industrial Chemistry*, Wiley-VCH Verlag GmbH & Co. KGaA, Weinheim; (b) Oppenheim, J.P. and Dickerson, G.L. (2003) *Kirk-Othmer Encyclopedia of Chemical Technology*, John Wiley & Sons, Inc.; (c) Mainhardt, H. (ICF Incorporated) (2001) N_2O emissions from adipic acid and nitric acid production, in *IPCC Good Practice Guidance and Uncertainty Management in National Greenhouse Gas Inventories*, Intergovernmental Panel on Climate Change (IPCC); (d)US Environmental Protection Agency – Office of Air and Radiation (2001) U.S. Adipic Acid and Nitric Acid Nitrous Oxide Emissions 1990–2020 – Inventories, Projections and Opportunities for Reductions; Jupp, C., Nakhutin, A., and Cianci, V.C.S. (2006) *2006 IPCC Guidelines for National Greenhouse Gas Inventories*, vol. 3: Industrial Processes and Product Use, Chapter. 3: Chemical Industry Emissions, Intergovernmental Panel on Climate Change.
4. (a) Bart, C.J. and Cavallaro, S. (2015) *Ind. Eng. Chem. Res.*, **54** (1), 1–46; (b) Bart, C.J. and Cavallaro, S. (2015) *Ind. Eng. Chem. Res.*, **54** (2), 567–576.
5. Niu, W., Draths, K.M., and Frost, J.W. (2002) *Biotechnol. Progr.*, **18**, 201–211.
6. Botes, A.L. and Van Eck Conradie, A. (2014) Methods of Producing 6-Carbon Chemicals Via Methyl-Ester Shielded Carbon Chain Elongation. PCT Patent WO 2014/105805.
7. Raemakers-Franken, P.C., Schürmann, M., Trefzer, A.C., and De Wildeman, S.M.A. (2010) Preparation of Adipic Acid. PCT Patent WO 2010/104391.
8. Baynes, B. and Geremia, J.M. (2010) Biological Synthesis of Difunctional Alkanes from Carbohydrate Feedstocks. PCT Patent WO 2010/068944.
9. Picataggio, S. and Beardslee, T. (2012) Biological Methods for Preparing Adipic Acid. US Patent 8,241,879.
10. Burgard, P.A., Pharkya, P., and Osterhout, R.E. (2010) Micororganisms for the Production of Adipic Acid and Other Compounds. US Patent 7,799,545.
11. Miltenberger, K. (2000) *Ullmann's Encyclopedia of Industrial Chemistry*, Wiley-VCH Verlag GmbH & Co. KGaA, Weinheim.
12. Kiely, D.E., Carter, A., and Shrout, D.P. (1997) Oxidation Process. US Patent 5,599,977.
13. Donen, S. and Jensen, K. (2014) Nitric Acid Oxidation Process. US Patent 2014/0275622.
14. Rivertop http://www.rivertop.com (accessed 30 September 2015).
15. (a) Merbouh, N., Bobbitt, J.M., and Bruckner, C. (2002) *J. Carbohydr. Chem.*, **21**, 65–77; (b) Ibert, M., Marsais, F., Merbouh, N., and Bruckner, C. (2002) *Carbohydr. Res.*, **337**, 1059–1063; (c) Merbouh, N., Thaburet, F., Ibert, M., Marsais, F., and Bobbitt, J.M. (2001) *Carbohydr. Res.*, **336**, 75–78; (d) Merbouh, N., Bobbitt, J.M., and

Bruckner, C. (2002) Method for the Oxidation of Aldehydes, Hemiacetals and Primary Alcohols. US Patent 6,498,269.

16. Bin, D., Wang, H., Li, J., Wang, H., Yin, Z., Kang, J., He, B., and Li, Z. (2014) *Electrochim. Acta*, **130**, 170–178.
17. (a) Capan, E., Hahnlein, M.S., Prusse, U., Vorlop, K.-D., and Begli, A.H. (2005) Catalytic Process for the Modification of Carbohydrates, Alcohols, Aldehydes or Polyhydroxyl Compounds. US Patent 6,894,160; (b) Deller, K., Krause, H., Peldszus, E., and Despeyroux, B. (1992) Method for Preparation of Gluconic Acid by Catalytic Oxidation of Glucose. US Patent 5,132,452; (c) Saito, H., Shinji, O., and Fukuda, S. (1989) Process for Producing Gluconic Acid. US Patent 4,843,173; (d) Acres, G.J.K. (1971) Catalytic Oxidation of Glucose. US Patent 3,607,922.
18. (a) de Lederkremer, R.M. and Marino, C. (2003) *Adv. Carbohydr. Chem. Biochem.*, **58**, 199–306; (b) Besson, M. and Gallezot, P. (2001) in *Fine Chemicals through Heterogeneous Catalysis* (eds R.A. Sheldon and H. van Bekkum), Wiley-VCH Verlag Gmbh & Co. KGaA, Weinheim, pp. 507–518; (c) Dirkx, J.M.H. and van der Baan, H.S. (1981) *J. Catal.*, **67**, 1–13; (d) Dirkx, J.M.H. and van der Baan, H.S. (1981) *J. Catal.*, **67**, 14–20; (e) Önal, Y., Schimpf, S., and Claus, P. (2004) *J. Catal.*, **223**, 122–133; (f) Thielecke, N., Aytemir, M., and Prüsse, U. (2007) *Catal. Today*, **121**, 115–120; (g) Dimitratos, N., Lopez-Sanchez, J.A., and Hutchings, G.J. (2009) *Top. Catal.*, **52**, 258–268.
19. Diamond, G.M., Murphy, V., and Boussie, T.R. (2014) in *Modern Applications of High Throughput R&D in Heterogeneous Catalysis*, Chapter 8 (eds A. Hagemeyer and A. Volpe Jr.), Bentham Science Publishers, pp. 299–309.
20. Boussie, T.R., Dias, E.L., Fresco, Z.M., Murphy, V.J., Shoemaker, J., Archer, R., and Jiang, H. (2014) Production of Adipic Acid and Derivatives from Carbohydrate-Containing Materials. US Patent 8,669,397.
21. Pedersen, C., Bock, K., and Lundt, I. (1978) *Pure Appl. Chem.*, **50**, 1385–1400.
22. Rennovia http://www.rennovia.com/wp-content/uploads/2014/12/JM-Davy-Rennovia-Press-Release-March-20-20141.pdf (accessed 19 May 2015).
23. (a) Pavone, A. (2012) Bio-Based Adipic Acid. IHS Process Economics Program Report 284, IHS Chemical, Santa Clara, CA; (b) Faisca, N. and Ping, C.C.W. (2013) Adipic Acid, PERP report 2013-3, Nexant Inc.
24. (a) Determined using PRé SimaPro LCA Software http://www.pre-sustainability.com/ (accessed 19 May 2015); Ecoinvent Life Cycle Inventory (LCI) Database http://www.ecoinvent.org (accessed 19 May 2015); Thinkstep GaBi LCA Database http://www.thinkstep.com/software/gabi-lca/gabi-databases (accessed 19 May 2015); (b) ISO (2006) Methods described in ISO 14040 *Environmental Management – Life Cycle Assessment – Principles and Framework*, International Organization for Standardization, Geneva; ISO (2006) ISO 14044 *Environmental Management – Life Cycle Assessment – Requirements and Guidelines*, International Organization for Standardization, Geneva; Scientific Applications International Corporation (SAIC) (2006) Life Cycle Assessment Principles and Practice. *EPA/600/R-06/060*, U.S. Environmental Protection Agency, Washington, DC; Intergovernmental Panel on Climate Change (2006) 2006 IPCC Guidelines for National Greenhouse Gas Inventories, Published by Institute for Global Environmental Strategies (IGES), Hayama, http://www.ipcc-nggip.iges.or.jp/public/2006gl/ (accessed 19 May 2015).; British Standards Institution (2011) Specification for the assessment of the life cycle greenhouse gas emissions of goods and services, publicly available specification PAS 2050:2011, BSI, London, United Kingdom.

7
Industrial Production of Succinic Acid

Alfred Carlson, Bill Coggio, Kit Lau, Christopher Mercogliano, and Jim Millis

7.1
Introduction

Succinic acid, a four-carbon dicarboxylic acid (Figure 7.1), has been the subject of intense industrial activity over the past several years, with technologies being commercialized by several companies (Table 7.1). The reasons for the interest, first highlighted in the DOE Top Value Added Chemicals from Biomass report [1], are several:

- There are increasing disparities between feedstock costs for petrochemicals and biomass-based feedstocks and between C_4-C_6 versus lighter C_1-C_3 building blocks.
- There is an attractive theoretical yield of succinic acid from glucose of 1.12 kg kg^{-1}, which is the highest among bio-based chemicals. This factor leads to an efficient use of feedstocks and less volatility and lower raw material cost.
- There are emerging applications for succinic acid and its derivatives driven by the promise of lower cost supply compared with succinic produced by hydrogenation of maleic acid.
- There is a favorable environmental footprint for succinic acid manufacture as a result of the overall process sequestering carbon dioxide from the atmosphere, and, more generally, there is an increasing awareness of the benefits of bio-based chemicals.

7.2
Market and Applications

The historical market for succinic acid, mostly produced by the hydrogenation of nonrenewable maleic acid, has been estimated to be 30–50 000 mT annually. Projections for market growth are 25–30% annually, exceeding 500 000 mT by the year 2020 [2]. The production gap is expected to be filled by fermentation-derived succinic acid (Figure 7.2).

Chemicals and Fuels from Bio-Based Building Blocks, First Edition.
Edited by Fabrizio Cavani, Stefania Albonetti, Francesco Basile, and Alessandro Gandini.

Figure 7.1 Succinic acid [CAS 110-15-6].

Table 7.1 Companies with commercial-scale succinic acid manufacturing facilities.

Company	Plant location	Capacity (mT)	Operation
BioAmber	Pomacle, France	3 000	2010
	Sarnia, Canada (Figure 7.2)	30 000	2015
	North America	200 000	2018
Reverdia (JV of DSM, Roquette)	Cassano Spinola, Italy	10 000	2012
Myriant	Lake Providence, Louisiana, USA	13 600	2013
Succinity (JV of BASF, Corbion Purac)	Montmeló, Spain	10 000	2014

Figure 7.2 30 kT BioAmber plant in Sarnia, Canada.

The growth in the succinic acid market is driven by two factors; (i) new applications, including substitution for adipic acid, and (ii) the use of succinic acid as an intermediate for the production of other chemicals (Figure 7.3). Current investment in succinic acid production facilities is spurred by the expectation of less volatility and more competitive pricing associated with the bio-based product.

7.2.1 Hydrogenation of Succinic Acid

Approximately one-third of the bio-based succinic produced by 2020 is expected to be used as an intermediate for the production of 1,4-butanediol (BDO) and

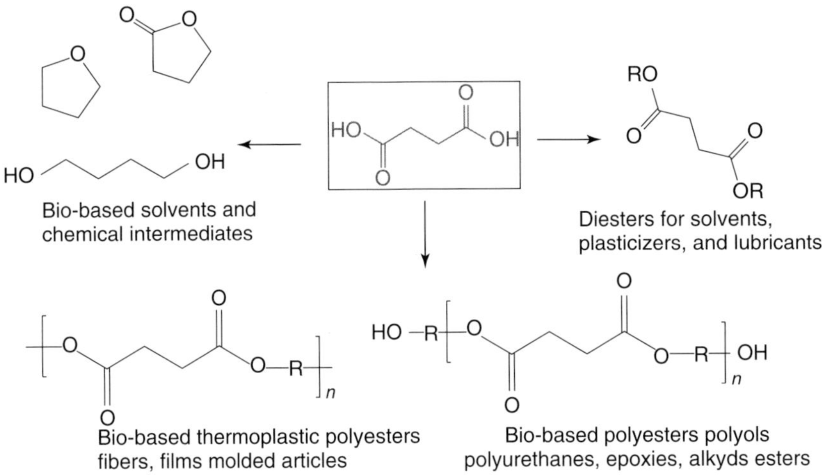

Figure 7.3 Succinic acid applications.

tetrahydrofuran (THF). Several catalytic technologies have been demonstrated for this conversion [3].

It is beyond the scope of this article to provide a technical summary of the hydrogenation process from bio-based succinic acid to BDO. However, since succinic acid is an intermediate in the reaction sequence from maleic to BDO already practiced in the chemical industry, known hydrogenation technologies are considered directly transferable to succinic acid or its esters without significant modification.

7.2.2 Polyester–Polyurethane Markets

Another one-third of the succinic market by 2020 is projected for use in polyesters, encompassing both thermoplastic polyesters and polyester polyols for use in polyurethane (PU) systems. The most immediate application of succinic acid in thermoplastic applications is polybutylene succinate (PBS), a polyester of succinic acid and BDO, currently being commercialized in Map Ta Phut, Thailand, by PTTMCC Biochem, a joint venture of PTT and Mitsubishi Chemical Company. The plant, scheduled for start-up in 2015, will have an annual capacity of 20 000 mT.

Succinic-based PUs enable a broader formulation and performance window compared with current materials [4]. Succinic is first converted to polyester polyols via a well-known condensation reaction between diacids and glycols (Figure 7.4). The resulting polyester polyols, having well-defined structures and molecular weights, can be easily incorporated into PUs via reaction between the polyester polyols, diisocyanates, and chain extenders (Figure 7.5). The segmented nature of the PU influences the final properties and applications of the material. When incorporated into the soft block of the PU, key performance properties such

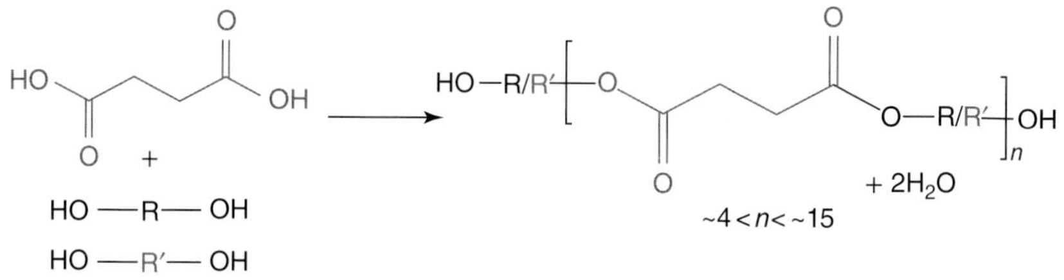

Figure 7.4 Formation of succinic-based polyester polyols.

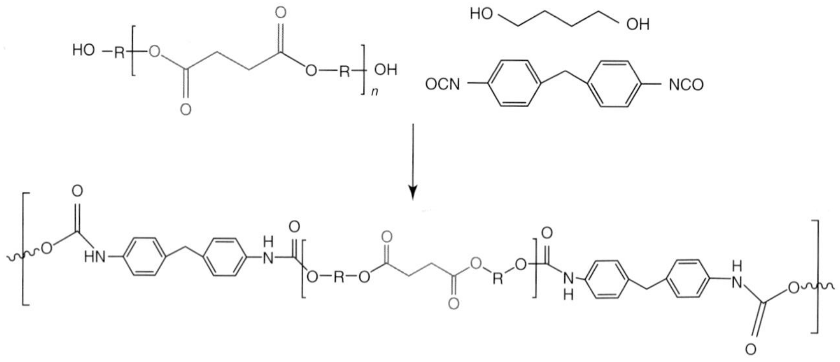

Figure 7.5 Polyurethane systems using succinic acid.

Table 7.2 Comparison of properties of polymers of bio-based succinic acid/PDO copolymers with conventional polymers containing adipic acid.

PU soft block	New carbon content	Break tensile (MPa)	Break elongation (%)	Solvent swell (%)		
				Oil	Ethanol	Toluene
SA-PDO	63	62	940	0.5	10	22
AA-PDO	20	55	777	2.2	13	41
AA-BDO	0	61	780	3.2	11	47
PTMEG	0	45	550	9.7	27	56

Succinic acid (SA), adipic acid (AA), 1,3-propanediol (PDO), 1,4-butanediol (BDO). Cured polyurethanes made using 56 OH number polyester polyols (approximate 2000 g mol^{-1}) or PTMEG. Made with the same MDI and chain extender content. Sixty-three percentage of new carbon content of PU when using bio-based 1,3-propanediol available from DuPont Tate and Lyle.

as durability and reliability are improved due to increased solvent and abrasion resistance. Table 7.2 compares two PUs made with 1,3-propanediol/succinate and 1,3-propanediol/adipate with commercially available PU systems based on polybutylene adipate or polytetramethylene glycol (PTMEG).

Other applications are under development that provide unique value propositions and will contribute to the growth of the succinic market shown in Table 7.3.

Table 7.3 Summary of the applications involving bio-based succinic acid expected by 2020.

Application area	Value proposition
Solvents	Up to 100% bio-based cleaners, degreasers, coalescent solvents for paints, potential to replace NMP
Deicers, heat transfer fluids	Noncorrosive and biodegradable
Lubricants, plasticizers	Good viscosity and low temperature pour points, replace phthalates in PVC
Thermoplastic elastomers, molded gaskets, textile coatings, industrial wheels and rollers synthetic leather, protective coatings, semirigid foams	High strength and elongation, improved abrasion, and solvent-resistant polyurethanes systems with >60% bio-based content
Gel coats, liquid, and powder polyester resins for anticorrosion and primers; coatings for metals used in industrial, automotive, energy, and appliance markets	Improve flexibility with good high temperature resistance. Improve UV and yellowing resistance when replacing aromatic acids

7.3 Technology

Over the last 30 years, the production of succinic acid by fermentation has been the subject of many academic research papers and reviews [5, 6] and industrial patent applications. Even earlier, more than 60 years ago, research papers and patents described the production of malic and fumaric acid, both immediate metabolic precursors to succinic, by fermentation processes similar to those for producing succinic acid.

The extent and diversity of the prior art provides many options in building a succinic acid process. Traditionally, processes have been designed around the choice of the host organism. However, a successful commercial process that minimizes capital and operating costs requires that all aspects of the manufacturing process – host microorganism, biochemical pathway, fermentation, and downstream processes – be considered and optimized as a system.

Commercial efforts by the companies listed in Table 7.1 have converged on two distinctly different solutions. Early efforts for succinic acid manufacture were constructed using bacterial fermentations conducted at neutral pH. While often capable of producing high succinic yields, these processes tend to have complex downstream processes that require splitting a succinate salt to form succinic acid and an inorganic salt coproduct. Later developments have focused on acidophilic yeast fermentations that operate below the lower pK_a of succinic acid (4.2), thus increasing the ratio of succinic acid to succinate salts and dramatically simplifying the downstream processes. These advancements have led to a decrease in the cost of manufacturing succinic acid and opened new markets. The following sections will first discuss the biochemical pathway considerations for the two options, followed by a comparison of the two processes and resulting consequences.

7.3.1 Biochemical Pathway and Host Microorganism Considerations

Three common pathways (Figure 7.6) from glucose to succinic acid have been identified; (i) the reverse TCA cycle or reductive pathway, (ii) the oxidative TCA pathway, and (iii) the glyoxylate pathway.

The pathway with the highest theoretical yield of succinate from glucose is the reverse TCA cycle or reductive pathway. In this pathway, glucose is converted to phosphoenolpyruvate (PEP) or pyruvate, which is then carboxylated to oxaloacetate (OAA), reduced to malate, and then further reduced to succinate. Stoichiometrically, 2 moles of PEP or pyruvate are formed from a mole of glucose in glycolysis and are available to produce succinate; however, a portion must be

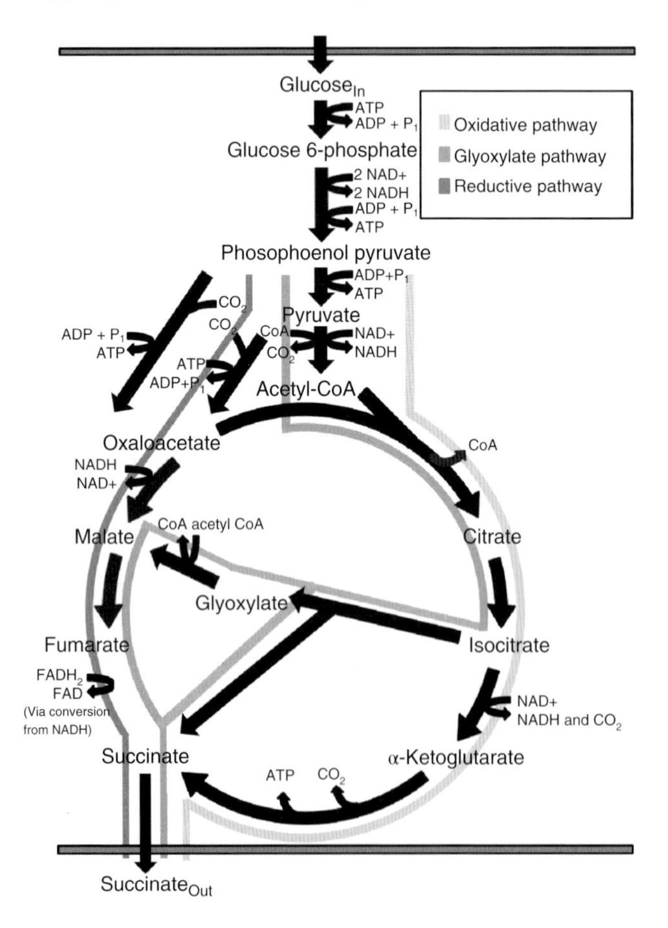

Figure 7.6 Biochemical pathways for succinic acid production.

diverted to the oxidative pathway for redox balancing, leading to the following equation:

$$C_6H_{12}O_6 + \frac{6}{7}CO_2 \rightarrow \frac{12}{7}C_4H_6O_4 + \frac{6}{7}H_2O$$

This stoichiometry, 1.714 moles of succinic acid per mole of glucose, or 1.12 g g^{-1}, exceeds 100% on a mass basis due to the incorporation of CO_2.

The oxidative TCA pathway involves glycolysis from glucose to pyruvate, and then conversion of pyruvate to CO_2 and acetyl-CoA through the action of pyruvate dehydrogenase. Acetyl-CoA reacts with OAA in the TCA cycle to form citrate that eventually is converted to succinate via two decarboxylation steps generating two additional molecules of CO_2 and one molecule each of NADH and succinic acid. As succinic acid is accumulated, anaplerotic replenishment of OAA takes place by direct carboxylation of PEP or pyruvate. The maximum yield of succinic acid on glucose by the oxidative route is 1 mole of succinate per mole of glucose or 0.655 g g^{-1}, not including the yield loss associated with cell mass production. Since this pathway produces excess NADH (and other reduced cofactors), oxygen must be supplied to the cells to oxidize these cofactors, or these cofactors must be used by other reductive pathways in the cell. Overall, this results in the stoichiometry for the oxidative TCA cycle of

$$C_6H_{12}O_6 + \frac{5}{2}O_2 \rightarrow C_4H_6O_4 + 2CO_2 + 3H_2O$$

The final pathway to succinate involves the glyoxylate cycle, in which OAA condenses with acetyl-CoA to form citrate that is then isomerized to isocitrate (as in the TCA cycle). Instead of being converted to α-ketoglutarate, isocitrate is converted to glyoxylate and succinic acid by the action of isocitrate lyase. Glyoxylate condenses with a second molecule of acetyl-CoA to form malic acid that is oxidized to OAA to complete the cycle. The stoichiometry of succinate formation by this process is the same as for the oxidative TCA pathway, 1 mole of succinate produced per mole of glucose. The pathway also requires oxygen to oxidize the excess reduced cofactors produced when acetyl-CoA is formed from pyruvate and malate is converted to OAA.

Many competing pathways can divert carbon from the succinate-producing pathways and lead to by-product accumulation in the fermentation. For example, in both bacteria and yeast, pyruvate can be decarboxylated to acetaldehyde (and subsequently is converted to acetic acid and CO_2) by pyruvate decarboxylase (*pdc*). Deletion of *pdc* reduces acetic accumulation and increases succinate yield. Pathway intermediates, such as malate and pyruvate, may also be secreted and accumulate as by-products.

These pathway strategies, alone or in combination, have been implemented differently in several organisms. The bacteria, *Actinobacillus succinogenes* [7], *Anaerobiospirillum succiniciproducens* [8], and *Mannheimia succiniciproducens* [9], all good succinic acid producers under anaerobic conditions, appear to use phosphoenolpyruvate carboxykinase (*pck*) to carboxylate PEP to OAA rather than PEP carboxylase, because the former enzyme produces 1 mole of ATP per mole of

OAA. At the same time, PEP carboxykinase requires a higher concentration of CO_2 to function than does PEP carboxylase. The additional ATP can then be used for other cellular functions, particularly for cell maintenance in the presence of high succinate concentrations.

The first commercial processes to manufacture succinic acid also used a genetically engineered *Escherichia coli* bacteria. BioAmber scaled a proprietary *E. coli* system derived from AFP111, first developed by Donnelly *et al.* [10], in their demonstration plant in Pomacle, France. Myriant uses an engineered *E. coli* strain based on the work of the Ingram lab [11] in their Lake Providence facility in Louisiana [12].

These same pathways have long been used in eukaryotic systems for C_4 carboxylic acid production. *Aspergilli,* especially *Aspergillus flavum,* was used in patented fermentation processes for malate production, and *Rhizopus* sp. were used in processes to produce fumaric acid, both prior to 1960. The main driving force for these low pH fermentation processes was ease of downstream processing. An additional benefit in large-scale fermentations of a low pH was that it also reduced the chance of contamination. An additional benefit of yeast is that they are not subject to bacteriophage. The fumaric process was later shown to use the same pathway (reverse TCA pathway) that current high yield succinate processes use [13]. Malic acid processes also use this pathway [14, 15].

In the mid-2000s, there was a renewed interest in developing eukaryotic systems, particularly the yeast *Saccharomyces cerevisiae,* to produce succinic and malic acids [16]. It was demonstrated that these cells, using a *pdc*$^-$ strain, were able to produce substantial malic and succinic acid through the reverse TCA cycle (i.e., at high yields) and at lower pH (<5.0) than bacterial systems. Commercial citric acid at that time was produced under aerobic and microaerobic conditions at pH ~2.0 and development of analogous, low pH processes for fumaric acid had been suggested [17].

The general biochemistry for production of succinate in eukaryotic cells is nearly the same as for bacteria, but there are important physiological differences and selected yeasts are particularly resistant to low pH. A leading difference between the eukaryotic reductive TCA pathway and the prokaryotic pathway is that yeast lack *pck* and thus carboxylate only pyruvate, not PEP, to supply OAA. The lack of *pck* means that the yeast pathway will produce less ATP than the bacterial system, so fermentations are generally run aerobically with oxygen uptake limited by mass transfer rather than under anaerobic conditions with CO_2 sparge as used in most bacterial systems. Indeed, the best results for succinic systems were obtained with microaerobic conditions in shake flasks similar to those previously observed in fumarate fermentations with *Rhizopus* sps. [18, 19]. For this reason, efficient pyruvate carboxylase activity is of key importance for high fluxes through the reverse TCA pathway in eukaryotic systems. Increasing the activity of pyruvate carboxylase while still allowing some O_2 transfer, by adding some CO_2 (but not 100%) or by using carbonate salts, is a critical issue in the use of the reverse TCA pathway in eukaryotic systems.

Another difference between the biochemistry of yeasts and bacteria is that the pathways are compartmentalized into mitochondrial and cytoplasmic compartments in yeast. While the oxidative TCA cycle operates exclusively in the mitochondria, the reverse TCA cycle operates best in the cytoplasm. Each compartment has its own requirement for redox balance with mechanisms for shuttling reducing equivalents across the mitochondria membrane into the cytoplasm. The pentose phosphate cycle can also be used as a mechanism to increase reduced cofactors (NADPH) in the cytoplasm, which also has the benefit of increasing glucose flux to pyruvate, independent of the mitochondria when necessary for redox balance in this compartment.

Yet another pathway difference between yeast systems and many bacterial systems relates to the final step in the reductive pathway. Many bacteria have a membrane-bound fumarate reductase enzyme complex. These complexes are able to access reducing equivalents from the quinol pool of the electron transport chain [20]. In current succinic acid production in yeast strains, the fumarate reductase is a soluble enzyme that converts fumarate to succinate using reducing equivalents from reduced flavins. These soluble fumarate reductase enzymes are employed in nature by certain bacteria [20, 21] and trypanosomatids [22], as well as in other organisms such as yeasts [23]. Reduced flavins can be generated by either the enzyme, by other enzymatic pathways, or by flavin reductases that can access and transfer the reducing charges from other cellular cofactor pools such as NADH or NADPH [23].

Finally, as noted earlier, some yeasts are better suited to tolerate low pH fermentation conditions and the resulting pH gradients and product gradients across the cell membrane. Yeasts have been used traditionally to produce citric acid but only recently have commercial efforts been expanded to produce other organic acids. Cargill developed a novel crabtree-negative yeast, denoted CB1, to produce lactic acid, a process that has been practiced commercially since 2007. The lessons from this effort have been adopted by BioAmber, in collaboration with Cargill, to develop CB1 for succinic production. Figure 7.7 illustrates the ability of CB1 to grow in the presence of and tolerate high concentrations of succinic acid relative to *Saccharomyces* strains, such as CEN PK and Ethanol Red. Similarly, Reverdia has developed a process using *S. cerevisiae.*

Yeast systems benefit from the ability to use a highly efficient succinate/malate transporter for export of product out of the cell. The *Schizosaccharomyces pombe* transporter (SpMAE1) promotes transport and excretion of both succinic and malic [24, 25] but not fumarate [26]. Both of these facts influence the pathway flux in these cells, although a transporter more specific to succinate would be more advantageous.

Under the more severe conditions of high external organic acid concentration and low pH, the free acid form of the acids can passively diffuse back into the cells at significant rates [27, 28]. Upon ionization at the higher internal pH of the cells, the acids release protons. Unless these protons are shuttled out of the cytoplasm by H^+-ATP pumps [29], the internal pH of the cells will drop to unacceptably low levels. However, the H^+-ATP pumps consume ATP which increases

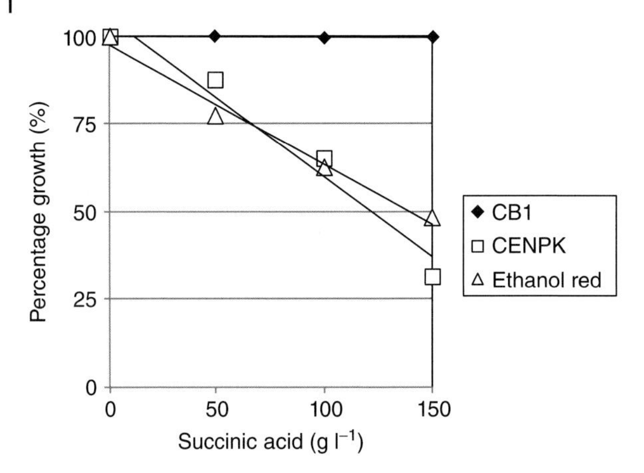

Figure 7.7 Growth effects of succinic acid on yeast strains.

the maintenance energy requirements of the cell. Such maintenance energy comes only at the cost of glucose, which ultimately competes for product formation and leads to yield drag [30]. Additional sugar costs for the acidophilic yeast fermentation have been shown to be offset by savings in the downstream process (described later). This ability to maintain internal cellular pH is a key benefit of these yeast systems. The greater ability of some yeast to limit passive back diffusion of acids into the cell and to generate ATP in the mitochondria, which is more protected from the external environment than in bacteria, gives these yeast systems an added advantage.

7.3.2
Fermentation Process Options

Of the numerous studies and patents on succinate production by fermentation, only a handful of commercially relevant production processes have been implemented (Table 7.4). Common features of most of the bacterial fermentations are operation under anaerobic conditions, at neutral pH, and with sparging of the fermentation with CO_2 to improve the performance and yield of the fermentation.

7.3.2.1 *E. coli* Systems

Early work, performed at Argonne National Laboratory and funded by the Department of Energy, led to the development of *E. coli* NZN111 (APF111), from which an improved strain was commercialized by BioAmber at a demonstration plant in Pomacle, France. Yields approaching theoretical value (1.12 g succinate per g glucose) have been reported with engineered *E. coli* strains in laboratory shake flask experiments; however, attempts to achieve these results in bioreactors have generally failed to reproduce the flask experiment results, though yields as high as $1.1\,g\,g^{-1}$ and productivities as high as $1.3\,g\,l^{-1}\text{-}h^{-1}$ with a titer of $99\,g\,l^{-1}$

Table 7.4 Commercial and semiworks processes for succinic acid production.

	Host	Company
Bacterial	*Escherichia coli AFP111*	BioAmber (first generation)
		Myriant
	Basfia succiniciproducens	Succinity (BASF/Corbion Purac)
	Corynebacterium glutamicum	Mitsubishi Chemical Corporation
	Actinobacillus succinogenes	Michigan Biotechnology Institute
Yeast	*Saccharomyces cerevisiae*	Reverdia (DSM, Roquette)
	CB1	BioAmber

using AFP111 was reported in a 1-l bioreactor system [31]. Various attempts at further engineering *E. coli* line have been tried with limited success. Dole and Yocum [32] observed a succinate yield from glucose of 0.70 $g\,g^{-1}$, a titer of about 60 $g\,l^{-1}$ and a productivity of about 0.6 $g\,l^{-1}$-h^{-1}. Lin *et al.* [33, 34] demonstrated similar results in a bioreactor system using an aerobic culture.

7.3.2.2 *Corynebacterium glutamicum* Systems

C. glutamicum has also been shown to be capable of industrially relevant succinic acid production under either anaerobic or aerobic conditions [35]. An engineered strain produced 146 $g\,l^{-1}$ of succinic acid from glucose in only 46 h (productivity of 3.17 $g\,l^{-1}$-h^{-1}) at a yield of 0.92 $g\,g^{-1}$ under anaerobic conditions [36]. By adding formate (for reducing equivalents) to the fermentation media, a productivity of 2.5 $g\,l^{-1}$-h^{-1} and a yield of 1.1 $g\,g^{-1}$, again using glucose, was observed with *C. glutamicum* engineered strains (yield does not count formate consumption) under aerobic conditions [37]. Mitsubishi, in an early collaboration with Ajinomoto, engineered a strain of *Brevibacterium flavum* (or *C. glutamicum*) for succinate production in their version of the succinate process [38]. A succinic acid concentration of 71 $g\,l^{-1}$ was reported with an 81% yield on sugar for this process.

7.3.2.3 Other Bacterial Systems

A. succinogenes has been reported to reach succinate titers of up to 106 $g\,l^{-1}$ in 40–60 h (productivity of 1.7–2.5 $g\,l^{-1}$-h^{-1}) with yields of >0.8 $g\,g^{-1}$ from glucose in 1-l fermenters [39]. Substantial side products especially acetic acid were formed (16 $g\,l^{-1}$), and soluble residues from the yeast extract and corn steep liquor, added to the fermentation media, add to the downstream purification costs. Other succinic acid producers, such as the rumen bacteria *M. succiniciproducens* and *Basfia succiniciproducens*, have been developed. *M. succiniciproducens* does not appear to give extremely high titers (25–35 $g\,l^{-1}$ range) in batch culture either on glucose or glucose/glycerol mixtures, but yields are competitive, 0.7–1.0 $g\,g^{-1}$ [9, 40, 41]. More recently, Succinity has deployed an engineered strain of *B. succiniciproducens* [42] grown on mixtures of glycerol and carbohydrates,

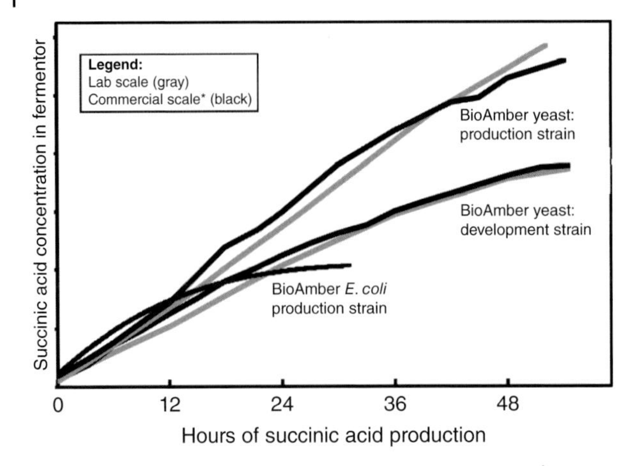

Figure 7.8 Performance and scalability of BioAmber yeast process.

reaching concentrations of $\gg$60 g l^{-1} and mass yields of at least 1.0 g g^{-1} on these substrates.

7.3.2.4 Yeast Systems

BioAmber and Reverdia have each developed and commercialized yeast processes to produce succinic acid. Reverdia's process [43] utilizes a highly modified *S. cerevisiae* operated at a fermentation pH of 3.0. They have reported 80 g l^{-1} succinic in 48 h and 94 g l^{-1} in 74 h [44]. The BioAmber process uses an engineered variant of the Cargill CB1 yeast under proprietary conditions. Fermentation performance of their process has not been disclosed; however, the scalability of the yeast fermentation from 2-liter laboratory fermentors to 180 m^3 production fermentors has been successfully demonstrated (Figure 7.8).

7.3.2.5 Media and pH Control

Some bacterial fermentations can use minimal or defined media containing glucose (or other carbohydrates), glycerol, or mixtures as the carbon source supplemented with salts and vitamins. Other bacteria, such as the rumen bacteria, require complex nutrients such as yeast extract or corn steep liquor [9, 45]. Although corn steep liquor is sometimes considered to be an inexpensive source of nutrients, it adds complexity to the downstream process, and residues can contribute to color or performance issues in product applications.

Succinic acidifies the fermentation medium as it accumulates and thus requires that the bacterial fermentations employ pH control with alkali metal hydroxides (NaOH, KOH), ammonia (NH_3), or alkaline earth metal (Ca^{+2}, Mg^{+2}) carbonates, bicarbonates, hydroxides, or mixtures. Mitsubishi reports using a variety of carbonates and bicarbonates for succinic fermentations using *C. glutamicum* or *B. flavum* to maintain the pH in the range 6–9.5 [46], and Litsanov *et al.* [47] describe using KOH for pH control with an aerobic culture of *C. glutamicum.*

Myriant [32] used a mixture of K_2CO_3 and KOH to maintain the pH of their *E. coli* fermentation process at 7.0. Fermentations with both *A. succinogenes* and *B. succiniciproducens* fermentations favor magnesium hydroxides and carbonates to control pH [9, 43]. Further discussion of the impact of the choice of base for pH control is discussed later with respect to downstream processing.

In contrast, the acidophilic yeast processes use a minimum of base for pH control, leaving the organic acid predominantly in the acid form [48, 49]. Low pH fermentation in combination with defined media also minimizes the risk of contaminating organisms, allowing for further simplifications in hardware design and lower capital.

7.3.2.6 Aeration and Gas Systems

Aeration conditions and equipment, with either air or CO_2, affect both fixed capital and variable costs. Most bacterial systems produce succinic acid best under anaerobic conditions and require CO_2 addition to achieve high yield. At production scale, a supply of oxygen-free CO_2 gas is needed, either as an overlay or sparge gas. Generally, a flow of CO_2 at 0.1 VVM is sufficient to meet stoichiometric needs. The CO_2 supplies bicarbonate for the PEP (or pyruvate) carboxylation to OAA, an important first reaction for the reverse TCA cycle. Alternately, carbonates can be added but are more costly and challenging to practice at scale.

A better process is one that uses no exogenous CO_2 or bicarbonate addition but rather produces a minimum amount of net CO_2. CO_2 produced by the cell is reconsumed in the succinate pathway. Litsanov *et al.* [47] describe an aerobic process using *C. glutamicum* that had been deleted in succinate dehydrogenase to deactivate the TCA cycle. This manipulation resulted in production of succinic using the glyoxylate cycle. This fermentation required no equipment for supplying CO_2 and emits only minimal CO_2 and substantially less than a normal aerobic fermentation.

Yeast systems do not generally produce substantial succinic acid under anaerobic conditions even when sparged with CO_2. Mixtures of O_2, CO_2, and N_2 or air have been used to produce malic and succinic acids in 1-l fermentations [50, 51]. Similarly, succinic acid production with *S. cerevisiae* has been described using a sparge gas with between 20 and 35% v/v O_2 and less than 0.1% CO_2, but operated in a manner that causes the partial pressure of the CO_2 in the fermenter to be above 0.35 bar [52]. Cargill described an optimal oxygen uptake rate (OUR) of a *pdc*$^-$ yeast producing any desired fermentation product [53, 54]. More production of succinic acid specified an OUR range >5 mmol l^{-1}-h^{-1} and showed reasonable yields (0.4 g g^{-1}) even at 25 mmol l^{-1}-h^{-1} in fermentations operated at pH 3.0.

7.3.3 Downstream Process Options

Following fermentation, the first step in downstream processing is typically the removal of biomass by well-established unit operations, such as microfiltration or centrifugation. At this point, the approaches that have been developed for isolating succinic acid diverge significantly. Isolation of an organic acid from its corresponding salt does not seem to be a difficult technical challenge until faced with economic constraints and the purity requirements of the chemical industry. Accordingly, early commercial efforts have struggled with purifying a succinate salt in a crude fermentation broth to achieve polymer-grade product.

Early work with calcium succinate utilized concentrated H_2SO_4 to acidify and resolubilized succinic acid, leading to calcium sulfate (gypsum) as a by-product. Disposal issues related to gypsum production, however, rendered this process uneconomical.

An alternative process is using desalting electrodialysis to split sodium succinate into NaOH and succinic acid, coupled with recycle of the NaOH to the fermentation process [55]. A version of electrodialysis with ammonium succinate was practiced by BioAmber at demonstration scale in Pomacle, France, to recycle the base to fermentation. Electrodialysis was found to have some limitations; (i) membrane fouling due to numerous saccharides and peptides commonly found in fermentation broth led to significant operational interruptions and costs and (ii) low conductivity at higher conversion resulted in higher electricity costs.

Liquid–liquid extraction using extraction agents, such as trialkylamines, has also been proposed for succinic acid [56]; however, this technology has only been commercialized for citric acid due to production of low levels of amine by-products when applied to other products.

In order to overcome the challenges associated with electrodialysis, BioAmber developed novel base (ammonia) recycling technology utilizing steam cracking of ammonia from diammonium succinate. BioAmber, working with MATRIC, published a series of patents on the technology. A two-step production method for succinic acid via monoammonium succinate [57] and a single-step production method for succinic acid were disclosed [58].

More recently, ThyssenKrupp has developed a process that uses simulated moving bed chromatography to purify and separate succinic acid [59], a step that requires that the fermentation broth first be acidified to produce stoichiometric ammonium sulfate, which can then be sold as a fertilizer by-product. Unfortunately, the stoichiometry dictates that the production of fertilizer exceeds that of succinic acid in these schemes.

The desire to avoid large by-product streams, such as gypsum and fertilizer, has led to other approaches that attempt to recover and recycle the base to the fermentation process. Alternate processes which produce magnesium succinate using $Mg(OH)_2$, $MgCO_3$, or $Mg(HCO_3)_2$ have been described. These processes involve acidification of fermentation broth with strong mineral acid, either HCl [60] or H_2SO_4 [61], from which succinic acid can be crystallized. In the former case, magnesium chloride is then reconverted to MgO through a high temperature, thermal decomposition step that releases HCl. The MgO is then quenched with water to form $Mg(OH)_2$ for reuse in the fermentation.

Regardless of the separation process used to remove the counterion, additional steps are needed to achieve a crystalline succinic acid product. Typically, polishing steps such as carbon treatment or ion exchange are employed, followed by evaporation, crystallization, and drying. Nevertheless, the complexity of the alternatives summarized previously has resulted in high capital investment for these processes and product purity issues when practiced at scale. It was in anticipation of these issues that producing succinic acid with an acidophilic yeast was developed.

With acidophilic yeast, succinic acid purity can be achieved with two crystallization steps following cell separation [62]. The two crystallization steps may also supplemented by polishing steps such as activated carbon for reducing certain impurities, especially those associated with color. The disclosure confirms that downstream processes coupled to lower pH yeast fermentations are substantially simpler compared to those associated with neutral pH bacterial fermentations. The simplicity in combination with the elimination of base recycle is a compelling argument for yeast-based technologies with respect to operability and lower costs.

7.4 Life Cycle Analysis

Two recent life cycle analyses related to succinic acid manufacturing processes have validated the positive impact of this technology relative to alternative processes. A comparison of an acidophilic yeast process with the petrochemical process to succinic showed near elimination of greenhouse gas emissions and a 64% reduction in energy consumption [63]. Separately, a comparison of a yeast process with a bacterial process producing an ammonium sulfate by-product predicted 40% lower greenhouse gas emission and 28% lower nonrenewable energy use [64].

7.5 Conclusion

Increasing market interest in sustainability, coupled with the recent dynamics of the petrochemical industry and the maturing of biotechnology, have created strong interest in bio-based chemicals, leading to strong competition of ideas for the production of bio-based succinic acid. These ideas cover a variety of philosophies to integrate biotechnology strategies – pathway, host properties, and fermentation – with process engineering. The two dominant strategies described herein are (i) the bacterial production of succinic salts with the associated downstream and (ii) the direct production of succinic acid with yeast leading to a simpler isolation and purification. The economics of these two routes have been independently assessed with the acidophilic yeast technology offering lower capital and operating costs and fewer by-products [65]. These developments are expected to lower the price of succinic acid to drive new applications in materials and uses of the intermediate for the production of commodity chemicals, such as BDO and THF.

References

1. Werpy, T. and Petersen, G. (eds) (2004) *Top Value Added Chemicals From Biomass*, US Department of Energy.
2. Roland Berger (2012) Bioplastics Market Study.
3. Delhomme, C., Weuster-Botz, D., and Kühn, F. (2009) *Green Chem.*, **11** (1), 13–26.
4. Coggio, W., Brouwer, F., Roche, X., and Alarcon, E. (2015) *Paints Coat. Ind.*, **31** (4), 56–62.
5. McKinlay, J., Vieille, C., and Zeikus, J. (2007) *Appl. Microbiol. Biotechnol.*, **76**, 727–740.
6. Bechthold, I., Bretz, K., Kabasci, S., Kopitzky, R., and Springer, A. (2008) *Chem. Eng. Technol.*, **31** (5), 647–654.
7. Van der Werf, M., Guettler, M., Jain, M., and Zeikus, J. (1997) *Arch. Microbiol.*, **167** (6), 332–342.
8. Samuelov, N. *et al.* (1991) *Appl. Environ. Microbiol.*, **57** (10), 3013–3019.
9. Lee, P., Lee, S., Hong, S., and Chang, H. (2002) *Appl. Microbiol. Biotechnol.*, **58** (5), 663–668.
10. Donnelly, M., Millard, C., Clark, D., Chen, M., and Rathke, J. (1998) *Biotechnol. Fuels Chem.*, 187–198.
11. Jantama, K., Haupt, J., Svoronos, S., Zhang, X., Moore, J.C., Shanmugam, K.T., and Ingram, L. (2008) *Biotechnol. Bioeng.*, **99** (5), 1140–1153.
12. Werpy, T. and Petersen, G., Editors (2012) US Department of Energy Publication DOE/EE-0824, December 2012.
13. Kenealy, W., Zaady, E., Du Preez, J., Stieglitz, B., and Goldberg, I. (1986) *Appl. Environ. Microbiol.*, **52** (1), 128–133.
14. Battat, E., Peleg, Y., Bercovitz, A., Rokem, J., and Goldberg, I. (1991) *Biotechnol. Bioeng.*, **37** (11), 1108–1116.
15. Zelle, R., Hulster, E., van Winden, W., de Waard, P., Dijkema, C., Winkler, A., Geertman, J., van Dijken, J., Pronk, J., and van Maris, A. (2008) *Appl. Environ. Microbiol.*, **74** (9), 2766–2777.
16. Winkler, A., DeHulster, A., van Dijken, F., and Pronk, J. (2006) US Patent 2006042754.
17. Gangl, L., Wiegand, W., and Keller, F. (1990) *Appl. Biochem. Biotechnol.*, **24**, 663–667.
18. Rhodes, R., Moyer, A., Smith, M., and Kelley, S. (1959) *Appl. Microbiol.*, **7**, 74–80.
19. Rhodes, R., Lagoda, A., Misenheimer, T., Smith, M., Anderson, R., and Jackson, R. (1962) *Appl. Microbiol.*, **10**, 9–15.
20. Miura, A., Kameya, M., Arai, H., Ishii, M., and Igarashi, Y. (2008) *J. Bacteriol.*, **190** (21), 7170–7177.
21. Bertsova, Y., Kostyrko, V., Baykov, A., and Bogachev, A. (2014) *Biochim. Biophys. Acta*, **1837** (7), 1122–1129.
22. Besteiro, S., Biran, M., Biteau, N., Coustou, V., Baltz, T., Canioni, P., and Bringaud, F. (2002) *J. Biol. Chem.*, **277** (41), 38001–38012.
23. Enomoto, K., Arikawa, Y., and Muratsubaki, H. (2002) *FEMS Microbiol. Lett.*, **215** (1), 103–108.
24. Grobler, J., Bauer, F., Subden, R., and van Vuuren, H. (1995) *Yeast*, **11** (15), 1485–1491.
25. Casal, M., Paiva, S., Querios, O., and Soares-Silva, I. (2008) *FEMS Microbiol. Rev.*, **32**, 974–994.
26. Saayman, M., van Vuuren, H., van Zyl, W., and Viljoen-Bloom, M. (2000) *Appl. Microbiol. Biotechnol.*, **54**, 792–798.
27. Jamalzadeh, E., Verheijen, P., Heijnen, J., and van Gulik, W. (2012) *Appl. Environ. Microbiol.*, **78** (3), 705–716.
28. Delcourt, F., Taillandier, P., Vidal, F., and Strehaiano, P. (1995) *Appl. Microbiol. Biotechnol.*, **43**, 321–324.
29. Holyoak, C., Stratford, M., McMullin, Z., Cole, M., Crimmins, K., Brown, A., and Coote, P. (1996) *Appl. Microbiol. Biotechnol.*, **62** (9), 3158–3164.
30. van Maris, A., Konings, W., von Dijken, J., and Pronk, J. (2004) *Metab. Eng.*, **6** (4), 245–255.
31. Vemuri, G., Eiteman, M., and Altman, E. (2002) *J. Ind. Microbiol. Biotechnol.*, **28** (6), 325–332.
32. Dole, S. and Yocum, R. (2014) US Patent 8,778,656.
33. Lin, H., Bennett, G., and San, K.-Y. (2005) *Biotechnol. Bioeng.*, **90**, 775–779.

34. Lin, H., Bennett, G., and San, K.-Y. (2005) *Metab. Eng.*, 7, 116–127.
35. Wieschalka, S., Blombach, B., Bott, M., and Eikmanns, B. (2012) *Microb. Biotechnol.*, **6**, 87–102.
36. Okino, S. and Yukawa, H. (2008) *Appl. Microbiol. Biotechnol.*, **81** (3), 459–464.
37. Litsanov, B., Brocker, M., and Bott, M. (2012) *Appl. Environ. Microbiol.*, **78** (9), 3325–3337.
38. Yoshikawa, K., Kido, D., and Murase, M. (2014) Method of producing succinic acid. US Patent 8,647,843.
39. Guettler, M., Jain, M., and Rumler, D. (1996) Method for making succinic acid, bacterial variants for use in the process, and methods for obtaining variants. US Patent 5,573,931.
40. Lee, P., Lee, S., Hong, S., and Chang, H. (2003) *Bioprocess. Biosyst. Eng.*, **26** (1), 63–67.
41. Oh, I., Kim, D., Oh, E., Lee, S., and Lee, J. (2009) *J. Microbiol. Biotechnol.*, **19** (2), 167–171.
42. Schroder, H., Haefner, S., von Abendroth, G., Hollmann, R., Raddatz, A., Ernst, H., and Gurski, H. (2014) Novel Microbial Succinic Acid Producers and Purification of Succinic Acid. US Patent US 20140127764.
43. Jansen, M. and Verwaal, R. (2015) Dicarboxylic acid production by fermentation at low pH. US Patent 9,012,187.
44. Jansen, M., Verwaal, R., Segueilha, L., van de Graaf, M., and Geurts, T. (2012) Low pH fermentation to succinic acid, the basis for efficient recovery. Bio4Bio Conference, Copenhagen, February 29, 2012.
45. Zhu, L., Wang, C., Liu, R., Li, H., Wan, D., and Tang, Y. (2012) *J. Biomed. Biotechnol.*, **2012**, 9.
46. Murase, M., Yonekura, M., Kido, D., Aoyama, R., Yunomura, S., and Koike, S. (2008) Method for production of succinic acid. European Patent PCT/JP2008/065030.
47. Litsanov, B., Kabus, A., Brocker, M., and Bott, M. (2012) *Microb. Biotechnol.*, **6** (1), 116–128.
48. Rush, B. and Fosmer, A. (2012) Methods for Succinate Production. WO Patent 2013/112939.
49. Jansen, M., August, L., and Verwaal, R. (2009) Dicarboxylic Acid Production by Fermentation at low pH. WO Patent 2010/003728.
50. Winkler, A., de Hulster, A., van Dijken, J., and Pronk, J. (2005) Malic Acid Production in Recombinant Yeast. WO Patent 2007/061590.
51. Winkler, A., de Hulster, A., van Dijken, J., and Pronk, J. (2006) Malic Acid Production in Recombinant Yeast. US Patent 2008/0090273.
52. Jansen, M., Heijnen, J., and Verwaal, R. (2011) Process for Preparing Dicarboxylic Acids Employing Fungal Cells. WO Patent 2013/004670.
53. van Hoek, P., Aristidou, A., and Rush, B. (2007) Fermentation process using specific oxygen uptake rates as a process control. US Patent 7,939,298.
54. van Hoek, P., Aristidou, A., and Rush, B. (2011) Fermentation process using specific oxygen uptake rates as a process control. US Patent 8,828,707.
55. Glassner, D. and Datta, R. (1992) Process for the production and purification of succinic acid. US Patent 5,143,834.
56. Hong, Y. and Hong, W. (2000) *Bioprocess. Eng.*, **23** (5), 535–538.
57. Fruchey, O., Keen, B., Albin, B., Clinton, N., Dunuwila, D., and Dombek, B. (2014) Processes for producing monoammonium succinate from fermentation broths containing diammonium succinate, monoammonium succinate and/or succinic acid, and conversion of monoammonium succinate to succinic acid. US Patent 8,624,059.
58. Fruchey, O., Manzer, L., Dunuwila, D., Keen, B., Albin, B., Clinton, N., and Dombek, B. (2013) Processes for Producing NH4+ -OOC-R-COOH Compounds from Fermentation Broths Containing NH4+ -OOC-R-COO- NH4+ Compounds and/or HOOC-R-COOH Compound Acids, and Conversion of NH4+ -OOC -R-COOH Compounds to HOOC-R-COOH Compound Acids. US Patent 20130140169.
59. Tietz, W. and Schulze, J. (2013) Process for Removing, Isolating and Purifying Dicarboxylic Acids. US Patent 20130096343.

60. De Haan, A., Van Breugel, J., Van Der Weide, P., Jansen, P., Lancis, J., and Daro, A. (2014) Acid/Salt Separation. US Patent 20140200365.
61. Hanchar, R., Kleff, S., and Guettler, M. (2014) Production of carboxylic acid and salt co-products. US Patent 8,829,237.
62. Van De Graaf, M., Valianpoer, F., Fiey, G., Delattre, L., and Schulten, E. (2011) WO Patent 2011/064151.
63. Riffel, B. (2012) Field-to-Gate Energy and Greenhouse Gas Emissions Associated with Succinic Acid Produced at BioAmber's Facility in Sarnia, Ontario. Riffel Consulting, Prepared for BioAmber, http://www.bio-amber.com/ignitionweb/data/media_centre_files/804/Ontario_Canada_LCA_04.16.2013.pdf (accessed 29 September 2015).
64. Benjamin, C., Ioannis, T., Alexander, L.R., Martin, K.P. (2013) ICIS Chemical Business (Dec. 9–15), p. 27.
65. Bell, S. (2015) IHS Chemical Process Economics Program, Report 292, September, 2014.

8
2,5-Furandicarboxylic Acid Synthesis and Use

Ying-Ting Huang, Jinn-Jong Wong, Cheng-Jyun Huang, Chia-Ling Li, and Guang-Way Bill Jang

8.1
Introduction

There has been extensive investigation on the synthesis and production of 2,5-furandicarboxylic acid (FDCA) [1–5] due to its potential as a substitute for a wide variety of petrochemicals, such as adipic acid (AA) and terephthalic acid, as well as for biobased intermediates such as levulinic acid and succinic acid [6, 7]. Purified terephthalic acid (PTA) is a monomer used in the synthesis of polyethylene terephthalate (PET) for the fabrication of a wide range of plastic materials, including bottles, films, and fibers. PET bottles can be made partly biobased by using ethylene glycol (EG) derived from renewable resources. At present, synthesis of biomass-derived PTA [8, 9] remains a challenge. Fully biobased bottles can be realized by introducing FDCA to polyester synthesis. AA is a key precursor for the synthesis of another important plastic, nylon. In 2004, the US Department of Energy identified FDCA as 1 of 12 priority chemicals [7] for establishing the "green" chemistry industry of the future. FDCA is a versatile biobased building block synthesized by the catalytic oxidation of 5-hydroxymethylfurfural (HMF), esters and ethers of HMF, as well as 5-alkyl furfurals. Recent improvements in the production of HMF, FDCA, and their derivatives provide the potential for replacing petrochemicals and polymers with renewable materials [7]. Potential applications for FDCA include polyesters [10, 11], polyamides [12], polycarbonates, and plasticizers. Commercially viable production employing FDCA has yet to be realized due to the challenge of biomass feedstock supply, instability of HMF, and the difficult purification of both the final product FDCA and its intermediate. Despite several suppliers' plans to provide FDCA for scientific and customized needs, the current focus is on Avantium's announcement to produce the biobased building block at 1000 dollars per ton with a capacity between 30 000 and 50 000 ton per annum from 2016 using a reactive separation and catalyst.

Chemicals and Fuels from Bio-Based Building Blocks, First Edition.
Edited by Fabrizio Cavani, Stefania Albonetti, Francesco Basile, and Alessandro Gandini.

8.1.1
2,5-Furandicarboxylic Acid and Terephthalic Acid

Biomass-derived FDCA is a desirable alternative to oil-based PTA for the production of biobased polyesters to replace PET and related plastics. The differences and similarities in the molecular structures of FDCA and PTA are illustrated in Table 8.1. Products made of PET are commonly seen in consumer products and their components, such as plastic bottles for water and soft drinks, optical films for display, and fibers for clothing and carpets. A preliminary study indicates that FDCA-derived polyethylene 2,5-furandicarboxylate or polyethylene furanoate (PEF) has comparable mechanical properties and barrier properties several times higher than those of PET (Table 8.2). Special attention was placed on the use of PEF for soft drink and beer bottle production to extend product shelf life without a barrier coating. Polyesters with 100% biomass origin can be realized by polymerization of bioethanol-derived EG and FDCA. A strategic working group of five global companies, Coca-Cola, Ford Moto, Heinz, Nike, and

Table 8.1 Molecular structures of FDCA and PTA.

Monomer	Structure	M_w	Symmetry	Angle between the two –COOH
FDCA	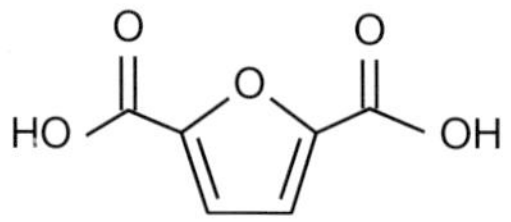	156.09	C_{2v}	130°
PTA	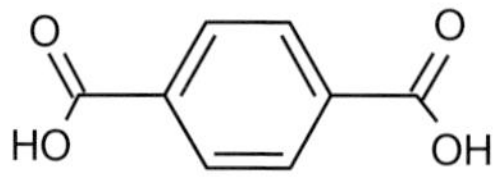	166.13	C_{2h}	180°

Table 8.2 Comparison of PEF and PET polyester physical properties [13–15].

Polymer	Structure	ρ (g cc^{-1})	T_g (°C)	T_m (°C)	T_d (°C)	Gas barrier properties		
						O_2	CO_2	H_2O
Poly(ethylene furanoate)	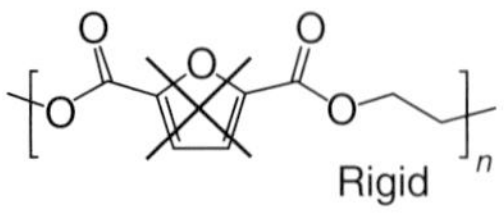 Rigid	1.4299	88	225	389	0.891	10.154	X2
Poly(ethylene terephthalate)	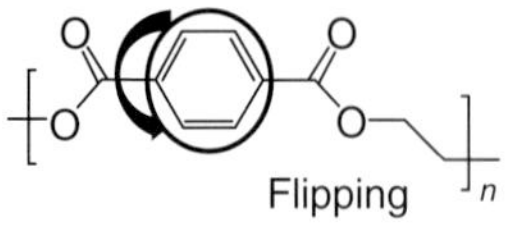 Flipping	1.3346	76	247	413	0.095	4.449	1

Procter & Gamble, was established in 2012 to develop 100% bioplastics using biomass-derived PTA or FDCA, with the two monomers typically obtained by oxidation of *para*-xylene (PX) and HMF. The foundation for this collaboration is Coca-Cola's PlantBottle technology. According to Grand View Research's estimate [6], synthesis of biobased polyesters has the highest market potential in the context of PET-related applications and will account for over 60% of the global FDCA consumption by 2020.

8.2 Synthesis of 2,5-Furandicarboxylic Acid by Oxidation of HMF

8.2.1 Aqueous Phase Oxidation of HMF

Water is a polar and low-cost solvent for HMF. Currently, the research focus is on the search for efficient catalysts and processing conditions to prepare FDCA in aqueous media (Table 8.3). Performing reactions in an environmentally benign solvent such as water is a critical step toward establishing sustainable manufacturing processes for FDCA production. Supported gold (Au) nanoparticles were observed to be very active in catalyzing aerobic oxidation of organic compounds [35]. Studies suggest that a strong basic condition was critical for the formation of the corresponding acid [36]. This poses a challenge for the process because the substrate HMF is unstable in alkaline aqueous solution. Bases are believed to increase the solubility of FDCA and prevent precipitation of the oxidation product onto the catalyst surface. Other difficulties for the synthesis of FDCA in water include structural changes and leaching of metallic catalysts.

Oxidation of HMF in basic condition (Scheme 8.1a) favors the selectivity of acids including 5-hydroxymethyl-2-furancarboxylic acid (HMFCA) and FDCA. On the other hand, oxidation proceeds through the oxidative dehydrogenation of the alcohol group of HMF to form 2,5-diformylfuran (DFF) under low pH conditions (Scheme 8.1b). The oxidation pathway and intermediates are governed by the oxidation conditions and catalyst use. Isotopic labeling investigation of the

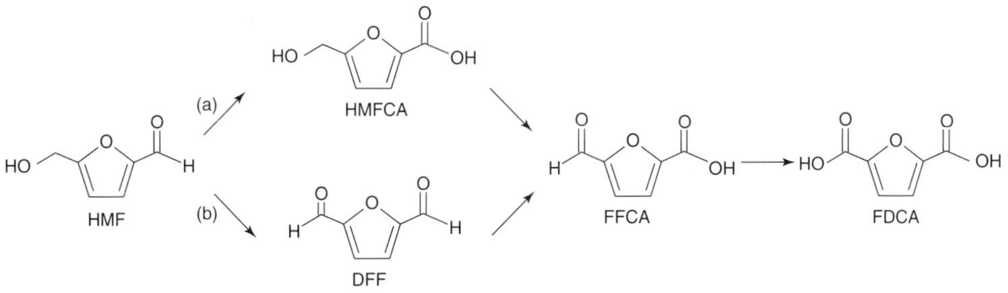

Scheme 8.1 Oxidation of HMF.

Table 8.3 Aqueous phase oxidation of HMF.

Catalyst	Base	Temperature (°C)	Time (h)	Oxidant	Pressure (bar)	Yield (%)				References
						DFF	FFCA	HMFCA	FDCA	
Ru/C	HT	150	1	O_2	20	0	3.6	—	78.2	[16]
Au/CNF–N	NaOH	22	24	O_2	6.9	0	0	100	0	[17]
						—	—	—	62	
Au/TiO_2	NaOH	22	22	O_2	20	0	0	21	79	[18]
Au/HY	NaOH	60	6	O_2	3	—	—	0	99	[19]
Au/HY-400								52	37	
Au–Cu/TiO_2	NaOH	95	4	O_2	10	—	—	—	90–99	[20]
Au–Cu/TiO_2	NaOH	60	4	O_2	10	—	—	—	57	[21]
Au–Cu/CeO_2									47	
Pt–Bi	Na_2CO_3	100	2.5	Air	40	—	<2	—	>90	[22]
Au/HT	—	95	7	O_2	1.013	—	0	0	>99	[23]
				Air			1	11	81	
Au–Pd/MgO	—	100	12	O_2	5	0	0.7	—	99	[24]
Au–Pd/CNT						0	1.7	—	94	
Au–Pd/HT						0	7.9	—	91	
Pt/C	NaOH	22	6	O_2	6.9	—	—	21	79	[25]
Au/C(sol)								93	7	
Pd/ZrO_2/La_2O_3	NaOH	90	8	O_2	1.013	—	0	7	90	[26]
γ-Fe_2O_3@HAP–Pd	K_2CO_3	100	6	O_2	—	—	—	—	92.9	[2]
Pd/C@Fe_3O_4	K_2CO_3	80	6	O_2	—	1.3	—	—	86.7	[27]
nano-Fe_3O_4–CoO_x [a)]	—	80	12	*t*-BuOOH	—	7.7	—	—	68.6	[28]
Merrifield resin-Co–Py [b)]	—	100	24	*t*-BuOOH	—	2.4	—	—	90.4	[29]
Fe–POP-1	—	100	10	Air	10	4	7	—	85	[30]
$CuCl_2$ [b)]	—	RT	48	Air	1	—	—	—	50	[31]
Au/TiO_2	NaOH	50	4	O_2	10	≦2	0.13	17.77	80.9	[32]
Pt, AgO, CuO	NaOH	24	5	O_2	—	—	—	—	99	[33]
Pt–Bi/C	$NaHCO_3$	100	6	Air	51	—	1	—	98	[34]

a) Solvent: DMSO.
b) Solvent: CH_3CN.

reaction mechanism suggests that water inserts oxygen into the product while oxygen acts as an electron scavenger [16–18]. Studies of the mechanism indicate that the reaction proceeds through the oxidation of the aldehyde moiety of HMF to form HMFCA, followed by oxidation of the alcohol group of HMFCA to aldehyde and then to carboxylic acid [17]. The presence of oxygen was critical for the conversion of HMFCA to FDCA, but it was not essential for the oxidation of HMF to HMFCA under high pH over Au/TiO_2 catalysts [18]. Hydroxide ions on the catalyst surface facilitate the transformation of the HMFCA alcohol side chain to aldehyde to form the intermediate 5-formyl-2-furancarboxylic acid (FFCA). Thus, high pH, catalyst loading, and oxygen pressure are required to obtain a high FDCA yield. Basic groups on carbon nanofibers render supported Au able to facilitate the formation of FDCA from HMF under mild conditions. There was 100% selectivity of HMFCA at 50% HMF conversion and 62% FDCA selectivity at 100% HMF conversion in the presence of 0.3 M NaOH at 295 °C [17]. Xu and coworkers [19] reported that the Au particle size affected the oxidation of HMF. Reduction of the Au particle size favors the selectivity of FDCA. Au nanoclusters were encapsulated in Y zeolite and demonstrated a superior catalytic activity for the selective oxidation of HMF to FDCA. Y zeolite has a unique supercage with a 0.74 nm open aperture and a 1.2 nm inner space which limits the growth of Au nanoclusters to an average size of ~1 nm. Studies suggest that the hydroxyl groups in the supercage also influence the size and distribution of Au nanoclusters. They obtained 99% FDCA yield under mild conditions of 60 °C and 3 bar oxygen pressure using Au nanoclusters confined in Y zeolite. The supported Au particle size on TiO_2 and channel-type zeolites (ZSM)-5 (silica alumina ratio = 38) was 10 and 20 nm, respectively. The oxidation products included 85% FDCA, 6% HMFCA, and 9% by-products when Au/TiO_2 was used as a catalyst under the same conditions. In the case of the ZSM-5, HMF was converted to 85% HMFCA and 15% by-products without formation of FDCA.

The catalytic activities of supported Au nanoparticles deteriorated rapidly during the oxidation process due to competitive adsorption onto the active phase and leaching of Au. Cavani and coworkers [20] successfully applied a bimetallic approach of supported Au–Cu on TiO_2 with improved stability and achieved a high FDCA yield of greater than 90% at relatively mild conditions, 10 bar, and 95 °C. The optimal Au/Cu weight ratio was about 1.14 : 0.36. The intermediate HMFCA yield reached a maximum after about 1 h in an HMF:metal loading:NaOH molar ratio of 1 : 0.01 : 4. The rate-determining step of the process was found to be the subsequent oxidation of the hydroxyl group of HMFCA. The conversion of FFCA to FDCA was very fast and a 99% FDCA yield was obtained after 4 h. It was suggested that Cu played a critical role in assisting dispersion of Au and/or as a promoter for the selective oxidation of HMF. The synergetic effect of Au and Cu on the selectivity of FDCA was substrate dependent. Although Au–Cu bimetallic catalysts demonstrated improved stability compared to monometallic Au on CeO_2, the FDCA yields were about 50 and 90%, respectively [21]. It was reported that a complete conversion of HMF exclusively to FDCA occurred within 2.5 h using a bimetallic catalyst, Pt–Bi on active carbon, in combination with a weak base, Na_2CO_3, under 40 bar air pressure at 100 °C [22]. The rate-determining step under these conditions is the oxidation of the aldehyde moiety of FFCA, and other intermediates observed include DFF and HMFCA. Degradation product formation often proceeds through the ring opening of HMFCA (Scheme 8.1a,b). The intermediate was quickly converted to FFCA. This may explain a high selectivity of FDCA. Besson and coworkers [1] achieved an FDCA yield of greater than 99% over Pt–Bi/TiO_2 (Bi/Pt = 0.22) at HMF/Pt molar ratio of 100 using Na_2CO_3 as a weak base within 5 h.

Hydrotalcite (HT) was employed to support Au nanoparticles for the selective oxidation of HMF to FDCA without addition of base to reaction media [23]. The hydrotalcite-supported Au nanoparticle (Au/HT) catalyst afforded 99% FDCA yield at 95 °C under ambient oxygen pressure. The catalyst was also effective in achieving 81% FDCA selectivity at full HMF conversion under air atmosphere. High concentrations of HMF feed had limited influence on the catalytic activity of Au/HT. FDCA selectivity was 83 and 72%, while HMFCA selectivity was 12 and 22%, respectively, when the initial feed increased from an HMF/metal mole ratio of 40–150 and 200. The study indicated that not only basicity of support but also formation of metal active sites played important roles in catalyzing the transformation of HMF to FDCA. This is because switching from HT to the other basic solid support, MgO, did not provide the same level of activity and FDCA selectivity. Initial oxidation of HMF to HMFCA (Scheme 8.1a) proceeded smoothly even at room temperature. As the temperature increased, the FDCA yield raised gradually to 99% at 95 °C.

Au catalysts are also inactive for the aerobic oxidation of alcohol in the absence of base additives or basic supports. Bimetallic catalysts were developed for the aerobic oxidation of HMF to FDCA under base-free conditions. It was known that Au–Pd alloys were very effective in catalyzing the oxidation of alcohols. Wang and coworkers [24] studied Au–Pd nanoparticles on various metal oxides, carbon, and hydrotalcite (HT) supports for the aerobic oxidation of HMF in water without base additives. Very high FDCA selectivity of 99%, 94%, and 91% at 100% HMF conversion was achieved for MgO, carbon nanotube (CNT), and HT supports, respectively. The trade-offs of operating at low pH value include higher reaction temperature (100 °C) and longer reaction time (12 h). The solid base supports, MgO and HT, are not stable in the reaction solution containing FDCA. Dissolution of the supports occurred. On the other hand, Au–Pd on CNT support was effective with high initial HMF concentration (0.15 M) and remained active for at least six recycling uses. Studies suggest that carbonyl and quinone groups on the CNT surface enhanced the absorption of HMF and intermediates and facilitated FDCA production. Carboxyl groups had an adverse effect on the conversion of HMF to FDCA. The optimal catalyst loading is 1 wt% at Au/Pd ratio of 1/1. The Pd moiety facilitated oxidation of the alcohol side chain and thus altered the mechanism through DFF intermediate (Scheme 8.1b) to FFCA followed by the formation of FDCA. Incorporation of Pd into the supported Au catalyst limited the formation of the HMFCA intermediate, which often leads to ring opening and degradation products.

Davis and coworkers [25] studied supported Pt, Pd, and Au catalysts for the oxidation of HMF and noticed the differences in reaction pathways. The NaOH/HMF ratio was optimized to 0.3 M/0.15 M to minimize side reactions, and the oxidation was carried out at 22 °C under 6.9 bar oxygen pressure. Au was not ready to oxidize the alcohol side chain under the reaction conditions. The conversion of HMF reached 100% after 1 h over the supported Au on C(WGC), C(sol), and TiO_2. HMFCA was the major product with 92–93% yield in 6 h. There was a limited quantity of FDCA formed and FFCA was not observed in the reaction mixture. The oxidation of HMF over a Pd/C catalyst apparently also proceeded through HMFCA and resulted in obtaining FDCA as a major product with a yield of 71%. The oxidation route was somewhat different over a Pt/C catalyst. Both HMFCA and FFCA were produced at about the same level and peaked at a time slightly earlier than that taken by HMF to reach full conversion at about 2 h. The HMF conversion rate over Pt/C was the slowest among the three metal catalyst studies but had the highest FDCA yield of 79% after 6 h. Supported Au catalysts do produce FDCA with suitable adjustment of the reaction conditions. The authors reported that Au/C (sol) did produce FDCA in a 21% yield at 50% HMF conversion when catalyst loading was lowered to 5.48×10^{-4} (metal:HMF) in order to obtain rate data unaffected by gas–liquid transport effects. The FDCA yield also increased to 36%, at 50% HMF conversion, over Au/TiO_2 at a catalyst loading of 3.57×10^{-4} as oxygen pressure increased to 30 bar. It was observed that high base concentration is necessary to oxidize the alcohol side groups of HMF or HMFCA. The challenge is degradation of products and intermediates in the presence of high base concentration. In this study, the major product was found to be 2,5-bis(hydroxymethyl)furan (BHMF) at high pH [25]. A high Au catalyst loading approach was applied with some degree of success in limiting the side reactions.

Steinfeldt and coworkers [26] identified ZrO_2/La_2O_3 as the best support among the substrates studied to stabilize Pd nanoparticles for the aerobic oxidation of HMF to FDCA. They achieved 90% FDCA yield at full conversion of HMF at 90 °C under ambient oxygen pressure. There was about 6% Pd leaching after 8 h of reaction. Studies indicated that the leached Pd was inactive in oxidizing HMFCA and FFCA intermediates. HMF was fully consumed after 1.5 h after HMFCA yield peaked at a slightly earlier stage. In the case of TiO_2-supported Pd nanoparticles, HMFCA oxidation stopped after 2 h while FFCA was continuously oxidized to FDCA. The active Pd surface sites might be partially blocked by adsorbed organic acids produced during the oxidation. This result suggests that the oxidation of HMFCA and FFCA could occur at different Pd surface sites. A magnetic palladium nanocatalyst was prepared for the selective aerobic oxidation of HMF into FDCA [2]. The catalyst was synthesized by ion exchange of Pd^{2+} with Ca^{2+} in hydroxyapatite-encapsulated magnetic γ-Fe_2O_3 (γ-Fe_2O_3@HAP) followed by reduction with $NaBH_4$ to produce γ-Fe_2O_3@hydroxyapatite (HAP)-Pd(0). γ-Fe_2O_3@HAP-Pd(0) can be easily separated from reaction medium for reuse. Deng and coworkers [2] achieved 92.9% FDCA yield at 97% HMF conversion using the Pd nanocatalyst in K_2CO_3 solution at 100 °C after 6 h. In a separated study, the catalyst of immobilized Pd on carbon shell of core–shell structure C@Fe_2O_3 provided 98.4% HMF conversion and 86.7% FDCA yield at 80 °C. The Pd nanoparticles' size and content were determined to be 10 nm and 3.85 wt%, respectively. In addition to FDCA, DFF was detected as one of the oxidation intermediates. The ratio of FDCA/DFF increases with increase of base dosage and temperature. FDCA selectivity decreases due to side reactions when temperature increases from the optimal of 80 to 100 °C. Cobalt nanoparticles that demonstrated high catalytic activity in alcohol oxidation were evaluated as a heterogeneous catalyst for the preparation of FDCA from HMF [27]. CoO_x nanoparticles were deposited onto the surface of Fe_3O_4 using $CoCl_2 \cdot 6H_2O$ in solution to prepare recyclable nano-Fe_3O_4–CoO_x catalysts. Investigations were carried out to evaluate oxidants and solvents for the oxidation process. The optimal reaction conditions identified are 100 mg nano-Fe_3O_4–CoO_x and 70 mg HMF in dimethyl sulfoxide (DMSO) using 0.5 ml of 70% aqueous *t*-BuOOH as oxidant at 80 °C. The FDCA yield obtained by the process is 68.6% at 97.2% HMF conversion [28]. Both DFF and HMFCA were identified as reaction intermediates with DFF as the major one. Zhang and coworkers [29] also evaluated cobalt(II)-*meso*-tetra(4-pyridyl)-porphyrin as a catalyst for the oxidation of HMF into FDCA. For the ease of separation, the catalyst was grafted onto the surface of a commercially available Merrifield resin support, which is soluble in reaction solvents such as acetonitrile. They achieved 95.6% HMF conversion and 90.4% FDCA yield at 100 °C after 24 h. The catalyst can be reused without significant loss of catalytic activity after washing with ethanol and vacuum drying at 60 °C.

An activated carbon-supported ruthenium (Ru/C) was evaluated as a catalyst for the aerobic oxidation of HMF in the presence of HT (Mg/Al = 3/1) to enhance FDCA selectivity [16]. An FDCA yield of 78.2% was achieved at 150 °C. However, the trade-off is degradation and condensation of HMF at high temperature and pH value. A highly cross-linked porphyrin-based porous organic polymer was synthesized as a basic support of iron (III) catalyst for the aerobic oxidation of HMF into FDCA in water [30]. DFF was the sole product obtained under atmospheric oxygen pressure at 70 °C. On the other hand, FDCA selectivity of 85% with 100% HMF conversion was achieved under 10 bar oxygen pressure at 100 °C when the feed contained 1.58 mmol HMF and 6 mg catalyst in 10 ml water. There were 4% DFF and 7% FFCA in the solution after 10 h of reaction.

Riisager and coworkers [31] employed a nonprecious metal, CuCl, as a catalyst and *tert*-butylhydroperoxide (*t*-BuOOH) as an oxidant for the conversion of HMF to FDCA in acetonitrile under ambient conditions. The HMFCA yield reached a maximum at 100% HMF conversion within a couple of hours and decreased gradually until it was fully consumed. An FDCA yield of 50% was achieved after a period of 48 h. It pointed to the formation of undetected by-products because the carbon balance was very low.

8.2.2 Oxidation of HMF in Acetic Acid

The advantage of applying the commonly known Amoco Mid-Century (MC) catalyst, $Co(OAc)_2$/$Mn(OAc)_2$/HBr, for aerobic oxidation of HMF into FDCA in acetic acid (Scheme 8.1b) is direct integration into existing PTA production lines. Acetic acid is a good solvent for oxidation substrates, such as HMF and HMF esters, but is not a good solvent for FDCA. This characteristic simplifies the subsequent separation and recovery of FDCA.

Initial evaluation of the selective oxidation of HMF to FDCA using the industrial practice of the synthesis of terephthalic acid achieved an FDCA yield of 60.9% at 100–125 °C and 70 bar air (Table 8.4) [3]. DFF was the main product at 75 °C (57 and 63%) either at ambient or high pressures. It suggests that acetoxylation of the alcohol moieties of HMF and HMFCA contributed to 5–8% yield loss. Oxidation of HMF using homogeneous metal salts and a bromide source under

Table 8.4 Oxidation of HMF in acetic acid.

Catalyst	Catalyst ratio	Temperature (°C)	Time (h)	Pressure (bar)	Gas	HMF conversion (%)	Yield (%)			References
							DFF	FFCA	FDCA	
Co/Mn/Br/Zr	1/1/2/—	125	3	70	Air	—	—	0	60.9	[3]
Co/Zn/Br	40/2.5/3	90	4	1.013	O_2	91	90	—	—	[37]
	40/11/44		3			90	—	29	60	
Co/Mn/Br[a)]	1/3.3/0.15	150	3	3 to >30	Air	—	—	—	73.4	[38]
				30		—	—	—	57.6	
				30					52.3	
Co/Mn/Br	1/0.05/1.11	132	2	8.96	Air	—	—	0.58	89.4	[4]
		130						0.16	90.2	
	1/0.05/0.74	130						0.02	88.8	
Co/Mn/Br	2.2/0.033/1.1	180	0.5	60	CO_2/O_2	>99	0.1	0	83.3	[39]
				30			—	—	89.6	
Co/Mn/Br	1/1/2	100	2	70	Air	—	—	3.3	44.8	[40]
Co/Mn/Br/Zr		105	12					2.5	58.8	
Co/Mn/Br	1/5/20	180	—	20	Air	100	—	—	72.2	[41]
Co/Mn/Br	1/1/1	120/150	2	30	Air	—	—	—	64.8	[42]
	4/1/1								77.7	

a) 30 g molecular sieve.

atmospheric pressure of O_2 at 90 °C does not progress beyond the formation of DFF [37]. The cocatalyst $Zn(OAc)_2$ instead of $Mn(OAc)_2$ was evaluated in this study to minimize the loss of the active bromide promoter as indicated in the oxidation of *para*-xylene. HMF conversion of 91% and 90% DFF in the resulting products were reported for a very low Br^- content catalyst composition (40 mM $Co(OAc)_2$/2.5 mM $Zn(OAc)_2$/3 mM NaBr). An acidic additive, trifluoroacetic acid (HTFA), was introduced to facilitate the formation of higher oxidation state products. Oxidation of HMF provided 60% FDCA and 29% FFCA in the presence of 1% HTFA using 40 mM $Co(OAc)_2$/11 mM $Zn(OAc)_2$/44 mM NaBr as the catalyst. However, the HMF conversion and FDCA formation decreased as the molar ratio of total metal catalysts to NaBr increased. Other studies suggest that high temperature and increasing oxygen pressure favor FDCA selectivity.

A practice of applying low substrate concentration and high catalyst loading to avoid water deactivation of catalyst limits industrial production of FDCA. A method for removing water generated during the oxidation process was developed by controlling the pressure in a reaction vessel below the vapor pressure of water and using a dehydration agent to absorb water [38]. An FDCA yield of 52.3% was obtained in the presence of 0.10 g of cobalt acetate tetrahydrate, 0.32 g of manganese acetate tetrahydrate, and 0.01 g of sodium bromide at 150 °C under 30 bar air pressure. Water produced during the oxidation process can be evaporated by lowering the pressure to 3 bar during the first hour of the reaction or by introducing a molecular sieve into the reaction mixture, thereby increasing FDCA yields to 55.1% and 57.6%, respectively. The yield was further increased to 73.4%

by combining the two approaches in one reaction. There was less effect on FDCA yield by regulating water at a higher temperature.

Eastman Chemical was able to push the FDCA yield to near 90% by optimizing the Co/Mn and Co/Br ratios of the catalyst system and controlling the temperature between 110 and 160 °C to minimize carbon burning [4]. They extended the feed to HMF esters in addition to HMF and HMF ethers. The highest yield achieved for HMF feed was 89.4% with a catalyst composition of 2000 ppm Co, 93.3 ppm Mn, and 3000 ppm Br using aqueous HBr as a source in a semibatch reaction. The color b^* values and FFCA yield were higher and the yield reduced when NaBr was used as the bromide source instead of HBr. The FDCA yield was 90.2% using the same catalyst composition and 5-(acetoxymethyl)furfural (AMF) as a feed at 130 °C. The catalyst composition was modified to 2500 ppm Co, 116.8 ppm Mn, and 2500 ppm Br to obtain 88.8% FDCA yield when using 5-(ethoxymethyl)furfural (EMF) as a feed. The presence of monocarboxylic acid by-products, including FFCA in the resulting FDCA solid, is not desirable because it terminates the chain growth during a polymerization process, resulting in a low molecular weight polymer. A process for purifying crude FDCA by means of hydrogenation of FFCA to aqueous-soluble products was patented [43]. The process comprises steps of crystallization, solid–liquid separation, dissolution, and mild hydrogenation. Crude FDCA was purified by catalytic hydrogenation of FFCA with carefully controlled temperature, pressure, residence time, and catalyst loading to avoid reduction of FDCA. The study results indicated that FFCA was converted to water-soluble products including tetrahydrofuran-2,5-dicarboxylic acid (THFDCA), tetrahydrofuran-2-carboxylic acid (THCA), and furan-2-carboxylic acid (FCA). FDCA can be easily separated from these by-products using conventional techniques. Alternatively, crude FDCA can react with excess alcohol to produce the corresponding diesters. The resulting crude diester contains the desired product, dialkyl-furan-2,5-dicarboxylate (DAFD), and by-products with boiling points different from that of the product, including 5-(alkoxycarbonyl)furan-2-carboxylic acid (ACFC), alkyl furan-2-carboxylate (AFC), alkyl-5-formylfuran-2-carboxylate (AFFC), and others [44]. The process involves esterification and purging of by-products to obtain purified DAFD for the synthesis of a variety of polyesters.

Archer Daniels Midland (ADM) employed an innovative spray oxidation process for the preparation of FDCA to minimize the temperature-related yield losses of substrate to by-products and burning of solvent [39]. In addition to being a good solvent for Co/Mn/Br catalysts and substrates, acetic acid also has a desired boiling point of 10–30 °C above the preferred temperature range for carrying out the oxidation process at reasonable pressures. Acetic acid serves not only as a solvent but also an evaporative heat sink to limit the rise of exothermic heat during the oxidation of substrates by suitable controlling of the operating pressure. The substrates and oxidation catalysts were introduced into the reaction chamber as a stream of mist made of a large number of tiny solution droplets. The droplets function as microreactors that contain furanic substrates and Co/Mn/Br catalysts dissolved in acetic acid. Each droplet should be sufficiently small to ensure diffusion of oxygen at stoichiometric amounts throughout the droplet for selective oxidation. The evaporative cooling mechanism was achieved by maintaining a vapor/liquid equilibrium for the solvent in the reactor. A predetermined amount of solvent, such as acetic acid, was discharged to the reactor before introducing the sprayable feed to avoid vaporization of the droplets and precipitation of substrates and catalysts. The preferable molar ratio of Co:Mn:Br is 1 : 1 : 2 and total catalyst concentration in the range of 0.8 – 1.2 wt% of the sprayable feed for the spray oxidation process. The degree of solvent and substrate burning monitored by analyzing CO and CO_2 yield in the gas phase increased as the oxidation temperature increased. The FDCA yield peaked at 180 – 190 °C. They achieved 83.3% FDCA yield for a 13.2 mmol 99% purity HMF under 60 bar ($CO_2/O_2 = 1$) at Co:Mn:Br mole ratio of 2.2 : 0.33 : 1.1. Water plays a different role at a high reaction temperature. The presence of a small amount of water up to about 10% by volume in the sprayable feed enhances FDCA yield by inhibiting burning of solvent and substrate at 180 °C. Further increase of water content resulted in generating intermediate FFCA and a decrease of FDCA yield. The promoting effect of cocatalyst $ZrO(OAc)_2$ at low temperatures, <160 °C, was diminished at 180 °C due to substantial solvent and substrate burning. The production of FDCA favored a lower pressure process by limiting burning of solvent and substrate. The FDCA yield was further increased to 89.6% by lowering the pressure to 30 bar. In separated experiments, liquid with entrained solid particles was withdrawn from the spray reactor and 85.5% FDCA yield was achieved with a Co:Mn:Br ratio of 1.3 mmol : 1.3 mmol : 3.5 mmol at 200 °C under CO_2/O_2 pressure of 15 bar. The resulting FDCA was obtained as solid precipitate from the separator and as filtrates from the separator and reactor. Temperature-related yield losses to by-products and solvent loss to burning were limited by means of precursor addition modes, optimization of catalyst composition, water content, and reaction pressure.

A crude dehydration solution contains about 21 wt% HMF and 0.3 wt% levulinic acid after extraction with ethyl acetate and concentration value increased to 60 and 2.6 wt%, respectively. Very low FDCA yield was obtained when crude HMF was used in a batchwise addition method due to rapid deactivation of the catalyst. This is overcome by continuously pumping acetic acid solution of crude HMF into the reactor at a predefined rate. In some cases, over 100% FDCA yields were obtained. For example, a 21% purity crude HMF containing 3.15 mmol HMF produced 3.22 mmol FDCA. Studies indicated that HMF dimer, 5,5′-[oxy-bis(methylene)]bis-2-furfural (OBMF), afforded about 40% FDCA conversion efficiency. On the other hand, a crude HMF with a purity of 60% containing 8.08 mmol HMF produced 7.28 mmol FDCA at 180 °C using a Co/Mn/Br catalyst at ratio of 2.2 mmol/0.33 mmol/1.1 mmol. Using crude HMF for the production of FDCA minimizes the loss of the thermally unstable raw material during purification. Some portions of by-products, such as HMF dimers, may be recovered to the desired product FDCA by the oxidation process. In addition to HMF and its dimer, levulinic acid is also present in the dehydration solution and can be oxidized to a valuable diacid, succinic acid. Thus, a crude dehydration product from the dehydration of fructose, glucose, or other carbohydrates can be directly oxidized to FDCA and succinic acid using the processes described above [45]. Oxidation of levulinic acid at 180 °C under 30 bar pressure using the same oxidant composition resulted in 12% yield of succinic acid at 99% levulinic acid conversion. Succinic acid is more soluble than FDCA in acidic acid at the reaction temperature. FDCA will be precipitated out and recovered first at higher temperatures as a substantially pure product, while succinic acid will be precipitated out with additional cooling of the reaction mixture.

It is a challenge to selectively oxidize alcohol functionality in the presence of a reactive aldehyde group on the same compound. DFF often was obtained as a minor by-product during oxidation of HMF using Co/Mn/Br catalyst. Du Pont's research team achieved a relatively high DFF selectivity of 68.3% at 92.2% HMF conversion at 75 °C using 406 ppm Co, 378 ppm Mn, and 1102 ppm HBr [40]. HMF was converted to FDCA with a yield of 44.8% using the same catalyst composition but instead raising the temperature to 100 °C. In the presence of a small amount of Zr, 20 ppm, HMF conversion was 99.7%, but DFF selectivity was reduced to 61.6% under the same reaction conditions. The maximum FDCA yield of 58.8% was achieved by tripling Co/Mn/Br catalyst concentration in the presence of 20 ppm Zr at 105 °C. The HMF esters are more easily recovered by distillation and extraction. The isolated HMF esters that may contain residual HMF can together be oxidized to FDCA in the presence of Co/Mn/Br catalysts and oxygen.

Mugweru and coworkers [46] synthesized spinel $Li_2CoMn_3O_8$ as a heterogeneous catalyst for the preparation of FDCA from HMF in acetic acid. The catalyst can be separated from the reaction products by dissolving the solid residue in 1 N NaOH and centrifugation. The technology has the advantages of utilizing earth-abundant materials and ease of catalyst recovery. They achieved 80% isolated yield of FDCA by performing the reaction at 150 °C under 55 bar oxygen pressure in the presence of cocatalyst sodium bromide (NaBr) after 8 h. DFF was obtained as the oxidation product when the reaction was carried out under atmospheric pressure. On the other hand, FFCA was the major product at lower temperatures.

At present, HMF and FDCA are not yet commercially available for downstream processing and application evaluations. A large quantity, often in kilogram scale, of HMF and derivatives is needed to synthesize functional additives for plastics and furanic polyesters for fabrication of films and bottles. We evaluated the dehydration of fructose and glucose to produce HMF in dimethylacetamide (DMAc) and achieved yields of 95% or higher. Studies were also carried out for sucrose, starch, and sugars from lignocellulose. Scale-up production of HMF was performed in a 200 l reactor with maximum operation temperature of 250 °C using fructose as the raw material and polymeric acid as the catalyst. HMF was isolated and purified as light yellowish crystals with a purity of greater than 99%. The production capacity is about 10–12 kg per batch when the reactor is half filled. The reaction facilities for the dehydration and oxidation processes as well as the isolated products are shown in Figure 8.1. The oxidation of HMF to FDCA process in acetic acid using MC catalyst was evaluated in a 1 l zirconium autoclave using batch and fed batch methods. Although production yields are about the same, the fed batch

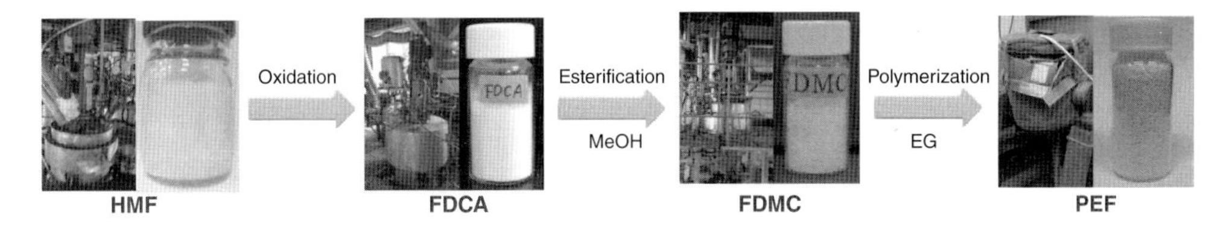

Figure 8.1 Facilities for the scale-up production of HMF and derivatives as well as related products.

method provides improved FDCA purity in the range of 95–97%, compared to 90–95% for batch-type synthesis. Pilot production of FDCA was performed in a 100 l titanium reactor with maximum pressure of about 50 bar. We achieved a 75% isolation yield at 92% purity and obtained 5–6 kg FDCA in 3 h. FFCA is identified as the major by-product with very small amount DFF based on high performance liquid chromatography (HPLC) studies. The crude FDCA is a yellowish powder. It is difficult to separate FDCA and FFCA by conventional recrystallization techniques. In order to synthesize high molecular weight fuanic polyesters, we adopted the dimethyl terephthalate (DMT) approach to obtain FDCA with high purity. Esterification of the crude FDCA with methanol resulted in a dimethyl ester of 2,5-furandicarboxylic acid (DMFDCA) with purity greater than 99%. DMFDCA can be used directly for the synthesis of polyesters through the transesterification method or converted to high-purity FDCA for polymerization.

8.2.3 Oxidative Esterification of HMF to 2,5-Furan Dimethylcarboxylate (FDMC)

FDCA is not soluble in most common solvents and it is difficult to isolate in the high purity required for polymerization. Its corresponding ester, 2,5-furan dimethylcarboxylate (FDMC) or 2,5-dimethylfuroate (DMF), is readily soluble in most solvents and can be obtained at high purities. FDMC can serve as an alternative to FDCA for the synthesis of polyesters. Several approaches for the conversion of HMF to FDMC were recently evaluated. Instead of using HMF as a feed for the production of FDCA, Furanix Technologies decided to utilize 5-alkoxymethylfurfural and/or 2,5-di(alkoxymethyl)furan as starting materials [41]. This resulted in producing a mixture of FDCA, mono- and/or diester of the compound. The mixture can be transformed to a diester of FDCA for the synthesis of biobased polyester by contact with a selected alcohol under appropriate conditions. Diesters of FDCA are ready to dissolve in most organic solvents and thus can be easily purified by conventional techniques. High purity is critical for obtaining high molecular weight polyesters. Ethers of HMF became desirable feeds for the synthesis of FDCA derivatives because of their stability in ambient environments and in oxidation media. In comparison studies, 5-methoxymethylfurfural (MMF) showed higher ester selectivity and overall furandicarboxylics selectivity than those of EMF in the presence of Co/Mn/Br catalyst with 60 bar air at 180 °C. The selectivity of furandicarboxylics increased with the decreasing of air pressure at the pressure range (60, 40, and 20 bar) studied. Metal/bromide catalysts containing low Br were most sensitive to air pressure change. In the case of MMF, the furandicarboxylics yields were 60.89%, 73.28%, and 76.66% with 0.78 mol% Co catalyst at Co/Mn/Br ratios of 1/5/5, 1/3/20, and 1/5/20, respectively. The highest yield, 82.27% (72.2% FDCA and 10.07% FDCA monomethyl ester), was achieved with a Co/Mn/Br ratio of 1/3/20 at 180 °C and 20 bar. Increased Br content enhanced the yield. However, corrosion can be a problem for commercial scale production if Br/(Co + Mn) ratio is greater than 1. The yield was about 77% (55% FDCA and 22% FDCA monomethyl ester) at a catalyst (Co + Mn) concentration of 4 mol% when Br/(Co + Mn) was set at 0.7 and Co/Mn in the range between 0.25 and 0.67. Undesired intermediate FFCA appeared and the yield of furandicarboxylics decreased when the feed concentration increased from 3.7 to 7.4% and 11.1%.

Corma and coworkers [47] reported that supported Au nanoparticle on nanoparticulated ceria (Au/CeO_2) is an effective catalyst for aerobic oxidation of HMF to FDMC with high selectivity using methanol as a solvent under mild conditions (65–130 °C, 10 bar oxygen) in the absence of base. Kinetics studies identified the oxidation of alcohol into aldehyde as the rate-determining step of the reaction. They observed a dependence of catalytic performance on the CeO_2 particle size due to the variation of chemical properties. The presence of oxygen vacancies and Ce^{III} on the surface renders nanoparticulated CeO_2 able to adsorb oxygen and thus enhance reoxidation processes. Stabilized positive gold atoms promote the initiation of the reaction, while the Lewis acid characteristic of unsaturated cerium atoms facilitates alcohol oxidation. Among the catalysts studied, Au/Fe_2O_3, Au/TiO, Au/C, and Au/CeO_2, supported Au on nanometer CeO_2 was the most active and afforded 99% FDMC yield after 5 h. Using a non-nanometer size CeO_2 support, it took 72 h to reach 85% FDMC yield. In the absence of oxygen, the Au/CeO_2 catalyst was capable of oxidizing HMF to the monoester alcohol and the acetal alcohol due to releasing and adsorbing oxygen through Ce^{+4}/Ce^{+3} redox processes. There were large quantities of organic deposits (>10 wt%) on the catalyst after the oxidation processes. However, catalytic activity can be recovered by oxidation at 250 °C for 12 h.

8.3
Synthesis of 2,5-Furandicarboxylic Acid from Carbohydrates and Furfural

The crude product mixture of a conventional acid dehydration of carbohydrates can be dissolved in acetic acid and converted FDCA using Co/Mn/Br oxidation catalysts. Because the dehydration step will produce water, the acetic acid would preferably be sufficiently concentrated so that the oxidation reaction mixture contains less than 10 wt% of water [39]. For this consideration, the water content of a cation exchange resin for the dehydration step also needs to be reduced. One-pot conversion of fructose into FDCA was performed via two consecutive steps of $Fe_3O_4@SiO_2{-}SO_3H$-catalyzed dehydration reaction using and nano-$Fe_3O_4{-}CoO_x$-catalyzed oxidation reaction in DMSO [28].

Dehydration of fructose proceeded rapidly in 2 h to achieve a 93.1% HMF yield at 90 °C. $Fe_3O_4@SiO_2-SO_3H$ was magnetically isolated prior to introducing nano-$Fe_3O_4-CoO_x$ and 70% aqueous *t*-BuOOH into the reaction mixture for the oxidation reaction at 80 °C. An FDCA yield of 59.8% based on the initial fructose feed was obtained after 15 h. There were 5.3% HMF and 2.6% DFF produced when both dehydration and oxidation catalysts were introduced into the starting fructose solution. This may be due to the oxidation of fructose to undesired by-products. Zhang and coworkers [48] developed a triphasic system for the direct conversion of sugars into FDCA using a modified H-shape reactor. Methyl isobutyl ketone (MIBK) serves as a solvent phase for extracting HMF. The solvents for dehydration and oxidation are tetraethylammonium bromide (TEAB) and water, respectively. Alternatively, water can be used for both reactions. For the dehydration of fructose, Amberlyst-15 was used as catalyst. In the oxidation chamber, Au_8Pd_2/HT and $NaCO_3$ were introduced. An overall FDCA yield of 78% was obtained with oxygen bubbling through oxidation chamber at 95 °C after 20 h. In the case of glucose, Amberlyst-15 and $CrCl_3$ were selected as catalysts. The dehydration reaction was first carried out at 120 °C for 30 min before the whole reactor was heated at 95 °C. An FDCA yield of 50% was achieved with full glucose conversion after 50 h.

Furfural is a commercially available chemical derived from agricultural by-products, including cane bagasse, corncobs, oat hulls, and sawdust, with good yield. Global production, 280 000 ton, is constrained due to the demand of the compound. Applications and value of furfural will be drastically improved by introducing functionality in the five-position. High yield was achieved for the key substitution processes. The challenge lies in the oxidation of disubstituted furan to diacid. Sanderson *et al.* [49] started with oxidation of furfural with oxygen and silver oxide in aqueous alkaline solution to produce furoic acid at an 87% yield after recrystallization. Esterification of furoic acid followed by region-selective chloromethylation of the resulting methyl furoate at the five-position resulted in formation of 5-chloromethylfuroate. FDCA was then obtained at 60% yield by oxidation of 5-chloromethylfuroate with concentrated nitric acid.

An alternative approach to the diacid starts with catalytic aerobic oxidation of furfural to furoate followed by catalytic disproportionation of furoate to furan and FDCA [50]. A previous report on the oxidation process was carried out in aqueous phase using heterogeneous catalysts and air or oxygen as an oxidant under alkaline conditions [51]. Pan *et al.* [50] achieved 93% selectivity of furoate at 100% furfural conversion using the nanoscale CuO at 60 °C for 30 min. The solid furoate was obtained by filtration to separate the catalyst and vacuum dried.

8.4
2,5-Furandicarboxylic Acid-Derived Surfactants and Plasticizers

A series of furan long-chain diesters, to be evaluated as an analog of the petroleum-based plasticizer di(2-ethylhexyl)phthalate (DEHP), was derived from crude FDCA synthesized from furfural. The high polarity of the furan ring as compared to the benzene ring in DEHP suggests improvement in compatibility with polyvinyl chloride (PVC) and possibly enhanced plasticizing efficiency. Raw material, dimethyl-2,5-furandicarboxylate, for the production of the biobased plasticizers was synthesized by acid-catalyzed diesterification of crude FDCA [49]. Other furan esters could be prepared by transesterifications of dimethyl-2,5-furandicarboxylate with 2-ethylhexanol, 2-octanol, hexanol, or butanol. Effectiveness of the furan diesters as plasticizers for PVC was evaluated based on their abilities in lowering glass transition temperature as a function of loading (ΔT_g/mass%). The values obtained for DEHP, di(2-ethylhexyl)furan-2,5-dicarboxylate, di(2-octyl)furan-2,5-dicarboxylate, dihexyl furan-2,5-dicarboxylate, and dibutyl furan-2,5-dicarboxylate were 2.45, 2.41, 2.23, 2.52, and 2.45 °C per mass% plasticizer, respectively. All four furan diesters exhibited comparable efficiency with improved compatibility with PVC.

The applications and processability of 2-ethylhexyl ester of FDCA are limited because the compound tends to crystallize at relatively low temperatures and 2-ethylhexanol is a hazardous substance. It is not suitable for manufacturing products intended for skin and/or food contact. In a series of patent applications, Evonik Oxeno GmbH demonstrated the use of isodecyl furan-2,5-dicarboxylate [52], isononyl furan-2,5-dicarboxylate [53], dipentyl furan-2,5-dicarboxylate [54], and diheptyl furan-2,5-dicarboxylate [55] as plasticizers for PVC with performance advantages. The diesters of FDCA were synthesized from FDCA via corresponding dichloride in two stages. The conversions of dichlorides to the corresponding esters are virtually quantitative. Alternatively, the diisononyl esters of FDCA can be obtained by transesterifying of an FDCA diester with an isononanol mixture. Differential scanning calorimetry (DSC) studies reveal a very different low-temperature behavior between bis(isononyl) furan-2,5-dicarboxylate and bis(2-ethylhexyl) furan-2,5-dicarboxylate. The isononyl ester compound shows a glass transition point at about −80 °C and no melting signals. This suggests that the isononyl ester compound does not turn solid and retains fluidity at low temperatures. On the other hand, the 2-ethylhexyl counterpart has a melting peak and is present in solid form even at temperatures above 0 °C. The diisononyl esters have a lower volatility and a lower increase in viscosity over time in plastisols when compared to those of FDCA esters based on 2-ethylhexanol. Plastisols containing the diisononyl esters exhibit higher gelling rates than those containing disononylphthalat (DINP). This indicates a lower processing temperature or a higher production rate when using the diisononyl esters as plasticizer. Thermally expandable plastisol formulations based on the diisononyl esters expand much faster than the formulations with DINP.

There is a trade-off in choosing optimal alkyl chain length for plasticizing efficiency. A rise in the incompatibility between plasticizer and polymer may occur due to an increase of the alkyl chain length of the esters. Nonuniform distribution of plasticizer in polymer matrices results in unforeseeable viscosity profiles, opaque appearance, and discoloration of the resulting films. Esters with short alkyl chains have high volatility and many esters of FDCA are room temperature crystalline solids, which can pose a difficulty for the production of liquid compositions. Dipentyl esters of FDCA were developed to overcome the above challenge. In addition to direct esterification of FDCA or transesterification of dimethyl esters of FDCA, dipentyl furan-2,5-dicarboxylate can be prepared by oxidative esterification of HMF. The DSC thermogram of dipentyl furan-2,5-dicarboxylate has a glass transition temperature of −14 °C and a melting point of +12 °C. The pentyl esters can be prepared and stored at room temperature in liquid form, which has advantages for polymer formulations. Transparent topcoat films for floor coverings use dipentyl furandicarboxylates in a PVC plastisol. The opacity of the above topcoat films is comparable to those comprising commercially available diisononyl (*ortho*)phthalate (DIPN, Vestinol®9 from Evonik Oxeno GmbH). This suggests superior compatibility between the dipentyl esters and the PVC matrices. Dipentyl furan-2,5-dicarboxylate also has the advantages of excellent gelation properties and a low dependence of the paste viscosity on the shear rate. Diheptyl furan-2,5-dicarboxylate was synthesized for the same purpose and has a glass transition temperature of −45 °C and a melting point of −19 °C. The plasticizer exhibits excellent compatibility with PVC. Topcoat films prepared from the formulation containing the heptyl esters have very low opacity values and yellowness indices.

8.5
2,5-Furandicarboxylic Acid-Derived Polymers

Polyesters are important polymers with applications in clothing and food packaging, such as beverage bottles. Global Industry Analysts estimated the global polyester market to reach 39.3 million ton by 2015. About 65% of polyester is produced for fibers, 30% for packaging, and 5% for film. Polyesters are typically synthesized by the esterification condensation of multifunctional alcohols and acids. The most familiar and commonly used polyester is PET. There is growing demand for polymeric materials from renewable resources due to concerns of crude oil shortage and global warming. Introduction of biodegradable biobased polymers for consumer and industrial applications provides solutions to not only dependency on fossil fuel-derived resources but also problems in waste management. A number of biobased polyesters, such as polylactic acid (PLA) and polyhydroxyalkanoates (PHAs), were developed, some at a commercial production scale, in the recent year. Biodegradable and compostable PLA is regarded as one of the most promising biopolymers with a large market potential, but its applications are limited to disposable products due to poor thermal stability and impact strength. PEF structure was revealed in the Celanese Corporation patent [10] shortly after introducing PET by the Calico Printers' Association [56]. Industrial production of PEF, however, was hindered by the supply of its raw material FDCA. Furanic polymers are derived from two basic furanic building blocks, furfural and HMF [57]. Studies suggest that PEF has comparable thermal properties and improved gas barrier properties to those of PET (Table 8.2) [13–15]. The ring-flipping motion in PEF is strongly hindered, which contributes to a reduction in diffusion coefficient and thus the large reduction in oxygen permeability for PEF according to chain mobility study [13].

8.5.1
Synthesis and Properties of Polyethylene Furandicarboxylate (PEF) and Related Polyesters

The first reported PEF was synthesized by transesterification of dimethyl ester of FDCA and 1.6 equiv of EG under ambient pressure between 160 and 200 °C followed by polycondensation under reduced pressure between 190 and 220 °C [10]. The resulting polyester has a melting point between 205 and 210 °C. An Osaka University team synthesized PEF with a melting point between 220 and 225 °C using lead as the catalyst [58]. Furanic polyesters consisting of long-chain diols have much lower melting temperatures, between 70 and 165 °C. In the patent application, Canon Kabushiki Kaisha disclosed the synthesis of PEF, polypropylene furandicarboxylate (PPF), and polybutylene furandicarboxylate (PBF) by reacting FDCA with the respective diol in the presence of the Sn/Ti catalyst [59]. Colored polymers were obtained after solid-state polymerization. Mitsubishi utilized the titanium tetrabutoxide/magnesium acetate mixed catalysts for the synthesis of PEF and PBF [60]. Moore and Kelly [61] obtained a white fibrous, film-forming polyester by reacting dimethyl FDCA with 1,6-hexane diol at a maximum temperature of 279 °C using the calcium acetate and antimony oxide catalyst.

Avantium employed a high-throughput film reactor technology to evaluate polycondensation of dimethyl FDCA and EG using the titanium isopropoxide catalyst. Strongly colored polymers were obtained when used FDCA for the polymerization [10, 14]. Drewitt and Lincoln suggested that this is related to decarboxylation of FDCA under the reaction conditions [10]. Transesterification was performed at 240 °C for 2.5 h for dimethyl FDCA and 210 °C for 5 h for FDCA, while polycondensation was carried out at 240 °C in both cases. The resulting PEFs have comparable molecular weight, $M_n \sim 13\,000$. Polycondensation was followed by solid-state polymerization at temperatures below the melting point of PEF to further increase molecular weight for selected applications. Scale-up production of PEF for the fabrication of fiber, film, and bottle was reported. Pilot-scale synthesized PEF was obtained by the direct esterification of FDCA and bio-EG for nonisothermal crystallization kinetics studies to avoid nucleating effects from residual transesterification catalysts [62].

Knoop *et al.* [11] synthesized FDCA-based PEF, PPF, and PBF for the evaluation of thermal and mechanical properties of the polyesters and comparison with PTA analogs, PET, polypropylene terephthalate (PPT), and PBT. PEF is obtained as a yellowish glass-like material, while PPF and PBF are colorless opaque. The glass transition temperature of PEF, PPF, and PBF are 77, 40, and 36 °C, respectively. In contrast to comparable T_gs, melting temperatures (T_ms) of the resulting polyesters are ~50 °C lower than the PTA-based analogs. PEF has a Young's modulus of 2450 MPa and maximum stress of 35 MPa, comparable to those of its industrial counterpart PET (2000 and 45 MPa, respectively). However, the polymer encounters poor impact strength and low elongation at break of 2.81%, similar to the case of PLA. Approaches for improving mechanical performance include increase of molecular weight without coloration, control of crystallinity, development of impact modifiers, and process optimization. Solid-state polymerization (SSP) that is carried out with polyesters in the crystalline state provides a viable method to increase the molecular weight of furanic polyesters under relatively mild conditions. Thus, crystallization kinetics and crystal morphology are critical factors governing the molecular weight increase of FDCA-based polyesters during SSP. Results of DSC studies indicate a dependence of melting temperature, T_m, with isothermal crystallization temperature. Multiple melting peaks with primary melting endotherms were located between 195 and 212 °C as the crystallization temperatures increased. Equilibrium melting temperature, T_m^0, measurements of the first and second melting endotherm revealed a set of very close values, 239.3 and 239.7 °C, respectively. This suggests the same crystal structure with the presence of imperfections and difference in crystal size. Knoop *et al.* carried out SSP for PEF after annealing at 140 °C. They observed a slow increase of molecular weight in the first 24 h of SSP and an increase in the rate of molecular weight increase in the second 24 h. The **defective** crystals rearranged into well-aligned larger crystals that concentrates the polymer chain ends in the amorphous phase during the first SSP stage. The chain end linkage reaction accelerates and a rapid increase in molecular weight occurs during the second 24 h of SSP.

In addition to 2,5-FDCA, 2,4-FDCA and a small quantity of 3,4-FDCA are obtained from solid-state disproportionation of 2-furoic acid [63]. The FDCA isomers 2,5-FDCA and 2,4-FDCA were isolated and purified to be used as building blocks for synthesis of biobased polyesters [15]. Thiyagarajan *et al.* [15] carried out polymerization of dimethyl esters of the FDCA isomers with linear aliphatic diols, including EG, 1,3-propanediol (1,3-PDO), 1,4-butanediol (1,4-BDO), and 2,3-butanediol (2,3-BDO). Melt polycondensation was performed at 160 °C for 12 h followed by further polymerization at 210–215 °C for 2 h under a reduced pressure of 0.02 mbar using titanium isopropoxide as the catalyst. Experimental results suggest that there is no correction between the position of the carboxylic acid groups on the furan ring and reactivity (yield, M_n) or selectivity (PDI, color). Diethylene glycol (DEG) residue in PEFs, about 1%, is lower than typically encountered during PET synthesis. They obtained a molecular weight (M_w) of 34 000–46 000 with high yields of 88–92% for EG series of polyesters. Polymerization of 2,5-FDCA and EG resulted in forming a semicrystalline polyester (2,5-PEF) with a glass transition temperature (T_g) of about 80 °C and melting temperature (T_m) of about 210 °C. The appearance of 2,5-PEF is opaque. On the other hand, a polymer based on the 2,4-isomer has T_g of about 72 °C and has no melting signal on DSC thermogram. The polymer 2,4-PEF is amorphous and thus appears translucent. Glass transition temperatures of the synthesized polyesters based on FDCA isomers decrease as the chain length of aliphatic diols increases. The polymers consisting of 2,4-isomers have comparable T_g with that of its 2,5-counterparts while T_g is significantly lower for the 3,4-analog. Translucent appearances of 2,4-PEF and 3,4-PEF suggest a low degree of crystallinity or very slow crystallization kinetics. The 2,4- and 3,4-isomer-based polyesters have comparable or even higher thermal stability compared to the 2,5-FDCA polyesters according to thermogravimetric analysis (TGA).

Synthesis of polyesters from FDCA often results in obtaining brownish colored polymers. Thus, we prepared PEF by transesterification of DMFDCA with EG followed by melt polymerization. The resulting polymer crystallized upon cooling when discharged from the reactor. DSC thermogram, Figure 8.2a, of the as-prepared polymer reveals a glass transition at above 80 °C followed by a melting peak at 217 °C. Crystallization of PEF through melting does not occur at a cooling rate of 10 °C min^{-1}. The peak intensity of the melting diminished sharply upon the second heating. The result suggests that the crystallization kinetic of PEF is relatively sluggish. A polymer film with superior optical properties was obtained when a PEF sheet was stretched 3 × 3 times. The resulting biaxially oriented polyethylene furanoate (BOPEF) film displays 91% transparency and 1.3 hazes. Solid-state polymerization was carried out at 200 °C for 48 h to obtain PEF with I.V. of 0.805. As shown in Figure 8.2b, both the glass transition temperature and melting point of PEF increase after SSP. The glass transition temperatures of PEF obtained after melt and solid-state polymerization are 87 and 90 °C, respectively.

8.5.2
Synthesis and Properties of Other Furanic Polyesters and Copolyesters

Moore and Kelly [61] carried out a systematic investigation in the reaction of FDCA and aliphatic or aromatic diols using melt transesterification

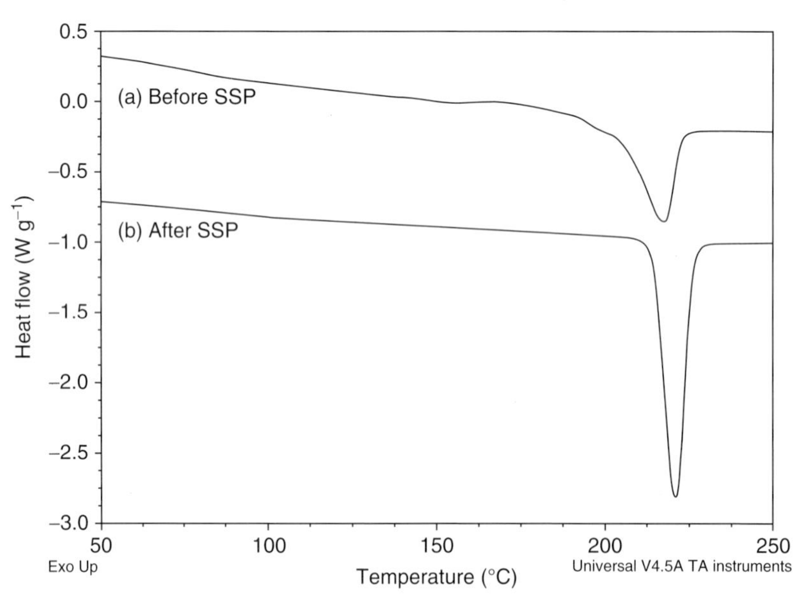

Figure 8.2 DSC thermograms of PEF from (a) melt polymerization and (b) solid-state polymerization.

and solution polymerization techniques. In addition to PEF, poly(2,5-furandiylcarbonyloxymethylene-2,5-furandiylmethyleneoxycarbonyl) and poly(2,5-furandiylcarbonyloxymethylene(*cis*-tetrahydro-2,5-furandiyl) methyleneoxycarbonyl) along with other furanic polyesters were prepared. They also studied the incorporation of furan rings into polyamides [64, 65]. Studies indicate that thermal properties and degradation behavior of furanic polymers differed widely according to their structures [57]. Furanic polyesters consisting of diphenylmethane are quite resistant to heat and atmospheric oxidation, while fully furanic polymers, poly(2,5-furandimethylene 2,5-furandicarboxylate), are quite fragile. Gandini *et al.* [66–68] extended the FDCA-based polyester series by introducing isosorbide, isoidide, bi(2,5-hydroxymethyl)-furan, bis-(1,4-hydroxymethyl)benzene, and hydroquinone for polymer synthesis. Isosorbide-based polyester was prepared by solution polycondensation in 1,1,2,2-tetrachloroethane (TCE) using 2,5-furandicarbonyl chloride as furan moiety [68]. The resulting polymer has a very high T_g of 180 °C due to rigid isosorbide structure. There was no melting detected in DSC thermograms up to 275 °C. Polyesters prepared from FDCA and hydroquinone display as highly rigid macromolecules that have a high degree of crystallinity but do not show any thermal transition up to 400 °C. On the other hand, a furanic polyester containing bis-(1,4-hydroxymethyl)benzene has a T_g not much higher than PEF.

Copolyesters derived from FDCA and PTA were prepared by using bis(hydroxyl)-2,5-furandicarboxylate and bis(2-hydroxyethyl)terephthalate as monomers and Sb_2O_3 as a catalyst [69]. Incorporation of both FDCA and PTA into the polymer chain disrupts the symmetry of the pristine homopolymers. At a 1/1 ratio, the copolyester has a glass transition temperature of 73.4 °C and displays no melting feature. When reducing the furanic units to 20%, the DSC thermogram of the copolyester shows a glass transition temperature at 62.4 °C, a cold crystallization at 125.2 °C, and a melting peak at 220.1 °C. Its thermal behavior is similar to that of PET with lower transition temperatures. AA and poly(tetramethylene glycol) (PTMG) were introduced to the polymer backbone to prepare a series of furanic elastomers and thermoplastics [70, 71]. PTA–AA-based copolyesters with tailored designs of biodegradability and physical properties have been commercialized by BASF and Eastman Chemical under the trademarks of Ecoflex® and Easter Bio®, respectively [70, 72]. Fully biobased counterparts can be realized by replacing PTA with FDCA. The poly (butylene adipate-*co*-butylene furandicarboxylate) could be synthesized by esterification at 190–210 °C under atmospheric pressure followed by polycondensation at 220–250 °C under a reduced pressure of 10 Pa. The resulting copolyesters have number average molecular weights between 26 400 and 41 300 and weight average molecular weights between 54 100 and 76 800. The copolyesters are amorphous at FDCA feed between 20 and 50 mol% while the crystal structure is the same as poly(butylene adipate) (PBA) at FDCA feed less than 20 mol% and the same as PBF at FDCA feed greater than 75 mol%. The amorphous copolyesters have elongation greater than 1000%. The tensile strength of samples with 40 and 50 mol% FDCA is about 20 MPa, which is better than that of PBA, 15 MPa. With a small amount of FDCA, 10 mol%, the copolyester is more ready to degrade than PBA due to reduced crystallinity. Copolyesters containing greater than 75 mol% of FDCA and PBF are not degradable by lipase under the testing conditions. Zhou *et al.* [71] designed biobased thermoplastic copolyester elastomers (TPEEs) using FDCA and PTMG for the synthesis of the hard and soft segments, respectively. The number average molecular weight achieved is in the range of 31 200 and 45 200. The PBF-PTMG copolyesters show typical characteristics of block copolymers. The biobased TPEEs have elongation at break in the range of 381–832% and are thermally stable at the temperatures below 280 °C. The mechanical properties can be improved by the thermal annealing process.

Copolyesters based on PEF and PLA were prepared for improved degradability [73]. At present, PLA is the most common biodegradable plastic derived from renewable resources. PEF-*co*-PLA copolyesters were synthesized by reacting bis(hydroxyethyl)-2,5-furandicarboxylate with PLA oligomer in bulk using Sb_2O_3 or a mixture of $SnCl_2{\cdot}2H_2O$ with p–toluenesulfonic (*p*-TSA) as a catalyst. Based on NMR studies, there are relatively higher quantities of furanic units incorporated into the copolyester than lactyl moieties. It is suggested that low relativity and depolymerization of PLA oligomer may be the causes. As lactyl moieties increase, water uptake increases, accompanied by the accelerated degradation of the copolyesters. However, PLA has a much lower degradation rate due to higher crystallinity.

A branched copolyester was prepared by reacting FDCA with glycerol in *N,N*-dimethyl formamide using Sb_2O_3 as a catalyst [74]. 2,5-Furan diacrylate was obtained by modifying 2,5-furan dimethanol with acryloyl chloride. Photocurable biobased resins were formulated by mixing 2,5-furan diacrylate with acrylated epoxidized soybean oil, acrylated castor oil, or acrylated 7,10-dihydroxy-8(*E*)-octadecenoic acid. Novel copolyesters were also synthesized from DFF using metal alkoxides [4]. The decomposition temperature of the resulting polymer is in the range of 100–120 °C.

8.6 Conclusion

Biomass-derived FDCA is a versatile renewable building block with great potential for a wide range of applications, including polymers, surfactants, and many specialty chemicals. FDCA is a desirable alternative to oil-based PTA for the synthesis of PET to produce soft drink bottles, fabrics for clothing, and optical films, as well as for making TPEEs to be used in air bag deployment doors, air intake ducting, tubing, and noise-absorbing connectors. Industrial applications are yet to be realized due to the challenges presented by biomass feedstock supply, instability of HMF, and a lack of a cost-effective synthetic process. Pilot production of FDCA in acetic acid using the MC catalyst is available for product development and manufacturing evaluations. It is possible to integrate the process into the existing PTA production line for commercial means. Academic studies focus on aqueous phase oxidation HMF using supported precious metals and many other catalysts. A high base concentration is necessary to oxidize the alcohol side groups of HMF and to prevent FDCA precipitation onto catalyst surface. This poses a challenge for aqueous phase process because HMF substrate is unstable in alkaline solution. FDCA is difficult to isolate in the high purity because it is not soluble in most common solvents. FDMC which can be obtained at high purities often serves as an alternative to FDCA for the synthesis of high molecular weight polyesters. Attention is currently focused on the synthesis of PEF due to its improved barrier performance and similar mechanical properties compared to PET. Strongly colored polymers were obtained due to decarboxylation of FDCA under the reaction conditions. Typical furanic polyesters were synthesized by transesterification of dimethyl ester of FDCA and a selected diol under ambient pressure followed by polycondensation under reduced pressure. For specific applications, such as blown bottles, SSP was performed in the crystalline state at temperatures below melting point to increase molecular weight of the biobased polyesters. PEF has a Young's modulus of 2450 MPa and maximum stress of 35 MPa, which are comparable to those of its industrial counterpart PET. To enhance its applications, impact strength of PEF will need to be improved. A wide range of furanic polymers with tailored properties can be derived from FDCA and diols, diamines, or others.

References

1. Ait Rass, H., Essayem, N., and Besson, M. (2015) *ChemSusChem*, **8**, 1206–1217.
2. Zhang, Z., Zhen, J., Liu, B., Lv, K., and Deng, K. (2015) *Green Chem.*, **17**, 1308–1317.
3. Partenheimer, W. and Grushin, V.V. (2001) *Adv. Synth. Catal.*, **343**, 102–111.
4. Janka, M.E., Parker, K.R., Shalkh, A.S., and Partin, L.R. (2014) Oxidation process to produce a crude dry carboxylic acid product. US Patent 2014/0256964 A1.
5. Carro, J., Ferreira, P., Rodríguez, L., Prieto, A., Serrano, A., Balcells, B., Ardá, A., Jiménez-Barbero, J., Gutiérrez, A., Ullrich, R., Hofrichter M., Martínez, A.T., *FEBS J.* DOI: 10.1111/febs.13177.
6. 2014) *Furandicarboxylic Acid (FDCA) Market Potential Analysis and Segment Forecasts to 2020*, Grand View Research. ISBN: 978-1-68038-043-9
7. Bozell, J.J. and Petersen, G.R. (2010) *Green Chem.*, **12**, 539–554.
8. Pacheco, J.J. and Davis, M.E. (2014) *Proc. Natl. Acad. Sci. U.S.A.*, **111**, 8363–8367.
9. Frost, J.W. (2014) Synthesis of biobased terephthalic acids and isophthalic acids. WO Patent 2014144843 A1.
10. James, G.N.D. and James, L. (1946) Improvements in polymers. GB Patent 621,971.
11. Knoop, R.J.I., Vogelzang, W., van Haveren, J., and van Es, D.S. (2013) *J. Polym. Sci., Part A: Polym. Chem.*, **51**, 4191–4199.
12. Chan, J.W., Nederberg, F., Rajagopalan, B., Williams, S. R., and Cobb, M.W. (2015) Furan based polyamides. US Patent 20150044927 A1.
13. Burgess, S.K., Leisen, J.E., Kraftschik, B.E., Mubarak, C.R., Kriegel, R.M., and Koros, W.J. (2014) *Macromolecules*, **47**, 1383–1391.
14. Jong, E.D., Dam, M.A., Sipos, L., and Gruter, G.J.M. (2012) *Bio-based Monomers, Polymers, and Materials*, ACS Symposium Series, American Chemical Society, Washington, DC, pp. 1–13.
15. Thiyagarajan, S., Vogelzang, W., Knoop, R.J., Frissen, A.E., van Haveren, J., and

van Es, D.S. (2014) *Green Chem.*, **16**, 1957–1966.

16. Xie, J., Nie, J., and Liu, H. (2014) *Chin. J. Catal.*, **35**, 937.
17. Davis, S.E., Benavidez, A.D., Cosselink, R.W., Bitter, J.H., de Jong, K.P., Datye, A.K., and Davis, R.J. (2014) *J. Mol. Catal. A: Chem.*, **388–389**, 123–132.
18. Davis, S.E., Zope, B.N., and Davis, R.J. (2012) *Green Chem.*, **14**, 143–147.
19. Cai, J., Ma, H., Zhang, J., Song, Q., Du, Z., Huang, Y., and Xu, J. (2013) *Chem. Eur. J.*, **19**, 14215–14223.
20. Pasini, T., Piccinini, M., Blosi, M., Bonelli, R., Albonetti, S., Dimitratos, N., Lopez-Sanchez, J.A., Sankar, M., He, Q., Kiely, C.J., Hutchings, G.J., and Cavani, F. (2011) *Green Chem.*, **13**, 2091.
21. Albonetti, S., Lolli, A., Morandi, V., Migliori, A., Lucarelli, C., and Cavani, F. (2015) *Appl. Catal., B: Environ.*, **163**, 520–530.
22. Ait Rass, H., Essayem, N., and Besson, M. (2013) *Green Chem.*, **15**, 2240–2251.
23. Gupta, N.K., Nishimura, S., Takagaki, A., and Ebitani, K. (2011) *Green Chem.*, **13**, 824–827.
24. Wan, X., Zhou, C., Chen, J., Deng, W., Zhang, Q., Yang, Y., and Wang, Y. (2014) *ACS Catal.*, 2175–2185.
25. Davis, S.E., Houk, L.R., Tamargo, E.C., Datye, A.K., and Davis, R.J. (2011) *Catal. Today*, **160**, 55–60.
26. Siyo, B., Schneider, M., Radnik, J., Pohl, M.M., Langer, P., and Steinfeldt, N. (2014) *Appl. Catal., A*, **478**, 107–116.
27. Liu, B., Ren, Y., and Zhang, Z. (2015) *Green Chem.*, **17**, 1610–1617.
28. Wang, S., Zhang, Z., and Liu, B. (2015) *ACS Sustainable Chem. Eng.*, **3**, 406–412.
29. Gao, L., Deng, K., Zheng, J., Liu, B., and Zhang, Z. (2015) *Chem. Eng. J.*, **270**, 444–449.
30. Saha, B., Gupta, D., Abu-Omar, M.M., Modak, A., and Bhaumik, A. (2013) *J. Catal.*, **299**, 316–320.
31. Hansen, T., Sadaba, I., Garcia-Suarez, E.J., and Riisager, A. (2013) *Appl. Catal., A*, **456**, 44–50.
32. Koteswara, R.V. and Peter, S. (2012) *Catal. Today*, **195**, 144–154.
33. Baak, W.L. and Wilmington, D. (1967) Method of producing dehydromucic acid. US Patent 3326944 A.
34. Hicham, A.R., Nadine, E., and Michèle, B. (2013) *Green Chem.*, **15**, 2240–2251.
35. Stratakis, M. and Garica, H. (2012) *Chem. Rev.*, **112**, 4469–4506.
36. Gorbanev, Y.Y., Klitgaard, S.K., Woodley, J.M., Christensen, C.H., and Riisager, A. (2009) *ChemSusChem*, **2**, 672.
37. Saha, B., Dutta, S., and Abu-Omar, M.M. (2012) *Catal. Sci. Technol.*, **2**, 79–81.
38. Yutaka, K., Miura, T., Eritate, S., and Komuro, T. (2012) Method of producing 2,5-furandicarboxylic acid. US Patent 8,242,292 B2.
39. Zuo, X., Subramaniam, B., Busch, D., and Venkitasubramaniam, P. (2013) Spray oxidation process for producing 2,5-furandicarboxylic acid from hydroxymethylfurfural. WO Patent 2013/033058 A1.
40. Grushin, V., Manzer, E., and Partenhelmer, W. (2014) Processes for preparing diacids, dialdehydes and polymers. US Patent 8,748,637 B2.
41. De Diego, C.M., Schammel, W.P., Dam, M.A., and Gruter, J.M. (2013) Method for the preparation of 2,5-furandicarboxylic acid and esters thereof. US Patent 8,519,167 B2.
42. Ebihara, Y. and Fujibayashi R. (2009) Process for producing 2,5-furandicarboxylic acid. JP Patent 5252969 B2.
43. Shaikh, A., Parker, K.R., Janka, M.E., and Partin, L.R. (2013) Process for purifying crude furan 2,5-dicarboxylic acid using hydrogenation. US Patent 8,748,479 B2.
44. Partin, L.R., Shaikh, A., Janka, M.E., and Parker, K.R. (2013) Method for producing purified dialkyl-furan-2,5-dicarboxylate vapor. US Patent 8,658,810 B2.
45. Subramaniam, B., Zuo, X., Busch, D.H., and Venkitasubramaniam, P. (2013) Process for producing both succinic acid and 2,5-furandicarboxylic acid. WO Patent 2013033081 A2.
46. Jain, A., Jonnalagadda, S.C., Ramanujachary, K.V., and Mugweru, A. (2015) *Catal. Commun.*, **58**, 179–182.
47. Casanova, O., Iborra, S., and Corma, A. (2009) *J. Catal.*, **265**, 109–116.

48. Yi, G., Teong, S.P., and Zhang, Y. (2015) *ChemSusChem*, **8**, 1151–1155.
49. Sanderson, R.D., Schneider, D.F., and Schreuder, I. (1994) *J. Appl. Polym. Sci.*, **53**, 1785–1793.
50. Pan, T., Deng, J., Xu, Q., Zuo, Y., Guo, Q.X., and Fu, Y. (2013) *ChemSusChem*, **6**, 47–50.
51. Harrisson, R.J. and Moyle, M. (1956) *Org. Synth.*, **36**, 36–37.
52. Michael, G. and Hinnerk, G.B. (2011) 2,5-furan dicarboxylates comprising isodecanols, and use thereof. WO Patent 2011023491 A1.
53. Michael, G. and Hinnerk, G.B. (2011) 2,5-furan dicarboxylate derivatives, and use thereof as plasticizers. WO Patent 2011023590 A1.
54. Michael, G. and Hinnerk, G.B. (2012) Pentyl esters of furandicarboxylic acid as softeners. WO Patent 2012113608 A1.
55. Hinnerk G.B., Michael, G., Andre, H., and Michael, W.-F. (2012) Heptyl esters of furan dicarboxylic acid as softeners. WO Patent 2012113609 A1.
56. John, R.W. and James, T.D. (1941) Improvements relating to the manufacture of highly polymeric substances. GB Patent 578,079.
57. Gandini, A. and Belgacem, M.N. (1997) *Prog. Polym. Sci.*, **22**, 1203–1379.
58. Hachihama, Y., Shono, T., and Hyono, K. (1958) *Technol. Rep. Osaka Univ.*, **8**, 475–480.
59. Matsuda, K., Matsuhisa, H., Horie, H., and Komuro, T. (2007) Polymer compound and method of synthesizing the same PCT patent application. WO Patent 2007052847 A1.
60. Kato, S. and Kasai, A. (2008) Method for producing polyester resin including furan structure. JP Patent 5233390 B2.
61. Moore, J.A. and Kelly, J.E. (1978) *Macromolecules*, **11**, 568–573.
62. Codou, A., Guigo, N., van Berkel, J., de Jong, E., and Sbirrazzuoli, N. (2014) *Macromol. Chem. Phys.*, **215**, 2065–2074.
63. Thiyagarajan, S., Pukin, A., van Haveren, J., Lutz, M., and van Es, D.S. (2013) *RSC Adv.*, **3**, 15678–15686.
64. Moore, J.A. and Partain, E.M. (1983) *Macromolecules*, **16**, 338–339.
65. Moore, J.A. and Kelly, J.E. (1984) *J. Polym. Sci. Polym. Chem. Ed.*, **22**, 863–864.
66. Gandini, A., Coelho, D., Gomes, M., Reis, B., and Silvestre, A. (2009) *J. Mater. Chem.*, **19**, 8656–8664.
67. Gandini, A. (2010) *Polym. Chem.*, **1**, 245–251.
68. Gomes, M., Gandini, A., Silvestre, A.J.D., and Reis, B. (2011) *J. Polym. Sci., Part A: Polym. Chem.*, **49**, 3759–3768.
69. Sousa, A.F., Matos, M., Freire, C.S.R., Silvestre, A.J.D., and Coelho, J.F.J. (2013) *Polymer*, **54**, 513–519.
70. Zhou, W., Wang, X., Yang, B., Xu, Y., Zhang, W., Zhang, Y., and Ji, J. (2013) *Polym. Degrad. Stab.*, **98**, 2177–2183.
71. Zhou, W., Zhang, Y., Xu, Y., Wang, P., Gao, L., Zhang, W., and Ji, J. (2014) *Polym. Degrad. Stab.*, **109**, 21–26.
72. Vroman, I. and Tighzert, L. (2009) *Materials*, **2**, 307–344.
73. Matos, M., Sousa, A.F., Fonseca, A.C., Freire, C.S.R., Coelho, J.F.J., and Silvestre, A.J.D. (2014) *Macromol. Chem. Phys.*, **215**, 2175–2184.
74. Amarasekara, A.S., Razzaq, A., and Bonham, P. (2013) *ISRN Polym. Sci.*, **2013**, Article ID 645169.

9
Production of Bioacrylic Acid

Benjamin Katryniok, Thomas Bonnotte, Franck Dumeignil, and Sébastien Paul

9.1
Introduction

With a growing demand of 4–5% per year and a forecasted annual production of almost 6 million tons for 2017, acrylic acid (AA), also called *2-propenoic acid* (Chemical Abstracts Service (CAS) number [79-10-7]), is one of the major chemicals issued from the organic chemical industry [1]. Today, while some processes of production are based on acetylene [2], acrylonitrile (in fact derived from propylene), or ethylene cyanohydrin [3], the most conventional route for producing acrylic acid starts from propylene as a raw material [4]. Note that some studies also report the seducing (but challenging) possibility of using propane as a starting material, but the actual yields reached after more than 20 years of research on the subject are still far from being sufficient to be able to envision the commercialization of the process [5].

Propene-based processes have been first developed in a one-step approach and then split in two steps, which is now the most largely spread-out technology at the industrial scale. In the two aforementioned approaches, propene is thus oxidized in one or two catalytic steps, but in any case via acrolein as an intermediate (Scheme 9.1).

The one-step process was replaced by the two steps one due to a better yield (85% vs. about 60%). Further, the catalyst used for the one-step process, mainly composed of molybdenum oxide supported on tellurium oxide, relatively rapidly deactivates. In the two-step process, propene is first oxidized to acrolein in a multitubular fixed-bed reactor at a temperature between 330 and 430 °C in the presence of a BiMo mixed oxide catalyst. The yield of this first step is more or less 85%. In the second step, the formed acrolein is very efficiently reacted with oxygen between 200 and 300 °C over a complex catalyst containing Mo and V oxides, with a yield close to 100%. This catalyst has been progressively improved by addition of elements such as Fe, Ni, Cu, Te, and so on, in order to activate the reaction at lower temperatures and to further improve its productivity [3]. Different formulations have been patented by Celanese [6], Nippon Kayaku [7], BASF [8], Nippon Shokubai [9], and so on. AA is then isolated and purified by absorption in

Chemicals and Fuels from Bio-Based Building Blocks, First Edition.
Edited by Fabrizio Cavani, Stefania Albonetti, Francesco Basile, and Alessandro Gandini.

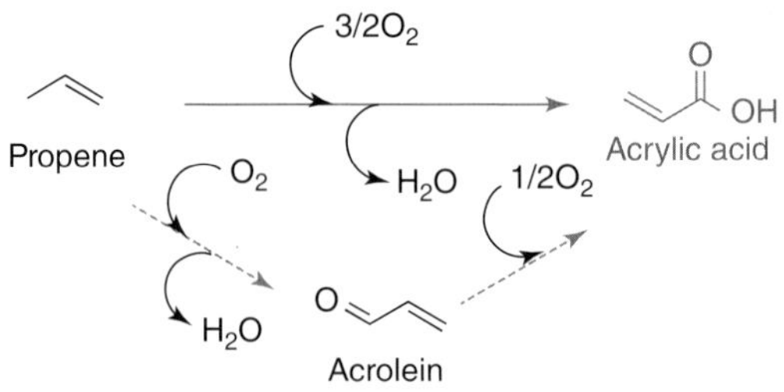

Scheme 9.1 Chemical routes for the one-step (solid arrow) and two-step (dashed arrows) propene oxidation process to acrylic acid [3].

water, extraction by a solvent, and a subsequent distillation, whereby the so-called glacial AA is obtained. Roughly, one third of the produced AA is polymerized to yield superabsorbents, for instance, used in babies' diapers. The two other thirds are esterified using methanol, ethanol, or butanol to yield acrylates, which are mainly used for paints and coatings production. The top AA producers are Dow Chemical, BASF, Arkema, Nippon Shokubai, Jiangsu Jurong Chemical Corp., and LG Chem.

Recently, academic and industrial researchers have paid a lot of attention to the development of a sustainable way for AA production either by a chemical or a biochemical route [10]. This chapter gives a synthetic view of the state of the art on the production of the so-called bioacrylic acid (bio-AA).

Bio-AA can be obtained following chemical routes starting from renewable compounds such as glycerol (GLY) (Section 9.2.1), lactic acid (LA) (Section 9.2.2) or C6 sugars, and ethanol (Section 9.2.3). GLY is the main by-product of the biodiesel production. LA can be easily obtained by a fermentation process based on compounds issued from biomass such as sugars or starch. LA can also be produced from GLY, as described in Section 9.2.2.1. GLY can be oxidized in one step to AA (Section 9.2.1.1) or using an indirect pathway, where acrolein, 3-hydroxypropionaldehyde, or acrylonitrile can be, respectively, used as intermediates (Section 9.2.1.2). LA is generally dehydrated to AA using a heterogeneously catalyzed process (Section 9.2.2.2). Another way of production of bio-AA is the biochemical route, which is described in Section 9.3.

9.2 Chemical Routes

9.2.1 Production of AA from GLY

Since the biodiesel boom at the end of the last century, GLY has come in the focus of research as a starting material for various chemicals such as 1,3-propanediol, acrolein, dihydroxyacetone, or epichlorohydrin. Starting from GLY, several pathways for the synthesis of AA are described in the literature: direct and indirect ones, meaning for these latter via intermediates such as acrolein, 3-hydroxypropionaldehyde, or acrylonitrile. These pathways will be discussed in detail in the following paragraphs.

9.2.1.1 Direct Pathway from GLY to AA

Various catalysts have been reported so far for the direct conversion of GLY to AA (Scheme 9.2), which is also referred as *oxidehydration* since it is generally assumed to proceed via the simultaneous dehydration of GLY to acrolein and its further oxidation to AA.

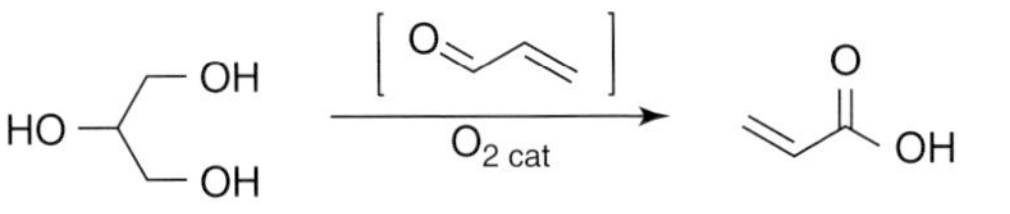

Scheme 9.2 Glycerol oxidehydration reaction to acrylic acid.

With respect to the supposed reaction mechanism of the oxidehydration reaction, the catalyst is bifunctional with acid properties for the dehydration reaction and oxidative properties for the successive oxidation of the intermediately formed acrolein to finally yield AA. Notably, mixed oxides from vanadium, molybdenum, tellurium, and tungsten were studied by several authors.

One of the first studies was performed by Ueda *et al.* [11] over molybdenum vanadate, tungsten vanadate, and a complex quaternary mixture of molybdenum, vanadium, tellurium, and niobium. The composition of the latter was quite close to those of catalysts used in the selective oxidation of acrolein to AA. Hence, it was not surprising that the best performance was observed over this complex catalyst, yielding 28% of AA. In all cases, the formation of acrolein remained very limited (<3%), whereby one cannot clearly conclude if the latter is really the intermediate in the reaction mechanism.

Following their initial work, Ueda *et al.* [12] modified a niobium–tungsten–vanadium mixed oxide catalyst by impregnation with phosphoric acid. On the one hand, the modification with phosphoric acid increased the acidity of the catalyst, as expected, and, on the other hand, the phosphoric acid is also supposed to moderate the redox properties of vanadium, thus inhibiting the overoxidation of the intermediately formed AA to carbon oxides (CO and CO_2). With this catalyst, an AA yield of 59% was obtained, which is so far the highest yield ever reported for the direct oxidation of GLY to AA.

The molybdenum–vanadium–tungsten system was also studied by Zhang *et al.* [13] who employed the binaries molybdenum–vanadium oxides and tungsten–vanadium oxides. Concerning these catalysts, the selectivity to AA and acrolein showed an inverse relationship: while the yield in acrolein increased with the vanadium content, the yield in AA increased when decreasing the vanadium content. In fact, high amounts of vanadium promoted the formation of carbon oxide, supposedly easily formed from AA degradation. Thus, the highest yield in

AA was 20% over the molybdenum–vanadium oxide catalyst and 26% over the tungsten–vanadium oxide catalyst.

Later, the molybdenum–vanadium–tungsten oxide system was deeply characterized and modeled by Yi *et al.* [14] using the Density Functional Theory (DFT) method. First, the authors modeled the catalyst before synthesizing and characterizing it. The as-obtained catalyst behaved identically to the modeled system and yielded 30% in AA. The authors established a correlation describing the influence of the tungsten content on the acidity as well as on the formation of AA and acrolein. Without any surprise, the yield in acrylic products increased with the tungsten content, notably due to the increased formation of acrolein over the acid sites of tungsten. Furthermore, an increase in the tungsten content also decreased the formation of carbon oxides, due to the larger amount of vanadium 4+ species.

Liu *et al.* [15] combined molybdenum vanadate with alumina-supported silicotungstic acid to prepare a bifunctional catalyst. The catalytic performance was significantly depending on the amount of molybdenum vanadate (Mo_3VO_x), since the latter was required for the oxidation step. Thus, the yield in oxidation products increased with the Mo_3VO_x content, whereas the formation of acrolein decreased. On the other hand, the selectivity to AA was the highest at a loading of 17 wt%, since at higher loadings the formation of carbon oxides (CO and CO_2) became predominant. Furthermore, the influence of the calcination temperature on the catalytic performance was studied, showing that calcination at high temperature (>450 °C) shifted the selectivity from AA to acrolein, which can be explained by the decomposition of silicotungstic acid to form silica and tungsten oxide, which is known as a *strong Lewis acid*. After all, the maximum observed yield in AA was 15%.

Possato *et al.* [16] also followed the idea of adding redox properties to an acid catalyst, using a zeolite doped with vanadium oxide (V_2O_5), but rather focused on the mechanistic aspects of the reaction rather than on the performance evaluation. The authors demonstrated by X-ray Photoelectron Spectroscopy (XPS) that the redox couple V^{4+}/V^{5+} was responsible for the oxidation, whereby a Mars and van Krevelen mechanism could be assumed.

Cavani *et al.* [17, 18] studied mixed oxides from molybdenum, vanadium, niobium, and tungsten with different relative ratios and structures as catalysts, whereby the catalytic performance was strongly depending on the reaction temperature. As a general trend, the formation of carbon oxides (CO and CO_2) significantly increased with the reaction temperature, whereby the yield in AA was in all cases between 20% (270 °C) and 55% (400 °C). On the other hand, at low temperature, the formation of AA was favored. Another important factor was the oxygen to GLY ratio. As expected, low oxygen concentrations (oxygen/GLY ratio of 1) favored the formation of acrolein, whereas high concentrations (oxygen/GLY ratio of 3) promoted the formation of carbon oxides. Thus, the best performance (50% yield in AA) was observed for an oxygen/GLY ratio of 2, which was also favorable to avoid the formation of heavy products leading to the deactivation of the catalyst [19].

9.2.1.2 Indirect Pathways from GLY to AA

With respect to the rather limited yields generally observed in the direct conversion of GLY to AA, the focus of the research moved more and more to indirect pathways. The most widely used intermediates in this case are acrolein, 3-hydroxypropionaldehyde, and acrylonitrile (Scheme 9.3).

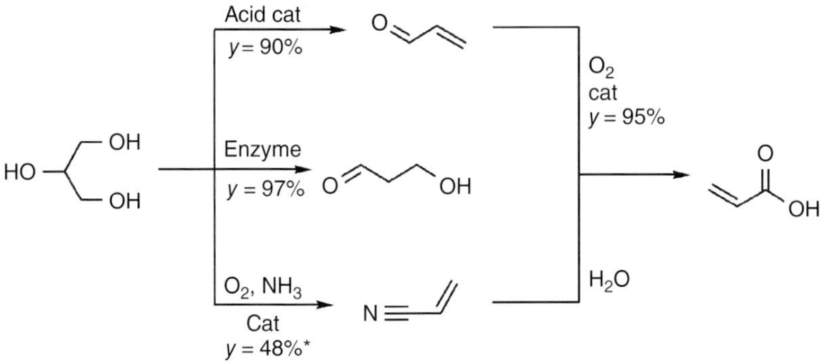

Scheme 9.3 Indirect pathways for acrylic acid synthesis from glycerol. *y stands for yield.

The formation of acrolein from GLY is well known and is the subject of over 100 articles and patents [20]. The reaction requires an acid catalyst and can be performed in the gas phase as well as in the liquid phase. Widely used catalysts for the gas-phase reaction are zeolites, inorganic acids (i.e., phosphorous acid, heteropolyacids, etc.) over various supports (silica, titania, alumina, etc.), and metal oxides. The latter can be pure (niobium oxide, tantalum oxide, tungsten oxide, etc.), or binary and even tertiary mixtures (tungsten oxide/zirconia, tungsten oxide/niobia, etc.).

The performance and the reaction conditions vary depending on the catalytic system. Generally speaking, inorganic acids show high performance from low temperature (starting from as low as 260 °C), whereas zeolites and metal oxide catalysts require increased temperature (>300 °C). Concerning the performance, with respect to the reaction mechanism, conventional Brønsted acid catalysts (phosphorous acid, heteropolyacids, and protonated zeolites) show very high selectivity to acrolein with more than 90%. On the other hand, catalysts based on conventional Lewis acids (niobium oxide, tungsten oxide) show slightly lower selectivity [21]. At the current point, the major drawback of such a technology lies in the deactivation of the solid acid catalysts, which suffer from coke deposition. Even though several techniques have been reported in the literature to decrease the formation of coke and thus extend the catalysts' life, the continuous regeneration of the deactivated catalysts is unavoidable to maintain high performance.

In 2012, the first report on the use of a tandem reaction for the conversion of GLY to AA via acrolein came from Witsuthammakul and Sooknoi [22], who combined a conventional solid acid catalyst, namely, a HZSM-5 zeolite in the first reactor, with a molybdenum–vanadium mixed oxide in the second. The advantage compared to the one-step oxidehydration reaction was notably the possibility to independently choose the reaction conditions for each step. In fact, while the dehydration of GLY achieves high yields at 280–310 °C, the oxidation of the intermediately formed acrolein requires rather lower temperature (220–300 °C). The oxygen concentration is also an important parameter, which can significantly impact the selectivity of the dehydration reaction. Thus, the adjustment of the oxygen concentration by intermediate feeding after the dehydration reaction is a possibility to independently generate optimal conditions in each reactor. Thereby, a yield of 46% in AA was obtained, which is already significantly higher than most of the results obtained by direct conversion of GLY (oxidehydration).

Following the work of Witsuthammakul *et al.*, Massa *et al.* [23] also studied the indirect conversion of GLY with acrolein as intermediate using a zirconium–tungsten mixed oxide as a catalyst for the dehydration reaction and a tungsten–vanadium–molybdenum mixed oxide for the oxidation reaction of the intermediately formed acrolein. The yield was comparable to that reported by Witsuthammakul *et al.* and reached 44%. In order to get a better understanding of the reaction mechanism, the authors also tested a physical mixture of the two catalysts in a single-bed reactor, whereby the corresponding configuration did not enable any AA formation.

Liu *et al.* [24] chose a combination of a cesium salt of phosphotungstic acid on niobium oxide and a silicon carbide-supported vanadium–molybdenum oxide as catalysts for the dehydration and oxidation steps, respectively. The authors evaluated the performance for a mechanical mixture as well as for the two catalytic bed processes. In the first case, the yield in AA remained limited to 25%, whereas the two-bed setup enabled a high and constant yield of 75% in AA, which is – so far – the highest performance ever reported and also the exact combination of the individual values observed when decoupling both steps: in fact, the individual yield in the dehydration of GLY to acrolein was 80%, whereas the oxidation of acrolein to AA was extremely efficient yielding more than 95%.

An alternative access to acrolein is based on 3-hydroxypropionaldehyde, a biotech-derived chemical easily obtained from glucose or GLY [25]. Yields higher than 97% have been reported [26]. This compound can be further efficiently dehydrated to AA over a solid acid catalyst [27]. In theory, the combination of the enzymatic and the chemical steps could lead to an overall yield in AA superior to 95% starting from GLY. In 2012, Cargill, BASF, and Novozymes joined their forces to produce AA via this route [28], but very recently BASF decided to exit the R&D program probably anticipating a problem of profitability [29]. With a similar approach, OPX Biotechnologies and Dow Chemical chose sugars (dextrose or sucrose) as a raw material fermented to 3-hydroxypropionic acid before subsequent step of dehydration over a bentonite clay [30].

In that kind of technology, the challenge consists in the development of an efficient and cost-effective fermentation and separation process to 3-hydroxypropionic acid, while further dehydration to AA is quite straightforward. In contrast, in technologies passing through LA formation (see

Section 9.2.2), while the fermentation step is well mastered, this is the chemocatalytic part of further oxidation to AA which is still a technological hurdle.

Another alternative to acrolein as an intermediate is acrylonitrile. In fact, the hydrolysis of acrylonitrile is well established and was formerly applied commercially for the synthesis of AA. Nowadays, the corresponding process is mainly abandoned, and Asahi Chemicals is the last remaining company producing AA from acrylonitrile [2].

Nevertheless, acrylonitrile is still a viable intermediate, since it can be obtained from GLY by ammoxidation in the gas phase in the presence of oxygen and ammonia. Several articles can be found in the literature reporting either the direct ammoxidation or the indirect ammoxidation [31]. Since the indirect ammoxidation is based on acrolein as the intermediate, the synthesis of AA would not make much sense in this case. On the other hand, a one-step direct ammoxidation can provide a viable alternative to the indirect pathway via acrolein.

The direct ammoxidation of GLY was reported by Banares and Guerrero-Perez [32] in the gas and the liquid phases using mixed oxides based on antimony, vanadium, and niobium. The claimed AA yield in gas phase was 48% with acrolein as major by-product (20% yield). However, these results are matter of a debate [33]. DFT calculations performed by Alliati *et al.* indicated that the cation-deficient (110) surface of vanadium–antimony oxide ($VSbO_4$) should be able to selectively activate ammonia and GLY by adsorption on the Lewis acid sites [34]. Hence, from the theoretical point of view, the direct gas-phase ammoxidation of GLY to acrylonitrile should be feasible.

Zhang *et al.* [35] also studied the direct ammoxidation of GLY in the gas phase over various metals supported on alumina, notably iron, cobalt, nickel, chromium, and zinc. All the catalysts promoted the formation of propionitrile and acetonitrile, whereby the latter was the main product (yield of 25–40%). Subsequently, the authors doped the catalyst with various alkalis, whereby the yield in nitriles (acetonitrile and propionitrile) reached 58% after optimization of the reaction conditions. Nonetheless, the formation of acrylonitrile was not observed in any case.

Banares *et al.* [36] also reported the use of their previously described catalytic system in the liquid phase under microwave irradiation. In that case, the combination of microwave irradiation with alumina-supported antimony–vanadium oxide as a catalyst enabled an acrylonitrile yield of 40%, which was twice higher than that observed in the absence of microwave irradiation.

Nevertheless, compared to the theoretical yield when using acrolein as an intermediate, the use of acrylonitrile is not a viable pathway at the current state of the art. This can be further underlined by the fact that the hydrolysis of acrylonitrile to AA with sulfuric acid generates stoichiometric amounts of monoammonium sulfate, which has only a limited commercial interest.

From the current point of view, it seems therefore that the most promising intermediate for the indirect pathway to AA from GLY is acrolein. However, when considering the catalyst deactivation issue in the dehydration reaction of GLY, one may even consider the three-step process including 3-hydroxypropionaldehyde.

Nonetheless, so far no process for the transformation of GLY to AA has been commercialized. It is well known that Arkema built a facility in France but stopped the project because of the volatility of the glycerine price, concluding that the bio-based process could not compete with that based on petro-sourced propylene [37]. Hence, the price of GLY currently remains the main hurdle.

9.2.2 Production of AA from LA

Since the first patent obtained by Holmen [38] in 1958 claiming the production of AA and acrylates from LA and alkyl lactates, respectively, the dehydration of LA to AA has been the source of a lot of interest. LA is mainly obtained by fermentation processes based on renewable raw materials such as starch or sugars, but it can also be produced from GLY. The next sections sum up the current state of the art on these topics of interest.

9.2.2.1 LA from GLY

LA salts can be directly obtained from GLY in the liquid phase in the presence of a catalyst and with or without the addition of a base. Scheme 9.4 shows the different pathways and mechanisms reported to get LA salts from GLY.

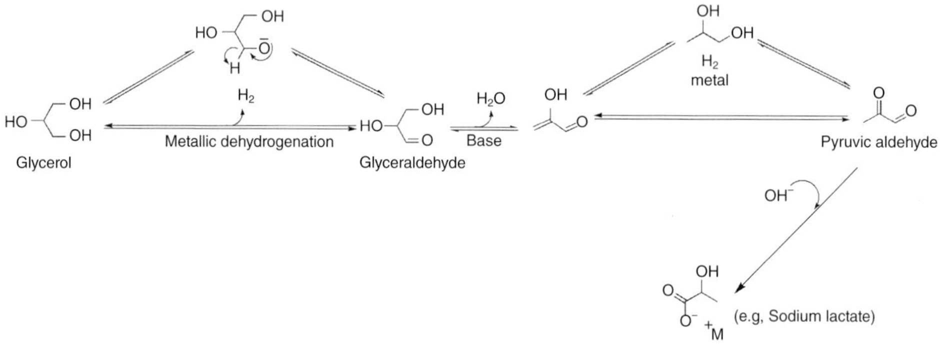

Scheme 9.4 Mechanism of lactate production from glycerol proposed by Dusselier *et al.* [39].

Following Montassier *et al.*'s work [40, 41] on GLY hydrogenolysis in the liquid phase in the presence of a metallic catalyst, Maris *et al.* have studied the formation of LA using Pt- and Ru-based catalysts supported on carbon in the presence of a base (NaOH, CaO). They reached a 58% yield using a Pt/C catalyst and 0.8 mol l^{-1} of CaO at 200 °C under 4 MPa of H_2. In the absence of a base, the authors did not observed any LA formation but rather 1,2-propanediol and ethylene glycol formation [42]. Auneau *et al.* [43] have shown the influence of the atmosphere on their iridium-based catalyst supported on $CaCO_3$. They obtained a 75% yield in LA under He (3 MPa) at 180 °C in the presence of NaOH (1.0 mol l^{-1}) [44]. Under H_2, the formation of 1,2-propanediol is even more favored. Roy *et al.* [45] obtained a similar yield using a copper oxide catalyst supported on alumina and placed under a nitrogen atmosphere (1.4 MPa) with a NaOH/GLY molar ratio of 1.1 at 240 °C. More recently, Ftouni *et al.* [46] succeeded in getting a very high yield in LA (80%) after 8 h using a Pt/ZrO_2 catalyst at 180 °C under He (3 MPa) in the presence of NaOH (1.0 mol l^{-1}).

In hydrothermal alkaline conditions, the reaction is possible at higher temperatures in the absence of a metallic catalyst. Kishida *et al.* [47] have reported a 90% yield in LA at 300 °C in the presence of NaOH. They have studied other alkaline solutions such as KOH and achieved also a 90% yield in LA with this base [48]. Even if the yield was already very high, another team has tried to improve the process because of its low productivity linked to the low concentration of GLY used in the feed. They obtained an 85% LA yield using a GLY concentration eight times higher than that in the previous study [49].

In oxidative conditions, it is also possible to get LA from GLY, as reported by Shen *et al.* [50] and Dusselier *et al.* [39] using a Au–Pt bimetallic catalyst supported on titania (Scheme 9.5). An 86% yield in LA was reached, but only after a very long reaction time of 176 h.

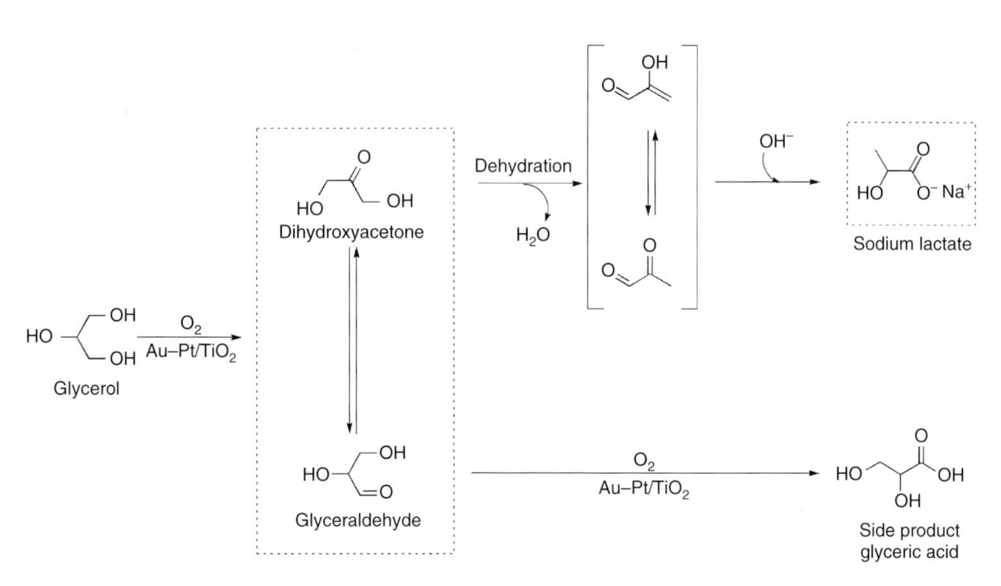

Scheme 9.5 Mechanism of glycerol oxidation to lactate over Au–Pt/TiO_2. (Adapted from [50].)

To conclude on this part, one can say that even if high lactate yields can be obtained from GLY, the productivities of the liquid-phase processes as well as the slow kinetics of the reaction are still detrimental to the commercialization of the process.

9.2.2.2 Direct Dehydration of LA to AA

As aforementioned, the first report about the direct dehydration of LA to AA (Scheme 9.6) is a patent attributed to Holmen in 1958. The process described in this work was carried out in the gas phase and led to AA and acrylates starting from LA or alkyl lactates, respectively [38]. A 68% yield in AA was obtained using a solid catalyst composed of Na_2SO_4 and $CaSO_4$ (1 : 25 molar ratio) tested in a Pyrex® reactor at 400 °C. A total of 10–15 ml h^{-1} of a 10 wt% LA solution was injected on the catalyst. Methyl lactate (ML) and butyl lactate were also used as reactants to directly get the corresponding acrylates. More concentrated LA aqueous solutions were also tested, but lower yields were then obtained. The patent also includes examples of use of phosphates, pyrophosphates, and sulfates as catalysts. Holmen underlined the bad results obtained on typical acid catalysts such as phosphoric and silicophosphoric acids, WO_3, W_2O_5/Al_2O_3, TiO_2, Na_2WO_4, Na_2MoO_4, $NaVO_3$, MoO_3, SiO_2, Al_2O_3, $NiMoO_2$, and $ZnMoO_3$.

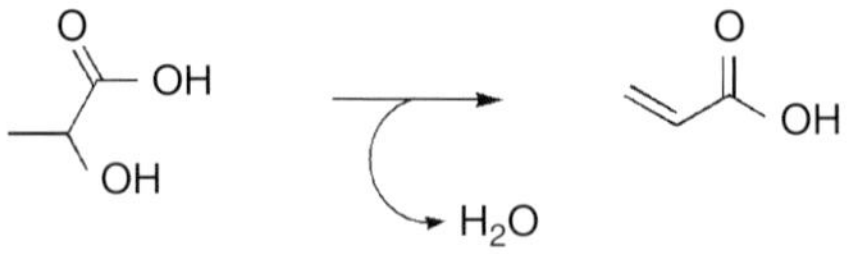

O OH OH → O OH
H_2O

Scheme 9.6 LA to AA dehydration.

Following this pioneer work, the LA dehydration to AA has been extensively studied in the gas phase as well as in sub- and supercritical water [51–54].

Irrespective of the pathway, a lot of by-products can be formed. Then, getting a high selectivity to AA becomes very challenging. LA decarbonylation and decarboxylation, leading to acetaldehyde, are the most important competitive reactions (Scheme 9.7). These reactions are favored on catalysts presenting medium and strong acid sites [55]. This is also the case in sub- and supercritical water, as extremely acid conditions take place in such media [51]. The hydrogen issued from the decarboxylation of LA can then further react with LA or AA to promote the formation of propionic acid. Wadley *et al.* [56] showed that the formation of acetaldehyde was favored compared to AA (activation energies of 115 and 137 kJ mol^{-1}, respectively). 2,3-Pentanedione and lactide are also typical by-products, respectively, obtained by autoesterification or condensation of LA.

Another difficulty in obtaining a high selectivity to AA starting from LA is the high reactivity of LA itself. At medium or high concentrations in water, it can be easily polymerized to form oligomers, hence limiting the yield in the desired product [57, 58].

Starting from this basis, a lot of researchers have tried to develop a catalyst able to dehydrate LA without enhancing the decarboxylation/decarbonylation process. A particular effort has been put on the understanding of the reaction mechanism. Three main types of catalysts have been identified from the literature: zeolites, phosphates, and hydroxyapatites (HAPs), which are nothing else than a special

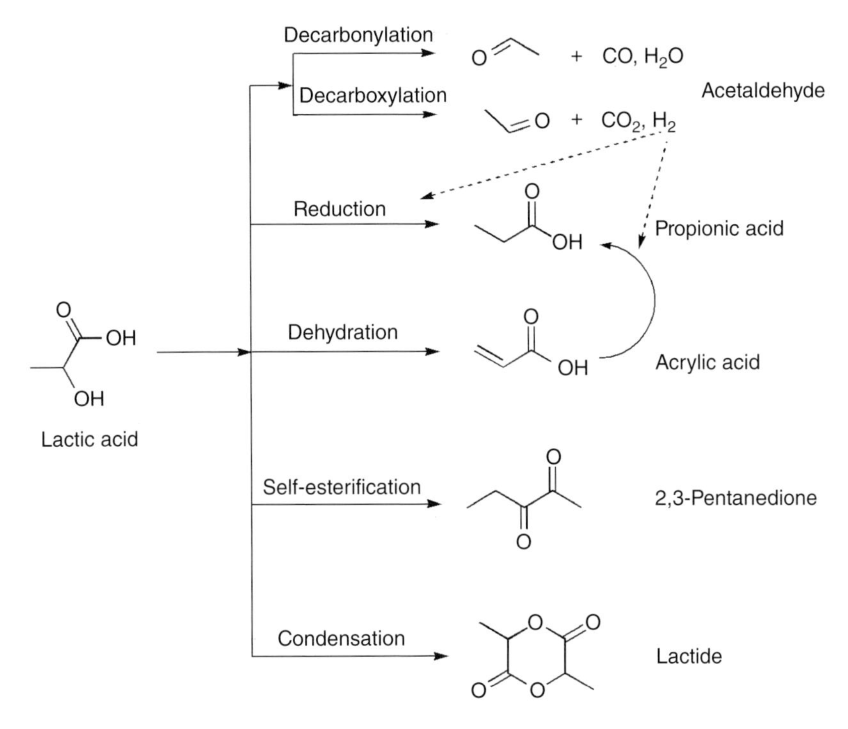

Scheme 9.7 Reactions competing with AA production from LA.

class of phosphates. The state of the art on the use of the three types of catalysts is exposed in the following sections.

LA Dehydration in the Gas Phase Using Zeolites as Catalyst In 1993, Takafumi and Hieda [59] were granted with a patent claiming the use of zeolites as catalysts for the production of unsaturated carboxylic acids and esters. The best results were obtained on zeolites of the Faujasite type (X or Y). Starting from ML, the authors reached a 99% conversion of ML and a selectivity to methyl acrylate (MA) of 93% at 240 °C. No example of direct dehydration of LA to AA is given, even if it is claimed in the patent.

Following this work, Huang *et al.* [60–65] studied the LA dehydration over modified zeolites. A first study on ML dehydration to MA on NaY zeolites modified by KCl to yield KNaY catalysts was published [66]. In very similar conditions compared to Takafumi's work, they obtained a slightly lower ML conversion (83% vs. 88.6%) but a much better selectivity to MA (45.7% vs. 32.1%). The authors explained this difference by the decrease in the strength of the active acid sites when the zeolites are modified by K cations.

Huang *et al.* [60, 62, 64] also modified NaY zeolites, but with lanthanides (La, Ce, Sm, and Eu). The best results are summarized in Table 9.1.

Table 9.1 Results of the catalytic tests realized by Huang *et al.* on lanthanide-impregnated zeolites [60, 64].

Catalyst	Conversion (%)	Selectivity (%)				Reference
		AA	AC	PA	Ot	
NaY	100	34.8	16.1	1.6	47.5	[60]
0.5%La/NaY	100	45.3	10.4	2.1	42.2	
1%La/NaY	100	43.8	6.8	1.7	47.7	
2%La/NaY	100	56.3	12.3	1.6	29.8	
3%La/NaY	100	50.2	11.9	2.3	35.6	
4%La/NaY	100	44.9	18.5	3.4	33.2	
2%Ce/NaY	100	45.8	9.5	2.0	42.7	
2%Sm/NaY	100	36.0	12.8	2.0	49.2	
2%Eu/NaY	100	40.2	16.1	3.3	40.4	
NaY	96.1	14.8	10.8	1.6	72.8	
2%La/NaY	97.6	36.2	19.3	3.8	40.7	[64]
2%La/NaY	98.5	56.3	12.3	1.6	29.8	
Test conditions						
T: 350 °C	Sol. Lac. A.: 38 wt%	LHSV: 3 h^{-1}	Gas flow (N_2): 30 ml min^{-1}		Catalyst weight: 1.5 g	TOS: 360 min

AA, acrylic acid; AC, acetaldehyde; PA, propionic acid; Ot, others (CO_x and unknown products); Lac.A Lactic Acid Solution or Solution of Lactic Acid.

The best results were obtained at 350 °C with the 2%La/NaY catalyst leading to an almost total conversion of LA and a 56.3% selectivity to AA. The authors also claimed that the formation of coke was strongly decreased on the La-modified catalyst and explained the best results by the presence of weak acid sites in the catalyst. Hence lanthanide-modified zeolites induced a decrease of the density and strength of the acid sites either by charge effects or structural modification of the sodalite cage of the zeolite.

The same group has pursued the study by modifying the zeolites by potassium and alkaline earth metals [61, 62]. The influence of the impregnation method (nature of the precursors) on the basic and acid properties of the catalysts was then put into evidence. The best results are summed up in Table 9.2.

It is clear from these results that the addition of K in the catalyst had a positive effect on the AA selectivity. The best yield in AA (viz., 66.2%) was obtained on KI/NaY (Si/Al = 4.5). The addition of K in the zeolite probably led to a better balance of the acid and basic sites strength by killing the strong acid and basic sites present in the initial zeolite. However, the influence of the precursor used for the impregnation of the parent zeolite proved to be very significant, and the authors also reported a strong deactivation of the catalyst by coke formation. Lately, Näfe *et al.* [67] solved the problem of deactivation by reducing the amount of reaction sites to avoid the accumulation and the formation of clusters, leading to the formation of coke. Thanks to a Na-ZSM-5 zeolite, they obtained a stable conversion

Table 9.2 Results of the catalytic tests made by Huang *et al.* on potassium-modified zeolites [61, 62].

Catalysts	Conversion (%)	Selectivity (%)						Reference
		AA	AC	PA	2,3P	Coke	Ot	
NaY (Si/Al = 2.5)	96.1	14.8	10.8	—	—	na	na	[61]
0.35K/NaY	95.9	20.6	10.2	—	3.5	na	na	
0.7K/NaY	96.4	31.2	8.3	2.1	4.8	na	na	
1.4K/NaY	97.5	39.8	6.6	4.1	9.4	na	na	
2.1K/NaY	98.2	40.2	4.7	4.0	10.1	na	na	
2.8K/NaY	98.2	50	2.8	3.8	10	na	na	
3.5K/NaY	98.8	41.3	1.4	3.5	10.6	na	na	
NaY (Si/Al = 4.5)	96.3	45	20.2	—	8	29.4	6.5	[62]
KF/NaY	91.2	39.9	17.1	—	—	19.3	23.7	
KCl/NaY	97.1	53.8	10	3.9	10.4	17.4	4.5	
KBr/NaY	97.1	59.9	9.2	—	8.1	15.7	7.1	
KI/NaY	97.6	67.9	—	—	7.9	13.0	11.2	
KNO_3/NaY	98	58.2	6.8	2.6	10	17.8	4.6	
KSO_4/NaY	94.8	51.6	20.2	—	—	18.1	10.1	
$KHPO_4$/NaY	94.5	49.7	6	—	6.5	20.9	16.9	
$K_2C_2O_4$/NaY	94.4	48.5	14.8	—	6.5	14.8	15.4	
K_2CO_3/NaY	95.1	46.7	14.7	—	4.9	13.8	19.9	
KOH/NaY	96.3	44.2	—	—	—	13.8	42	
Test conditions								
T: 325 °C	Sol Lac. A.: 29 wt%	Liquid flow: 4.5 ml h^{-1}		Gas flow (N_2): 30 ml min^{-1}				Catalyst weight: 1.5 g

AA, acrylic acid; AC, acetaldehyde; PA, propionic acid; Ot, others (CO_x and unknown products).

of 96% and a selectivity of 53% after almost 5 h. It would be however interesting to test the catalyst stability on a longer period.

Huang's group finally studied the modification of the zeolite by alkaline earth metals (Ca, Mg, Sr, and Ba). The best result was obtained on 2%Ba/NaY (Si/Al = 4.5) with a yield in AA of 44.6% [64]. They described the catalytic site as a cluster formed by the cation and three atoms of oxygen of the network, as represented in Scheme 9.8 [64, 68].

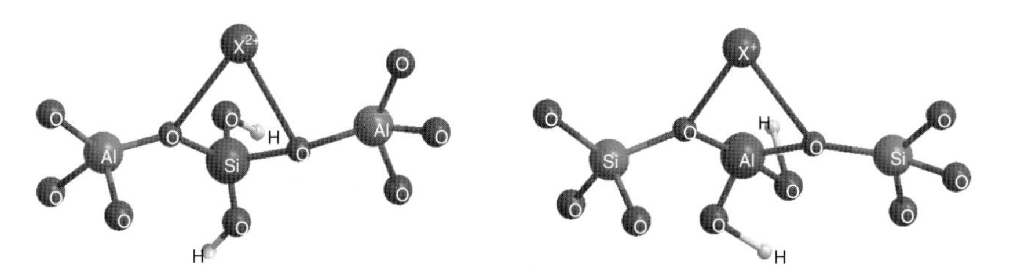

Scheme 9.8 Clusters formed by the alkali (X^+) and alkaline earth metals with the oxygen atoms from the zeolite NaY lattice. (Adapted from [64].)

LA Dehydration in the Gas Phase Using Phosphate-Based Catalysts Following Holmen's work [38], phosphate-based catalysts have been extensively studied for the dehydration of LA to AA. Sawicki's [69] and Paparizos' [70] groups were granted with two patents on this topic in 1988.

Paparizos *et al.* used an aluminum phosphate ($AlPO_4$) treated *in situ* by an ammonia solution just before reaction. A 43.3% yield in AA was observed starting from LA [68]. At the temperature of the reaction (340 °C), the ammonia was desorbed from the weak and medium acid sites, letting them available for the dehydration reaction. On the contrary, the strong sites were still occupied, hence preventing the decarboxylation/decarbonylation side reactions. The strong influence of the catalyst calcination temperature, of the reaction temperature, and of the pretreatment was underlined by the inventors.

In parallel, the Sawicki's group used various phosphates supported on SiO_2, TiO_2, and Al_2O_3 and treated by different bases before reaction. The best performances (58% yield in AA with 89% of LA conversion) were obtained using silica impregnated by sodium hydrogen phosphate as a support and a sodium bicarbonate pretreatment. These performances were attributed to the neutralization of the strong acid sites and to the formation of new sites, which are favorable to LA dehydration. However, the selectivity to acetaldehyde stayed high, and no information nor on the other by-products neither on the carbon balance was given in the patent, which makes it difficult to interpret the results in a more detailed manner.

Finally, Gunter *et al.* [71–73] studied the formation of AA and 2,3-pentanedione on phosphate-based catalysts. The authors suggest that the phosphates, which are already known as *good catalysts* for the dehydration of alcohols, could stabilize the carboxylic acid group, hence favoring the AA selectivity to the detriment of the acetaldehyde formation. Based on Fourier Transformed Infra-Red (FTIR) and ^{31}P-NMR results, these authors studied the interactions between LA and various sodium phosphates species such as $(NaPO_3)_n$, NaH_2PO_4, $Na_3P_3O_9$, Na_2HPO_4, Na_3PO_4, and $Na_4P_2O_7$. Their work shows that Na_2HPO_4 condensates into $Na_4P_2O_7$, which then exchanges a proton with LA to yield sodium lactate and $Na_3HP_2O_7$. Following the same mechanism, Na_3PO_4 would accept a proton from LA to yield Na_2HPO_4, which then condensates into $Na_4P_2O_7$. Actually, Na_2HPO_4 and Na_3PO_4 were the most selective species tested in their work. The authors showed that the lactate formation is a critical step in the mechanism.

These results are in good agreement with the work of Wadley *et al.* [56] who claimed the formation of a liquid layer at the surface of catalyst made of sodium nitrate supported on silica. They also underlined that the formation of sodium lactate is a key to reach a high AA selectivity.

Starting from ML on NaH_2PO_4/SiO_2 (80 : 20 in wt%), Zhang *et al.* [74] reached a cumulated yield in AA and MA of 52% (roughly 25% of each product). Silica has also been used by Lee *et al.* as a support for tricalcic phosphate-based catalysts [75]. The optimal weight ratio was here again 80 : 20 between $Ca_3(PO_4)_2$ and SiO_2. The yield did not exceed 60% since at low temperature (350 °C) the conversion was

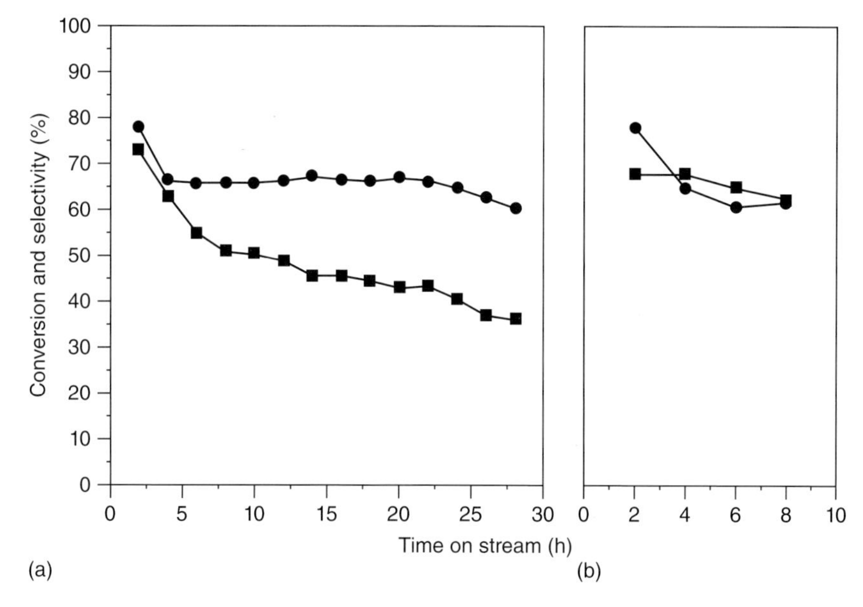

Figure 9.1 Evaluation of the catalyst stability and of its regeneration under air [77].

limited to 64% and by increasing the temperature the selectivity decreased sharply. The authors however claimed that this catalyst was better than the others because of its high specific surface area and its adequate porosity and appropriate acid and basic properties.

In another work by Zhang *et al.* [76, 77], zeolites were used as supports for sodium hydrogen phosphate. The *in situ* formation of sodium lactate was also presented here as a key step of the mechanism. The influence of the increase of the loading clearly showed that the strong acid sites progressively disappeared, hence decreasing acetaldehyde selectivity and enhancing the AA formation. However the conversion also decreased with the loading. Another problem was also faced by Zhang *et al.*: the progressive deactivation of the catalyst. Huang *et al.* [60, 61] had already encountered the same problem before. Zhang *et al.* studied this phenomenon as shown on Figure 9.1 where the LA conversion (•) and the AA selectivity (■) are plotted versus time.

It is clear from these results that not only the activity of the catalyst could be fully recovered after a regenerative treatment (calcination under air at 500 °C during 5 h) but also that the AA selectivity was much better then than during the first time on stream. The calcination was expected to eliminate the coke formed at the surface of the catalysts and maybe also to improve the phosphate species dispersion and hence their interaction with lactate. However the best AA yield obtained after regeneration was still inferior to the one observed on KI/NaY (i.e., 68%) [62].

A very good improvement of the performance was obtained using zeolite nanocrystallites, avoiding the formation of coke in the zeolite microstructure (Table 9.3) [77].

Table 9.3 Catalytic performance for LA dehydration obtained over sodium phosphate-based catalysts supported on zeolite nanocrystallites [77].

Catalysts	Conversion (%)	Selectivity AA (%)	Yield (%)			
			AA	AC	2,3P	PA
NaY_{com}	95.5	37.2	35.5	22	3.1	1.6
NaY-13.8[a)]	89.6	43.4	38.9	37.7	3	0.8
NaY-16.9	88.9	46.2	41.1	19.9	3.1	1.3
NaY-20.1	91.6	45.9	42	13.6	2.5	1.1
NaY-26.3	86.2	39.5	34	18.4	2.4	1.6
Na_2HPO_4/NaY_{com}	82.3	65.7	54	4.6	4.6	1
Na_2HPO_4/NaY-13.8	92.3	76.8	70.9	4.8	5.3	1.1
Na_2HPO_4/NaY-16.9	94.6	75.4	71.3	3.8	5	1.1
Na_2HPO_4/NaY-20.1	93.5	79.5	74.3	5.1	5.2	1
Na_2HPO_4/NaY-26.3	94.3	75	70.7	4.5	4.5	1
Test conditions						
T: 340 °C	Gas flow (N_2): 30 ml min^{-1}	Catalyst weight: 1.5 g		Sol. lac. A.: 34 wt%		Liquid flow: 6 ml h^{-1}

a) H_2O/S_iO_2 ratio used during the zeolite syntheses.
AA, acrylic acid; AC, acetaldehyde; 2,3P, 2,3-pentanedione; PA, propionic acid.

These results showed that the use of a zeolite nanocrystallite-based support led to a much better AA yield. For instance, 74.3% is reached on Na_2HPO_4/NaY-20.1, which is a much better result than when Na_2HPO_4/NaY_{com} was used (65.7%). The authors ascribed this difference to the incomplete structure of the nanocrystallites of zeolite, which did not present narrow channels where coke can be easily formed and deposited. A productivity in AA of 12 mmol g^{-1} h^{-1} was claimed by the authors of this study, which represents a four times higher productivity than the one reported by Peng *et al.* [78]. As before a progressive deactivation of the catalysts was observed, but a regeneration of the catalyst was possible.

The previous studies reported in this section concerned the used of supported phosphate-based catalysts, but most of the time they are used as bulk catalysts, that is, without any support [79–83].

Starting from ML, Lee *et al.* [79] have studied a catalyst made of calcium phosphate and pyrophosphate (50 : 50). A very good yield in MA and AA was obtained (72%).

More recently, Blanco *et al.* [82] published a detailed study of the reactivity of earth metal phosphates and interpreted the results by correlating the AA selectivity with the balance between acid and basic sites. The best molar ratio was close to 1 : 1 for the acid and basic sites, which comforted the hypothesis of the conjugated action of both types of sites for the formation of AA.

The best conversion was obtained for the catalysts presenting the highest specific surface area (calcium HAPs), but the best selectivity was reached on a barium phosphate.

Table 9.4 Catalytic performance for the dehydration of lactic acid to acrylic acid over phosphate-based catalysts [80].

Catalysts	Conversion (%)	Yield (%)		Selectivity AA (%)	B.E.T. ($m^2\,g^{-1}$)	Acidic site density ($mmol\,g^{-1}$)	Basic site density ($mmol\,g^{-1}$)
		AA	PA				
A	91	85	1.1	93	0.57	0.25	77.8
B	77	72	0	92	0.4	0.18	36.1
Test conditions							
T: 325–350 °C		Sol. lac. A.: 17–25 wt%		GHSV ≈ 3500 h^{-1}		Quartz reactor	

AA, acrylic acid; PA, propionic acid.

Published also very recently, the patent of Lingoes *et al.* [80] from Procter & Gamble contains very good catalytic results as well as very interesting remarks on the process operating conditions. Mixtures of phosphates presenting very low surface area were used in this work (<1 m^2). The authors reported excellent catalytic performances for potassium and barium phosphate mixture (K:Ba = 40 : 60). However, two batches of an identical catalyst were prepared and tested (batches A and B), and, as it can be seen in Table 9.4, while the results are very good in terms of AA selectivity (92–93%), the conversion is not reproducible.

For batch A, the best yield in AA ever claimed was reached (85%). However, as it can be seen in Table 9.4, the authors could not control the acid and basic sites densities, then leading to a strong variation in the catalyst performances. It is interesting to note that the acid and basic site densities are very high in this work, namely, two to six times higher than those reported on zeolites [60, 61, 64] and on phosphates [79, 80, 84].

Barium pyrophosphate has also been used by Tang *et al.* [81] who reported an AA yield of 76% starting from LA. The results were quite different from those obtained by Blanco *et al.* [82], but this can be explained by the very different operating conditions used for the tests in both works. Here also, a progressive deactivation of the catalyst was observed.

Among all the bulk phosphate-based catalysts, the best results were reported by Lingoes *et al.* [80]. Barium phosphates were identified as very interesting catalysts to get high AA selectivity.

LA Dehydration in the Gas Phase Using HAP-Based Catalysts Among the phosphate family, HAPs are a special class of solids because of their specific structure and versatility of composition. They are interesting in catalysis because of their basic or bifunctional properties. They have been used, for instance, to catalyze the Guerbet reaction [85], the Knoevenagel condensation reaction [86], or the Michael addition [87] or for dehydration [88], oxidation [89], and dehydrogenation reactions

[90]. The acid and basic properties of HAPs can be easily tuned by substituting elements in their structure (Ca^{2+}, PO_4^{3-}, OH^-) or by playing on their stoichiometry. These materials were used as catalysts for LA dehydration to AA by Matsuura *et al.* [91–93] and Umbarkar *et al.* [84, 94, 95].

Matsuura *et al.* [92] have studied the impact of the variation of the Ca/P ratio associated to the substitution of Ca by Na atoms, as well as of the Ca substitution by Pb or Sr and of P by V in HAPs prepared by a hydrothermal method [93]. A very good yield in AA (close to 70%) was obtained on Ca–Na–HAP (Ca/P = 1.55; (Ca + Na)/P = 1.65), which presented a remarkable stability after 50 h on stream. Matsuura *et al.* concluded – in good agreement with Blanco *et al.* [82] – on the importance of the good balance between acid and basic acid sites.

Umbarkar *et al.* [84] came to very different conclusions using HAPs prepared by coprecipitation. This group studied the modification of HAP properties not only by changing their Ca/P ratio without Na addition but also by changing the pH during the synthesis procedure. Their best catalyst presented a very high acid/base site ratio of 210, which is very far from the optimum found by Blanco *et al.* and Matsuura *et al.* However, it is probable that the HAP phase is not the only one in presence and that the catalyst is in fact a mixture of phosphates of different natures. Later, the same team obtained a very high yield in AA (78% at full LA conversion) on a calcium phosphate catalyst prepared in the presence of Na and for a Ca/P ratio of 0.76. The acid/base ratio of this catalyst is 11, and, here again, the authors claimed that the good result was correlated to the high value of this ratio [95].

Very recently, Yan *et al.* [96] have studied the influence of the temperature of calcination between 360 and 700 °C on the HAP catalytic performances. They have also carried out a detailed examination of the performance evolution as a function of the Ca/P ratio (between 1.58 and 1.69) and of the acid/base properties. The best yield was obtained at 360 °C (62%) for a Weight Hourly Space Velocity (WHSV) equal to 1.4 h^{-1} on a HAP catalyst (Ca/P = 1.62) calcinated at 360 °C. The authors presented the evolution of the basic and acid site densities versus the Ca/P ratio and the calcination temperature (Figure 9.2).

An almost linear increase in the basic site density versus the Ca/P ratio is observed, whereas an exponential decrease of the acid site density is obtained. The calcination temperature has no real influence on the basic site density. The acid site density versus the calcination temperature is a volcano-type curve with a maximum around 500 °C. This result is very interesting because it shows that it is possible to tune the acid/base ratio of the HAP by choosing the appropriate calcination temperature. The same authors also put the focus on the effect of the acid/base ratio on the AA and acetaldehyde-specific productivities. It is clear that acetaldehyde formation is favored when the acid/base ratio is increased, but the tendency for AA is less clear even if a maximum seems to be reached around 3–4.

Phosphates are clearly very interesting candidates to catalyze the LA dehydration to AA in the gas phase. The selectivity to AA is much higher over these catalysts compared to those observed over zeolites. Moreover, the phosphates seem

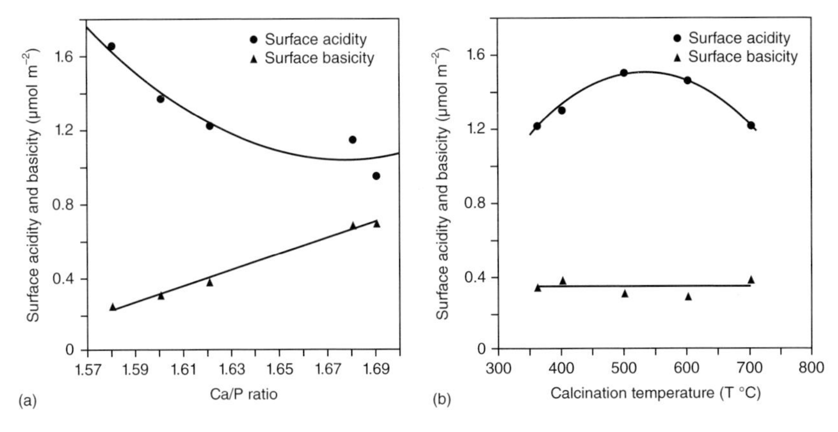

Figure 9.2 Acidic and basic site densities of HAPS as a function of the Ca/P ratio (a) or calcination temperature (b). For (a) all the HAPs have been calcined at 360 °C. For (b) all the HAPs have a Ca/P ratio of 1.62 [96].

to be more resistant to deactivation. However, it is still difficult to identify clearly which phases are responsible for the good performances or which types of sites are necessary. The conclusions of the authors are sometimes contradictory as far as the acid and base site densities are concerned. Lingoes *et al.* claim that the best performances are obtained with a high basic site density, whereas Umbarkar *et al.* claim the contrary. In between, Blanco *et al.*, Tsuchida *et al.*, and Yan *et al.* rather advise a good balance between both types of sites. It is clear that the scientific community still needs to deepen the study to get a better understanding and a clarification of this aspect.

To conclude on that part, the best results published in the literature so far for LA dehydration to AA are summed up in Table 9.5.

NaY zeolites modified by alkaline or metal earth elements, and more particularly by K, proved to be good catalysts for LA dehydration to AA. The LA conversion was always very high (close to 100% generally), but the selectivity to AA tended to decrease with time on stream. The performances have been linked to the interaction between the cations and the oxygen atoms of the network tuning the acid and basic properties (number and strength of the sites).

To conclude on this part, it is also of interest to note that an alternative synthetic route starting from LA to AA has been very recently proposed by Beerthuis *et al.* [97]. These authors actually designed an indirect heterogeneously catalyzed route proceeding via acetoxylation of LA to 2-acetoxypropionic acid, which is further pyrolyzed to AA. Acetic acid was used as a solvent and as a reagent, but could be recycled after the pyrolysis step. Y zeolites, sulfated zirconia, ion-exchange resins, sulfonated graphene, and various sulfonated silica gels and mixed oxides were tested as catalysts. The best results were obtained using ion-exchange resins such as Amberlyst 70, with a yield of 37% in 2-acetoxypropionic acid at low conversion. However, the authors expect an AA yield around 85% when the two steps are combined under optimized conditions.

Table 9.5 Main catalysts and their performances for LA dehydration in the gas phase.

Catalysts	Reactant	*T* (°C)	GHSV (h^{-1}) / WHSV (h^{-1})	Reactor ($Q^{a)}$/$SS^{b)}$)	C (%)	S (%) AA (AlA + AA)[c)]	Y (%)	Reference
Modified zeolites								
Mol.S.13X	MA	240	882	Q	99	95[c)]	94	[59]
2.8KNaY	LA	325	8906	Q	98.8	50	49.4	[61]
KI/NaY	LA	325	8906	Q	97.6	67.9	66.3	[62]
Supported phosphates								
Na_2HPO_4/NaY-20.1	LA	340	10780	Q	93.5	79.5	74.3	[76]
Nonsupported phosphates								
K- and Ba-mixed phosphates	LA	325–350	3500	Q	91	93	85	[79]
				SS	90	64	58	
$Ca_3(PO_4)_2-Ca_2(P_2O_7)$ (50 : 50)	LA	390	990	Q	100	54	54	[74]
	MA		975		91	80[c)]	72.8	
	EA		964		57	84[c)]	47.9	
$Ba_2P_2O_7$	LA	400	2801	Q	99.7	76	75.7	[80]
Calcium phosphate (Ca/P: 0.76)	LA	375	3[d)]	Q	100	74	74	[94]
Hydroxyapatites								
HAP (Ca/P: 1.3)	LA	375	3[d)]	Q	100	60	60	[83]
Ca–Na–HAP(Ca/P:1.55) (Ca + Na/P: 1.65)	LA	350	7471	Q	90	78	70.2	[90]
HAP. (Ca/P: 1.62)	LA	360	1.4[d)]	Q	84	74	62	[95]

C (%), conversion (%); S (%), selectivity (%); AA, acrylic acid; Y (%), yield (%).
a) Quartz.
b) Stainless steel.
c) AlA + AA: alkyl acrylates and acrylic acid
d) Contact time (s).

9.2.3 Production of AA from Biopropylene

In this section, some strategies of AA production from biopropylene obtained by C6 sugar fermentation or by bioethanol chemocatalytic conversion (bioethanol being also produced by sugar fermentation) are given. Some routes based on GLY will also be mentioned mainly for comparison/benchmarking, this raw material being fully described in Section 9.2.1 of the present chapter.

Answering the question of AA production via biopropylene synthesis is, at least from a technological point of view, equivalent to answering the question of the possible ways for producing biopropylene. Of course, a downstream process of AA production from such biopropylene will have to be compatible with the specifically generated process impurities, and the catalytic formulation and the process in itself might need some adaptation to tolerate such a new biopropylene quality.

To the best of our knowledge, the combination of biopropylene production followed by propylene oxidation to AA has not been reported yet.

Hereafter are listed the ways for preparing biopropylene from biomass, mostly from glucose and ethanol:

- *Ethanol to propylene*: Different catalytic formulations have been identified to realize this quite complex reaction [98]. In early studies, zeolites such as ZSM-5 and SAPO have shown a certain efficiency in this reaction, with, at that time, a proposed reaction pathway as follows: first (i) formation of ethylene by ethanol dehydration, followed by (ii) formation of C6 hydrocarbon by ethylene oligomerization reaction, and finally (iii) formation of propylene by β-cleavage. For this whole reaction process, acidic sites play an important role. Especially, in such a configuration, the key point for obtaining a maximal yield in propylene leads in the capacity of hindering overoligomerization during step (ii). However, when using pure zeolites as catalysts, overoligomerization occurs, which leads to low propylene selectivities and also to deactivation of the catalytic system. Nevertheless, recent developments show that the reaction pathway is more complex than initially imagined, with even more steps supposedly involving intermediates such as diethyl ether, acetaldehyde, acetic acid, ethyl acetate, and acetone. As reminded in Iwamoto's review [98], three kinds of efficient catalysts are now known from the literature, namely, Ni/MCM-41, Sc/In_2O_3, and $Y_2O_3-CeO_2$, which act through different reaction networks to yield propylene from ethanol. Research in this field is of growing interest, and enhanced catalytic formulations are being gradually proposed. Note that some industrial stakeholders have announced upscaling actions.
- *GLY to propylene*: While not being exactly within the scope of this part of the chapter, we would like also to mention here a recent study showing efficient production of propylene from GLY [99]. The group of Sato obtained at 240 °C and in the presence of H_2 a propylene selectivity of about 85% at full GLY conversion using a double-bed reactor containing a Cu/Al_2O_3 catalyst, followed by a $SiO_2-Al_2O_3$ catalyst.
- *Glucose to propylene*: In that case, biotechnologies are envisioned, and Global Bioenergies (GBE) has launched a research program to adapt a glucose fermentation process to directly yield propylene [100].

9.3 Biochemical Routes

In addition to cascade reactions with first biotechnological conversion followed by a chemocatalytic conversion, direct biochemical routes for AA production are also envisioned.

AA, under its acrylate form, is an intermediate (as an ester of coenzyme A (CoA)) [101–103] in the metabolic system of bacteria such as *Clostridium propionicum*, which can reduce lactate to propionate [100]. The enzymatic scheme established by Dalal *et al.* [103] is reported in Scheme 9.9.

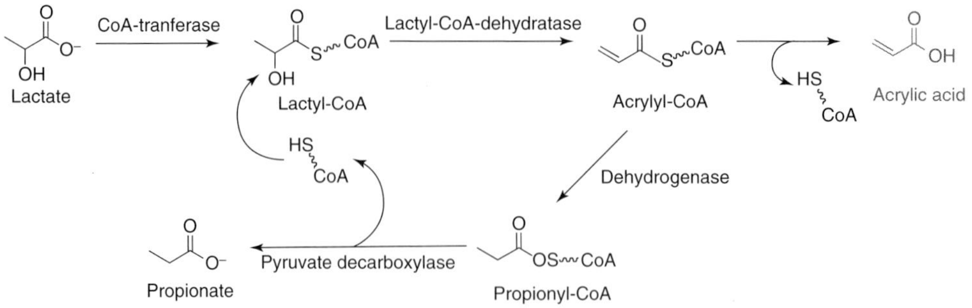

Lactate — CoA-tranferase → Lactyl-CoA — Lactyl-CoA-dehydratase → Acrylyl-CoA → Acrylic acid (HS-CoA)

Acrylyl-CoA — Dehydrogenase → Propionyl-CoA — Pyruvate decarboxylase → Propionate (HS-CoA)

Scheme 9.9 Simplified scheme of the metabolic pathway of direct reduction with *C. propionicum*. (Adapted from [103].)

By modifying the genes involved in the metabolic pathway of reduction of *C. propionicum* or by using blocking agents such as 3-butynoic acid, it is possible to inhibit to a certain extent the dehydrogenase action that transforms acrylyl-CoA to propionyl-CoA (Scheme 9.9) instead of giving the way to AA production. However, the obtained concentrations are low not only due to the actual difficulty of efficiently inhibiting the dehydrogenase action but also due to the presence of other parallel pathways [104]. Moreover, an increase in the concentration of AA can inhibit the strains' growth, and there is thus a need for selecting/developing more resistant mutant strains [105].

However, while AA has been identified as an intermediate in the metabolic pathway of bacteria, algae, and fungi – often under its form associated to CoA (acrylyl-CoA) – for reasons of feasibility and when keeping productivity as a target, the most studied route consists in the direct fermentation of sugars [104]. Straathof *et al.* [106] propose a schematic synthesis of the as-established or supposed routes for the production of acrylate from sugars (Scheme 9.10).

Based on the established metabolic pathways, on bacterial growth models, and on a model of ethanol production, Lunelli *et al.* [107, 108] have proposed a process of production of AA from glucose via pyruvate and lactate as intermediates. However, in 2013, Straathof *et al.* [109] pointed out that the processes based on enzymatic activity always suffer from the presence of undesired parallel routes of fermentation and also from the fact that an external electron acceptor is needed. Thus, no one has developed to date any fully biotechnological process for producing AA or its esters that would give yields enough high to enable envisioning a viable industrial application.

9.4 Summary and Conclusions

In this chapter, we have seen that bio-AA can be obtained using chemo- or biocatalyzed pathways starting from GLY, LA or C6 sugars, and ethanol (sometimes via biopropylene).

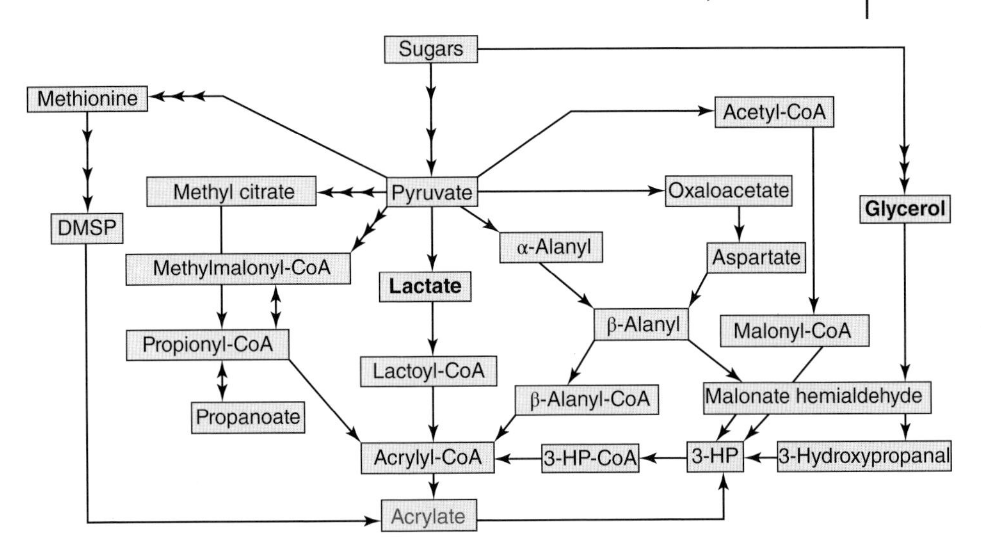

Scheme 9.10 Schematics of the metabolic routes for acrylate biosynthesis from sugars. (Adapted from [106].)

It has been shown that the best yield for the direct conversion of GLY to AA has been obtained by Ueda's group using phosphoric acid-modified niobium–tungsten–vanadium mixed oxide [12]. With this catalyst, an AA yield of 59% was obtained. Since this yield is still too low to envision commercialization, a two-step process remains an interesting alternative: within this frame, the most promising intermediate is acrolein (yield in AA of 75% starting from GLY). However, when considering the catalyst deactivation issue in the dehydration reaction of GLY, one may even consider the three-step process involving 3-hydroxypropionaldehyde. In that case, a yield of 95% in AA is theoretically reachable, but a lot of work has still to be performed to experimentally confirm these performances. Even if GLY is probably the most promising starting material, no process for the transformation of GLY to AA either by the direct or the indirect pathway has been commercialized so far. The volatility of the GLY price and the stability of the catalysts used for GLY dehydration remain the main hurdles.

Starting from LA could be an alternative, since the latter can be derived from GLY or by fermentation whereby high lactate yields can be obtained. For the following gas-phase dehydration of LA to AA, several catalysts were reported whereby phosphates (and more particularly HAPs) are then clearly the most interesting. Actually, the selectivity to AA is much higher over these catalysts compared to those observed over zeolites. Moreover, the phosphate catalysts seem to be more resistant to deactivation. However, it is still difficult to identify clearly which phases are responsible for the good performances or which types of sites are necessary, and further extended studies should give the answers in the future.

As far as the biochemical routes are concerned, a lot of progress have been recently made. However, the processes based on enzymatic reactivity always suffer from the presence of undesired parallel routes of fermentation and also from the fact that an external electron acceptor is needed. Thus, no one has developed to date any fully biotechnological process for producing AA or its esters that would give yields high enough to enable envisioning a viable industrial application.

The best options described so far are thus the gas-phase conversion of GLY to AA in two or three steps (i.e., involving acrolein and 3-hydroxypropionaldehyde as intermediates). Actually, following this pathway, a yield in AA of 95% should be reachable.

References

1. Tanya, L. (2014) Global Acrylic Acid Production to Reach 5.94 Mln Tonnes in 2017, According to In-demand Report by Merchant Research & Consulting, http://www.prweb.com/releases/2014/02/prweb11597545.htm (accessed 23 December 2014).
2. Ohara, T., Sato, T., Shimizu, N., Prescher, G., Schwind, H., Weiberg, O., Marten, K., and Greim, H. (2000) *Ullmann's Encyclopedia of Industrial Chemistry*, Wiley-VCH Verlag GmbH & Co. KGaA, pp. 1–18.
3. Bauer, W. (2000) *Kirk-Othmer Encyclopedia of Chemical Technology*, vol. 1, John Wiley & Sons, Inc., pp. 342–369.
4. Arpe, H.-J. (2010) *Industrial Organic Chemistry*, 5th edn, Wiley-VCH Verlag GmbH.
5. Duprez, D. and Cavani, F. (2014) *Handbook of Advanced Methods and Processes in Oxidation Catalysis*, Imperial College Press, London.
6. Allen, G. (1972) Oxidation of unsaturated aldehydes to the corresponding acids. US Patent 3644509.
7. Yamaguchi, G. and Takenaka, S. (1967) Production of unsaturated aliphatic acids US Patent 3567773.
8. Krabetz, R. and Engelbach, H. (1970) Production of acrylic acid by oxidation of acrolein. US Patent 3845120.
9. Nakajima, A. and Fukumoto, N. (2010) Process for producing acrylic acid. WO Patent 2009028292.
10. Beerthuis, R., Rothenberg, G., and Raveendran Shiju, N. (2015) *Green Chem.*, **17**, 1341–1361.
11. Deleplanque, J., Dubois, J.-L., Devaux, J.-F., and Ueda, W. (2010) *Catal. Today*, **157**, 351–358.
12. Omata, K., Matsumoto, K., Murayama, T., and Ueda, W. (2014) *Chem. Lett.*, **43**, 435–437.
13. Shen, L., Yin, H., Wang, A., Lu, X., and Zhang, C. (2014) *Chem. Eng. J.*, **244**, 168–177.
14. Yun, Y.S., Lee, K.R., Park, H., Kim, T.Y., Yun, D., Han, J.W., and Yi, J. (2015) *ACS Catal.*, **5**, 82–94.
15. Liu, L., Wang, B., Du, Y., Zhong, Z., and Borgna, A. (2015) *Appl. Catal., B: Environ.*, **174–175**, 1–12.
16. Possato, L.G., Cassinelli, W.H., Garetto, T., Pulcinelli, S.H., Santilli, C.V., and Martins, L. (2015) *Appl. Catal., A: Gen.*, **492**, 243–251.
17. Chieregato, A., Soriano, M.D., Garcia-Gonzalez, E., Puglia, G., Basile, F., Concepción, P., Bandinelli, C., López Nieto, J.M., and Cavani, F. (2015) *ChemSusChem*, **8**, 398–406.
18. Chieregato, A., Basile, F., Concepción, P., Guidetti, S., Liosi, G., Soriano, M.D., Trevisanut, C., Cavani, F., and López Nieto, J.M. (2012) *Catal. Today*, **197**, 58–65.
19. Chieregato, A., Soriano, M.D., Basile, F., Liosi, G., Zamora, S., Concepción, P., Cavani, F., and López Nieto, J.M. (2014) *Appl. Catal., B: Environ.*, **150–151**, 37–46.
20. Katryniok, B., Paul, S., and Dumeignil, F. (2013) *ACS Catal.*, **3** (8), 1819–1834.
21. Katryniok, B., Paul, S., Bellière-Baca, V., Rey, P., and Dumeignil, F. (2010) *Green Chem.*, **12**, 2079–2098.
22. Witsuthammakul, A. and Sooknoi, T. (2012) *Appl. Catal., A: Gen.*, **413–414**, 109–116.

23. Massa, M., Andersson, A., Finocchio, F., Busca, G., Lenrick, F., and Wallenberg, L.R. (2013) *J. Catal.*, **297**, 93–109.
24. Liu, R., Wang, T., Cai, D., and Jin, Y. (2014) *Ind. Eng. Chem. Res.*, **53** (21), 8667–8674.
25. Sheldon, R.A. (2014) *Green Chem.*, **16**, 950–963.
26. Vollenweider, S. and Lacroix, C. (2004) *Appl. Microbiol. Biotechnol.*, **64**, 16–27.
27. Thompson, B., Moon, T.S., and Nielsen, D.R. (2014) *Curr. Opin. Biotechnol.*, **30**, 17–23.
28. Novozymes (0000) http://www.novozymes.com/en/news/news-archive/Pages/BASF,-Cargill-and-Novozymes-achieve-milestone.aspx (accessed 24 September 2015).
29. Novozymes (0000) http://www.novozymes.com/en/news/news-archive/Pages/Novozymes-Cargill-continue-bio-acrylic-acid-partnership-BASF-exits.aspx (accessed 24 September 2015).
30. (a) OPX Biotechnologies (0000) http://www.opxbio.com/2012/09/the-commercialization-of-bioacrylic-acid/ (accessed 24 September 2015); (b) Bomgardner, M. (2011) *Chem. Eng. News Arch.*, **89** (16), 9; (c) Decoster, D., Hoyt, S., and Roach, S. (2013) Dehydration of 3-hydroxypropionic acid to acrylic acid. WO Patent 2103192451.
31. Liebig, C., Paul, S., Katryniok, B., Guillon, C., Couturier, J.-L., Dubois, J.-L., Dumeignil, F., and Hölderich, W.F. (2013) *Appl. Catal., B: Environ.*, **170**, 132–133.
32. Guerrero-Perez, M.O. and Banares, M.A. (2008) *ChemSusChem*, **1** (6), 511–513.
33. Liebig, C., Paul, S., Katryniok, B., Guillon, C., Couturier, J.-L., Dubois, J.-L., Dumeignil, F., and Hölderich, W.F. (2014) *Appl. Catal., B: Environ.*, **148–149**, 604–605.
34. Alliati, I.M. and Irigoyen, B.L., (2015) *Catal. Today*, **254**, 53–61.
35. Zhang, Y., Ma, T., and Zhao, J. (2014) *J. Catal.*, **313**, 92–103.
36. Calvino-Casilda, V., Guerrero-Pérez, M.O., and Banares, M.A. (2009) *Green Chem.*, **11**, 939–941.
37. Tullo, A.H. (2013) *Chem. Eng. News*, **91**, 18.
38. Holmen, R.E. (1958) Production of acrylates by catalytic dehydration of lactic acid and alkyl lactates. US Patent 2859240.
39. Dusselier, M., Van Wouwe, P., Dewaele, A., Makshina, E., and Sels, B.F. (2013) *Energy Environ. Sci.*, **6** (5), 1415–1442.
40. Montassier, C., Giraud, D., and Barbier, J. (1988) in *Polyol Conversion by Liquid Phase Heterogeneous Catalysis Over Metals* (eds Guisnet, M., Barrault, J., Bouchoule, C., Duprez, D., Montassier, C. and Pérot, G.), Elsevier, pp. 165–170.
41. Montassier, C., Ménézo, J.C., Hoang, L.C., Renaud, C., and Barbier, J. (1991) *J. Mol. Catal.*, **70** (1), 99–110.
42. Maris, E.P. and Davis, R.J. (2007) *J. Catal.*, **249** (2), 328–337.
43. Auneau, F., Sadr Arani, L., Besson, M., Djakovitch, L., Michel, C., Delbecq, F., Sautet, P., and Pinel, C. (2012) *Top. Catal.*, **55** (7–10), 474–479.
44. Auneau, F., Noël, S., Aubert, G., Besson, M., Djakovitch, L., and Pinel, C. (2011) *Catal. Commun.*, **16** (1), 144–149.
45. Roy, D., Subramaniam, B., and Chaudhari, R.V. (2011) *ACS Catal.*, **1** (5), 548–551.
46. Ftouni, J., Villandier, N., Auneau, F., Besson, M., Djakovitch, L., and Pinel, C. (2015) *Catal. Today*, **257** (2), 267–273.
47. Kishida, H., Jin, F., Zhou, Z., Moriya, T., and Enomoto, H. (2005) *Chem. Lett.*, **34** (11), 1560–1561.
48. Shen, Z., Jin, F., Zhang, Y., Wu, B., Kishita, A., Tohji, K., and Kishida, H. (2009) *Ind. Eng. Chem. Res.*, **48** (19), 8920–8925.
49. Ramírez-López, C.A., Ochoa-Gómez, J.R., Fernández-Santos, M., Gómez-Jiménez-Aberasturi, O., Alonso-Vicario, A., and Torrecilla-Soria, J. (2010) *Ind. Eng. Chem. Res.*, **49** (14), 6270–6278.
50. Shen, Y., Zhang, S., Li, H., Ren, Y., and Liu, H. (2010) *Chem. Eur. J.*, **16** (25), 7368–7371.
51. Aida, T.M., Ikarashi, A., Saito, Y., Watanabe, M., Lee Smith, R. Jr., and

Arai, K. (2009) *J. Supercrit. Fluids*, **50** (3), 257–264.

52. Mok, W.S.L., Antal, M.J., and Jones, M. (1989) *J. Org. Chem.*, **54** (19), 4596–4602.
53. Hatada, K., Hakuta, Y., and Ikushima, Y. (2005) Multi-organic compounds synthetic system with high temperature and high pressure water. JP Patent 2005089428.
54. Hatada, K., Ikushima, Y., Sato, O., and Saito, I. (2004) Method for synthesizing acrylic acid and/or pyruvic acid. JP Patent 3873123.
55. Katryniok, B., Paul, S., and Dumeignil, F. (2010) *Green Chem.*, **12** (11), 1910–1913.
56. Wadley, D.C., Tam, M.S., Kokitkar, P.B., Jackson, J.E., and Miller, D.J. (1997) *J. Catal.*, **165** (2), 162–171.
57. Montgomery, R. (1952) *J. Am. Chem. Soc.*, **74** (6), 1466–1468.
58. Vu, D.T., Kolah, A.K., Asthana, N.S., Peereboom, L., Lira, C.T., and Miller, D.J. (2005) *Fluid Phase Equilib.*, **236** (1–2), 125–135.
59. Takafumi, A. and Hieda, S. (1993) Process for preparing unsaturated carboxylic acid or ester thereof. US Patent 5250729.
60. Wang, H., Yu, D., Sun, P., Yan, J., Wang, Y., and Huang, H. (2008) *Catal. Commun.*, **9** (9), 1799–1803.
61. Sun, P., Yu, D., Fu, K., Gu, M., Wang, Y., Huang, H., and Ying, H. (2009) *Catal. Commun.*, **10** (9), 1345–1349.
62. Sun, P., Yu, D., Tang, Z., Li, H., and Huang, H. (2010) *Ind. Eng. Chem. Res.*, **49** (19), 9082–9087.
63. Yan, J., Yu, D., Li, H., Sun, P., and Huang, H. (2010) *J. Rare Earths*, **28** (5), 803–806.
64. Yu, D., Sun, P., Tang, Z., Li, Z., and Huang, H. (2011) *Can. J. Chem. Eng.*, **89** (3), 484–490.
65. Yan, J., Yu, D., Sun, P., and Huang, H. (2011) *Chin. J. Catal.*, **32** (3–4), 405–411.
66. Shi, H.F., Hu, Y.C., Wang, Y., and Huang, H. (2007) *Chin. Chem. Lett.*, **18** (4), 476–478.
67. Näfe, G., Lóper-Martínez, M.-A., Dyballa, M., Hunger, M., Traa, Y., Hirth, T., and Klemm, E. (2015) *J. Catal.*, **329**, 413–424.
68. Deka, R.C., Kinkar Roy, R., and Hirao, K. (2000) *Chem. Phys. Lett.*, **332** (5–6), 576–582.
69. Sawicki, R.A. (1988) Catalyst for dehydration of lactic acid to acrylic acid. US Patent 4729978.
70. Paparizos, C., Dolhyj, S.R., and Shaw, W.G. (1988) Catalytic conversion of lactic acid and ammonium lactate to acrylic acid. US Patent 4786756.
71. Gunter, G.C., Miller, D.J., and Jackson, J.E. (1994) *J. Catal.*, **148** (1), 252–260.
72. Gunter, G.C., Langford, R.H., Jackson, J.E., and Miller, D.J. (1995) *Ind. Eng. Chem. Res.*, **34** (3), 974–980.
73. Gunter, G.C., Craciun, R., Tam, M.S., Jackson, J.E., and Miller, D.J. (1996) *J. Catal.*, **164** (1), 207–219.
74. Zhang, J., Lin, J., Xu, X., and Cen, P. (2008) *Chin. J. Chem. Eng.*, **16** (2), 263–269.
75. Lee, J.-M., Hwang, D.-W., Hwang, Y.K., Halligudi, S.B., Chang, J.-S., and Han, Y.-H. (2010) *Catal. Commun.*, **11** (15), 1176–1180.
76. Zhang, J., Zhao, Y., Pan, M., Feng, X., Ji, W., and Au, C.-T. (2010) *ACS Catal.*, **1** (1), 32–41.
77. Zhang, J., Zhao, Y., Feng, X., Pan, M., Zhao, J., Ji, W., and Au, C.-T. (2014) *Catal. Sci. Technol.*, **4** (5), 1376–1385.
78. Peng, J., Li, X., Tang, C., and Bai, W. (2014) *Green Chem.*, **16** (1), 108–111.
79. Hong, J.H., Lee, J.-M., Kim, H., Hwang, Y.K., Chang, J.-S., Halligudi, S.B., and Han, Y.-H. (2011) *Appl. Catal., A: Gen.*, **396** (1–2), 194–200.
80. Lingoes, J., Villalobos, C., and Dimitris, I. (2013) Catalytic conversion of lactic acid to acrylic acid. WO Patent 2013155245.
81. Tang, C., Peng, J., Fan, G., Li, X., Pu, X., and Bai, W. (2014) *Catal. Commun.*, **43**, 231–234.
82. Blanco, E., Delichere, P., Millet, J.-M.M., and Loridant, S. (2014) *Catal. Today*, **226**, 185–191.
83. Tang, C., Peng, J., Li, X., Zhai, Z., Jiang, N., Bai, W., Gao, H., and Liao, Y. (2014) *RSC Adv.*, **4** (55), 28875–28882.

84. Ghantani, V.C., Lomate, S.T., Dongare, M.K., and Umbarkar, S.B. (2013) *Green Chem.*, **15** (5), 1211–1217.
85. Tsuchida, T., Kubo, J., Yoshioka, T., Sakuma, S., Takeguchi, T., and Ueda, W. (2008) *J. Catal.*, **259** (2), 183–189.
86. Sebti, S., Tahir, R., Nazih, R., Saber, A., and Boulaajaj, S. (2002) *Appl. Catal., A: Gen.*, **228** (1–2), 155–159.
87. Gruselle, M., Kanger, T., Thouvenot, R., Flambard, A., Kriis, K., Mikli, V., Traksmaa, R., Maaten, B., and Tõnsuaadu, K. (2011) *ACS Catal.*, **1** (12), 1729–1733.
88. Lan, J. and Zhang, Z. (2015) *J. Ind. Eng. Chem.*, **23**, 200–205.
89. Zhao, K., Qiao, B., Zhang, Y., and Wang, J. (2013) *Chin. J. Catal.*, **34** (7), 1386–1394.
90. Hara, T., Mori, K., Mizugaki, T., Ebitani, K., and Kaneda, K. (2003) *Tetrahedron Lett.*, **44** (33), 6207–6210.
91. Matsuura, Y., Onda, A., and Yanagisawa, K. (2014) *Catal. Commun.*, **48**, 5–10.
92. Onda, A., Matsuura, Y., and Yanagisawa, K. (2012) Method for synthesizing unsaturated carboxylic acid and/or derivative of same. US Patent 8772539.
93. Matsuura, Y., Onda, A., Ogo, S., and Yanagisawa, K. (2014) *Catal. Today*, **226**, 192–197.
94. Dongare, M.K., Umbarkar, S., and Lomate, S.T. (2014) An improved process for catalytic dehydration of lactic acid to acrylic acid. WO Patent 2012156921.
95. Ghantani, V.C., Dongare, M.K., and Umbarkar, S.B. (2014) *RSC Adv.*, **4** (63), 33319–33326.
96. Yan, B., Tao, L.-Z., Liang, Y., and Xu, B.-Q. (2014) *ACS Catal.*, **4** (6), 1931–1943.
97. Beerthuis, R., Granollers, M., Brown, D.R., Salavagione, H.J., Rothenberg, G., and Shiju, R.N. (2015) *RSC Adv.*, **5**, 4103–4108.
98. Iwamoto, M., (2015) *Catal. Today*, **242** (Part B), 243–248.
99. Sun, D., Yamada, Y., and Sato, S. (2015) *Appl. Catal., B: Environ.*, **174**, 13–20.
100. Global-Bioenergies (0000) http://www.global-bioenergies.com/wp-content/uploads/2015/08/141208_cp_fr.pdf.
101. Kuchta, R.D. and Abeles, R.H. (1985) *J. Biol. Chem.*, **260** (24), 13181–13189.
102. O'Brien, D.J., Panzer, C.C., and Eisele, W.P. (1990) *Biotechnol. Progr.*, **6** (4), 237–242.
103. Dalal, R.K., Akedo, M., Cooney, C.L., and Sinskey, A.J. (1980) *Biosources Dig.*, **2**, 89–97.
104. Danner, H., Ürmös, M., Gartner, M., and Braun, R. (1998) *Appl. Biochem. Biotechnol.*, **70–72** (1), 887–894.
105. Luo, W., Cai, J., Zhu, L., Zhu, X., Huang, L., Xu, Z., and Cen, P. (2012) *Eng. Life Sci.*, **12** (5), 567–573.
106. Straathof, A.J., Sie, S., Franco, T.T., and Van Der Wielen, L.M. (2005) *Appl. Microbiol. Biotechnol.*, **67** (6), 727–734.
107. Lunelli, B.H., Duarte, E.R., Vasco de Toledo, E.C., Wolf Maciel, M.R., and Maciel Filho, R. (2007) *Appl. Biochem. Biotechnol.*, **137–140** (1–12), 487–499.
108. Lunelli, B.H., Rivera, E.C., Vasco de Toledo, E.C., Wolf Maciel, M.R., and Maciel Filho, R. (2008) in *Biotechnology for Fuels and Chemicals* (eds W.S. Adney, J.D. McMillan, J. Mielenz, and K.T. Klasson), Humana Press, pp. 693–705.
109. Straathof, A.J.J. (2013) *Chem. Rev.*, **114** (3), 1871–1908.

10
Production of Ethylene and Propylene Glycol from Lignocellulose

Anna Katharina Beine, Peter J.C. Hausoul, and Regina Palkovits

10.1
Introduction

10.1.1
Motivation

Ethylene and propylene glycols are petrochemical-based compounds that are produced on a large scale. Their main areas of application relate to the manufacturing of polymers, antifreeze, cosmetics, and food additives. In recent years, renewed interest has been devoted to suitable catalytic processes for the preparation of ethylene and propylene glycol based on renewable feedstocks. Especially glycerol, a major by-product of biodiesel production formed in amounts as high as 226 kg per ton of biodiesel produced, presents an interesting substrate [1]. Numerous studies addressed a selective transformation of glycerol into 1,2- and 1,3-propylene glycol as well as ethylene glycol. Several recent reviews allow a comprehensive overview about the current state of the art with regard to catalyst development and process design for the valorization of glycerol [2]. However, vegetable oils utilized in biodiesel manufacturing result in a competition with food production and are critically discussed with regard to sustainable land use.

Polysaccharides present alternative renewable feedstocks for production of propylene and ethylene glycol. In this regard, especially cellulose and hemicellulose as major components of lignocellulose bear a high potential as raw materials. The annual worldwide biomass production on land is estimated to sum up to $170-200 \times 10^9$ kton with around 70% of lignocellulose as major component [3]. Lignocellulose does not compete with food industry and is readily available as a part of various waste streams of, for example, agriculture, forestry, and paper industry. Lignocellulose itself is a composite of the natural polymers cellulose, hemicellulose, and lignin. Cellulose and hemicellulose are polysaccharides consisting of hexoses and pentoses, while aromatic alcohols present the building blocks of lignin. Dependent on the type and source of lignocellulose, the amount of the three components can strongly vary. Nevertheless, average plant cell walls are composed of 50–70% cellulose, 10–40% hemicellulose, and 10–30% lignin

Chemicals and Fuels from Bio-Based Building Blocks, First Edition.
Edited by Fabrizio Cavani, Stefania Albonetti, Francesco Basile, and Alessandro Gandini.

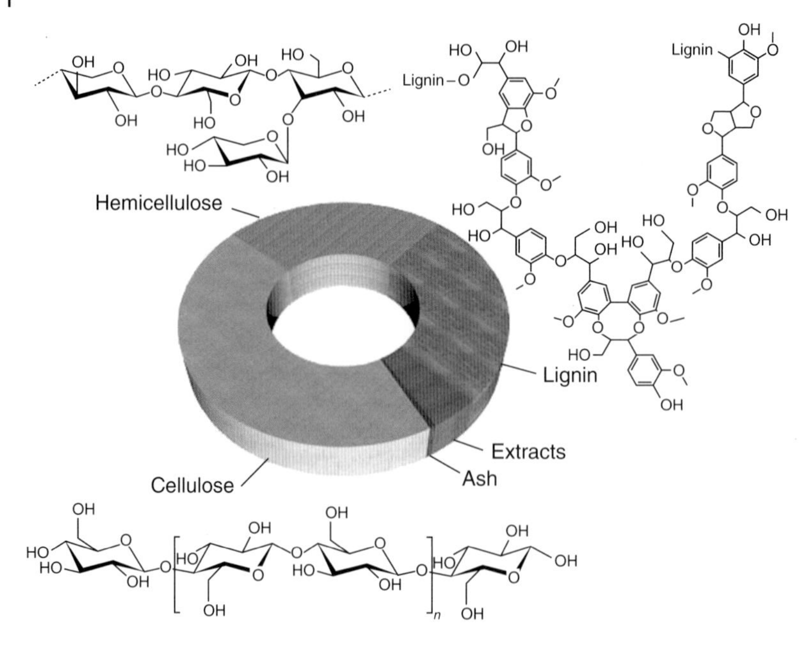

Figure 10.1 Illustration of the main constituents of lignocellulose [2c]. Agnieszka et al. Reproduced with kind permisison from John Wiley & Sons.

(Figure 10.1). Consequently, polysaccharides are available in significant amounts and enable potential access to a CO_2-neutral production of chemicals.

10.1.2
Early Examples

Already around one century ago, the capability of cellulosic raw materials as feedstocks for production of glycols has been pointed out. In 1933, a patent by Lautenschläger *et al.* [4] described glycerol and 1,2-propylene glycol as major products of the hydrogenolysis of cellulose over nickel- and copper-based catalysts at 200–300 °C. A process reported by Gürkan in 1949 emphasized the potential of glycol production based on a treatment of cellulose in water in the presence of hydrogen pressure and a suitable catalyst [5]. Therein, cellulosic material was first dissolved in an ammoniacal copper oxide solution (Schweizer's reagent) and subjected to hydrogen pressure. The treatment caused formation of finely dispersed copper nanoparticles within the cellulose matrix. In a second step, he proposed treatment of the material in an aqueous methanol solution at 250–270 °C and 69 bar hydrogen pressure, resulting in full conversion of cellulose into a mixture of short-chain products including propanol, hydroxyacetone, and 1,2-propylene glycol. Also other early studies on wood and cellulose liquefaction via hydrogenation name glycols as part of a broad product spectrum. Concerning a technical application, Russian scientists described a one- and a two-step process for the transformation of agricultural and municipal waste into hexitols, pentitols, and short-chain polyols including glycols [6]. The two-stage process relied on hydrolysis of purified cellulose to glucose in the first step catalyzed by concentrated mineral acids. After acid removal, a base was added to the aqueous glucose solution, followed by hydrogenation in the presence of a nickel catalyst at 120 °C and 60 – 150 bar hydrogen pressure. The reaction products under those comparatively mild reaction temperatures were predominantly hexitols. In contrast, at temperatures of 220 – 240 °C, glucose underwent hydrogenolysis with glycerol, propylene, and ethylene glycol as major products. The one-step process was catalyzed by small amounts of mineral acids and a ruthenium catalyst at 160 °C combining cellulose hydrolysis and hydrogenation of released glucose to mainly yield hexitols. Despite the technical potential of these early examples, comprehensive discussions concerning the involved reaction pathways and mechanisms were not available up until recently.

10.2 Reaction Mechanism

10.2.1 Possible Transformation Schemes

In general, one may assume that a transformation of polysaccharides such as cellulose and hemicellulose into glycols covers their depolymerization to the corresponding monosaccharides followed by hydrogenolysis enabling the necessary C–C and C–O bond cleavage. Therein, the term *hydrogenolysis* does not comprise a certain type of reaction mechanism but summarizes all involved processes. As mentioned in Section 10.1.2, processes to transform polysaccharides into glycols can be carried out in two steps, separating hydrolysis and subsequent hydrogenolysis (Scheme 10.1). Even a further breakdown into a process sequence involving (i) hydrolysis of polysaccharides into monosaccharides and (ii) their hydrogenation into the corresponding sugar alcohols followed by (iii) hydrogenolysis to glycols is possible. These sequential strategies bear the major advantage that reaction conditions and catalysts of the individual steps can be tailored enabling high selectivity. Considering the broad set of possible products, such an approach – despite

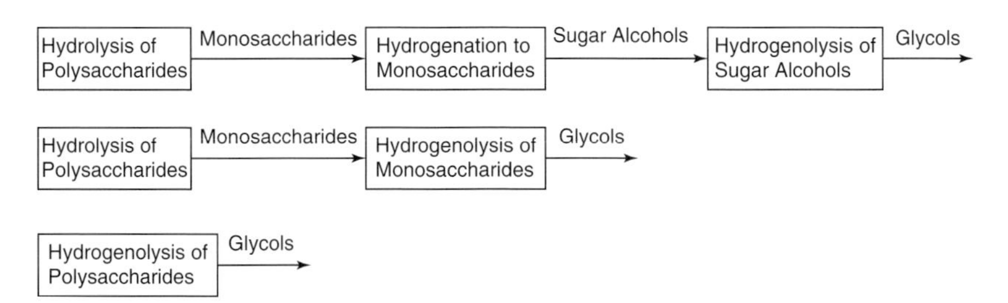

Scheme 10.1 Different process configurations of the transformation of polysaccharides into glycols covering three (top)-, two (middle)-, and one (bottom)-step processes.

higher capital costs – clearly facilitates selectivity control. Alternatively, one-pot strategies are possible. They are attractive due to simplicity and significantly lower investment costs. However, the advantages are opposed to the need for highly selective catalysts and precise control of all process parameters.

Recently, Negahdar *et al.* [7] provided a comprehensive investigation of sorbitol formation based on cellobiose and cello-oligomers as model substrates for cellulose. In general, most studies on hydrogenation or hydrogenolysis of cellulose consider a sequential process. First hydrolysis of polysaccharides to monosaccharides occurs, followed by hydrogenation or hydrogenolysis of these monosaccharides to sugar alcohols and glycols, respectively. Interestingly, hydrogenation of cellobiose, a dimer of glucose, in the presence of an acid catalyst and a metal catalyst for hydrogenation confirmed an alternative reaction pathway to play a crucial role. Therein, cellobiose underwent hydrogenation to cellobitol (3-β-D-glucopyranosyl-D-glucitol) as main reaction intermediate [8]. A thorough kinetic analysis confirmed a facilitated hydrolysis of cellobitol compared to cellobiose, furthering this alternative reaction pathway. Both trisaccharides based on α-1,4- and β-1,4-glycosidic linkages emphasized reaction pathways via these hydrogenated intermediates to be dominant at lower temperature or low acid concentration compared to simple hydrolysis. A study using maltodextrins and cello-oligomers with a degree of polymerization of 2–7, substantiated these observations further. Consequently, a hydrogenation or even hydrogenolysis of a polysaccharide chain end enables direct release of sorbitol or glycols into the reaction solution without prior depolymerization into monosaccharides.

10.2.2 Undesired Side Reactions

To gain an understanding of parameters influencing the efficient transformation of polysaccharides into glycols, the main reactions occurring in the course of hydrogenolysis need to be discussed. A key point with regard to undesired reaction pathways relates to the hydrolysis of polysaccharides to monosaccharides (Scheme 10.2). Hydrolysis is a well-established acid-catalyzed process. Hydrolysis of cellulose has even been employed on a technical scale in the form of wood saccharification at the beginning of the twentieth century [9]. Different processes were used including the Bergius–Rheinau and the Scholler–Tornesch process, which mainly differ in terms of type of acid, temperature, and acid concentration. In recent years, hydrolysis has been revisited intensively to optimize reaction conditions and to explore the potential of solid acids to reduce salt formation associated with traditional wood saccharification processes [10]. Nevertheless, a major challenge of the acid-catalyzed polysaccharide hydrolysis concerns a subsequent acid catalyst dehydration of the formed monosaccharides to, for example, furfural and 5-hydroxymethylfurfural. These compounds inhibit not only subsequent fermentation to ethanol but present main sources of undesired polymerization reactions toward humins. Another challenge arises in hydrogenating monosaccharides in the presence of acids or bases. Dependent on the

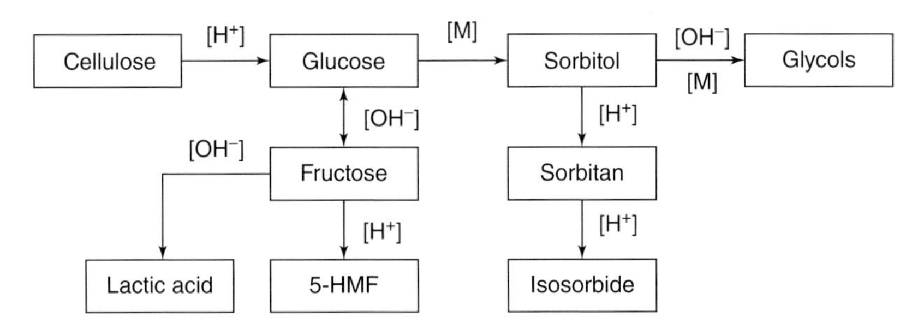

Scheme 10.2 Simplified representation of major undesired reaction pathways for a transformation of cellulose to glycols as selected example (5-HMF – 5-hydroxymethylfurfural). (Please note that also further isomerization reactions can occur, delivering mannose in addition to fructose and mannitol/iditol together with the corresponding dehydration products for sorbitol.)

relative rate of monosaccharide hydrogenation, again undesired side reactions can occur. In particular, the base-catalyzed isomerization of glucose to fructose and mannose and subsequent base-catalyzed transformation to lactic acid are major side-reactions. Lactic acid is assumed to form via β-elimination of the hydroxyl group and a benzylic acid-like rearrangement (or an intramolecular Cannizzaro-type reaction) of dihydroxyacetone and glyceraldehyde products of retro-aldolization of fructose [11]. In addition, acid-catalysed dehydration of sugar alcohols leads to cyclic products (e.g., sorbitan and isosorbide in the case of sorbitol). Especially isosorbide is highly stable in the presence of acids but also in the presence of supported metal catalysts and hydrogen presenting a dead end for glycol formation.

10.2.3
C–C and C–O Bond Cleavage for Selective Glycol Formation

The selective formation of glycols via hydrogenolysis of cellulosic biomass is highly challenging due to the large number of possible reaction pathways available under typically employed reaction conditions (high T, high p, and H_2O). To understand the influence of the substrate, metal catalyst, and acid/base cocatalyst on the resulting product selectivities, it is imperative to be able to discern between different reaction mechanisms and identify the crucial selectivity-determining steps. Until now various reaction mechanisms have been proposed to account for the observed product mixtures of hydrogenolysis reactions [12]. In what follows only the most commonly accepted routes will be discussed, and a distinction will be made between metal-catalyzed and acid/base-catalyzed reactions.

First, it should be clear that polyols are relatively stable compounds that can be heated to high temperatures (200–250 °C) in aqueous solutions without any reaction occurring. As discussed in Section 10.2.2, the addition of catalytic amounts of acid leads, via dehydration, to cyclization products such as anhydroxylitol, sorbitan, and isosorbide. Thus under metal-free conditions, no C–C cleavage reactions occur. In contrast when carbohydrates (i.e., aldoses and ketoses) such as xylose, glucose, or fructose are heated under acidic, neutral, or basic conditions, various reactions occur. Besides isomerization, dehydration, and polymerization reactions, also fragmentation reactions (i.e., retro-aldolization) are typically observed [13]. These reactions are essentially enabled by the aldehyde/ketone functionality of the carbohydrate. The metal catalysts employed for hydrogenolysis reactions are in many cases suitable hydrogenation catalysts [2c, 14]. Since hydrogenolysis is typically performed under high H_2 pressure, it would be expected that no reaction occurs as the equilibrium is shifted toward the hydrogenated products. However, via micro reversibility, the metal catalyst enables the formation of minor amounts of aldose and ketose sugars and as such facilitates subsequent irreversible degradation reactions. Hence, the first and foremost function of the metal catalyst is dehydrogenation and hydrogenation. Scheme 10.3 shows that the dehydrogenation proceeds via initial adsorption of the alcohol to yield a surface-bound alkoxide. Subsequent activation of the C–H bond results in the formation of a surface-bound carbonyl that depending on metal and temperature can detach.

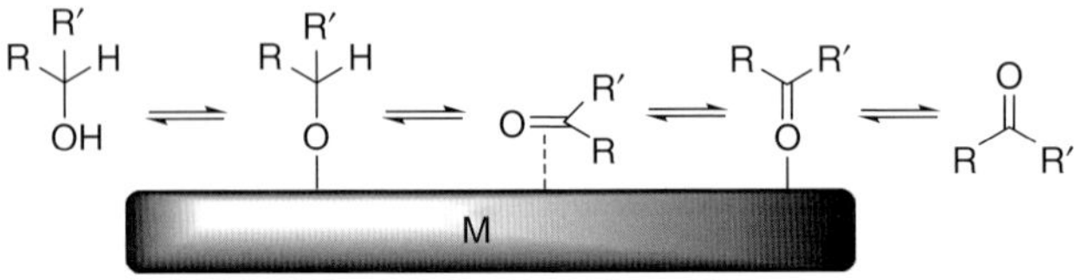

Scheme 10.3 Simplified representation of (de)hydrogenation mechanism on metal surfaces.

Following carbonyl formation, C–C cleavage via retro-aldol addition can occur [15]. This reaction can be catalyzed by both acids and bases but proceeds more easily and selectively in the presence of bases at temperatures between 150 and 250 °C. Under neutral or acidic conditions, the retro-aldol reaction only occurs at very high temperatures. Prerequisite for the retro-aldol reaction is a β-hydroxy carbonyl structure motif (Scheme 10.4). Cleavage occurs between the

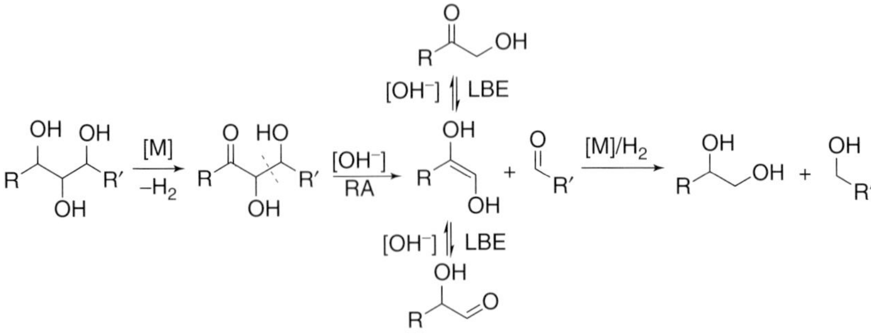

Scheme 10.4 Retro-aldol addition and Lobry de Bruyn–Alberda van Ekenstein (LBE) isomerization under basic conditions.

α- and β-carbon, leading to the formation of an aldehyde and a 1,2-enediol. The 1,2-enediol will in turn isomerize to either an aldehyde or ketone via the Lobry de Bruyn–Alberda van Ekenstein (LBE) transformation [16]. Finally the newly formed carbonyls are hydrogenated to the respective polyols.

As an example, Scheme 10.5 shows the reaction network of hexitol hydrogenolysis under basic conditions via a combination of dehydrogenation, retro-aldol, and LBE reactions. Depending on the site of dehydrogenation, aldose, 2-ketose, and 3-ketose intermediates can be obtained. Even if the metal catalyst has a high preference for the dehydrogenation of a certain position (e.g., 1-position), it is still possible to obtain the other isomers via LBE isomerization. Also in the presence of Lewis acidic catalysts, isomerization can occur via a 1,2-hydride shift [17]. Retro-aldol cleavage of the aldose leads to a C_2/C_4 split, whereas cleavage of the 2-ketose leads to a C_3/C_3 split. Cleavage of the 3-ketose can follow two different courses, that is, either resulting in a C_5/C_1 split or C_2/C_4 split. Thus it is clear that the retro-aldol reaction can account for the formation of smaller-chain polyols such as ethylene glycol, glycerol, tetritols, and pentitols. Via the same reaction pathway, pentitols, tetritols, and glycerol can be further converted to ultimately give ethylene glycol and C_1 compounds (e.g., formaldehyde, methanol, and CO_2).

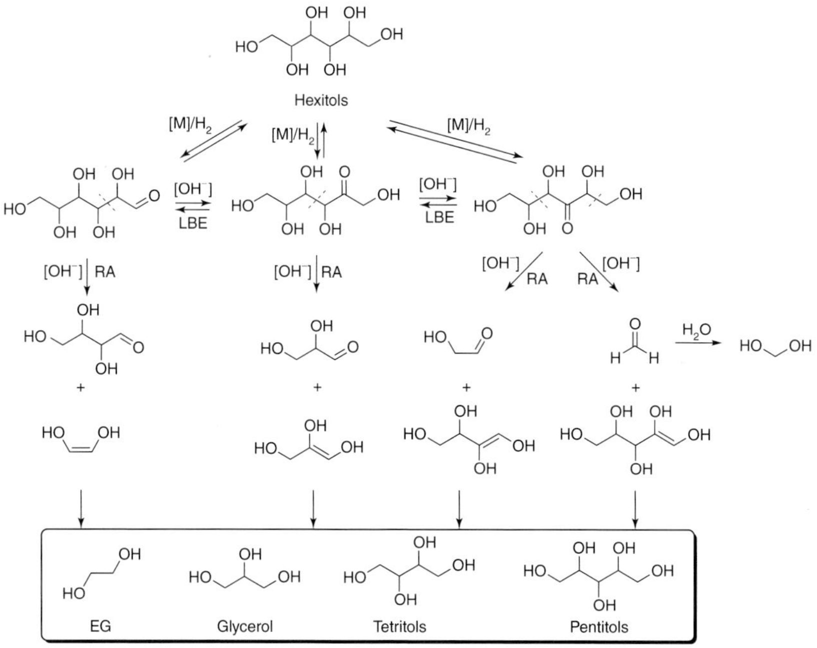

Scheme 10.5 Reaction network of the retro-aldol-based hydrogenolysis of hexitols.

Besides retro-aldol, also metal-catalyzed decarbonylation is readily observed under hydrogenolysis conditions [18]. This reaction involves initial aldose formation, activation of the aldehydic proton leading to a surface-bound acyl species, and C–C bond cleavage to give an adsorbed CO and an alkyl species (Scheme 10.6). Addition of hydrogen to these species ultimately leads to the formation of the corresponding C_{n-1} polyol and methane. Particularly Ru is known to catalyze decarbonylation under relatively mild conditions. Nevertheless, under more severe conditions, this reaction can also occur on various other metals (e.g., Rh, Ir, Pd).

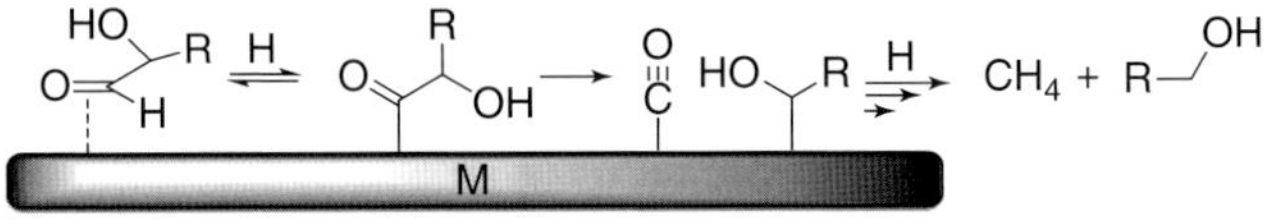

Scheme 10.6 Simplified representation of the decarbonylation mechanism on metal surfaces.

When applied to hexitols, decarbonylation results in the successive formation of pentitols, tetritols, glycerol, and ethylene glycol (Scheme 10.7). Thus although a different mechanism is at play, the same series of compounds as for the retro-aldol reaction is obtained. The major difference between these pathways is the evolution of the product spectrum. Namely, in the case of retro-aldol cleavage, ethylene glycol and glycerol are obtained directly from C_2/C_4 and C_3/C_3 splits, whereas for decarbonylation these compounds are only obtained after the initial formation of pentitols and tetritols. Furthermore decarbonylation does not end with ethylene glycol, and the reaction can proceed to full conversion to methane. Therefore selectivity control can only be achieved by optimization of reaction time and pressure. In contrast, the retro-aldol condensation does not convert ethylene glycol, and the selectivity of the reaction is mostly dependent on the site of dehydrogenation (i.e., aldehyde formation vs. ketone formation). In principle retro-aldol-based hydrogenolysis of hexitols can give rise to 3 equiv of ethylene glycol via exclusive aldehyde formation (e.g., C_6 to C_2/C_4 and C_4 to C_2/C_2). Therefore the selectivity of the reaction will be strongly dependent on the dehydrogenation selectivity of the catalyst and the possible occurrence of LBE isomerization or 1,2-hydride shift reactions.

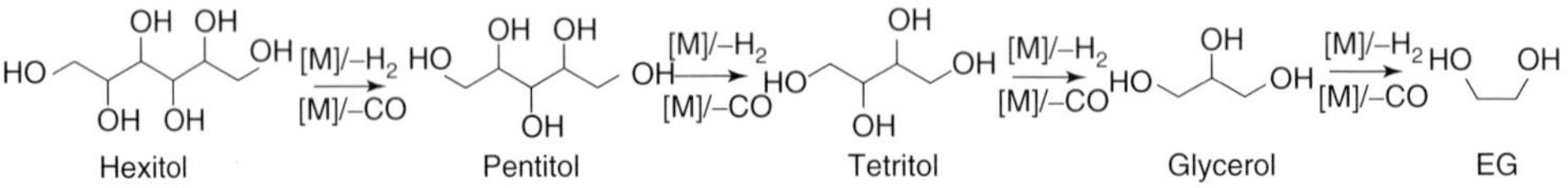

Scheme 10.7 Successive decarbonylation of hexitols toward ethylene glycol.

In addition to products originating from C–C cleavage, also products indicative of C–O cleavage are typically observed. In particular 1,2-propylene glycol is often formed in significant amounts. Most likely, glyceraldehyde is dehydrated via acid or base catalysis to give 2-hydroxyacrolein (Scheme 10.8). This is then rapidly isomerized to pyruvaldehyde and can give rise to either 1,2-propylene glycol via hydrogenation or lactic acid via hydration and base-catalyzed intramolecular hydride shift. Controlling the selectivity of the hydrogenolysis reaction toward

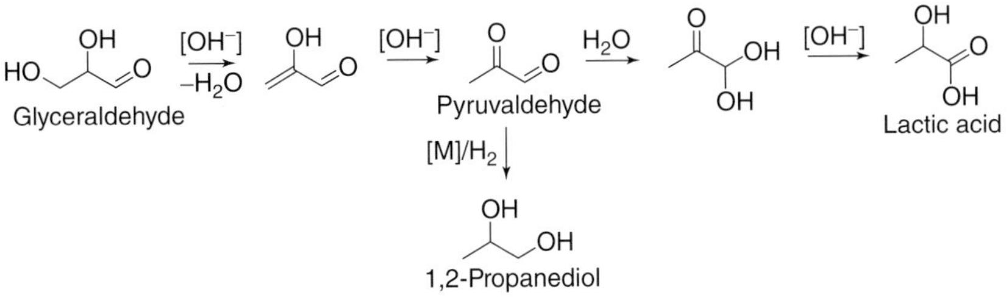

Scheme 10.8 Formation of 1,2-propylene glycol and lactic acid from glyceraldehyde.

1,2-propylene glycol will therefore require a catalyst that is able to favor the C_3/C_3 split of hexitols. However, as discussed earlier, retro-aldol addition is most efficient under basic conditions, and this will also favor lactic acid formation. Thus to enable selective formation of 1,2-propylene glycol, a catalyst system is required that is capable of selective ketose formation and retro-aldol reaction under neutral to acidic conditions.

There are also strong indications that C–O cleavage may also be catalyzed by the metal catalyst or the support. The surfaces of oxidic supports possess both Lewis acid and Lewis base properties that can cause dehydration reactions. However deoxygenation is also observed for carbon-supported catalysts, highlighting the role of the metal catalyst. As indicated in Section 10.2.2, Brønsted acidic conditions lead to cyclodehydration via a S_N2-type mechanism. However in the presence of metal catalysts, deoxygenation generally follows a different course. More specifically the terminal hydroxyl groups of the polyol chain are sequentially removed with high selectivity suggesting an E_2/E_1 type mechanism. The reaction sequence is generally considered to proceed via dehydration, enol/keto tautomerization, and hydrogenation steps (Scheme 10.9). Since both retro-aldol reaction and decarbonylation give rise to series of $C_{n<6}$ polyols, hydrodeoxygenation can render a large variety of dehydroxylation products. Depending on the reaction conditions, alkanes can also be obtained from the polyols.

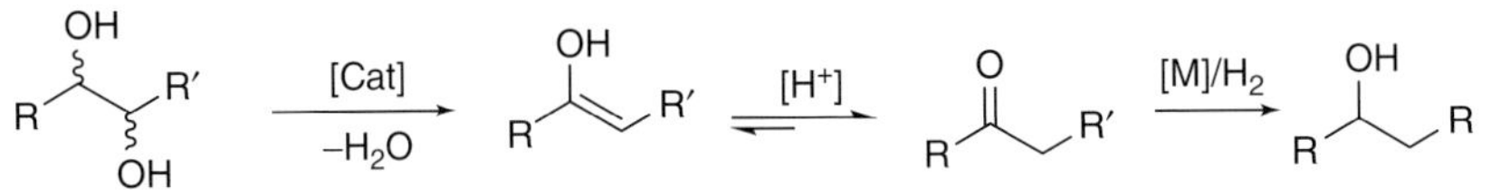

Scheme 10.9 Hydrodeoxygenation via dehydration and hydrogenation.

10.3 Glycol Production

Most catalysts employed in the transformation of mono- and polysaccharides to glycols are typical hydrogenation catalysts including supported noble metals such as ruthenium and platinum as well as base metal catalysts with major focus on nickel and copper. Essential steps of a selective glycol formation comprehend dehydrogenation of sugar alcohols as substrates, C–C and C–O bond cleavage, and subsequent rehydrogenation. To enhance the retro-aldol reaction, base promoters or additional metals were most frequently used in literature. Metal catalysts were supported on materials such as activated carbon, silica, alumina, and zeolites. However, one has to point out that a lot of conventional support materials are not stable under the hydrothermal reaction conditions usually applied in glycol production. The dissolution of silica, boehmite formation for alumina carriers, and insufficient stability of several zeolites have to be stressed [19]. In contrast, carbon materials exhibit excellent stability under those reaction conditions. The following sections provide an overview of potential catalyst systems facilitating access to glycol formation based on mono- and polysaccharides as well as starting from sugar alcohols such as xylitol and sorbitol. The discussion is structured into noble and base metal catalysts. For a comprehensive literature overview, we would like to refer the reader to available literature reviews, while herein only selected examples will be discussed illustrating recent developments and remaining challenges [2c, 14, 20].

10.3.1 Ruthenium Catalysts

10.3.1.1 C_5 and C_6 Sugar Alcohols and Monosaccharides

Sorbitol and glucose can be used in a one-step reaction to produce glycols. Using ruthenium-based catalysts on the conversion of sorbitol into glycols, mainly carbon-based supports are discussed in literature. Reactions were preferentially carried out in water as solvent adding a suitable base to achieve high selectivity. Base addition accelerates retro-aldolization as crucial step in the selective formation of glycols. Applying Ru/C as catalyst with a metal loading of 3 wt%, a conversion of 71% could be achieved within 4 h at 220 °C [21]. CaO served as basic promoter, and selectivities toward ethylene and propylene glycol of 11 and 16% were possible. Although the reactions appeared to occur rather slow, one needs to stress the low catalyst to substrate ratio with 1 : 44. Using a 5 wt% Ru/C catalyst at a catalyst to substrate ratio of 1 : 5 and a reaction temperatures of 200 °C, Soták *et al.* [22] reported full conversion within 45 min. $Ba(OH)_2$ was used to adjust the pH value of the reaction solution producing ethylene and propylene glycol with 18 and 34% selectivity, respectively. Also carbon nanofibers (CNFs) proved to be promising support materials for ruthenium catalysts. A 3 wt% ruthenium catalyst provides higher selectivity than the corresponding catalyst on activated carbon [23]. Overall, selectivities to ethylene and propylene glycol of 19% and 32% could be reached at a conversion of sorbitol of 86% [21]. Enhanced dispersion and electronic effects were made responsible for this observation.

Investigating the influence of base promoters, the group of Zhou *et al.* [24] found CaO to be the best suited promoter for the reaction. It not only provides OH^- to support the retro-aldol reaction, but Ca^{2+} also stabilizes formed aldehydes that are the starting point for C–C bond cleavage. In comparsion, NaOH, KOH, $Ba(OH)_2$, and $Mg(OH)_2$ did not perform as good. Additionally, the amount of base turned out to be crucial for high selectivity and conversion. With increasing amounts of base, the conversion increased, however, associated by a decrease in selectivity to glycols. A potential explanation relates to the increasing formation of side products such as lactic acid for high base concentrations. In the study of Zhou et al., the best results were obtained using CaO reaching a conversion of 19% and a selectivity toward ethylene and propylene glycol of 31 and 41%.

Taking a look at glucose as substrate, which can also be used to produce ethylene glycol in a one-step reaction, $Ru/NbOPO_4$ appeared to be a promising catalyst candidate [25]. Ethylene glycol yields of 37% were obtained within 6 h at 220 °C. Additionally, the monoether of ethylene glycol was formed with 28% yield. The only other product was propylene glycol with a yield of 5%. Table 10.1 summarizes the mentioned results.

Other interesting substrates for glycol production comprehend xylose as major monomer of hemicellulosic biomass together with the corresponding sugar alcohol xylitol. Montassier *et al.* [15b] used Ru/C for the conversion of xylose and also doped the catalyst with sulfur. Sulfur is a catalyst poison decreasing the activity of the catalyst by a factor of 4. However, this decrease of catalytic activity caused a clear increase in selectivity. The authors reported 84% combined yield of glycerol, ethylene glycol, and propylene glycol at conversion of 100%. Obviously, optimizing the relative reaction rates of the associated reactions within the sequence to glycols is important for high selectivity. Although catalyst poisoning is not the potential optimum situation, it offers a means to properly control catalytic activity for individual reaction steps. Using xylitol as starting material, Lee *et al.* [26] screened a huge number of catalysts supported on Al_2O_3 at 200 °C. They found 1 wt% Ru/Al_2O_3 to be most active providing an initial TOF of $92 \times 10^{-3}\ s^{-1}$. The selectivity toward ethylene glycol was found to be 10%. Propylene glycol was formed with 11% selectivity. The fact that besides these products also methanol, butanol, butanediol, and pentanol were found in the reaction mixture indicates that the reaction was not taking place over a retro-aldol mechanism but rather via decarbonylation and deoxygenation reactions.

Table 10.1 Results for ruthenium-based catalysts in the conversion of sorbitol.

Reference	Substrate	Catalyst	*m*(base) (g)	X (%)	S(ethylene glycol) (%)	S(propylene glycol) (%)
[21][a)]	Sorbitol	Ru/C (3 wt%)	10	71	11	16
[22][b)]	Sorbitol	Ru/C (5 wt%)	0.25	100	1	34
[21][a)]	Sorbitol	Ru/CNF (3 wt%)	10	9	19	34
[24][a)]	Sorbitol	Ru/CNF (3 wt%)	0.5	19	31	41
[25][c)]	Glucose	$Ru/NbOPO_4$	—	69	54	7

a) $T = 220\,°C$, $p(H_2) = 80$ bar, $t = 4$ h.
b) $T = 200\,°C$, $p(H_2) = 40$ bar, $t = 45$ min.
c) $T = 220\,°C$, $p(H_2) = 30$ bar, $t = 6$ h.

Also for xylose and xylitol, base addition was reported as main factor for high selectivity. Changing the support material and working with the addition of base, Sun et al. and Liu et al. [27] were able to increase both selectivity and activity of supported ruthenium catalysts. Screening different supports including TiO_2 and Al_2O_3, ruthenium supported on activated carbon showed better performance. The authors explained this observation by the different electronic properties of ruthenium on the supports and the modification of the adsorption behavior of xylitol on the catalyst surface. Additionally, one may consider the high specific surface area of activated carbon and the potentially resulting high metal dispersion. However, further investigations clarifying these points were not provided. Comparable to the observations made for sorbitol as substrate, ethylene and propylene glycol were only formed in trace amounts under neutral conditions. Instead of the retro-aldol reaction, isomerization to arabitol occurred as well as decarbonylation and deoxygenation emphasized by the formation of erythritol and threitol. Comparing base additives, $Ca(OH)_2$ was again found to be a more active base than $Mg(OH)_2$ and $CaCO_3$. After 1 h at 200 °C, a 4 wt% Ru/C catalyst allowed a xylitol conversion of 20% delivering ethylene and propylene glycol with 32 and 25% selectivity, respectively. However, also lactic acid was formed in significant amounts ($S = 17\%$). As mentioned in Section 10.2.5, lactic acid is a typical by-product of carbohydrate degradation in the presence of bases and certainly one of the major challenges to tackle for selective glycol formation. A promising result relates to good stability of the reported catalyst achieving stable activity and selectivity over six recycling runs.

10.3.1.2 **Polysaccharides**

In the case of cellulose and hemicellulose as substrates, a transformation to glycols comprehends depolymerization and hydrogenolysis to the desired products. Consequently, suitable catalysts and the selected reaction conditions need to facilitate both steps. Using supported ruthenium catalysts without modification or addition of acids or bases, rather high temperatures over 200 °C are necessary for an efficient depolymerization of untreated cellulose. For example, for Ru on HUSY, a yield of sugar alcohols of 30% could be achieved after 24 h at 190 °C illustrating the difficult conversion of the substrate [28]. In contrast, Hilgert *et al.* [29] demonstrated 100% conversion of a pretreated cellulose at 160 °C within 1 h. The main products were C_6 alditols, suggesting that no further C–C bond cleavage occurred under those mild conditions. The pretreatment step covered ball milling of cellulose in the presence of small amounts of an acid resulting in depolymerization to water-soluble cello-oligomers significantly facilitating a subsequent transformation. Also Xi *et al.* [25] used Ru/C as catalyst with methanol as solvent identifying sorbitol as major product, while only traces of ethylene or propylene glycol were formed. Also the influence of cellulose crystallinity was pointed out. The authors varied both the degree of crystallinity of cellulose from 85% to 33% applying a phosphoric acid treatment and the surface modifications of carbon nanotubes as catalyst support via oxidation with nitric acid [30]. Interestingly, they found Ru supported on strongly modified carbon nanotubes to be the most effective catalyst for the formation of sorbitol reaching up to 40% yield of hexitols for microcrystalline cellulose (85% crystallinity) and even up to 73% for pretreated cellulose. However, the applied reaction conditions (185 °C, 50 bar, 3 h) did not enable significant glycol formation. Luo *et al.* [12a] used Ru/C and demonstrated that cellulose can be efficiently converted at a temperature of 245 °C reaching 85–87% conversion within 30 min with and without a catalyst. Obviously, the high temperature enables fast hydrolysis of cellulose. In addition to sorbitol, sorbitan, and mannitol as main products with a combined yield of about 40%, further hydrogenolysis occurred, resulting in various other C_1–C_5 products including xylitol, erythritol, glycerol, propylene glycol, ethylene glycol, methanol, and methane. Facilitating an efficient conversion of cellulose at lower temperatures, the groups of Sels *et al.* and Palkovits *et al.* presented different concepts combining supported noble metal catalysts with the addition of molecular and solid acids [12d, 31]. These studies demonstrated efficient conversion of cellulose already at 160 °C. Though these systems could even be expanded to lignocellulose in the form of spruce as feedstock, hexitols remained the predominate reaction products.

Zheng *et al.* [32] introduced a promising bimetallic catalyst providing a significantly shifted selectivity. They used tungsten-promoted transition metals supported on activated carbon with 5% metal and 25% tungsten and water as solvent. For Ru–W/AC a conversion of 100% was reached after 30 min reaction at 245 °C. Therein, ethylene glycol was built with 62% selectivity. The authors explained this good catalytic behavior by the unique interaction of tungsten and transition metals of the groups 8, 9, and 10 in the periodic table of the elements. Nevertheless, it has to be taken into account that the catalyst to substrate ratio with 1 : 3 was high; therefore, a huge amount of catalyst is needed for the reaction. Interestingly, certain support materials facilitate a significant shift of the selectivity. $Ru/NbOPO_4$ presents such an example, already highlighted for the outstanding ethylene glycol selectivity with glucose as substrate (Table 10.1). A catalyst exhibiting 3 wt% of ruthenium loading allowed 96% conversion of ball-milled cellulose within 20 h at 220 °C and 30 bar hydrogen. The built products were again mainly ethylene glycol ($S = 27\%$) and its monoether ($S = 30\%$). The reason for this good performance might indeed be the synergetic effect between the two metals, where niobium is suggested to support the C–C bond cleavage while ruthenium is in charge of hydrogenation. Additionally, the increased acidity of the support needs to be considered potentially facilitating hydrolysis. Nevertheless, also the solvent has a great influence on the reaction selectivity. Using the same catalyst in water, the selectivity changed drastically, yielding mainly isosorbide emphasizing the acidic character of the support material. To form the desired products, primary alcohols were identified as optimum solvents being capable of stabilizing the carbonyl group formed as an intermediate to enable retro-aldolization. Adding nickel to this catalyst further improved selectivity. The resulting $3\%Ru-1.8\%Ni/NbOPO_4$ catalyst facilitated a selectivity to ethylene glycol of 30% together with 35% selectivity of the monoether.

10.3.2 Platinum Catalysts

10.3.2.1 C_5 and C_6 Sugar Alcohols and Monosaccharides

Platinum is a rarely used metal in the conversion of sorbitol and has nearly not been used for glucose. In one of the rare examples, 5 wt% Pt/C yielded $0.2\,mol_{glycols}\,g_{Pt}^{-1}\,h^{-1}$ at 20% conversion of sorbitol. Pressurizing the reactor with nitrogen instead of hydrogen, selectivity shifted completely, delivering eight times more lactic acid [33]. Using platinum on a silica–alumina support with acid sites, Li et al. and Huber et al. [34] were able to increase the selectivity toward the desired products. The optimized catalyst consisted of 4 wt% platinum and a ratio of $SiO_2 : Al_2O_3$ of 4 : 1. In a tubular flow reactor at 245 °C and 30 bar hydrogen pressure, a conversion of 19% could be reached producing ethylene and propylene glycol with 19% and 38% selectivity, respectively. The other products were mainly isosorbide and follow-up products such as CO_2. The mass balance did not close with around 30% of carbon missing. A reference experiment in the presence of the support material only showed a significant conversion of sorbitol to isosorbide with 95% selectivity. This led to the conclusion that mostly the metal centers were responsible for C–C bond cleavage and retro-aldol reactions. Because of the acidic support, a concurrence between dehydration and retro-aldol reaction occurs, causing insufficient selectivity to the desired products.

Using 8 wt% Pt/C, Sun et al. and Liu et al. [27] found a four times greater activity in the conversion of xylitol than for Ru/C, but the selectivity toward the desired products decreased. Ethylene glycol was now built with 25% and propylene glycol with 23% selectivity. Therein, lactic acid formed as main by-product under the applied basic reaction conditions. A system without basic promoter was investigated by Lee *et al.* [26] observing the opposite trend. When 3 wt% Pt/Al_2O_3 was used, a very low TOF of $4 \times 10^{-3}\,s^{-1}$ compared to the ruthenium-based catalyst was obtained. The selectivity to ethylene glycol of 9% and propylene glycol of 6% was also not significant. For the system of Sun *et al.*, decarbonylation and deoxygenation are the main reaction pathways. In contrast, for Pt/Al_2O_3 nearly no intermediates appeared, and the metal seems inactive for these reaction pathways.

Changing hydrogen to nitrogen as gas to pressurize the reaction using 5 wt% Pt/C, it becomes obvious that the formation of acids really is a side reaction that has to be taken seriously [33]. At 160 °C 20% of the used xylitol was converted, but the selectivity to glycols increases drastically to just 10%, while lactic acid was formed with a selectivity of 55%.

Xylose is faster converted compared to xylitol as substrate. Using 3.5 wt% Pt/Al_2O_3, 94% conversion of xylose occurs within 15 min at 160 °C [35]. Therein, glycols were formed with 29% and xylitol with 27% selectivity emphasizing a competition between retro-aldolization and direct hydrogenation. Furthermore, various other products could be observed, confirming the challenges associated with monosaccharides as substrates.

10.3.2.2 Polysaccharides

In contrast to ruthenium for monometallic platinum catalysts supported on activated carbon even under neutral pH, a promising selectivity to glycols occurs. Zheng *et al.* [32] reached a conversion of 64% of cellulose in 30 min at 245 °C. Ethylene glycol formed with a selectivity of 19%, while propylene glycol was a minor product with only 5% selectivity. Using a bimetallic Pt/$NbOPO_4$ catalyst, selectivity increased to 22% for ethylene and 14% for propylene glycol at 90% conversion; however, a reaction time of 20 h at 220 °C was necessary to achieve these results [25].

The described synergistic effect for a combination of ruthenium and tungsten could also be observed for supported platinum catalysts. Using Pt–W/AC led to the production of ethylene glycol ($S = 59\%$) rather than propylene glycol ($S = 3\%$) [32]. Changing to a platinum catalyst supported on tungstated alumina, selectivity changed completely. For a reaction temperature of 190 °C and 24 h reaction time, propylene glycol formed with 29% selectivity at 70% conversion of cellulose [36]. Additionally, acetol contributed with a selectivity of 40% and was suggested as crucial intermediate in the formation of propylene glycol. Only small amounts of lactic acid were observed. Obviously, the reaction conditions and catalyst favor C–C bond cleavage between position 3 and 4 preferentially yielding C_3 molecules over C_2 products. However, the significant fraction of acetol indicates a slow relative rate of hydrogenation over the platinum catalyst.

Supported platinum catalysts also possess activity for the one-pot transformation of hemicellulose. Hemicellulose itself mainly consists of xylan, a polymer of xylose. Tathod and Dhepe [35] used the 3.5 wt% Pt/Al_2O_3 catalyst already discussed for the conversion of xylose and xylitol under the same conditions but elevated the reaction time to 16 h. Therein, 93% of hemicelluloses were converted, yielding glycols ($Y = 23\%$), xylose ($Y = 2\%$), and xylitol ($Y = 17\%$). Additionally, furfural, furfuryl alcohol, and oligomers were formed, pointing to slow hydrogenation of xylose, which can undergo dehydration to furfural, giving rise to oligomer formation or hydrogenation to furfuryl alcohol. Consequently, a proper balance of relative reaction rates is of major importance in integrating the transformation of polysaccharides into glycols in one process step.

10.3.3 Other Noble Metal Catalysts

Interestingly, only few examples covering other noble metal catalysts with significant activity for glycol formation exist. Palladium and rhodium were investigated by Sun *et al.* and Pd/Al_2O_3 by Lee *et al.* All catalysts showed a low activity without addition of base promoters [26]. Nevertheless, the metals exhibit reasonable catalytic activity for a reaction in basic medium utilizing activated carbon as support material [27]. The selectivity to glycols is comparable to ruthenium- or platinum-based catalysts, but the metals exhibit an approximately three times lower activity than ruthenium. Therefore, these metals can be considered active in the conversion of xylitol but do not provide any advantages over ruthenium or platinum catalysts discussed previously.

10.3.4 Nickel-Based Catalysts

10.3.4.1 C_5 and C_6 Monosaccharides and Sugar Alcohols

Already the patent filed in 1933 by Lautenschläger *et al.* and a report by Clark in 1958 illustrate the potential of nickel as active metal for the conversion of sorbitol into glycols. In the contribution by Clark, temperatures of 215 °C and high hydrogen pressures above 130 bar enabled a conversion of 100% with a selectivity toward ethylene glycol of 18% and 22% selectivity for propylene glycol after 6.5 h [37]. Approximately 70% of the products were identified. Recently, renewed interest in this reaction occurred. Banu *et al.* synthesized nickel-based catalysts on a NaY zeolite to make use of the basic sites of the support. Catalysts were prepared by incipient wetness impregnation rather than by ion exchange which led to decreased catalyst activity and the formation of polyols. Reactions were carried out at 220 °C under a hydrogen pressure of 60 bar over 6 h [38]. Considering that the cage opening of the zeolite is just slightly bigger than sorbitol, a limitation by diffusion is likely to occur, and reactions will mainly take place at the outer surface of the catalyst. A loading of 6 wt% nickel proved to be most effective enabling 66% conversion with a selectivity of ethylene and propylene glycol of 7% and 62%, respectively. Addition of $Ca(OH)_2$ even facilitated a higher conversion of 75%, and the selectivity to propylene glycol could be increased to 69%. Interestingly, the selectivity toward ethylene glycol remained comparable with or without $Ca(OH)_2$ addition. In the presence of base, ion exchange occurred, exchanging sodium by calcium in the zeolite framework.

As an alternative to zeolite-supported nickel catalysts, Soták *et al.* [22] introduced 2% nickel phosphide supported on activated carbon as active catalyst, resulting in 99% sorbitol conversion within 45 min at 200 °C. $Ba(OH)_2$ was chosen as basic promoter leading to 17% and 28% selectivity toward ethylene and propylene glycol. Besides glycerol, ethanol, and methanol were also formed as by-products. Recycling experiments showed a 6% decrease of conversion after three runs. Therein, XRD analyses confirmed that the Ni_2P phase is not stable in the reaction but decomposes into $Ni(PO_3)_2$ and metallic nickel. Ye *et al.* [39] demonstrated a promoting effect of cerium for Ni/Al_2O_3. The optimum loadings of Ce and Ni were about 0.5 and 20%, respectively. Investigating the conversion of a basic ($Ca(OH)_2$) 30 wt% aqueous sorbitol solution at 240 °C, not only above 95.6% conversion but also selectivities to glycerol, ethylene, and propylene glycol of 25, 18 and 36% were achieved in 8 h. Recycling experiments confirmed a positive effect of Ce on catalyst stability with minor deactivation within 10 recycling runs.

It is also possible to access glycols directly based on glucose as substrate. However, one has to consider the higher reactivity of glucose compared to sorbitol due to the carbonyl group. At the same time, glucose is also more prone to side reactions like the formation of soluble and insoluble polymers. Consequently, Ooms *et al.* [40] were able to convert 100% of glucose within 3 h but failed to reach high selectivity or to close the mass balance. When changing the process from a batch to a semibatch configuration, they achieved a significant increase of glycol formation reaching up to 47% ethylene glycol selectivity. The applied catalyst was 2%Ni–30%W_2C/AC that, despite promising selectivity, suffered from tungsten leaching upon recycling and an associated loss of selectivity. A multimetallic catalyst composed of Ni (30%), Mo (7%), and Cu (1%) supported on Kieselguhr presents another suitable catalyst for the conversion of sucrose, a disaccharide composed of glucose and fructose [41]. Under relatively mild conditions (150 °C, 135 min), full conversion with a selectivity to ethylene and propylene glycol of 22% and 13% could be reached. Additionally, glycerol formed as a major by-product with around 28% yield.

10.3.4.2 **Polysaccharides**

Most nickel-based catalysts used in the conversion of cellulose yield hexitols. Glycols are only achieved in small amounts. The group of Liang *et al.* [42] reported various nickel catalysts on different supports including ZSM-5, Al_2O_3, SiO_2, bentonite, TiO_2, and Kieselguhr. Using cellobiose as model compound, Ni/ZSM-5 mainly yielded hexitols. Interestingly, all other support materials gave rise to the formation of glycerol, ethylene, and propylene glycol. A comparable trend occurred for microcrystalline cellulose as substrate. The authors suggested a synergistic effect of Ni active species and acid/base sites accelerating dehydrogenation of sorbitol as important factor for glycol formation. Clearly, nickel allows efficient hydrogenation, but for glycol formation a suitable support or addition of a second metal is required.

Xi *et al.* [25] used a Ni/$NbOPO_4$ catalyst reaching 66% conversion of cellulose over 20 h and a yield toward ethylene glycol of 7%. This result stands far behind the performance of the ruthenium–niobium catalysts presented in Section 10.3.1. Nevertheless, other catalytic systems used in literature confirm that nickel-based catalysts can perform as good as noble metal catalysts. Pang *et al.* [43] used nickel–tungsten carbide catalysts and achieved full conversion of cellulose within 30 min at 245 °C. A selectivity of 55% toward ethylene glycol and 14% toward propylene glycol was reached. The group of Zhang *et al.* also worked on the development of nickel-based tungsten-containing catalysts. In 2008 they presented a 2%Ni–30%W_2C/AC catalyst facilitating full cellulose conversion under similar conditions [44]. Surprisingly, the selectivity toward ethylene glycol was with 61% comparable to noble metal catalysts. The used amount of catalyst was very high (catalyst to substrate ratio 1 : 3), and recycling experiments showed a slight decrease in selectivity explained by oxidation of W_2C. Further optimizing the catalyst, the group presented Ni5%–W15% and Ni5%–W25% catalysts on SBA-15 that even showed better performance at the same conditions [32]. At full conversion, a selectivity toward ethylene glycol of 75% was reached. The authors suggested that tungsten supports depolymerization and retro-aldol reaction, while nickel provides hydrogenation activity. In subsequent contributions, the group investigated catalyst stability of the presented tungsten carbide catalysts, revealing that a decrease of selectivity from 73% to 58% is found after three recycling runs [45]. Because the reaction was conducted at full conversion, this might not only indicate a decrease in selectivity but also in catalyst activity. ICP analysis made clear that leaching of the active components nickel and tungsten appeared. After three runs 19% of the initial amount of nickel and 11% of the initial tungsten were lost. Nevertheless, a yield of 75% of ethylene glycol starting from cellulose certainly presents a highly promising result.

10.3.5 Copper and Other Base Metal Catalysts

10.3.5.1 C_5 and C_6 Monosaccharides and Sugar Alcohols

Also other base metals can be used in the conversion of sorbitol. Here especially other hydrogenation catalysts such as Cu, Fe, and Co were investigated. Using a combined CuO–ZnO catalyst as usually used in the production of methanol from syngas, mainly deoxygenation occurred [46]. The main products were deoxyhexitols, and only a low selectivity toward ethylene and propylene glycol of 6–8% could be found. For Cu_3P supported on activated carbon, a conversion of 75% was achieved within 45 min using basic promoters. However, ethylene and propylene glycol are not the main products reaching only a selectivity of 8–9% [22]. The same study also covered iron- and cobalt-based catalysts. Fe_2P on activated carbon enabled a conversion of 70% but low selectivity to the desired products of only 3–4%. Cobalt-based phosphide catalysts were even less active.

Focusing on xylitol as substrate, Huang *et al.* [47] used Cu–SiO_2 catalysts of different copper loadings and $Ca(OH)_2$ as basic promoter. At 200 °C and 2 h reaction 29% of the used xylitol was converted, obtaining a selectivity to the two glycols of about 20% each. Also after three recycling runs, selectivity stayed constant with only slightly decreasing activity to 24% conversion. In addition to glycols, also lactic acid and glycolic acid were found in the product spectrum. Cu supported on MgO showed a higher activity with around 40% of xylitol conversion [47]. Nevertheless, the selectivity to glycols was lower, and MgO is not stable under hydrothermal reaction conditions.

The same authors also varied the amount of base, confirming increasing conversion for higher amounts of base in the system. Additionally, the selectivity to glycols increased, however, associated by the formation of acids. Eight millimole of base and a pH value of 12.4 were found to provide the optimal selectivity and activity. Under such reaction conditions, a conversion of 56% and a selectivity to ethylene and propylene glycol of 21% and 25% could be reached. Investigating the influence of copper dispersion, a particle size of 20–35 nm appeared to provide the best activity. Overall, the group concluded that the reaction rate is primarily determined by the speed of (de)hydrogenation, while selectivity relies on an efficient retro-aldol reaction. Interestingly, also CuO/ZnO/Al_2O_3 facilitated direct

Table 10.2 Conversion of xylose and xylitol over $CuO/ZnO/Al_2O_3$ ($T = 245\,°C$, $p(H_2) = 50$ bar).

Substrate	Reaction time (h)	X (%)	S(ethylene glycol) (%)	S(propylene glycol) (%)
Xylose	1	100	4	9
Xylitol	1	31	32	35
Xylitol	2	59	31	34
Xylitol	4	63	33	37

glycol formation based on xylose and xylitol (Table 10.2) [48]. The catalyst was used at a catalyst to substrate ratio of 1 : 2.5 at 245 °C without a basic promoter.

Again, it can be noticed that xylose is far more reactive than xylitol, but side reaction hinder a selective transformation. The absence of erythritol or threitol in the reaction mixtures confirms that decarbonylation did not significantly contribute to product formation. In contrast, some decarbonylation occurred for nickel and platinum but only presented a significant reaction pathway in the case of ruthenium-based catalysts.

10.3.5.2 Polysaccharides

Only few examples for the conversion of cellulose by copper-based catalysts exist. Therein, $CuO/ZnO/Al_2O_3$ was investigated facilitating 100% conversion at 245 °C after 5 h [49]. However, ethylene and propylene glycol were only minor products with yields of 5 and 8%. Instead, a significant amount of methanol was formed pointing toward extensive C–C and C–O bond cleavage [48].

10.4 Direct Formation of Glycols from Lignocellulose

In some cases, also the conversion of lignocellulose directly has been considered. However, most literature examples report hexitols as main products [40, 41]. A catalyst that is suitable to both convert cellulose and hemicellulose is a bimetallic catalyst composed of nickel and tungsten. A 4%Ni–30%W_2C/AC catalyst was used to convert different lignocellulosic biomass obtained from hard and soft wood [50]. Reactions were carried out at 235 °C, under 60 bar hydrogen pressure over 4 h. The used amount of catalyst was very high with a catalyst to substrate ratio of 1 : 2.5. The results are summarized in Table 10.3.

Obviously, the composition of lignocellulose exhibits a significant influence on the selectivity. This observation has certainly to be taken into consideration envisaging future biorefineries with flexible sources of renewable feedstocks.

Another approach to convert lignocellulose directly was made by Pang *et al.* [43]. They converted corn stalk that has a content of lignin of approximately 13%

Table 10.3 Conversion of woody biomass over Ni–W_2C/AC catalyst ($T = 235\,°C$, $t = 4\,h$, $p(H_2) = 60\,bar$).

Substrate	Y(ethylene glycol) (%)	Y(propylene glycol) (%)	Amount of lignin (%)
Birch	51	14	20
Poplar	49	13	15
Basswood	49	12	15
Ash tree	53	12	18
Beech	35	11	25
Xylosma	36	14	23

Table 10.4 Conversion of corn stalk over different catalysts ($T = 245\,°C$, $t = 2\,h$, $p(H_2) = 60\,bar$).

Catalyst	X(corn stalk) (%)	Y(ethylene glycol) (%)	Y(propylene glycol) (%)
2%Ni/AC	92	6	7
2%Pt/AC	90	6	10
2%Ni–30%W_2C/AC	96	18	14

[50]. Again the best results were obtained using a nickel–tungsten carbide catalyst, but also other carbon-supported catalysts were tested as shown in Table 10.4. The used amount of catalyst is again high and comparable to the reaction introduced earlier.

10.5 Technical Application of Glycol Production

For the technical application of the process from xylitol to ethylene and propylene glycol next to the chosen catalyst, also process parameters, workup, and reactors have to be discussed. Numerous patents are dealing with these or other challenges on the way to industry. Because just a few examples can be mentioned, the reader is revised to the published literature in the field [51].

As mentioned in Section 10.3.4 the design of the process can have a significant influence on the selectivity and activity of a reaction. Ooms *et al.* [40] compared the performance of a nickel–tungsten carbide catalyst in a batch and a semibatch reaction. Both reactions were carried out under the same conditions – the only difference was that for the semibatch process the starting material was slowly added to the reactor over a time of 3 h. Table 10.5 gives an overview about glucose

Table 10.5 Conversion of glucose over Ni–W_2C/AC (T = 245 °C, t = 3 h, $p(H_2)$ = 60 bar).

	Batch	Semibatch
X (%)	100	100
S(ethylene glycol) (%)	8	47
S(propylene glycol) (%)	5	4

conversion and selectivity comparing these process configurations. Obviously, slow glucose addition suppresses undesired side reactions facilitating higher glycol selectivity.

Another possibility is the application of a two-step process as investigated by Xiao *et al.* Therein, glucose was first converted to sorbitol or mannitol under mild conditions. Afterward the reaction mixture was transformed further to glycols in a second step under different reaction conditions (see Section 10.3) [52]. The used catalyst was Cu–Cr based. A glucose concentration of 333 g l^{-1} was fed and $Ca(OH)_2$ used as basic promoter. Both steps were carried out subsequently in one pot reaching 100% conversion of the starting material and a final selectivity of propylene glycol of 53%. The formed C_2-C_4 components summed up to a combined yield of 77% (Scheme 10.10).

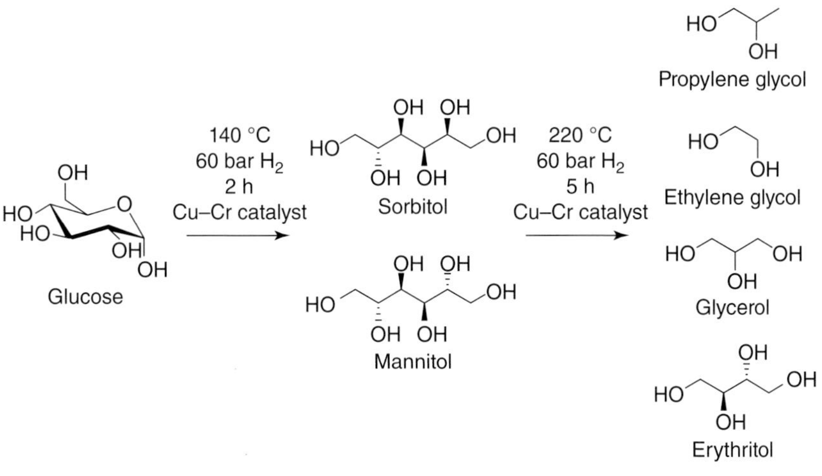

Glucose

140 °C
60 bar H_2
2 h
Cu–Cr catalyst

Sorbitol

Mannitol

220 °C
60 bar H_2
5 h
Cu–Cr catalyst

Propylene glycol

Ethylene glycol

Glycerol

Erythritol

Scheme 10.10 Two-step reaction of glucose to C_2-C_4 components over a Cu–Cr catalyst.

Considering process design, a patent by Chopade *et al.* [53] discussed a trickle-bed reactor to convert xylitol into glycols over a Ru/C catalyst (see Scheme 10.11). The substrate solution is first heated to the desired reaction temperature of 190–240 °C before being pressurized with hydrogen. The reaction then takes place in the fixed-bed reactor, and the product mixture is worked off

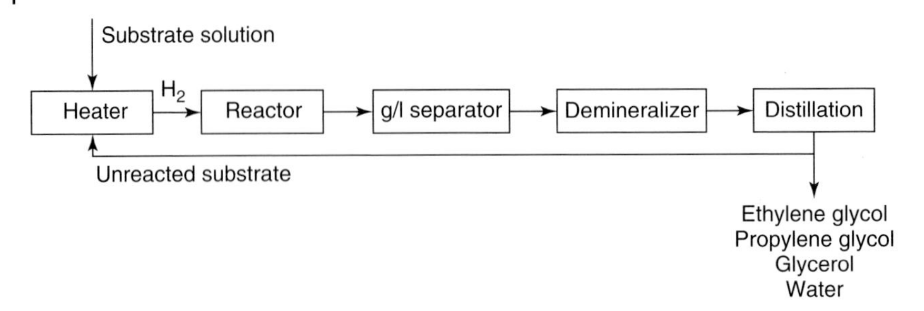

Scheme 10.11 Valorization of xylitol in a trickle-bed reactor.

first by a gas–liquid separator and second by demineralization. Finally, the end products are separated by means of distillation. The three products – ethylene glycol, propylene glycol, and glycerol – are separated from water and unreacted substrate that is led back into the reactor.

10.6
Summary and Conclusion

Cellulose and hemicellulose as major constituents of lignocellulose certainly present promising feedstocks for production of propylene and ethylene glycol. Potential catalyst systems mainly consist of supported noble metals such as ruthenium and platinum or base metals such as nickel and copper. Crucial steps of the reaction pathway are dehydrogenation of sugar alcohols, subsequent C–C cleavage via retro-aldol reaction, and hydrogenation of the formed products hindering further side reactions. Despite numerous publications covering glycol formation based on monosaccharides, sugar alcohols, and polysaccharides, selectivity control remains a challenge. Base addition proved to be advantageous for increasing activity and selectivity. The common explanation for this observation relates to an enhanced reaction rate of retro-aldol reactions in the presence of a base. However, the interaction of the base with the applied metal catalysts has not been fully clarified yet.

Future process design aiming for an efficient glycol production based on the mentioned substrates will indispensably rely on a comprehensive understanding of the individual factors governing the complex reaction network. In addition, suitable methods reducing lactic acid formation as major by-product of base-promoted reactions are required. Finally yet importantly, separation of high-boiling glycols from the mainly aqueous reaction solutions is energy intense and needs to be considered in discussing a technical application of such technologies. Nevertheless, glycol production based on lignocellulosic feedstocks bears the potential to serve as crossing technology connecting biorefinery approaches with established petrochemical value chains.

References

1. Quispe, C.A.G., Coronado, C.J.R., and Carvalho, J.A. Jr. (2013) *Renewable Sustainable Energy Rev.*, **27**, 475–493.
2. (a) Behr, A., Eilting, J., Irawadi, K., Leschinski, J., and Lindner, F. (2008) *Green Chem.*, **10**, 13–30; (b) Tan, H.W., Abdul Aziz, A.R., and Aroua, M.K. (2013) *Renewable Sustainable Energy Rev.*, **27**, 118–127; (c) Ruppert, A.M., Weinberg, K., and Palkovits, R. (2012) *Angew. Chem. Int. Ed.*, **51**, 2564–2601.
3. Pauly, M. and Keegstra, K. (2008) *Plant J.*, **54**, 559–568.
4. Lautenschläger, K.L., Bockmühl, M., and Ehrhart, G. (1933) Process of hydrogenating polyhydroxy compounds and substances obtainable thereby. US Patent 1 915 431.
5. Gürkan, H.H. (1949) Catalytic hydrogenation of cellulose to produce oxygenated compounds. US Patent 2 488 722.
6. (a) Sharkov, V.I. (1963) *Angew. Chem. Int. Ed. Engl.*, **2**, 405–409; (b) Sharkov, V.I. (1963) *Chem. Ing. Tech.*, **35**, 494–497.
7. (a) Negahdar, L., Oltmanns, J.U., Palkovits, S., and Palkovits, R. (2014) *Appl. Catal., B*, **147**, 677–683; (b) Negahdar, L., Hausoul, P.J.C., Palkovits, S., and Palkovits, R. (2015) *Appl. Catal., B*, **166–167**, 460–464.
8. (a) Deng, W., Liu, M., Tan, X., Zhang, Q., and Wang, Y. (2010) *J. Catal.*, **271**, 22–32; (b) Yan, N., Zhao, C., Luo, C., Dyson, P.J., Liu, H., and Kou, Y. (2006) *J. Am. Chem. Soc.*, **128**, 8714–8715; (c) Li, J., Spina, A., Moulijn, J.A., and Makkee, M. (2013) *Catal. Sci. Technol.*, **3**, 1540–1546.
9. Rinaldi, R. and Schüth, F. (2009) *ChemSusChem*, **2**, 1096–1107.
10. (a) Huang, Y.-B. and Fu, Y. (2013) *Green Chem.*, **15**, 1095–1111; (b) Hu, L., Lin, L., Wu, Z., Zhou, S., and Liu, S. (2015) *Appl. Catal., B*, **174–175**, 225–243.
11. (a) Harris, J.F. (1972) *Carbohydr. Res.*, **23**, 207–215; (b) Delidovich, I.V., Simonov, A.N., Taran, O.P., and Parmon, V.N. (2014) *ChemSusChem*, **7**, 1833–1846; (c) Dusselier, M., Van Wouwe, P., Dewaele, A., Makshina, E., and Sels, B.F. (2013) *Energy Environ. Sci.*, **6**, 1415–1442; (d) Pescarmona, P.P., Janssen, K.P.F., Delaet, C., Stroobants, C., Houthoofd, K., Philippaerts, A., De Jonghe, C., Paul, J.S., Jacobs, P.A., and Sels, B.F. (2010) *Green Chem.*, **12**, 1083–1089; (e) Delidovich, I. and Palkovits, R. (2014) *Catal. Sci. Technol.*, **4**, 4322–4329.
12. (a) Luo, C., Wang, S., and Liu, H. (2007) *Angew. Chem. Int. Ed.*, **46**, 7636–7639; (b) Ding, L.-N., Wang, A.-Q., Zheng, M.-Y., and Zhang, T. (2010) *ChemSusChem*, **3**, 818–821; (c) Van de Vyver, S., Geboers, J., Dusselier, M., Schepers, H., Vosch, T., Zhang, L., Van Tendeloo, G., Jacobs, P.A., and Sels, B.F. (2010) *ChemSusChem*, **3**, 698–701; (d) Palkovits, R., Tajvidi, K., Procelewska, J., Rinaldi, R., and Ruppert, A. (2010) *Green Chem.*, **12**, 972–978.
13. (a) Kabyemela, B.M., Adschiri, T., Malaluan, R.M., and Arai, K. (1999) *Ind. Eng. Chem. Res.*, **38**, 2888–2895; (b) Kabyemela, B.M., Adschiri, T., Malaluan, R.M., and Arai, K. (1997) *Ind. Eng. Chem. Res.*, **36**, 1552–1558; (c) Kabyemela, B.M., Adschiri, T., Malaluan, R., and Arai, K. (1997) *Ind. Eng. Chem. Res.*, **36**, 2025–2030.
14. Van de Vyver, S., Geboers, J., Jacobs, P.A., and Sels, B.F. (2011) *ChemCatChem*, **3**, 82–94.
15. (a) Sohounloue, D.K., Montassier, C., and Barbier, J. (1983) *React. Kinet. Catal. Lett.*, **22**, 391–397; (b) Montassier, C., Ménézo, J.C., Hoang, L.C., Renaud, C., and Barbier, J. (1991) *J. Mol. Catal.*, **70**, 99–110; (c) Wang, K., Hawley, M.C., and Furney, T.D. (1995) *Ind. Eng. Chem. Res.*, **34**, 3766–3770; (d) Andrews, M.A. and Klaeren, S.A. (1989) *J. Am. Chem. Soc.*, **111**, 4131–4133.
16. (a) de Bruyn, C.A.L. and van Ekenstein, W.A. (1895) *Recl. Trav. Chim. Pays-Bas*, **14**, 203–216; (b) de Bruyn, C.A.L. and van Ekenstein, W.A. (1896) *Recl. Trav. Chim. Pays-Bas*, **15**, 92–96.
17. (a) Sowden, J.C. and Schaffer, R. (1952) *J. Am. Chem. Soc.*, **74**, 505–507; (b) Harris, D.W. and Feather, M.S. (1973) *Carbohydr. Res.*, **30**, 359–365.

18. Deutsch, K.L., Lahr, D.G., and Shanks, B.H. (2012) *Green Chem.*, **14**, 1635–1642.
19. (a) Xiong, H., Pham, H.N., and Datye, A.K. (2014) *Green Chem.*, **16**, 4627–4643; (b) Ravenelle, R.M., Schüßler, F., D'Amico, A., Danilina, N., van Bokhoven, J.A., Lercher, J.A., Jones, C.W., and Sievers, C. (2010) *J. Phys. Chem. C*, **114**, 19582–19595.
20. (a) Alonso, D.M., Wettstein, S.G., and Dumesic, J.A. (2012) *Chem. Soc. Rev.*, **41**, 8075–8098; (b) Liu, X., Wang, X., Yao, S., Jiang, Y., Guan, J., and Mu, X. (2014) *RSC Adv.*, **4**, 49501–49520.
21. Zhao, L., Zhou, J.H., Sui, Z.J., and Zhou, X.G. (2010) *Chem. Eng. Sci.*, **65**, 30–35.
22. Soták, T., Schmidt, T., and Hronec, M. (2013) *Appl. Catal., A*, **459**, 26–33.
23. Zhou, J.H., Zhang, M.G., Zhao, L., Li, P., Zhou, X.G., and Yuan, W.K. (2009) *Catal. Today*, **147**, 225–229.
24. Zhou, J., Liu, G., Sui, Z., Zhou, X., and Yuan, W. (2014) *Chin. J. Catal.*, **35**, 692–702.
25. Xi, J., Ding, D., Shao, Y., Liu, X., Lu, G., and Wang, Y. (2014) *ACS Sustainable Chem. Eng.*, **2**, 2355–2362.
26. Lee, J., Xu, Y., and Huber, G.W. (2013) *Appl. Catal., B*, **140–141**, 98–107.
27. Sun, J. and Liu, H. (2011) *Green Chem.*, **13**, 135–142.
28. (a) Fukuoka, A. and Dhepe, P.L. (2006) *Angew. Chem. Int. Ed.*, **45**, 5161–5163; (b) Dhepe, P. and Fukuoka, A. (2007) *Catal. Surv. Asia*, **11**, 186–191; (c) Fukuoka, A. and Dhepe, P. (2009) Catalyst for Cellulose Hydrolysis and/or Reduction of Cellulose Hydrolysis Products and Method of Producing Sugar Alcohols From Cellulose. US Patent 0 217 922.
29. Hilgert, J., Meine, N., Rinaldi, R., and Schuth, F. (2013) *Energy Environ. Sci.*, **6**, 92–96.
30. Deng, W., Tan, X., Fang, W., Zhang, Q., and Wang, Y. (2009) *Catal. Lett.*, **133**, 167–174.
31. (a) Palkovits, R., Tajvidi, K., Ruppert, A.M., and Procelewska, J. (2011) *Chem. Commun.*, **47**, 576–578; (b) Geboers, J., Van de Vyver, S., Carpentier, K., de Blochouse, K., Jacobs, P., and Sels, B. (2010) *Chem. Commun.*, **46**, 3577–3579; (c) Geboers, J., Van de Vyver, S., Carpentier, K., Jacobs, P., and Sels, B. (2011) *Chem. Commun.*, **47**, 5590–5592; (d) Geboers, J., Van de Vyver, S., Carpentier, K., Jacobs, P., and Sels, B. (2011) *Green Chem.*, **13**, 2167–2174.
32. Zheng, M.-Y., Wang, A.-Q., Ji, N., Pang, J.-F., Wang, X.-D., and Zhang, T. (2010) *ChemSusChem*, **3**, 63–66.
33. Jin, X., Roy, D., Thapa, P.S., Subramaniam, B., and Chaudhari, R.V. (2013) *ACS Sustainable Chem. Eng.*, **1**, 1453–1462.
34. Li, N. and Huber, G.W. (2010) *J. Catal.*, **270**, 48–59.
35. Tathod, A.P. and Dhepe, P.L. (2014) *Green Chem.*, **16**, 4944–4954.
36. Chambon, F., Rataboul, F., Pinel, C., Cabiac, A., Guillon, E., and Essayem, N. (2013) *ChemSusChem*, **6**, 500–507.
37. Clark, I.T. (1958) *Ind. Eng. Chem.*, **50**, 1125–1126.
38. Banu, M., Sivasanker, S., Sankaranarayanan, T.M., and Venuvanalingam, P. (2011) *Catal. Commun.*, **12**, 673–677.
39. Ye, L., Duan, X., Lin, H., and Yuan, Y. (2012) *Catal. Today*, **183**, 65–71.
40. Ooms, R., Dusselier, M., Geboers, J.A., Op de Beeck, B., Verhaeven, R., Gobechiya, E., Martens, J.A., Redl, A., and Sels, B.F. (2014) *Green Chem.*, **16**, 695–707.
41. Saxena, U., Dwivedi, N., and Vidyarthi, S.R. (2005) *Ind. Eng. Chem. Res.*, **44**, 1466–1473.
42. (a) Liang, G., He, L., Cheng, H., Li, W., Li, X., Zhang, C., Yu, Y., and Zhao, F. (2014) *J. Catal.*, **309**, 468–476; (b) Liang, G., Cheng, H., Li, W., He, L., Yu, Y., and Zhao, F. (2012) *Green Chem.*, **14**, 2146–2149.
43. Pang, J., Zheng, M., Wang, A., and Zhang, T. (2011) *Ind. Eng. Chem. Res.*, **50**, 6601–6608.
44. Ji, N., Zhang, T., Zheng, M., Wang, A., Wang, H., Wang, X., and Chen, J.G. (2008) *Angew. Chem.*, **120**, 8638–8641.
45. Ji, N., Zheng, M., Wang, A., Zhang, T., and Chen, J.G. (2012) *ChemSusChem*, **5**, 939–944.
46. Blanc, B., Bourrel, A., Gallezot, P., Haas, T., and Taylor, P. (2000) *Green Chem.*, **2**, 89–91.

47. Huang, Z., Chen, J., Jia, Y., Liu, H., Xia, C., and Liu, H. (2014) *Appl. Catal., B*, **147**, 377–386.

48. Tajvidi, K., Hausoul, P.J.C., and Palkovits, R. (2014) *ChemSusChem*, **7**, 1311–1317.

49. Tajvidi, K., Pupovac, K., Kükrek, M., and Palkovits, R. (2012) *ChemSusChem*, **5**, 2139–2142.

50. Li, C., Zheng, M., Wang, A., and Zhang, T. (2012) *Energy Environ. Sci.*, **5**, 6383–6390.

51. (a) Zhang, T., Zheng, M., Ji, N., Wang, A., and Shu, Y. (2010) A process for preparing ethylene glycol using cellulose. WO Patent 2010045766; (b) Zhang, T., Ji, N., Zheng, M., Wang, A., Shu, Y., Wang, X., and Chen, J. (2014) Methods of using tungsten carbide catalysts in preparation of ethylene glycol. US Patent 8 692 032.

52. Xiao, Z., Jin, S., Sha, G., Williams, C.T., and Liang, C. (2014) *Ind. Eng. Chem. Res.*, **53**, 8735–8743.

53. Chopade, S.P., Miller, D.J., Jackson, J.E., Werpy, T.A., Frye, J.G. Jr., and Zacher, A.H. (2001), Catalyst and process for hydrogenolysis of sugar alcohols to polyols. WO Patent 2001066499.

Part III
Polymers from Bio-Based building blocks

Chemicals and Fuels from Bio-Based Building Blocks, First Edition.
Edited by Fabrizio Cavani, Stefania Albonetti, Francesco Basile, and Alessandro Gandini.

11
Introduction

Alessandro Gandini

Interest in polymers from renewable resources has been witnessing an incessant growth in both academia and industry with these blooming activities covering a progressively wider domain of sources and approaches worldwide. The association of these innovative scientific and technological investigations with the need of introducing a growing dose of green chemistry connotations in their conception is opening the way to an all-inclusive sustainability for this new generation of polymers, which justifies their vision as macromolecular materials for the twenty-first century.

After the first comprehensive monograph dealing with these materials, published in 2008 [1], numerous reviews and books have appeared, mostly dealing with specific topics or brief analyses of the overall state of the art [2, 3], with a very recent update [4], which was however aimed at highlighting trends rather than at detailing the numerous papers covering different aspects of the vast field.

Given the structure of this book, the present chapter concentrates exclusively on the use of bio-based building blocks as potential monomers and on the recent studies dealing with their polymerizations and copolymerizations, as well as on the properties of the ensuing materials. Interestingly, moreover, the rate at which new exciting results are being published, amply justifies the coverage given here, since it also provides a "further update" of the mentioned "recent update," [4] at least within its specific scope. Natural polymers (polysaccharides, lignin, suberin, proteins, natural rubber, etc.) and their modifications fall outside that scope, and the reader will find information about the multitude of equally exciting and equally incessant studies and applications of these renewable resources in such materials as thermoplastics, thermosets, blends, and composites in other publications [1–4].

The topics discussed here are divided into two major sections, one dealing with polymers from pristine and chemically modified *natural monomers*, notably terpenes, rosin, sugars, and vegetable oils, and the other dealing with polymers from monomers obtained from various vegetable biomass components, notably furans, diacids, diols, hydroxy acids, and glycerol. The polymerizations involving classical fossil-derived monomers, such as ethylene and terephthalic acid, now also prepared from renewable resources, are obviously not discussed here since

Chemicals and Fuels from Bio-Based Building Blocks, First Edition.
Edited by Fabrizio Cavani, Stefania Albonetti, Francesco Basile, and Alessandro Gandini.

the novelty of these systems only resides in the alternative mode of monomer synthesis.

No attempt was made to cover each family of monomers exhaustively but rather to concentrate on contributions that appeared to provide originality and hence potential follow-ups. This personal choice did not hinder a critical approach whenever it was deemed constructively appropriate. Given the modest amount of space allotted to this chapter, even some of the numerous interesting studies had to be sacrificed to avoid a list of cursory mentions and privilege instead a reasoned treatment of carefully selected examples.

References

1. Belgacem, M.N. and Gandini, A. (eds) (2008) *Monomers, Polymers and Composites from Renewable Resources*, Elsevier, Amsterdam.
2. Gandini, A. (2008) *Macromolecules*, **41**, 9491–9504.
3. Gandini, A. (2011) *Green Chem.*, **13**, 1061–1083.
4. Gandini, A. and Lacerda, T.M. (2015) *Prog. Polym. Sci.* **48**, 1–39.

12
Polymers from Pristine and Modified Natural Monomers

Annamaria Celli, Alessandro Gandini, Claudio Gioia, Talita M. Lacerda, Micaela Vannini, and Martino Colonna

12.1
Monomers and Polymers from Vegetable Oils

12.1.1
Introduction

Nature is generous about the number of plant oils available for mankind, and their importance as a source of food is substantial. Recently, equally relevant has become their industrial applications as raw material for the preparation of biodiesel, chemicals, and polymers, as partial replacements of fossil feedstock. Approximately 90% of all the oils extracted for those purposes are burned as fuel and the rest used for industrial production, ranging from polymers to fine chemicals for the pharmaceutical industry [1]. Since vegetable oils correspond to an abundant and democratically available bioresource, their different uses can easily coexist. In this section, we will focus on the more recent advances related to the great potential of vegetable oils to serve as building blocks for the development of novel polymeric materials.

According to the United States Department of Agriculture (USDA), only in the last decade, the world production of major oilseeds has increased from 331 to 529 million tons, and the harvested areas from 186 to 234 million acres. Soybean oilseeds represent over half of the total production and are mainly grown in Brazil, the United States, and Argentina, followed by rapeseed (grown in European Union (EU)-27, Canada, and China), cottonseed (China and India), and sunflower oilseeds (Ukraine, EU-27, and Russia) [2].

The increasing usage of oils as the main source of polymeric materials is related to their chemical composition. They are all triglycerides in which the fatty acid structures vary as a function of the source from which they were extracted and are responsible for the final features of the oil, determining their possible applications as renewable feedstock. The number of double bonds, as well as their positions within the aliphatic chain, and the presence of other reactive moieties such as hydroxyl groups, strongly affects the oil properties. One main characteristic of

Chemicals and Fuels from Bio-Based Building Blocks, First Edition.
Edited by Fabrizio Cavani, Stefania Albonetti, Francesco Basile, and Alessandro Gandini.

Table 12.1 Common fatty acids found in general vegetable oil compositions [4, 5].

Fatty acid	Systematic name	Formula	CN/DB	Type
Caprylic (I)	Octanoic acid	$C_8H_{16}O_2$	8 : 0	Saturated
Capric (II)	Decanoic acid	$C_{10}H_{20}O_2$	10 : 0	Saturated
Lauric (III)	Dodecanoic acid	$C_{12}H_{24}O_2$	12 : 0	Saturated
Myristic (IV)	Tetradecanoic acid	$C_{14}H_{28}O_2$	14 : 0	Saturated
Palmitic (V)	Hexadecanoic acid	$C_{16}H_{32}O_2$	16 : 0	Saturated
Stearic (VI)	Octadecanoic acid	$C_{18}H_{36}O_2$	18 : 0	Saturated
Arachidic (VII)	Eicosanoic acid	$C_{20}H_{40}O_2$	20 : 0	Saturated
Palmitoleic (VIII)	Hexadec-9-enoic acid	$C_{16}H_{30}O_2$	16 : 1	Monounsaturated
Oleic (IX)	Octadec-9-enoic acid	$C_{18}H_{34}O_2$	18 : 1	Monounsaturated
Erucic (X)	Docos-13-enoic acid	$C_{22}H_{42}O_2$	22 : 1	Monounsaturated
Linoleic (XI)	9,12-Octadecadienoic acid	$C_{18}H_{32}O_2$	18 : 2	Polyunsaturated
α-Linolenic (XII)	Octadeca-9,12,15-trienoic acid	$C_{18}H_{30}O_2$	18 : 3	Polyunsaturated
α-Eleostearic (XIII)	Octadeca-9,11,13-trienoic acid	$C_{18}H_{30}O_2$	18 : 3	Polyunsaturated
Ricinoleic (XIV)	(9*Z*,12*R*)-12-Hydroxyoctadec-9-enoic acid	$C_{18}H_{34}O_3$	18 : 1	Monounsaturated
Vernolic (XV)	*cis*-12,13-Epoxy-*cis*-9-octadecenoic acid	$C_{18}H_{32}O_3$	18 : 1	Monounsaturated

CN = carbon number and DB = number of double bonds at the carbon chain.

the chemistry of vegetable oils is the heterogeneity and variability of the triglycerides used, due to the statistical distribution of fatty acids per triglyceride. When used for polymer synthesis, it is not possible to set a fine correlation between the material properties and the oil structures, contrary to conventional petrochemical sources that can provide polymers with specific properties through well-optimized processes [3].

A list of the most common fatty acids found in various vegetable oils in the form of the corresponding triglycerides is given in Table 12.1, whereas Table 12.2 provides their typical compositions.

It is important to mention that these contents listed below can be modified by appropriate crop breeding or genetic modifications [6, 7]. For instance, erucic acid is the main fatty acid in standard rapeseed oil (43%), but several of the rapeseed varieties in cultivation are based on zero content of erucic acid [8], given its toxicity in humans if ingested at high doses [9].

The iodine value is used as an index of the extent of C=C unsaturations of a given plant oil, and three conventional categories are derived, ranging from *drying*, if the iodine value is >130, *semidrying*, if it is 90–130, and *nondrying*, if it is <90. Exposure of a spread of drying oil to air for a period of time turns it into a solid film because of the cross-linking between fatty acid chains through their free radical polymerization induced by atmospheric oxygen [10]. As a classical example to illustrate the association between the iodine value and the properties of the oils, coconut oil has an average iodine value of 7–12, which means that it is essentially saturated and therefore good for making soap. On the other hand, linseed oil has an iodine value of 136–178, which makes it a drying oil, well suited for making paints and inks. However, in order to precisely

Table 12.2 Common vegetable oils and their fatty acid content [4, 5].

Oil	I	II	III	IV	V	VI	VII	IX	XI	XII	XIV
Palm	—	—	—	1.2	41.8	3.4	—	41.9	11.0	—	—
Soybean	—	—	—	—	14.0	4.0	—	23.3	52.2	5.6	—
Coconut	6.2	6.2	51.0	18.9	8.6	1.9	—	5.8	1.3	—	—
Sunflower	—	—	—	—	6.5	2.0	—	45.4	46.0	0.1	—
Rapeseed	—	—	—	—	4.0	2.0	—	56.0	26.0	10.0	—
Castor	—	—	—	—	1.5	0.5	—	5.0	4.0	0.5	87.5
Linseed	—	—	—	—	5.0	4.0	—	22.0	17.0	52.0	—
Nahar seed	—	—	—	—	15.9	9.5	—	52.3	22.3	—	—
Corn	—	—	—	—	10.0	4.0	—	34.0	48.0	—	—
Olive	—	—	—	—	6.0	4.0	—	83.0	7.0	—	—
Sesame	—	—	—	0.1	8.2	3.6	—	42.1	43.4	—	—
Safflower	—	—	—	0.1	6.8	2.3	0.3	12.0	77.7	0.4	—

characterize a vegetable oil sample and forecast its potential applications, Fourier transform infrared (FTIR) and proton nuclear magnetic resonance (^{1}H-NMR) spectroscopies are fundamental. Scheme 12.1 illustrates a generic structure of a triglyceride bearing C=C unsaturations, highlighting the sites that can undergo reactions or chemical modifications, that is, useful tools to prepare polymers, either directly or after preliminary conversions into suitable monomers: (i) unsaturated bonds, (ii) methylene moieties next to one of them (allylic carbons), (iii) ester groups, and (iv) methylene sites next to one of them (α-carbons). Another possibility, not illustrated here, is the presence of hydroxyl or oxirane groups, which allow an even wider variety of chemical manipulations.

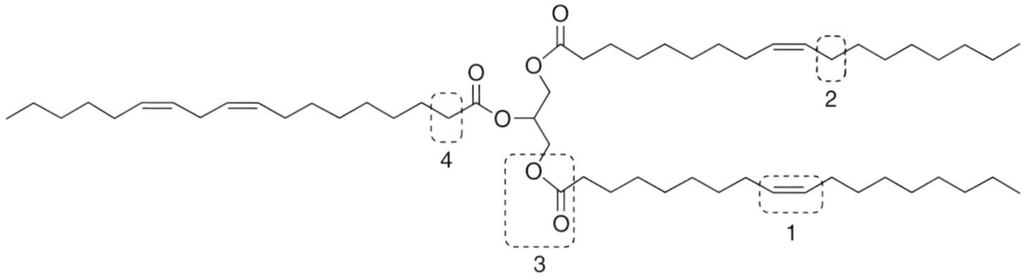

Scheme 12.1 Schematic representation of the reactive sites in a general unsaturated triglyceride.

The huge potential of vegetable oils for the polymer industry is reflected by the increasing number of scientific and patent output about the preparation of linear polymers with well-defined structures, cross-linked thermosets, and matrices for composites and hybrid materials.

Cádiz and coworkers [11] recently reviewed the main uses of vegetable oil-derived polymers, classified the ensuing materials into groups according to their applications, and discussed their properties. *Composites* (or biocomposites) derived from plant oils exhibit typical long methylene chain features, that is, high elongation at break and relatively low stiffness [11, 12]. The interfacial compatibility between the filler/fiber and the hydrophobic nature of the plant-based matrix is of great importance, and it is common to treat or functionalize the surface of the reinforcing material to achieve better final mechanical properties for the composites. The exploitation of vegetable oils for *paints, coatings, and adhesives* dates back to the beginning of civilization, and cave paintings were all based on drying oils. This evolved to the development of renewable resources-based coating formulations with improved performance and properties [11, 13], for example, solubility in water (waterborne coatings) and UV-cured systems. Adhesives synthesized from soybean oil-based polymers exhibit thermal and mechanical properties that are competitive with those of petrochemical-based counterparts [11]. *Biomedical applications* are also favored, since vegetable oils can be metabolized in the human body, which makes the materials derived from them potentially biocompatible. They include surgical sealants and glues, pharmacological patches, wound healing devices, drug carriers, and scaffolds for tissue engineering [11].

The purpose of this section is to overview the most recent progress in the preparation of polyesters and polyurethanes derived from vegetable oils, and a brief final topic will describe some interesting work on the synthesis of polyamides.

12.1.2 Polyesters

There are two main reasons that put fatty acid-based polyesters in a prominent position: the first one is the fact that plant oils (and their isolated fatty acids) are substrates that already provide aliphatic acidic or ester functions. The second is also related to the ester linkage, known to be hydrolysable and degradable, which makes the final polyesters susceptible to degradability by the action of naturally occurring microorganisms such as bacteria, fungi, and algae.

One class of interesting derivatives of unsaturated fatty acids is ω-hydroxy fatty acids [$HO–(CH_2)_n–COOH$], since they have both alcohol and ester groups (AB-type monomers), from which aliphatic polyesters can be readily prepared. Narine and colleagues [14] reported the preparation of three different ω-hydroxy fatty acids and the corresponding ester counterparts [$HO–(CH_2)_n–COOCH_3$], with varying lengths of the aliphatic chain ($n = 8$, 12, and 17), to serve as monomers for the synthesis of polyesters and copolyesters. The methyl ester homologs ($n = 8$ and 12) were synthesized from methyl oleate and methyl erucate, respectively, by ozonolysis–reduction reactions at the fatty acid double bonds, and subsequent saponification reactions led to their acid homologs. The long-chain homolog ($n = 17$) was obtained by cross-metathesis of methyl oleate and oleyl alcohol using a Grubbs catalyst. The melt polymerization reactions were conducted using catalytic amounts of titanium(IV) isopropoxide. The authors successfully obtained polymers that reached a maximum value of molecular weight (M_n) of

35 000 when the monomer with the longest aliphatic chain was used. Copolymers were also prepared and the association of the ester monomers with $n = 13$ and 9 (80 : 20 M ratio) led to a polyester with $M_n = 18\,500$. A subsequent study dealt with the mechanical and thermal properties of these materials [15].

The metathesis reaction is known to be an important tool to generate a vast array of monomers from unsaturated fatty acids [9]. Studies in this field are frequent, and a recent one [16] described the synthesis of unsaturated dicarboxylic acids prepared from oleic and linoleic acids through metathesis reactions, to serve as monomers to produce polyesters with aliphatic and aromatic diols. The best results were achieved when the diacid synthesized from oleic acid (octadec-9-enedioic) and 1,6-hexanediol were combined to form a polyester with $M_w = 3500$, a glass transition temperature of 143 °C, and a melting point of 36 °C.

Cross-linked and hyperbranched polymers are also interesting approaches to convert plant oil derivatives into functional materials. Unsaturated polyester resins were synthesized [17] by alcoholysis of palm oil, producing palm oil monoglyceride (POMG), further treated with maleic anhydride (MA) at different ratios. The curing procedure was carried out with 35 wt% of styrene, 0.3 wt% of cobalt naphthenate, and 1 wt% of methyl ethyl ketone peroxide. The activation energy of the reaction POMG/MA was studied, as well as the swelling properties of the cured resin. A nonedible vegetable oil (pongamia oil) was used to prepare a hyperbranched polyester–styrene copolymer, with potential applications as antimicrobial and corrosion-resistant coating materials via the introduction of silver nanoparticles. In this study, a hydroxyl derivative of the oil was esterified with linolenic acid to form a hyperbranched polyester. A further styrenation was performed at the conjugated double bond in the chains of linolenic acid, and in a final step, silver nanoparticles were added in different amounts [18]. Enzyme-catalyzed polymerization reactions are also an interesting tool to obtain high-molecular-weight polymers using mild conditions without metal-based catalysts, a positive step toward the global environmental concern. This concept was recently presented by Galià and Cádiz's group [19], by the synthesis of a vinylsulfide-containing hydroxy acid (VSHA, Scheme 12.2) from the alkyne-derivatized fatty 10-undecynoic acid, via hydrothiolation. 10-Undecynoic acid was prepared by successive bromination and dehydrobromination of 10-undecenoic acid, which is the major product of castor oil pyrolysis. VSHA was polymerized using metallic and enzymatic catalysts, and the best results arose from the enzymatic catalysis ($M_w = 75\,000$ with Novozymes-435 and $M_w = 19\,000$ with $Sn(Oct)_2$, both after 48 h reaction). Some relevant examples of the direct application of fatty acid-derived polymers can be illustrated by the preparation of pressure-sensitive adhesives (PSAs), which are adhesives that form a bond when pressure is applied to connect the adhesive with the adherent surface without the need of activation by water, solvents, heat, or radiation. Wu and Li [20] described the synthesis of hydroxyl-containing polyesters from epoxidized oleic acid, derived from the epoxidation reaction of methyl oleate, followed by selective hydrolysis of the ester group. This AB-type monomer containing both carboxylic acid and epoxy functions (Scheme 12.3) was tested to serve as a PSA.

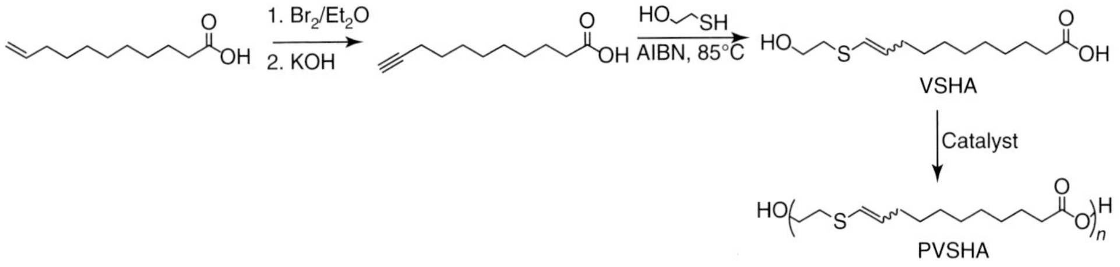

Scheme 12.2 Synthesis and polymerization of VSHA [19].

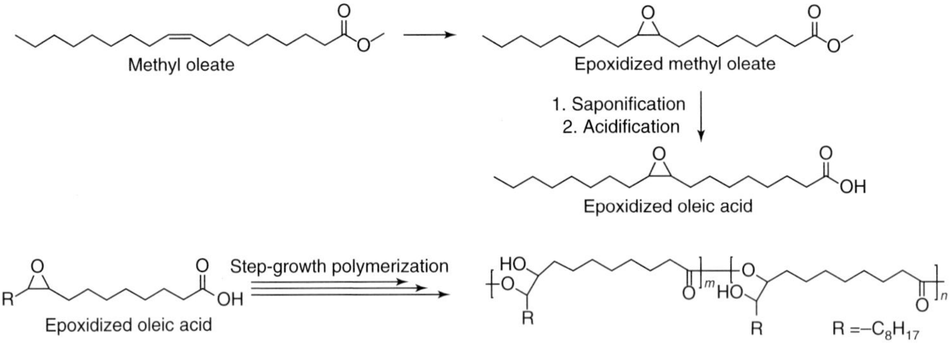

Scheme 12.3 Polycondensation reaction of epoxidized oleic acid [20].

The authors obtained a polyester with $M_n = 3600$ and $M_w = 53\,800$, and the high value of polydispersity was attributed to side reactions at the epoxide backbone.

Another practical example that is worth mentioning is related to the preparation of a polyethylene-like polyester containing long aliphatic chains between the ester functions [21]. Self-metathesis of methyl oleate and methyl erucate led to, respectively, C_{19} and C_{23} diesters with high purity (>99%), and their reduction reactions gave the subsequent long-chain aliphatic diols. The polyesters (PE19–19 from methyl oleate and PE23–23 from methyl erucate) were synthesized using 0.04 mol% of $Ti(OBu)_4$ as catalyst, at 200 °C, and 0.01 mbar for 16 h. PE19–19 was isolated with an M_n value of 30 000 (PDI = 2.6) and a T_m of 102 °C; PE23–23 had an M_n of 39 000 (PDI = 2.4) and a T_m of 107 °C. The materials were processed by injection molding, film extrusion, and electrospinning, and the good properties obtained indicate a high potential as polyethylene-like materials with the important additional advantage of the typical *biodegradability* of aliphatic polyesters.

Finally, an interesting study was devoted to jojoba oil, extracted from the seeds of the jojoba plant (*Simmondsia chinensis*). The seed contains 42–58% of a liquid wax that is uniquely composed of two unsaturated hydrocarbon chains linked by an ester moiety with the (*Z*-) carbon–carbon double bond present in the ninth carbon of each side counting from the end (Scheme 12.4) [22]. When submitted to cross-metathesis reactions, jojoba oil displayed a good potential to serve as a source of *both* biofuels, by the formation of long-chain

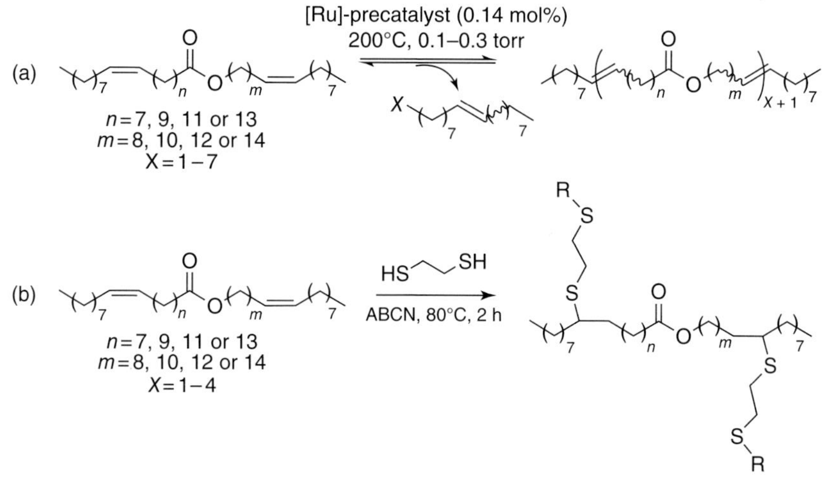

(a) [Ru]-precatalyst (0.14 mol%) 200°C, 0.1–0.3 torr

n=7, 9, 11 or 13
m=8, 10, 12 or 14
X=1–7

(b) HS–CH$_2$CH$_2$–SH, ABCN, 80°C, 2 h

n=7, 9, 11 or 13
m=8, 10, 12 or 14
X=1–4

Scheme 12.4 (a) General scheme for ADMET reaction of jojoba oil and (b) oligomerization of jojoba oil with 1,2-ethanedithiol.

hydrocarbons (e.g., 9-octadecene), *and* renewable polyesters. Butilkov and Lemcoff [22] described the acyclic diene metathesis polymerization (ADMET) of neat jojoba oil using different reaction conditions. The best results were achieved when Hoveyda–Grubbs second generation (HG2) was used as a precatalyst, and two potentially useful materials were obtained, that is, polyesters up to nonamers (3400–3500) and a hydrocarbon distillate of variable composition. In parallel, the authors tested the thiol–ene reaction of jojoba oils with 1,2-ethanedithiol and 1,1′-diazene-1,2-diyldicyclohexanecarbonitrile (ABCN) as the radical initiator. ESI-MS analysis indicated the formation of oligomers up to tetramers. This preliminary study demonstrates the intrinsic potential of vegetable oils for polymer synthesis, even when it is hidden behind exotic specimens.

12.1.3 Polyurethanes

Polyurethanes constitute an important class of versatile materials, mainly manufactured in the form of flexible and rigid foams, whose applications can largely vary from mattresses to automotive parts and insulating materials. The urethane linkage can be synthesized by various methods, and the most used route is the classical condensation reaction between hydroxyl and isocyanate functions [3]. Vegetable oils offer innumerous possibilities to be converted into polyurethane precursors, although the vast majority of investigations reported up to the present are directed to the synthesis of hydroxyl-containing structures (polyols), with fatty acid-derived polyisocyanates occupying a less important position. This is readily explained by the much more straightforward ways to arrive at OH groups by different chemical routes involving the oil unsaturations than to generate NCO moieties. Additionally, the abundant supply of castor oil, which naturally corresponds to a polyol with its three hydroxyl functions, allows the direct production of biobased polyurethanes, and BASF commercializes a castor oil-based polyether polyol (Lupranol Balance® 50) since 2009 [23].

Vegetable oils also offer a third route to polyurethanes by using the biodiesel-derived crude glycerol as a hydroxyl-containing raw material. This strategy has been gaining growing attention in the last several years [24, 25], since it corresponds to the application of a renewable molecule that is, additionally, the side product of the biofuel industry.

The following paragraphs summarize the most recent advances in the preparation of polyurethanes from vegetable oil derivatives. As already mentioned, most of successful works on the synthesis of plant oil-based polyurethanes are grounded on the search for novel hydroxyl-containing structures. The corresponding polyurethanes that arise from this strategy are however only partially renewable, since they are prepared with conventional fossil-based diisocyanates, for example, 4,4′-methylenediphenyldiisocyanate (MDI) and toluene diisocyanate (TDI). The pursuit of materials that are made of 100% renewable sources has encouraged some researchers to convert fatty acids into isocyanate building blocks [26]. Scheme 12.5 illustrates the routes to plant oil-based isocyanates that were successfully reported thus far.

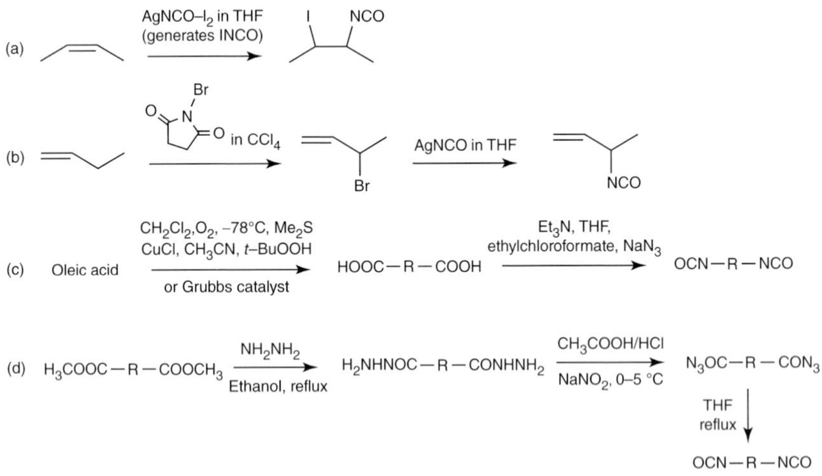

(a) $AgNCO–I_2$ in THF (generates INCO) → I, NCO

(b) NBS in CCl_4 → Br; AgNCO in THF → NCO

(c) Oleic acid —(CH_2Cl_2, O_2, −78°C, Me_2S; CuCl, CH_3CN, *t*-BuOOH; or Grubbs catalyst)→ HOOC—R—COOH —(Et_3N, THF, ethylchloroformate, NaN_3)→ OCN—R—NCO

(d) $H_3COOC—R—COOCH_3$ —(NH_2NH_2, Ethanol, reflux)→ $H_2NHNOC—R—CONHNH_2$ —(CH_3COOH/HCl, $NaNO_2$, 0–5 °C)→ $N_3OC—R—CON_3$ —(THF reflux)→ OCN—R—NCO

Scheme 12.5 Alternative routes explored to convert triglycerides and their fatty acids into polyisocyanates. (Adapted from [26].)

This challenging subject, pioneered by Cramail's group [27], was recently reported by Avérous and colleagues [28] in an interesting study about the application of 2-heptyl-3,4-bis(9-isocyanatononyl)-1-pentylcyclohexane (abbreviated as DDI, Scheme 12.6), a biobased diisocyanate obtained from the chemical modification of a hydrogenated dimer from fatty acids. The authors obtained a 100% biobased polyurethane from DDI, isosorbide (a starch derivative), and

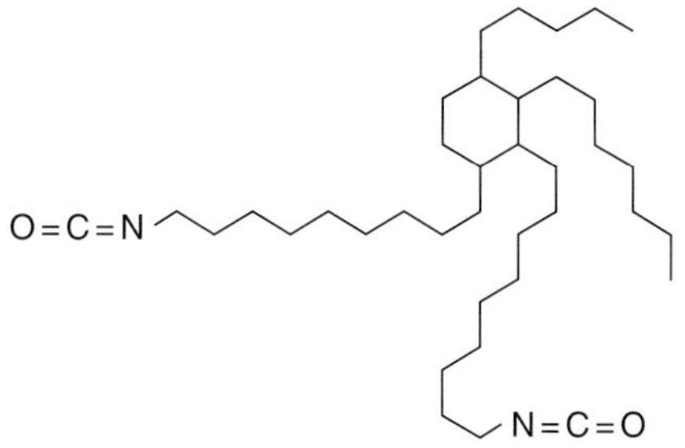

Scheme 12.6 The structure of DDI, a fatty acid-based diisocyanate [28].

1,4-butanediol. High-molecular-weight polyurethane was isolated ($M_w = 45\,000$) when no isosorbide was added. This value progressively decreased, and with 20% isosorbide, the molecular weight reached its minimum value ($M_w = 11\,400$). The glass transition of the materials ranged from −15 to 60 °C, and the melting point from −1 to 82 °C. Partially renewable polyurethanes were prepared from palm, soy, and sunflower oils, converted into their corresponding monoglycerides, and further reacted with phthalic anhydride (2 : 1 monoglyceride-to-phthalic anhydride ratio) to prepare alkyd diols, a new class of polyols (Scheme 12.7) [29]. The alkyd diols were then reacted with MDI, resulting in the formation of poly(alkyd-urethane)s. The materials exhibited good chemical, mechanical, and thermal resistances.

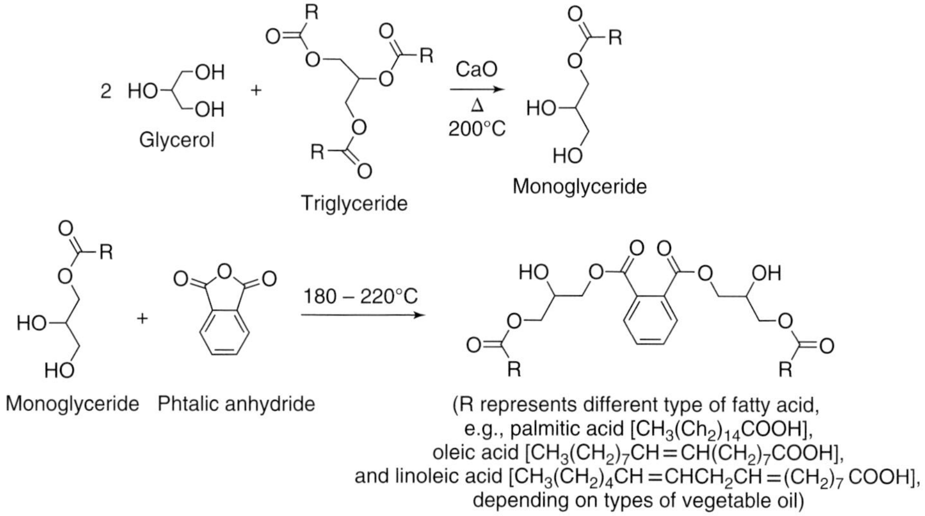

Glycerol + Triglyceride → (CaO, Δ, 200°C) Monoglyceride

Monoglyceride + Phtalic anhydride → (180 – 220°C)

(R represents different type of fatty acid, e.g., palmitic acid [$CH_3(Ch_2)_{14}COOH$], oleic acid [$CH_3(CH_2)_7CH{=}CH(CH_2)_7COOH$], and linoleic acid [$CH_3(CH_2)_4CH{=}CHCH_2CH{=}(CH_2)_7COOH$], depending on types of vegetable oil)

Scheme 12.7 Synthesis route of (a) palm oil monoglyceride and (b) palm oil alkyd diols [29].

The path for the development of new polymeric materials from plant oils requires the exhaustive evaluation of the wide range of alternative sources that are naturally available. The oil extracted from *Jatropha curcas* is called *jatropha oil*, a nonedible oil that is mainly converted into biodiesel. The residual material of the biodiesel production of jatropha oil has many applications that range from biomass feedstock to power electricity plants and to medicinal purposes. An additional advantage of this oil is the fact that *J. curcas* grows well in arid and hot regions such as the desert areas. Aung and colleagues [30] recently reported a jatropha oil-based alternative to polyurethane wood adhesives derived from palm oil, which frequently exhibit poor performances due to their high contents of saturated fatty acids. Jatropha oil has approximately 78.5% of unsaturated fatty acids, which would in principle lead to better properties of these materials. Polyols were prepared by esterification and alcoholysis or epoxidation of crude jatropha oil and then polymerized with TDI, to give a material with comparable shear strength to the palm oil-based one, but with superior chemical resistance.

Another important field of application of polyurethanes is related to paints and coating materials. The worldwide tendency to reduce and monitor the release of volatile organic compounds greatly interests this industrial sector, since solvent-based polyurethanes correspond to 18% of the total value of global consumption of coating materials [31, 32]. Aiming to develop alternatives to solvent-based polyurethanes, the focus turned to waterborne polyurethane coatings as an effective means to tackle the problem. This switch is also associated with the reduction of costs, fire, and health risks. Anionic waterborne polyurethanes are prevalent, and they are typically prepared by incorporating a diol or polyol that also contains a carboxylic acid group, such as dimethylolpropionic acid, into the polymer backbone. The polyurethane is then neutralized with a tertiary amine such as triethylamine and dispersed in water [32]. Associated with the absence of organic solvents, the use of renewable building blocks for polyurethane synthesis is also advantageous, since it makes an important part of the whole process highly sustainable. Larock and coworkers [33] described the synthesis of anionic waterborne polyurethanes using polyols prepared from peanut, corn, soybean, and linseed oils (degree of unsaturation varying from 0.4 to 3.5 C=C double bonds per triglyceride), with the main goal of evaluating the effects of residual unsaturation in vegetable oil-based polyols on the thermal and mechanical properties of the final materials. The crude oils were partially epoxidized and subsequently ring opened to prepare polyols with M_n values near 1000. Polyurethane films were synthesized with isophorone diisocyanate (IPDI), and the evaluation of their tensile properties indicated that the modulus and toughness increased and the % strain decreased as the residual unsaturation increased. Glass transition temperatures were also a function of this parameter, as the polyols synthesized from peanut and linseed oil had degrees of unsaturation of 0.4 and 3.5 and T_g values of -9.2 and 13 °C, respectively.

Motivated by the lack of antimicrobial activity of anionic waterborne polyurethanes, a key inherent disadvantage of these materials, the same group had previously reported novel cationic polyurethane dispersions from vegetable oil-based polyols [34]. They are typically prepared by incorporating a tertiary amine diol or polyol followed by treatment with an acid and exhibit excellent adhesion properties. In a more recent contribution, Larock and colleagues [32] described the preparation of soy oil-based cationic aqueous polyurethane dispersions from the reaction of methoxylated soybean polyols (MSOL), *N*-methyl diethanolamine (MDEA), and IPDI (Scheme 12.8). The authors found that polyurethane films with higher amounts of MDEA, which acts as a chain extender, improved mechanical properties and that increasing the hydroxyl numbers of the soy polyol, while keeping the overall molar ratios between NCO and OH groups constant, increased the T_g from 3.0 to 17.8 °C as the hydroxyl number went from 140 to 174 mg KOH g^{-1}. All polyurethane films displayed inhibitory activity against three foodborne pathogens.

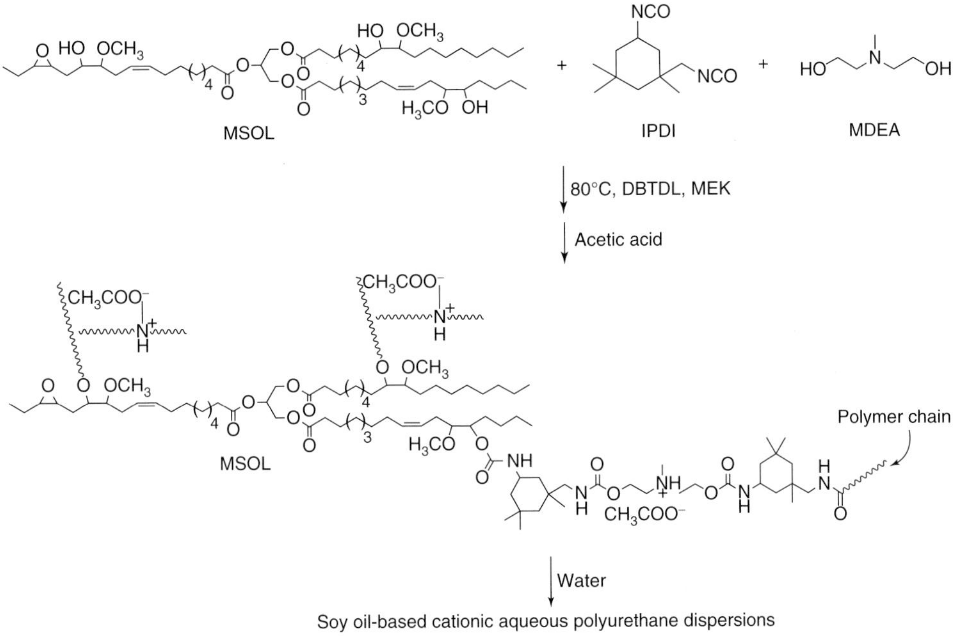

MSOL + IPDI + MDEA

80°C, DBTDL, MEK

Acetic acid

MSOL

Polymer chain

Water

Soy oil-based cationic aqueous polyurethane dispersions

Scheme 12.8 Reaction pathway for the preparation of soy oil-based cationic aqueous polyurethane dispersions [32].

One of the disadvantages of using polyurethane foams is their high flammability, and flame-retardant additives are used in their formulation to attenuate the burning process. An interesting application of vegetable oil-based polyurethanes, described by Petzhold and colleagues [35], is based on the preparation of phosphorylated oil-derived polyols in the formulation of rigid polyurethane foams and their properties as flame retardants. The foams were synthesized from epoxidized soybean oil, phosphoric acid in two different amounts (5% and 10% weight of phosphoric acid), castor oil, glycerin, and polymeric N,N-diphenylmethane diisocyanate (pMDI). The polyurethane flammability was evaluated by means of oxygen index (OI), and all polyurethanes that contained the phosphorylated polyols exhibited OI values similar to, or better than, those of polyurethanes that contained commercial flame retardants.

It is also possible to apply oil-based polyurethanes in composite materials. Polyurethanes were successfully synthesized from epoxidized soybean–castor oil fatty acid-based polyols and pMDI, and used as surface modifier of graphene oxide, synthesized via a modified pressurized oxidation method [36]. Another remarkable study describes the preparation of a polyurethane foam from a rapeseed oil-derived polyol and pMDI, reinforced with micro-/nanocellulose, (MNC) in order to increase the amount of biobased components in the final material [37]. Glycerol was also used to serve as a reactive modifier that allowed to increase the cross-linking density in foamed PUs and to improve their mechanical properties. The authors reported an increase in parameters like compression strength, rigidity, and glass transition temperatures of the materials incorporating MNC and glycerol to the foam formulation. The association of oleic acid and lignin gave rise to a novel macropolyol (Scheme 12.9) using a solvent-free and catalyst-free method [38]. A series of polyurethanes were prepared by a two-step procedure: the first involved the synthesis of linear isocyanate prepolymers of MDI and poly(propylene)glycol of varying molecular weights and then used in the second step to obtain different polyurethane macromolecular architectures in the presence of the lignin–fatty acid-based macropolyol. The originality of this

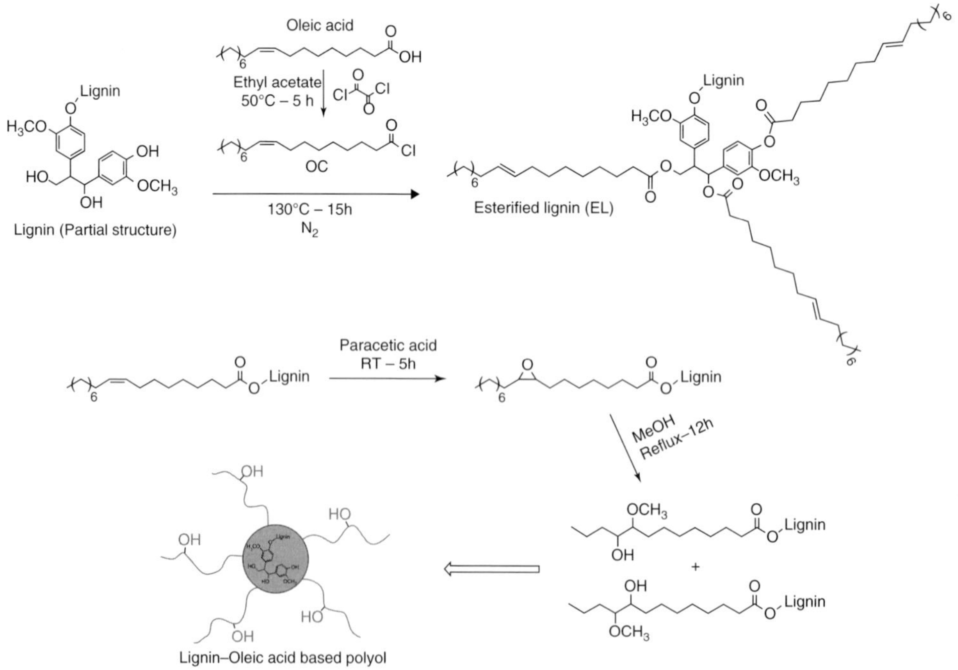

Scheme 12.9 Synthetic pathway to the lignin–oleic acid macropolyol [38].

study is to have combined the valorization of two different renewable resources to prepare a family of viable materials.

12.1.4 Polyamides

Polyamides (PAs) from ω-aminodecanoic acid, a castor oil derivative prepared from 10-undecenoic acid, have already been used for a large-scale production of PA11 for decades [39, 40]. Among the most recent contributions to the preparation of oil-based PAs, those of Meier's research team predominate [41, 40, 42, 43]. AB-type polyamide monomers were obtained from methyl oleate and methyl erucate [41] via a classic hydrobromination reaction with 33% hydrobromic acid solution in acetic acid at room temperature (RT). The fatty acid methyl ester bromides, respectively, methyl 9(10)-bromostearate and methyl 13(14)-bromodocosanoate, were converted into azides and subsequently reduced to the corresponding amines. The polycondensation reactions were conducted with 1,5,7-triazabicyclo[4.4.0]dec-5-ene (TBD) as catalyst, giving branched polymers with M_n of 6500 and 7900, and melting points of 40 and 89 °C for the PA synthesized from methyl oleate and methyl erucate, respectively. PAs were also prepared from biochemically derived fatty acid derivative followed by chemical derivatization [40]. The copolymerization of ω-amino fatty acid methyl esters and a long-chain diester led to the formation of renewable polyesters and polyamides with interesting thermal properties, that is, melting points of 109 °C for the polyester and 166 °C for the polyamide. The same group described the oxy-functionalization of methyl oleate and methyl erucate [42], employing molecular oxygen as oxidizing agent, for the preparation of keto fatty acid esters to serve as intermediates for the synthesis of renewable PAs. Amine-functionalized AB-type monomers were obtained by reductive amination of the keto fatty acids, and their homo- and copolymerization reactions were studied with hexamethylenediamine and dimethyl adipate to modify the properties of conventional nylon 6,6. These polycondensations were catalyzed by TBD and gave PAs with M_n values up to 7500 PDI = 1.65 and T_m = 45 °C and copolyamides with M_n values up to 21 000, PDI = 2.3, and T_m = 250 °C. Meier and Winkler [43] also described more recently the application of olefin cross-metathesis of methyl oleate- and methyl erucate-derived benzyl carbamates with methyl acrylate (Scheme 12.10). This strategy was a novel approach to obtain linear AB-type monomers, used for the preparation of renewable polyamides PA11, PA12, and PA15, obtained with M_n values of 14 900, 22 600, and 15 200 and T_m values of 186, 182, and 169 °C, respectively.

An innovative approach called upon the ring-closing metathesis (RCM) reaction for the synthesis of PA11, PA12, and PA13 from oleic acid (Scheme 12.11) [44]. In this three-step mechanism, oleic acid was converted to alkenyl amides, which were then subjected to RCM, generating unsaturated ene-lactams, subsequently hydrogenated to saturated lactams, the precursors for PA12 and PA13. The precursor of PA11 was a linear amino ester, which could be readily accessed by

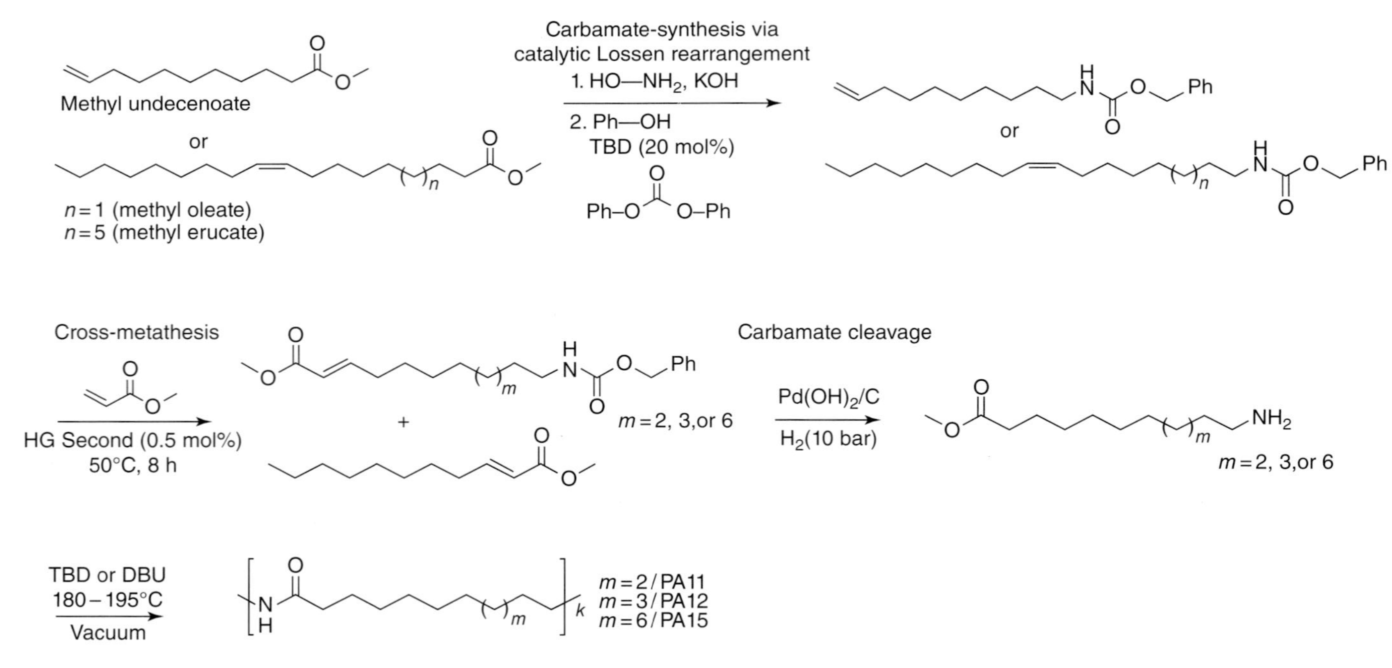

Scheme 12.10 Synthesis of AB-type monomers for the preparation of renewable PAs [43].

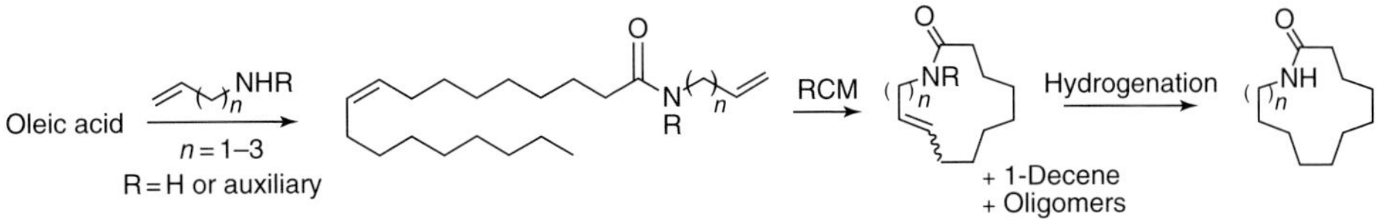

Scheme 12.11 General procedure for the synthesis of nylon precursors from oleic acid [44].

solvolysis of the lactam. The authors state that, comparing with the current methods used for the cyclic precursors of polyamides, this approach has the important advantage of the direct introduction of amine groups, avoiding additional reaction steps. It is also less susceptible to undesirable isomerization or self-metathesis, which complicates the isolation process of the final material.

12.2 Sugar-Derived Monomers and Polymers

12.2.1 Introduction

Carbohydrates can be considered one of the most promising feedstocks for the preparation of sustainable chemicals and materials [45]. They are fully renewable, widely spread compounds, easily accessible from a huge number of different natural sources. A great variety of sugars can be readily obtained from crops (edible and nonedible sources) or through extraction and enzymatic hydrolysis from lignocellulosic wastes. Sugars are actually the most exploited bioresources since they are the source of strategic monomers, such as ethylene glycol, succinic acid, and lactic acid. When properly modified and functionalized, carbohydrates constitute a platform for new substances [46], often incorporating structures that, introduced into macromolecular chains, provide materials displaying interesting properties, such as enhanced hydrophilicity, biodegradability, or low toxicity [4]. The main occurrence of hydroxyl groups, acid, and amino moieties in the carbohydrate structure makes sugar-based monomers particularly suitable for polycondensation reactions in the synthesis of materials such as polyesters, polyamides, polyurethanes, and other polycondensates. The purpose of this section is to report the latest advances which involve the most promising monomers derived from the chemical modification of carbohydrates.

12.2.2 Polymers from 1,4:3,6-Dianhydrohexitols

Bicyclic compounds such as isosorbide, isomannide, and isoidide (Scheme 12.12) have recently asserted their particularly promising role as precursors of macromolecular materials. The vigorous interest in these compounds is mainly due to

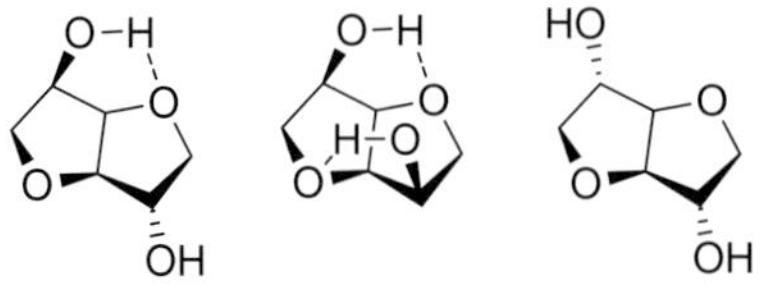

Scheme 12.12 Isosorbide, isomannide, and isoidide.

their rigidity, chirality, and nontoxicity, coupled with the fact that isosorbide is now a chemical commodity. Indeed, the two cis-fused V-shaped tetrahydrofuran rings can confer a significant rigidity to the polymeric structure, producing materials with high glass transition temperature and good thermomechanical resistance. The chirality inherited from the sugar source can furthermore affect the properties as well as transfer specific optical properties to the material. Finally, since these compounds are not toxic, they can be used in the synthesis of materials for biomedical applications.

Despite the resembling chemical structure, these three diastereomeric compounds display different reactivities and therefore different ensuing properties when introduced into a polymer chain. The reason for this behavior must be referred to the stereoisomerism of the hydroxyl groups, the *exo* configuration, less hindered and not involving intramolecular hydrogen bonds, being therefore more reactive than the *endo* counterpart. As a consequence, isoidide is the most reactive monomer while isomannide shows the poorest reactivity toward polycondensation. Furthermore, only the more symmetric structure of isoidide can produce semicrystalline materials. The synthetic pathway of 1,4:3,6-dianhydrohexitols starts with the enzymatic degradation of starch for the production of monomeric sugars (D-glucose and D-mannose). These sugars are hydrogenated to obtain D-sorbitol and D-mannitol and subsequently isosorbide and isomannide are synthesized through a dehydration process, whereas isoidide can be obtained from iditol via a similar procedure, but it does not derive from starch. Up to now, isosorbide is the only compound produced on an industrial scale due to the large availability of glucose. Isomannide is usually less applied to materials synthesis due to its lower reactivity, while isoidide shows poor industrial applicability because of its scarce sources. Given the strong academic and industrial interest related to these renewable compounds, much effort has been devoted in the last decade to the synthesis of a large variety of macromolecular structures [47–49]. The most relevant use of 1,4:3,6-dianhydrohexitols, and isosorbide in particular, concerns the field of copolyesters since the rigid structure together with the low reactivity of secondary alcohols usually produce stiff and fragile homopolymers with low molecular weights. Different polymerization partners have been studied, from both the fossil-based and renewable biobased pools of monomers. Kricheldorf's group, for example, developed biodegradable copolyesters involving isosorbide and lactide, with aromatic monomers [50, 51]. These materials were synthesized with a two-step one-pot procedure involving a ring-opening polymerization of lactide with different ratios of isosorbide, followed by polycondensations with terephthaloyl and isophthaloyl chloride (Scheme 12.13).

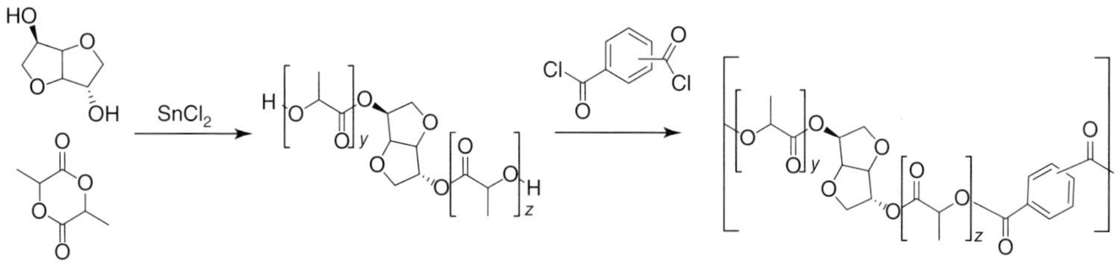

Scheme 12.13 Biodegradable copolyesters involving isosorbide, lactide, and aromatic monomers [50, 51].

The copolyesters obtained displayed an amorphous character and a range of glass transition temperatures between 64 and 180 °C according to the isosorbide content. Since the marriage between sugar-derived monomers and aromatic compounds appeared to be particularly promising in terms of obtaining materials with high thermomechanical properties, many efforts were then directed toward the study of fully biobased aliphatic/aromatic polyesters. Allais *et al.* [52] studied sustainable alternating aliphatic/aromatic copolyesters based on ferulic acid as the aromatic source with aliphatic isosorbide or 1,4-butanediol. The materials obtained showed tunable properties, in particular glass transition temperatures from 0 to 75 °C. Interesting possible industrial applications were recently studied for the pilot-scale synthesis of poly(butylene succinate) (PBS) copolymers with isosorbide and 2,5-furanedicarboxylic acid [53]. Due to the lower reactivity of its *endo* hydroxyl group, the presence of isosorbide slowed down the polycondensations. The glass transition temperature however increased from −30 °C (PBS) to −11 °C (PBS with 14 mol% of isosorbide). As expected, a progressive reduction of crystallinity was observed with increasing amounts of the sugar-based comonomer. A different approach was reported by Colonna *et al.* [54, 55], who developed and studied a procedure where biobased compounds, such as isosorbide, succinic acid, and glycerol, were combined with chemical recycling. Postconsumer PET (polyethylene terephthalate) was successfully depolymerized by isosorbide in order to obtain oligomers which were subsequently repolymerized with bio-succinic acid (Scheme 12.14). The new polyester exhibited properties suitable for powder coating applications, namely, an amorphous character, a glass transition of 55 °C, low molecular weights, and acid terminal groups. The introduction of isosorbide rigid structure into the polymer chain resulted to be suitable for coating properties such as thermal and chemical stability as well as a good retention of these properties after accelerated aging. The process is currently being upgraded to a pilot scale.

Although dianhydrohexitols have demonstrated a remarkable potential in the field of polyesters, they are actually modestly employed, without sensible chemical modifications, to other polycondensation materials. Recently, however, some interesting investigations reported the use of isosorbide for the production of polyethers [58], polycarbonates [59], and polyurethanes [56, 57]. One such study was reported by Avérous *et al.* [28] who described the synthesis and properties of

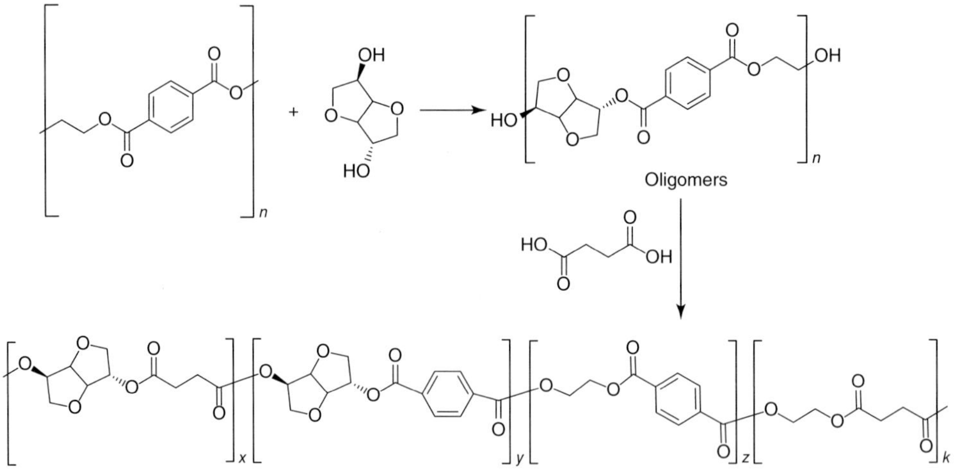

Scheme 12.14 Chemical recycling of PET with isosorbide and succinic acid for powder coating applications [54, 55].

fully biobased thermoplastic polyurethane obtained from isosorbide, butanediol, and a diisocyanate based on biobased building blocks (Scheme 12.15). These polyurethanes displayed a semicrystalline character resulting from different segregate hard and soft domains. The introduction of isosorbide into the polymeric chain deeply affected the molecular weight (from 45 000 to 11 400) and the glass transition temperature increased from −15 to 2 °C while the melting temperature increased from 60 to 84 °C, even if associated with a progressive reduction of crystallinity. In order to overcome the previously discussed problems of reactivity associated with dianhydrohexitols, as well as to enlarge the horizon of applications of such compounds, many efforts have been directed toward chemical functionalization. This approach consists in designed modifications of 1,4:3,6-dianhydrohexitols in order to introduce specific functional groups without affecting the main structure. A remarkable example was developed

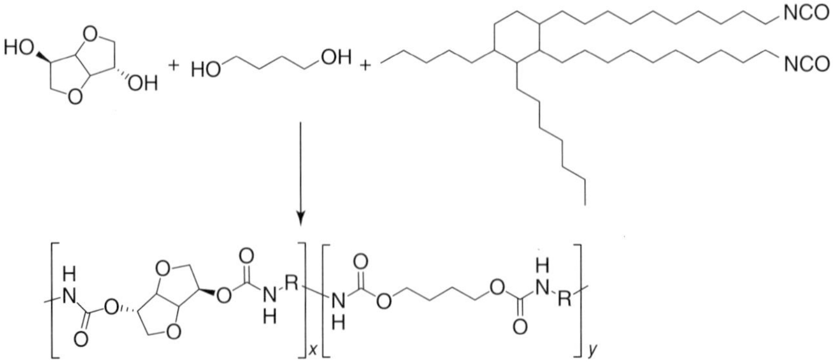

Scheme 12.15 Fully biobased thermoplastic polyurethanes incorporating isosorbide [28].

by van Es's group [61] who developed the first synthesis of a new family of dianhydrohexitols, using different synthetic routes designed to obtain primary diols, primary diamines, or diacids from isomannide (Scheme 12.16). These new building blocks showed a high reactivity since the reactive groups are no longer hindered secondary groups but primary glycol, diamine, and diacid. Furthermore, the choice of isomannide as starting material led, after inversion of configuration of the chiral center, to symmetric isoidide-like compounds, suitable for the synthesis of semicrystalline materials. Different polyesters involving isoidide dicarboxylic acid and isoidide dimethanol were synthesized along with different comonomers in order to determine their potential applications. Both isoidide dicarboxylic acid and isoidide dimethanol were shown to confer rigidity to the macromolecular structure with enhanced reactivity and semicrystalline features [62]. Fully isohexide-based polyesters were also reported [63]. Isoidide dimethyleneamine, moreover, was used for the synthesis of new biobased semicrystalline polyamides with melting points up to 250 °C [64]. More recently, the same authors developed the synthesis of new isohexide hydroxyesters (Scheme 12.17) [60]. This new class of biobased AB-type building blocks was employed for the synthesis of stereoregular polyesters. The results demonstrated that the orientation of the hydroxyl group has a tremendous influence on the properties of the ensuing materials. While *endo*-OH monomers (Scheme 12.17, compounds 2 and 4) displayed the expected reduction of reactivity and low molecular weights, surprisingly, homopolymers obtained from compounds 1 and 2 had low glass transition temperatures (15–20 °C), while copolyesters from compounds 1 and 3 gave T_g values up to 80 °C. Clearly, these singular properties could be related to particular interactions created by the stereoregularity of the various materials. Further investigations are required in view of future applications.

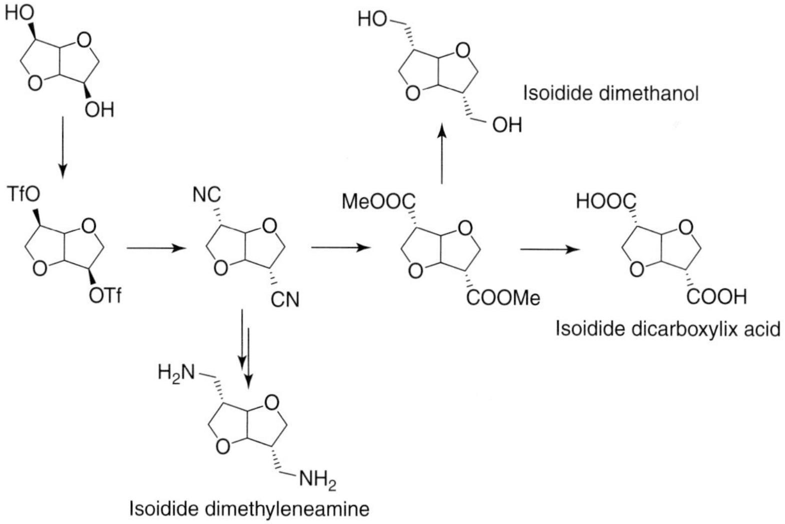

Isoidide dimethanol

Isoidide dicarboxylix acid

Isoidide dimethyleneamine

Scheme 12.16 Synthesis of new biobased monomers from isomannide [61].

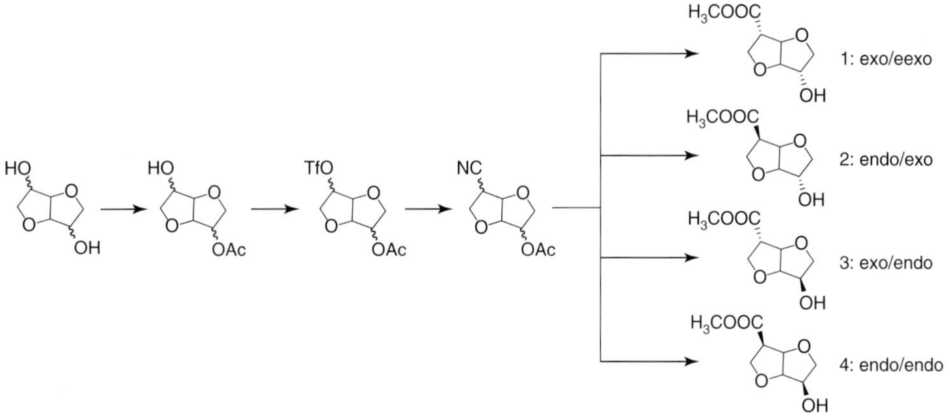

Scheme 12.17 Synthesis of biobased AB monomers [60].

12.2.3
Polymers from Diacetals Derived from Sugars

Acetalized carbohydrates constitute an independent class of bicyclic monomers which demonstrate a vast potential, although no examples of industrial application are currently reported. The most remarkable examples, shown in Scheme 12.18, were developed by Lavilla and Muñoz-Guerra [65–70], for the synthesis of aliphatic/aromatic biobased polyesters with enhanced thermomechanical properties. The reported configurations were chosen from the natural pool of carbohydrates in order to exploit simple, commercial, and accessible starting materials (respectively, galactaric acid, mannitol, and D-gluconolactone). The synthetic routes for gluco- and galacto-derived monomers (respectively, Glux and Galx) were designed through specific protections of the secondary hydroxyl moieties with paraformaldehyde for the formation of cycloacetals, followed by the reduction or oxidation of terminal groups in order to obtain glycols or diesters. Manno-derived monomers (Manx) required a preliminary protection of the primary alcohols followed by diacetylation. All the compounds reported were

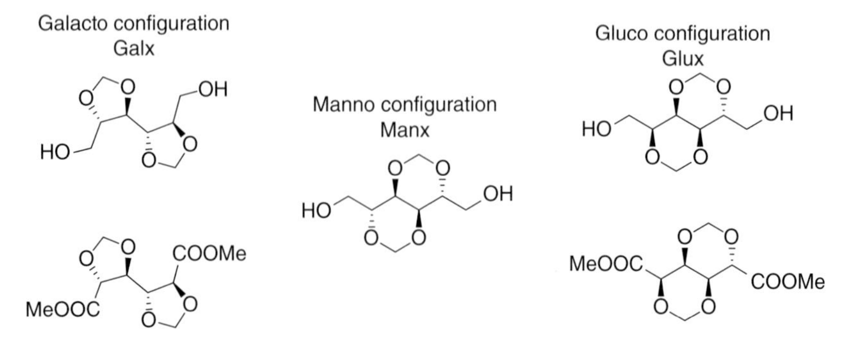

Scheme 12.18 Diacetal monomers derived from sugars.

crystalline, easily isolated and purified. The properties of the materials involving these monomers are highly affected by their structure, chirality, and symmetry. All of them are bicyclic structures, but Manx and Glux contain two fused rings that confer high stiffness compared to the two independent rings of Galx compounds. Moreover, Galx and Manx bear, respectively, a centrosymmetric structure and a C_2 axis, whereas Glux has no symmetric element. These differences are reflected by a different stereoregularity introduced into the polymer chain, as well as a dissimilar reactivity toward esterification of Glux moieties.

Several remarkable studies have been recently reported involving these monomers. Galacto derivatives, for example, were copolymerized with different monomers to determine the effect of their structure on the properties of the ensuing materials [66, 67]. In particular, copolyesters were synthesized from galactitol and galactaric acid with 1,4-butanediol and dimethyl terephthalate in different proportions (Scheme 12.19) [66]. This set of experiments produced

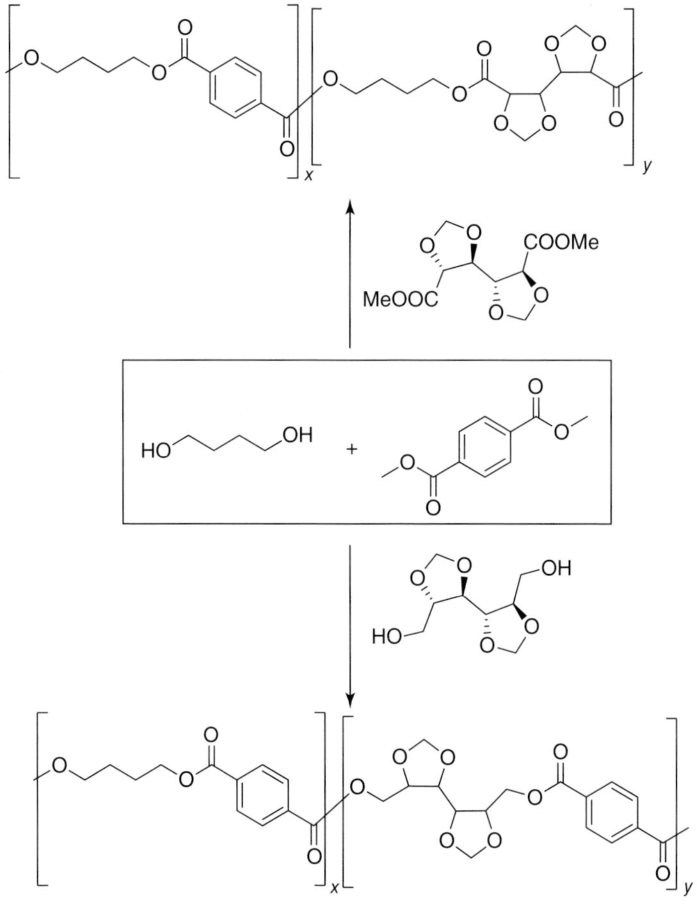

MeOOC COOMe HO OH HO OH + x y

Scheme 12.19 Copolyesters based on galactitol and galactaric acid derivates [66].

different polymers with satisfactory molecular weights (32 000–41 000) and randomly distributed microstructures. All such materials displayed a semicrystalline character with T_m ranging from 115 to 210 °C and T_g between 20 and 70 °C. The authors furthermore demonstrated that the value of T_g decreased with increasing content of galactarate, while they increased with increasing content of galactitol.

Additionally, mannitol-based fused-ring monomers (Manx) have been employed in the synthesis of polyesters [68, 65]. Remarkably, in their most recent investigations, Lavilla *et al.* reported the effect of such monomers for the development of aliphatic polyesters with 1,4-butanediol and succinic acid (Scheme 12.20). Properties such as stiffness and thermomechanical resistance were comparable with those of isosorbide-based copolyesters with the additional advantage of desirable molecular weights related to the enhanced reactivity of Manx. The homopolymer had a melting temperature of 127 °C and a T_g of 68 °C, which is much higher compared to those of PBS. All the copolyesters were semicrystalline materials with high molecular weights (from 30 800 to 49 100) and thermal and mechanical properties strongly related to the increasing content of the bicyclic monomer (T_g values from −29 to +51 °C).

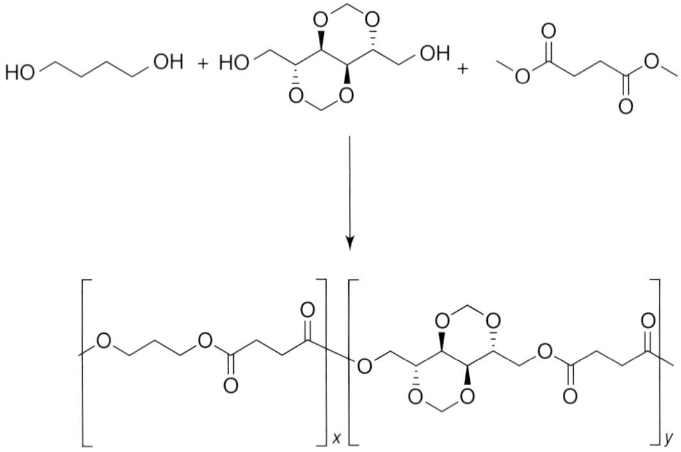

HO OH + HO OH + O O O O

Scheme 12.20 Fully aliphatic polyesters based on mannitol derivatives [65].

A different approach was finally studied in order to prepare partially renewable copolyesters of PBT with acetalized carbohydrates by means of a solid-state modification [67, 69]. The resulting materials had a unique blocklike microstructure with homogeneous crystalline sequences of PBT attached to random amorphous phases enriched into the carbohydrate-based monomer. This unusual structure gave rise to an unexpected enhancement of the thermal properties of the materials. Because of the low rate of randomization of the phases, these blocklike copolyesters displayed higher melting and crystallization temperatures and a considerably higher crystallinity compared with materials obtained by melt polymerization. Moreover, Glux monomer produced enhanced melting points

Table 12.3 Comparison among different aromatic polyesters based on carbohydrates [70].

Entry	Polyester	M_w	PDI	T_g (°C)	T_m (°C)	T_d (°C)
1	PET	32 100	2.5	81	250	411
2	PBT	41 300	2.4	31	223	371
3	PIsT	25 600	not reported	205	—	300
4	PIiT	14 500	not reported	209	261	not reported
5	PGalxT	30 500	2.5	87	—	382
6	PManxT	30 200	2.3	137	—	378
7	PGluxT	12 800	2.3	154	272	377
8	PEGlux	9 000	2.4	112	—	262
9	PBGalx	30 800	2.4	18	122	328

and glass transition temperatures compared to those of the other sugar-based polymers. All the reported studies demonstrate the huge interest and therefore the potential impact of carbohydrate-based monomers for the synthesis of partially biobased polyesters with enhanced properties. This topic is thoroughly discussed by Muñoz-Guerra's group in their latest review [70], which describes the recent achievements regarding terephthalate homopolymers and copolymers giving a clear-cut comparison on the two families of carbohydrate-based bicyclic monomers, namely, isohexide and diacetals (Table 12.3).

The results reported in Table 12.3 show that the addition of carbohydrate-based bicyclic monomers, in general, slightly reduces the thermal stability of the corresponding materials and deeply affects their crystallinity, with only PIiT and PGluxT retaining some semicrystalline character due to their configuration. On the contrary, they can increase significantly the glass transition temperature of the polymers. As a consequence, these materials exhibit higher mechanical modules and are more brittle compared with their fossil-based counterparts. Furthermore, diacetalized compounds (Manx, Galx, Glux) demonstrated to be more reactive compared with the dianhydrohexitols isosorbide and isoidide [71], thus providing polymers with higher molecular weights. Finally, biodegradability was enhanced as the content of sugar-based monomers was increased, with hydrolysis taking place preferably at the sugar moieties.

12.3 Polymers from Terpenes and Rosin

12.3.1 Introduction

Terpenes, terpenoids, and rosins are classes of natural molecules obtained from biomass that contain unsaturated cycloaliphatic and/or aromatic structures. They are major components of tree resin, which is an exudate obtained from trees (e.g., pine trees and conifers) [72]. Turpentine, the volatile fraction of the resin, is mainly composed of a mixture of terpenes (mainly mono- and sesquiterpenes), which bear a cycloaliphatic structure made of isoprene units as building block of the molecular skeleton [73]. Monoterpenes have the general isoprene-dimer formula $C_{10}H_{16}$ (Scheme 12.21), while sesquiterpenes consist of three isoprene units with the molecular formula $C_{15}H_{24}$. Terpenoids are modified terpenes, in which methyl groups have been isomerized or removed, or oxygen atoms were added. Rosin is the nonvolatile solid part of the resin and is produced by heating the liquid resin in order to vaporize the volatile terpene components. Rosin is composed of diterpenic acids with the chemical formula $C_{19}C_{29}COOH$. It consists mainly of abietic- and pimaric-type resin acids (or rosin acids) with characteristic hydrophenanthrene structures (Scheme 12.22).

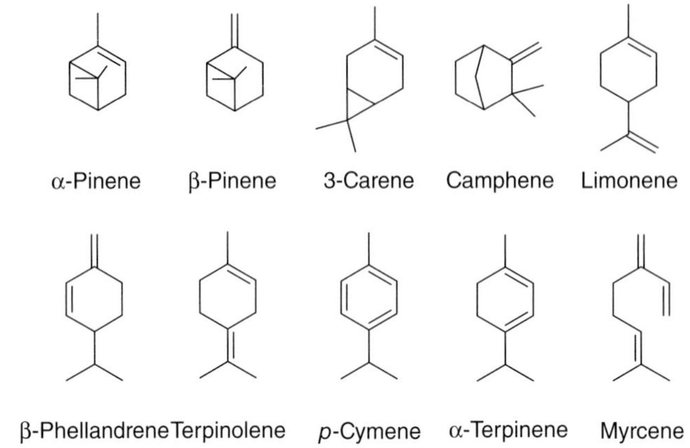

α-Pinene
β-Pinene
3-Carene
Camphene
Limonene
β-Phellandrene
Terpinolene
p-Cymene
α-Terpinene
Myrcene

Scheme 12.21 Chemical structure of the most common monoterpenes.

COOH
Abietic acid
COOH
Neoabietic acid
COOH
Levopimaric acid
COOH
Palustric acid
COOH
Dehydroabietic acid

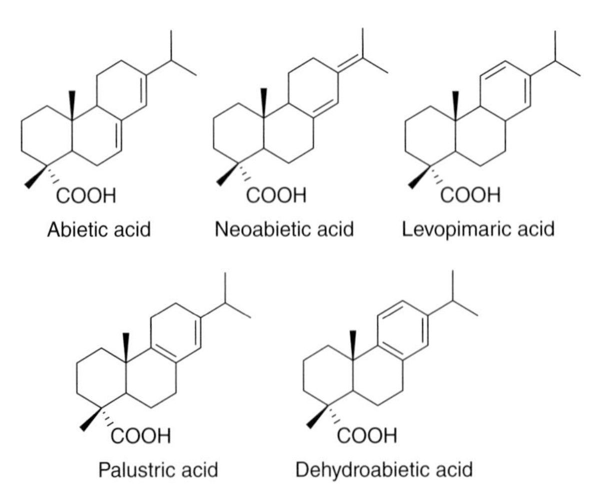

Scheme 12.22 Chemical structure of the most common rosin components.

12.3.2
Terpenes and Rosin Production and Application

Terpenes are considered as secondary metabolites of plants and of some fungi and insects [74]. They are widely distributed in higher plants, in particular in softwoods, and are present in roots, stems, foliage, and seeds [75]. Terpenes have been used for centuries as a source of resins for adhesive and sealants, as well as for other industrial applications. Turpentine is by far the main source of terpenes with a global production of more than 300 000 tons per year (70% as sulfate turpentine) and its major constituents are α-pinene (45–97%) and β-pinene (0.5–28%) with smaller amounts of other monoterpenes [76]. D-Limonene is the main constituent of citrus skin and is mainly obtained as by-product of citric fruit juice processing, mostly by a cold process involving centrifugal separation or by steam distillation [77]. Its worldwide production in 2013 was estimated to exceed 70 000 tons. On the other hand, the orange production for 2014 was projected to rise by 5% from the previous year, reaching a global amount of 51.8 million tons. Brazil is by far the largest producer, with an output approaching 40% of the world's orange production. About 40% of the 51.8 million tons global orange production in 2014 was used for orange juice production [78]. This left about 13 million tons of waste orange peel, from which, assuming a D-limonene content of 3.8 wt% [79], 520 000 tons of D-limonene could be obtained [78]. Terpenes, especially pinene and limonene, are mainly used as starting material for the synthesis of flavors, fragrances, solvents, pesticides, pharmaceuticals, and chiral intermediates [80]. Cleaning products for industrial and household applications originally represented the largest market for D-limonene. Nowadays, limonene is widely used as solvent to degrease and clean vessels and equipment encrusted by heavy oil or in the extraction of petroleum from oil sands [81]. Caryophyllene and humulene are the most abundant and cheapest available sesquiterpenes. They are found in many plants and fungi. For example, more than 10^5 tons of clove oil is produced annually by the clove tree *Eugenia caryophyllata.* Clove oil contains 7–12% caryophyllene and 1–4% humulene [82]. Another potential source of sesquiterpenes is hop oil, with up to 25% caryophyllene and 45% humulene [82]. Rosin has a production of more than 1 million tons per year in the world. This pine resin has been widely used in the past for the waterproofing of wooden ships and for this reason is also known as *naval stores.* Rosin and its derivatives have been traditionally widely used as paper seizers, emulsifiers, tackifiers, and additives for printing inks and varnishes [83]. Due to its hydrophobicity, rosin has also been used for decades as a marine antifouling material.

12.3.2.1 Isomerization Reactions to Obtain Different Terpenes

Pinenes can be used as a source of other less common terpenes through selective catalyzed or heat-induced isomerization processes (Scheme 12.23) [84]. Several processes are reported in the literature for the isomerization/rearrangement of terpenes [85], for example, α-pinene has been isomerized to β-pinene in the liquid phase at 160 °C using titanic acid as catalyst and calcium amide has also been used

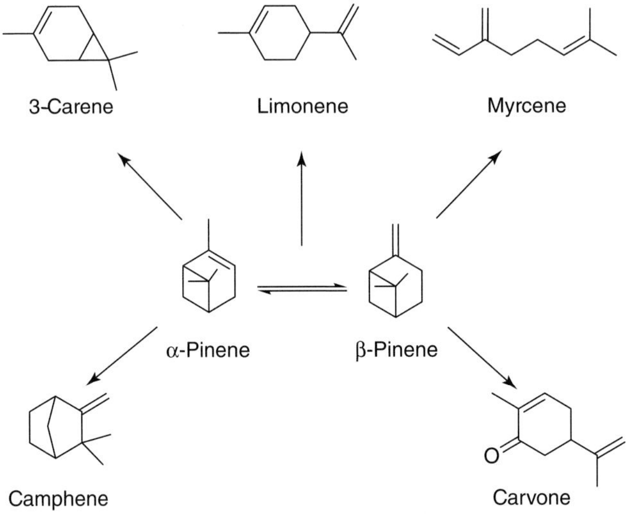

3-Carene Limonene Myrcene

α-Pinene β-Pinene

Camphene Carvone

Scheme 12.23 Isomerization and oxidation processes for converting pinenes into other terpenes and a terpenoid.

as basic catalyst for this isomerization. Alkaline-earth metal oxides can also be used, with strontium oxide being the most active [86]. The isomerization of pinene to camphene is an industrially important process for the preparation of camphor [87]. This reaction is catalyzed by several solid acid catalysts such as zeolites [88] and titania [89]. The isomerization of α-pinene to limonene in high yields has been performed using zeolites as catalysts [88], and myrcene has been produced from β-pinene with a yield of 83% [89].

12.3.3 Terpenes as Monomers for Polymer Synthesis without Any Modification

The first reference to the production of a macromolecular material from turpentine, and indeed the *very first mention* of a clear-cut polymerization, dates back from 1798, when Bishop Watson obtained a "sticky material" treating turpentine with sulfuric acid. Nowadays, the most important process for the production of polymers from terpenes is cationic polymerization, followed by free radical polymerization The main applications of polymers from terpenes are in the field of PSAs, hot melt adhesives (e.g., for cardboard sealing), and construction adhesives (e.g., tile adhesives).

12.3.3.1 Cationic Polymerization of Pinenes

Most terpenes have a methyl group or other electron-donating groups on the double bond, and for this reason, cationic polymerization has been the most suitable polymerization process of terpenes. Among terpenes, β-pinene is by far the preferred substrate for cationic polymerization because the reactivity of its double bond is not hindered [90]. The cationic polymerization of β-pinene has been extensively studied, first with Lewis acid initiators [91] and then using living systems [92]. On the contrary, the structure of α-pinene lacks the highly reactive *exo*-methylene double bond that is present in β-pinene, thus making α-pinene highly inert toward cationic polymerization, and only oligomers can be obtained [93]. The first example of the living cationic polymerization of β-pinene was reported by Sawamoto *et al.* [94]. The mechanism passes through the addition of the cationic initiator on the monomer yielding a tertiary carbocation, which then undergoes an isomerization reaction to produce the propagating isobutylene tertiary carbocation. However, only low-molecular-weight polymers were obtained using aluminum chloride as catalyst (Scheme 12.24), but high-molecular-weight poly(β-pinene)s were obtained by Keszler *et al.* [96] and Kamigaito *et al.* [97] using $H_2O/EtAlCl_2$. Temperatures below −80 °C were required in order to achieve high-molecular-weight polymers since chain transfer reaction competes with propagation at higher temperatures. These polymers exhibit a glass transition temperature close to 90 °C. The low reaction temperature, dilute reaction conditions, and high initiator concentrations limit the large-scale production of poly(β-pinene). More recently, a Schiff-based nickel complex catalyst in combination with methylaluminoxane (MAO) as activator has been used to produce poly(β-pinene) by cationic polymerization at 40 °C [98]. Random and block copolymers of β-pinene with styrene have been obtained by both living cationic polymerization and radical polymerization [95].

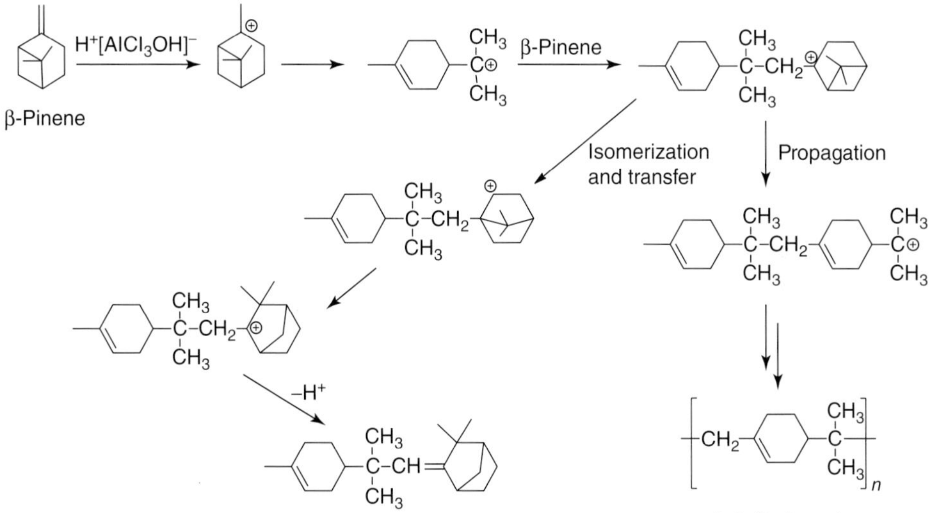

Scheme 12.24 The cationic polymerization of β-pinene initiated by the $AlCl_3/H_2O$ complex [94].

12.3.3.2 Polymyrcene

Myrcene is one of the few noncyclic terpenes and can be polymerized through a double mechanism that involves in the first step a cyclization performed through an RCM reaction that employs a second-generation Grubbs catalyst, followed in the second step by its cationic activation to give high-molecular-weight polymers [99]. The emulsion polymerization technique was also used to prepare a biobased polymer from β-myrcene [100]. The persulfate-initiated polymyrcene possesses 3,4 and 1,2 vinyl defects along with 1,4 microstructures, while the redox analog contains only 1,4 addition moieties (Scheme 12.25). The polymerization of myrcene has also been performed using a cationic β-diimidosulfonate lutetium catalyst affording isotactic 3,4-polymyrcene with a low T_g (−42 °C) [102], as well as random and block copolymers with isoprene.

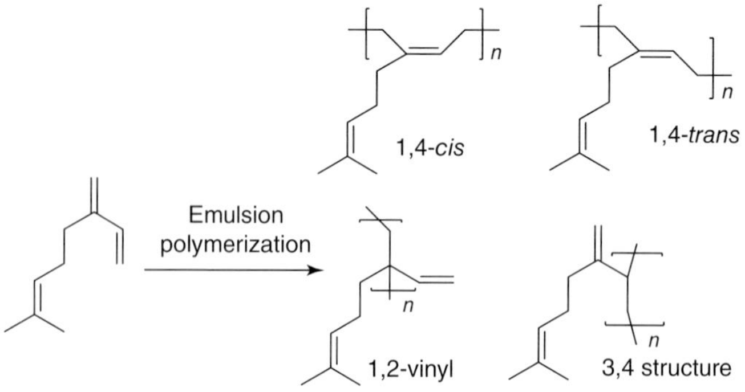

Emulsion polymerization

1,4-*cis*

1,4-*trans*

1,2-vinyl

3,4 structure

Scheme 12.25 Emulsion polymerization of myrcene [100].

12.3.4 Terpenes as Monomers after Chemical Modification

12.3.4.1 Limonene Modified by the Thiol–Ene Reaction

The thiol–ene reaction is a very useful technique to convert unsaturated carbon–carbon double bonds into a wide range of functional groups [103]. Several routes are reported in the literature regarding the insertion of thiol groups into terpenes, for example, by reaction with hydrogen sulfide (Scheme 12.26) [105]. The insertion of thiol groups can be also performed by the thiol–ene reaction on the double bonds present on terpenes (Scheme 12.27) [101]. Thiol–ene reactions can be used to transform terpenes into alcohols, esters, and amines. In particular, monomers bearing terminal primary OH and methyl ester moieties have been used for the preparation of polyesters by polytransesterification reactions [101]. For example, Meier and coworkers [108] used thiol–ene click chemistry to produce limonene-based homopolymers (Scheme 12.28). TBD was used as catalyst due to its high transesterification activity. Polymers with molecular weight of 10 000, a PDI as low as 1.65, and a T_g around −10 °C were obtained with this catalyst. These monomers have also been copolymerized with

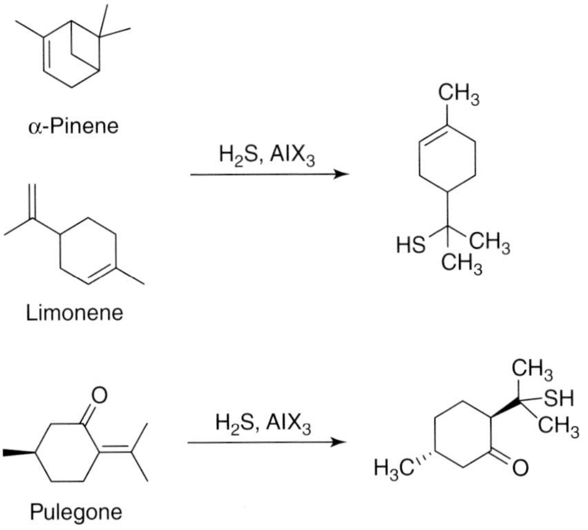
α-Pinene
Limonene
Pulegone
H2S, AlX3
H2S, AlX3
CH3
HS
CH3
CH3
CH3
SH
CH3
H3C
O
O

Scheme 12.26 Preparation of terpene-based thiols by the reaction of hydrogen sulfide with monoterpenes [105].

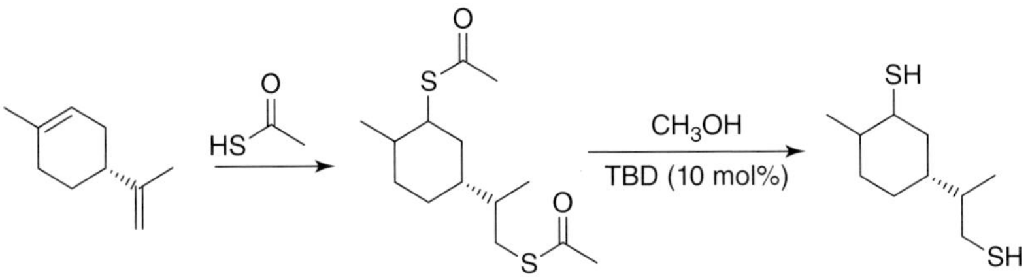
HS
O
S
O
O
S
CH3OH
TBD (10 mol%)
SH
SH

Scheme 12.27 Thiol–ene click chemistry between limonene and thiols [101].

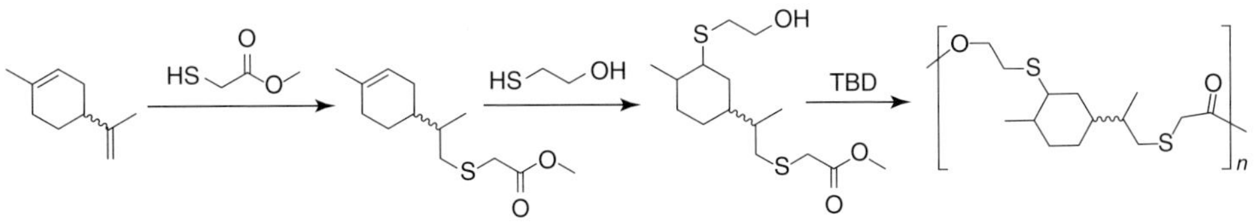
HS
O
O
HS
OH
S
OH
TBD
S
O
O
S
O
O
O
S
S
O
n

Scheme 12.28 Polyester synthesis through the thiol–ene reaction [108].

diols and diesters derived from vegetable oils, such as castor oil [101]. Polyamides and polyurethanes have also been prepared from limonene-based diamines and diamides synthesized by a click reaction involving thiol amine hydrochlorides [109]. This reaction can be also exploited for the preparation of thermoset polymers. For example, Johansson and coworkers [104] showed that thermoset polymers synthesized via the thiol–ene reaction applied to D-limonene are potentially useful in a wide variety of applications including sealants, adhesives, and organic coatings.

12.3.4.2 Dimethylstyrene from Limonene

Owing to their chemical structure, cyclic terpenes can be used as starting materials for the preparation of aromatic monomers, for example, dimethylstyrene (DMS) can be obtained by the selective dehydrogenation of limonene at 120 °C with $Pd(OAc)_2$ as catalyst, 2,6-lutidine as a noncoordinating base, and $CuCl_2$ as oxidant [106].

12.3.4.3 Terephthalic Acid Synthesis from Terpenes

Limonene has been used for the production of *p*-cymene that is an intermediate for the production of terephthalic acid, the monomer used to produce poly(alkylene terephthalate)s such as PET [107]. *p*-Cymene has been oxidized to terephthalic acid (Scheme 12.29) whose polycondensation with glycols (e.g., butanediol), leading to the first synthesis of a terephthalate polyesters completely derived from renewable resources [107]. The method reported has not been yet optimized for large-scale production, and moreover, the amount of limonene available on the market is not enough to cover the terephthalic acid market. This suggests that other terpenes should also be tested, for example, the more widely available pinene that could be isomerized to limonene or directly converted to *p*-cymene. Several catalytic procedures have been reported in the literature for the preparation of *p*-cymene from pinene. For example, Roberge *et al.* [111] reported that the dehydrogenation of α-pinene to *p*-cymene over carriers impregnated with palladium involves a reaction mechanism in which the catalyst has a dual functionality with the acid sites in charge of isomerization and the metallic sites responsible for the hydrogenation/dehydrogenation. This study also showed that crude sulfate turpentine could be used as a raw material for this purpose, giving high yields of *p*-cymene. However, the authors did not report the complete process to prepare polyesters directly from turpentine.

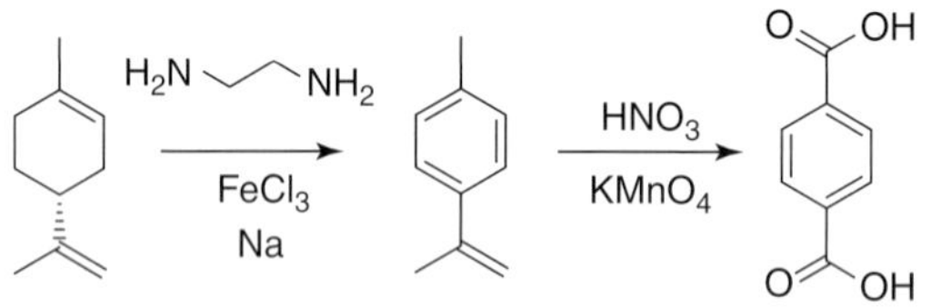

H_2N–CH₂CH₂–NH_2, $FeCl_3$, Na → ; HNO_3, $KMnO_4$ →

Scheme 12.29 Terephthalic acid synthesis from limonene via *p*-cymene [107].

12.3.4.4 Epoxidation of Limonene for the Synthesis of Polycarbonates and Polyurethanes

Limonene oxide is another important monomer derived from terpenes. Several routes are reported in the literature for its epoxidation reaction, like the use of hydrogen peroxide along with a tungsten-based catalyst [112]. Manganese-based catalytic systems have also been extensively investigated for the epoxidation of limonene [113], mostly manganese porphyrins with either hydrogen peroxide or sodium periodate as oxidant. Limonene dioxide is commercially available and used industrially as a diluent for epoxy resin and as a solvent. It can be copolymerized with CO_2 using β-diiminate zinc complexes to yield polycarbonates [110]. Mulhaupt *et al.* [114] employed limonene dioxide/CO_2 reactions to produce isocyanate-free polyurethanes and thermosetting polymers (Scheme 12.30).

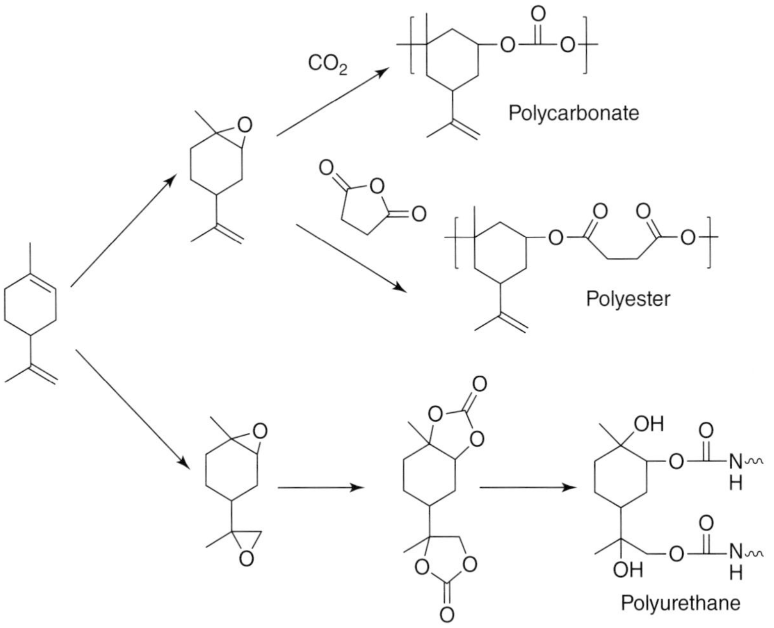

Scheme 12.30 Limonene oxides for the synthesis of polyurethanes, polycarbonates, and polyesters [114].

12.3.4.5 Copolymers Containing Terpenes

The free radical and cationic copolymerization of different terpenes or with comonomers derived from fossil resources, such as styrene, has been used to produce random and block copolymers. For example, hyperbranched structures can be obtained from dicyclopentadiene and terpenes using ring-opening metathesis polymerization (ROMP) (Scheme 12.31) [117].

12.3.5 Sesquiterpenes

Sesquiterpenes such as caryophyllene and humulene can be polymerized by ROMP with appropriate ruthenium alkylidenes (Scheme 12.32) [82]. Their 9- and 11-membered rings are opened with selective reaction of only one of the different types of double bonds present and the reaction yields non-cross-linked polymers with a well-defined microstructure. These polymers are soft materials characterized by a glass transition in the range of −15 to −50 °C [82] that renders

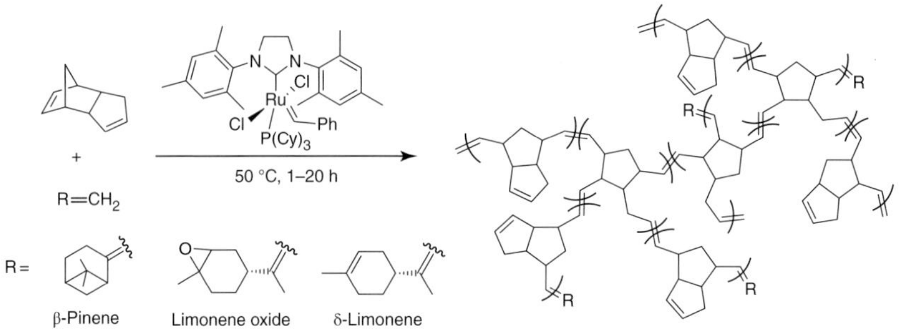

Scheme 12.31 Hyperbranched polymers obtained from dicyclopentadiene and terpenes by ROMP [117].

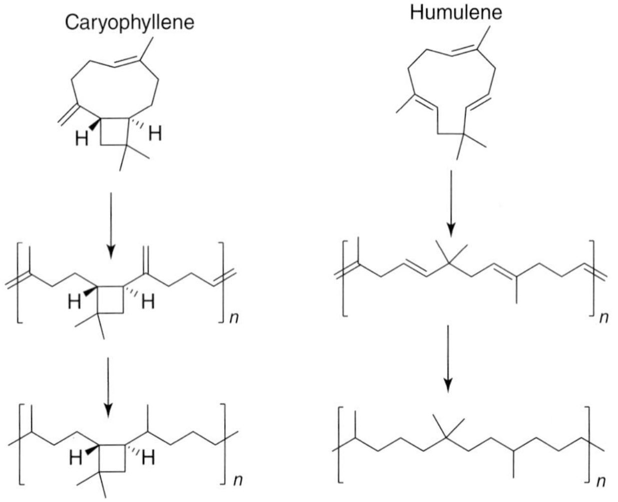

Scheme 12.32 ROMP of sesquiterpenes [82].

them very attractive for the preparation of environmentally friendly hydrophobic films and coatings based on renewable resources.

12.3.6 Terpenoids

Terpenoids are molecules with structures similar to those of terpenes but with functional groups such as OH or carbonyl moieties. The two most important terpenoids are carvone and menthol (Scheme 12.33). Dihydrocarvone has been oxidized to an epoxylactone and copolymerized with caprolactone to give shape-memory copolyesters [119]. Menthone was converted into menthide, the corresponding lactone, which was then polymerized using a zinc alkoxide catalyst at room temperature in toluene, giving high-molecular-weight materials

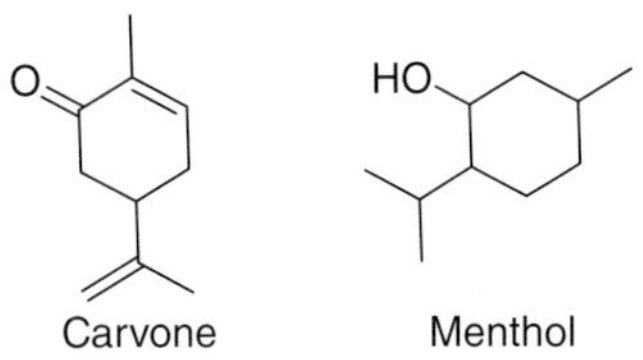

Carvone Menthol **Scheme 12.33** The two most common terpenoids.

(Scheme 12.34) [115]. Menthide was also used as comonomer in the synthesis of triblock copolymers, to be employed as pressure-sensitive adhesives [116]. A methacrylic monomer derived from myrtenol was polymerized using free radical initiation and the material tested for antibacterial properties, thanks to the antimicrobial activity of myrtenol [118]. δ-Decalactone, which is available in large quantities for flavor and fragrance applications, has been polymerized with an organocatalyst to give an aliphatic polyester and copolymerized to prepare poly(lactide)–poly(δ-lactone)–poly(lactide) triblock copolymers [121].

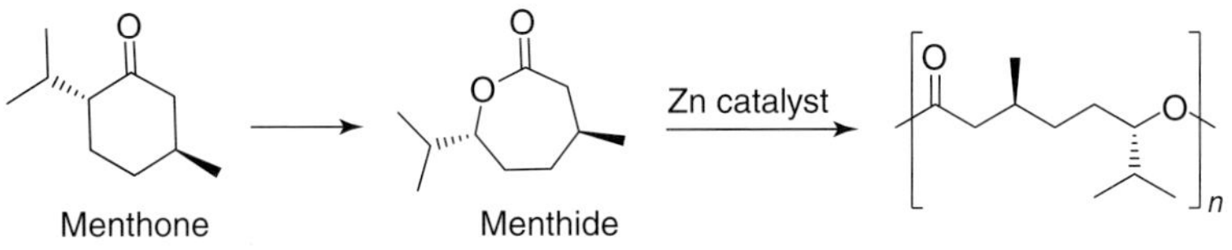

Menthone Menthide Zn catalyst

Scheme 12.34 Ring-opening polymerization of menthone [115].

12.3.7 Rosin

Several chemical modifications have been applied to rosin in order to obtain monomers for the preparation of both linear and cross-linked materials [83]. The modification reactions can be performed on the conjugated C–C double bonds or on carboxylic groups. Reactions involving the olefin moiety include oxidation, hydrogenation, dehydrogenation, isomerization, Diels–Alder (DA) couplings, and reactions with formaldehyde and phenol [83]. Oxidation reactions are generally avoided since they produce colored materials. Hydrogenation and dehydrogenation reactions give rise to structures more stable against oxidation. The reactions of the carboxylic functional groups include esterification, alkoxylation, and the synthesis of salts with several types of cations. The conjugated double-bond system can easily undergo acidic or thermal isomerization. Many of these modifications are aimed at the synthesis of monomers for both polycondensation and polyaddition reactions. Both thermosetting and thermoplastic polymers have been obtained from rosin. The main applications of rosin-based materials are in the fields of paper sizing agents, soaps and surfactants for emulsion polymerization (for rosin salts), as coating, and varnishes and printing inks [83].

12.3.7.1 Thermoset Polymers from Rosin

Rosin has been derivatized with anhydrides, carboxylic, or epoxy groups in order to prepare curing agents. In this way, it is possible to replace the petroleum-derived curing agents that are widely used in industry. As an example, abietic acid, which is the major component of rosin, can undergo isomerization at high temperatures to produce levopimaric acid which can then be reacted with maleic anhydride through the DA reaction to yield acrylopimaric acid or with acrylic acid to yield maleopimaric acid (MPA) (Scheme 12.35) [120]. The isomerization is necessary since only levopimaric acid is able to participate in the DA reactions. MPA-based polymers have higher T_g due to the bulky structure of rosin compared with those cured with 1,2,4-benzenetricarboxylic anhydride that is a typical curing agent derived from fossil-based resources [120]. Zhang and coworkers [123] have also placed short polymer segments (e.g., polycaprolactone or polyethylene glycol) between two rosin molecules in order to decrease the T_g and increase flexibility of the ensuing thermosetting polymers. DA adducts can be also useful monomers for the synthesis of poly(amide-imide) materials [124] and for the derivatization with acrylic functionalities [125]. Abietic acid reacts with formaldehyde in a similar way as phenol and can be used for the preparation of rosin–phenol–formaldehyde resins [122].

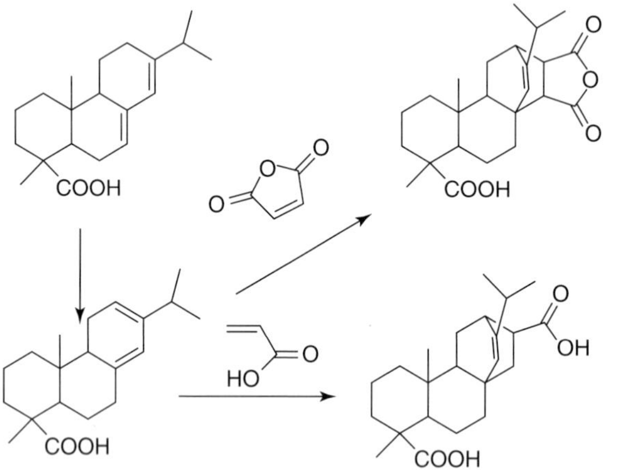

Scheme 12.35 Derivatization of abietic acid for the synthesis of acrylopimaric acid and maleopimaric acid [120].

12.3.7.2 Thermoplastic Polymers from Rosin

Polymers with rosin groups along the main polymeric chain can be prepared by various condensation methods, but only low-molecular-weight polymers can be obtained due to steric hindrance and monomer impurities [83]. On the contrary, the synthesis of polymers with rosin in the side chain permits to avoid the problems associated with their main-chain counterparts. Radical polymerization is used to produce polymers with rosin in the side chain. For this purpose, vinyl, acrylic, or allyl ester groups have been appended to rosin, although steric hindrance tends to limit their reactivity and hence the molecular weight of the final polymers. Acrylic monomers give the best results in terms of molecular weights in the preparation of side-chain rosin-based polymers [83]. Both ATRP and RAFT polymerizations can be used for the preparation of high-molecular-weight polymers derived from rosin. For ATRP, dehydroabietic acid is the preferred starting material due to its higher stability [126].

12.4 Final Considerations

Within the general context of the growing interest in polymers from renewable resources, vegetable oils and their fatty acids are of great importance because they offer a wide range of possibilities for polymer syntheses, and the ensuing plant oil-based materials are progressively gaining increasing relevance and evolving each day. Moreover, the large and ubiquitous availability of carbohydrates, terpenes, and rosins, generally as wastes of the food and wood industries, is an attractive key point for the development of green polymers with enhanced properties. Indeed, the cyclic structure of such feedstocks makes them suitable for the preparation of materials with rigid polymer backbones and high thermomechanical properties. Additionally, the presence of double bonds and of other functional groups provides the opportunity to produce a wide range of polymers with unique properties and tailored performances.

Acknowledgment

Talita M. Lacerda acknowledges São Paulo Research Foundation (FAPESP) for her postdoctoral fellowship (2012/00124-9).

References

1. US Energy Information Administration http://www.eia.gov (accessed 24 September 2015).
2. USDA http://www.fas.usda.gov .psdonline/circulars/oilseeds.pdf (accessed 24 September 2015).
3. Maisonneuve, L., Lebarbé, T., Grau, E., and Cramail, H. (2013) *Polym. Chem.*, **4**, 5472–5517.
4. Belgacem, M.N. and Gandini, A. (eds) (2008) *Monomers, Polymers and Composites from Renewable Resources*, Elsevier, Amsterdam.
5. Barnwal, B.K. and Sharma, M.P. (2005) *Renewable Sustainable Energy Rev.*, **9** (4), 363–378.
6. Cahoon, E.B., Shockey, J.M., Dietrich, C.R., Gidda, S.K., Mullen, R.T., and Dyer, J.M. (2007) *Curr. Opin. Plant Biol.*, **10** (3), 236–244.
7. Dyer, J.M. and Mullen, R.T. (2008) *Physiol. Plant.*, **132** (1), 11–22.
8. Carlsson, A.S. (2009) *Biochimie*, **91** (6), 665–670.
9. Gandini, A. and Lacerda, T.M. (2015) Metathesis reactions applied to plant oils and polymers derived from the

ensuing products, in *Polymers from Plant Oils* (A. Gandini and T.M. Lacerda), Smithers Rapra, Shawbury, chap. 5, pp. 83–108
10. Islam, M.R., Beg, M.D.H., and Jamari, S.S. (2014) *J. Appl. Polym. Sci.*, **131** (18), 40787. doi: 10.1002/app.40787
11. Lligadas, G., Ronda, J.C., Galià, M., and Cádiz, V. (2013) *Mater. Today*, **16** (9), 337–343.
12. Frederick, T., Wallenberger, T., and Norman, E. (2004) *Natural Fibers, Plastics and Composites*, Springer, New York.
13. Derksen, J.T.P., Cuperus, F.P., and Kolster, P. (1996) *Prog. Org. Coat.*, **27** (1–4), 45–53.
14. Jose, J., Pourfallah, G., Merkley, D., Li, S., Bouzidi, L., Leão, A.L., and Narine, S.S. (2014) *Polym. Chem.*, **5**, 3203–3213.
15. Jose, J., Li, S., Bouzidi, L., Leão, A.L., and Narine, S.S. (2014) *J. Appl. Polym. Sci.*, **131** (13), 40492. doi: 10.1002/app.40492
16. Rahim, N.F.A., Watanabe, K., Ariffin, H., Andou, Y., Hassan, M.A., and Shirai, Y. (2014) *Chem. Lett.*, **43** (9), 1517–1519.
17. Lai, C.M., Rozman, H.D., and Tay, G.S. (2013) *Polym. Eng. Sci.*, **53** (6), 1138–1145.
18. Alam, M., Shaik, M.R., and Alandis, N.M. (2013) *J. Chem.* doi: 10.1155/2013/962316, Article ID 962316, 11 pages.
19. Beyazkilic, Z., Lligadas, G., Ronda, J.C., Galià, M., and Cádiz, V. (2014) *Macromol. Chem. Phys.*, **215** (22), 2248–2259.
20. (a) Wu, Y. and Li, K. (2015) *J. Am. Oil Chem. Soc.*, **92** (1), 111–120; (b) Wu, Y., Li, A., and Li, K. (2014) *J. Appl. Polym. Sci.*, **131** (23). http://onlinelibrary.wiley.com/doi/10.1002/app.41143/epdf
21. Stemple, F., Ritter, B.S., Mülhaupt, R., and Mecking, S. (2014) *Green Chem.*, **16**, 2008–2014.
22. Butilkov, D. and Lemcoff, N.G. (2014) *Green Chem.*, **16**, 4728–4733.
23. Basf https://www.basf.com/documents/corp/en/sustainability/management-and-instruments/quantifying-sustainability/eco-efficiency-analysis/examples/lupranol-balance-50/Lupranol_BALANCE_EEA.pdf (accessed 24 September 2015).
24. Piszczyk, L., Strankowski, M., Danowska, M., Hejna, A., and Haponiuk, J.T. (2014) *Eur. Polym. J.*, **57**, 143–150.
25. Hu, S., Luo, X., and Li, Y. (2015) *J. Appl. Polym. Sci.* http://onlinelibrary.wiley.com/doi/10.1002/app.41425/abstract.
26. Miao, S., Yong, P., Su, Z., and Zhang, S. (2014) *Acta Biomater.*, **10** (4), 1692–1704.
27. More, A.S., Lebarbé, T., Maisonneuve, L., Gadenne, B., Alfos, C., and Cramail, H. (2013) *Eur. Polym. J.*, **49** (4), 823–833.
28. Charlon, M., Heinrich, B., Matter, Y., Couzigné, E., Donnio, B., and Avérous, L. (2014) *Eur. Polym. J.*, **61**, 197–205.
29. Ling, J.S., Mohammed, I.A., Ghazali, A., and Khairuddean, M. (2014) *Ind. Crops Prod.*, **52**, 74–84.
30. Aung, M.M., Yaakob, Z., and Kamarudin, S. (2014) *Ind. Crops Prod.*, **60**, 177–185.
31. Meier-Westhues, U. (2007) *Polyurethanes Coatings, Adhesives and Sealants*, Vincentz Network, Hannover.
32. Garrison, T.F., Zhang, Z., Kim, H.-J., Mitra, D., Xia, Y., Pfister, D.P., Brehm-Stecher, B.F., Larock, R.C., and Kessler, M.R. (2014) *Macromol. Mater. Eng.*, **299** (9), 1042–1051.
33. Garrison, T.F., Kessler, M.R., and Larock, R.C. (2014) *Polymer*, **55** (4), 1004–1011.
34. (a) Lu, Y. and Larock, R.C. (2010) *ChemSusChem*, **3** (3), 329–333; (b) Xia, Y., Zhang, Z., Kessler, M.R., Brehm-Stecher, B., and Larock, R.C. (2012) *ChemSusChem*, **5** (11), 2221–2227.
35. Heinen, M., Gerbase, A.E., and Petzhold, C.L. (2014) *Polym. Degrad. Stab.*, **108**, 76–86.
36. Zhang, J., Zhang, C., and Madbouly, S.A. (2015) *J. Appl. Polym. Sci.* http://onlinelibrary.wiley.com/doi/10.1002/app.41751/abstract.
37. Mosiewicki, M.A., Rojek, P., Michalowski, S., Aranguren, M.I.,

and Prociak, A. (2015) *J. Appl. Polym. Sci.* http://onlinelibrary.wiley.com/doi/10.1002/app.41602/abstract

38. Laurichesse, S., Huillet, C., and Avérous, L. (2014) *Green Chem.*, **16**, 3958–3970.
39. Genas, M. (1962) *Angew. Chem.*, **74** (15), 535–540.
40. Kolb, N., Winkler, M., Syldatk, C., and Meier, M.A.R. (2014) *Eur. Polym. J.*, **51**, 159–166.
41. Winkler, M., Steinbiß, M., and Meier, M.A.R. (2014) *Eur. J. Lipid Sci. Technol.*, **116** (1), 44–51.
42. Winkler, M. and Meier, M.A.R. (2014) *Green Chem.*, **16**, 1784–1788.
43. Winkler, M. and Meier, M.A.R. (2014) *Green Chem.*, **16**, 3335–3340.
44. Mudiyanselage, A.Y., Viamajala, S., Varanasi, S., and Yamamoto, K. (2014) *ACS Sustainable Chem. Eng.*, **2** (12), 2831–2836.
45. Wool, R. and Sun, S. (2005) *Biobased Polymers and Composites*, Academic Press, Amsterdam.
46. Werpy, T. and Petersen, G. (2004) *Top Value Added Chemicals from Biomass*, US Department of Energy.
47. Fenouillot, F., Rousseau, A., Colomines, G., Saint-Loup, R., and Pascault, J.-P. (2010) *Prog. Polym. Sci.*, **35** (5), 578–622.
48. Rose, M. and Palkovits, R. (2012) *ChemSusChem*, **5** (1), 167–176.
49. Vilela, C., Sousa, A.F., Fonseca, A.C., Serra, A.C., Coelho, F.J., Freire, C.S.R., and Silvestre, A.J.D. (2014) *Polym. Chem.*, **5**, 3119–3141.
50. Kricheldorf, H.R. and Weidner, S.M. (2013) *Macromol. Chem. Phys.*, **214** (6), 726–733.
51. Kricheldorf, H.R. and Weidner, S.M. (2013) *Eur. Polym. J.*, **49** (8), 2293–2302.
52. Pion, F., Ducrot, P.-H., and Allais, F. (2014) *Macromol. Chem. Phys.*, **215** (5), 431–439.
53. Jacquel, N., Saint-Loup, R., Pascault, J.-P., Rousseau, A., and Fenouillot, F. (2015) *Polymer*, **59**, 234–242.
54. Gioia, C., Vannini, M., Marchese, P., Minesso, A., Cavalieri, R., Colonna, M., and Celli, A. (2014) *Green Chem.*, **16** (4), 1807–1815.
55. Gioia, C., Minesso, A., Cavalieri, R., Marchese, P., Colonna, M., and Celli, A. (2015) *J. Coat. Technol. Res.* **12**, 555–562.
56. Park, H.S., Gong, M.S., and Knowles, J.C. (2013) *J. Mater. Sci. - Mater. Med.*, **24** (2), 281–294.
57. Li, Y., Noordover, B.A.J., Van Benthem, R.A.T.M., and Koning, C.E. (2014) *Eur. Polym. J.*, **59**, 8–18.
58. Belgacem, C., Medimagh, R., Fildier, A., Bulete, A., Kricheldorf, H.R., Romdhane, H.B., and Chatti, S. (2014) *Des. Monomers Polym.*, **18** (1), 64–72.
59. Nelson, A.M. and Long, T.E. (2012) *Polym. Int.*, **61** (10), 1485–1491.
60. Thiyagarajan, S., Wu, J., Knoop, R.J.I., van Haveren, J., Lutz, M., and van Es, D.S. (2014) *RSC Adv.*, **4** (89), 47937–47950.
61. Wu, J., Eduard, P., Thiyagarajan, S., van Haveren, J., van Es, D.S., Koning, C.E., Lutz, M., and Guerra, C.F. (2011) *ChemSusChem*, **4** (5), 599–603.
62. Wu, J., Eduard, P., Thiyagarajan, S., Jasinska-Walc, L., Rozanski, A., Guerra, C.F., Noordover, B.A.J., van Haveren, J., van Es, D., and Koning, C.E. (2012) *Macromolecules*, **45** (12), 5069–5080.
63. Wu, J., Eduard, P., Jasinska-Walc, L., Rozanski, A., Noordover, B.A.J., van Es, D.S., and Koning, C.E. (2013) *Macromolecules*, **46** (2), 384–394.
64. Jasinska, L., Villani, M., Wu, J., van Es, D., Klop, E., Rastogi, S., and Koning, C.E. (2011) *Macromolecules*, **44** (9), 3458–3466.
65. Lavilla, C., Alla, A., Martinez de Ilarduya, A., and Muñoz-Guerra, S. (2013) *Biomacromolecules*, **14**, 781–793.
66. Lavilla, C., Alla, A., Martínez de Ilarduya, A., Benito, E., García-Martín, M.G., Galbis, J.A., and Muñoz-Guerra, S. (2012) *Polymer*, **53** (16), 3432–3445.
67. Gubbels, E., Lavilla, C., Martinez de Ilarduya, A., Noordover, B.A.J., Koning, C.E., and Muñoz-Guerra, S. (2014) *J. Polym. Sci., Part A: Polym. Chem.*, **52** (2), 164–177.
68. Lavilla, C., Martinez de Ilarduya, A., Alla, A., García-Martín, M.G., Galbis,

J.A., and Muñoz-Guerra, S. (2012) *Macromolecules*, **45** (20), 8257–8266.
69. Lavilla, C., Gubbels, E., Mart, A., Noordover, B.A.J., and Koning, C.E. (2013) *Macromolecules*, **46**, 4335–4345.
70. Muñoz-Guerra, S., Lavilla, C., Japu, C., and Martínez de Ilarduya, A. (2014) *Green Chem.*, **16** (4), 1716–1739.
71. Storbeck, R., Rehahn, M., and Ballauff, M. (1993) *Makromol. Chem.*, **194**, 53–64.
72. Gandini, A. (2011) *Green Chem.*, **13**, 1061–1083.
73. Corma, A., Iborra, S., and Velty, A. (2007) *Chem. Rev.*, **107**, 2411–2502.
74. Gandini, A. and Silvestre, A.J.D. (2008) in *Monomers, Polymers and Composites from Renewable Resources* (eds M.N. Belgacem and A. Gandini), Elsevier Ltd., Amsterdam, pp. 17–38.
75. Wilbon, P.A., Chu, F., and Tang, C. (2013) *Macromol. Rapid Commun.*, **34**, 8–37.
76. Gandini, A. (2011) in *Biocatalysis in Polymer Chemistry* (ed P.K. Loos), Wiley-VCH Verlag GmbH & Co., Weinheim, pp. 1–29.
77. Ciriminna, R., Lomeli-Rodriguez, M., Demma Cara, P., Lopez-Sanchez, J.A., and Pagliaro, M. (2014) *Chem. Commun.*, **50**, 15288–15296.
78. US Department of Agriculture (2015) Citrus: World Markets and Trade, January 2015, http://www.fas.usda.gov/data/citrus-world-markets-and-trade (accessed 24 September 2015).
79. Pourbafrani, M., Forgacs, G., Horvath, I.S., Niklasson, C., and Taherzadeh, M. (2010) *Bioresour. Technol.*, **101**, 4246–4250.
80. Teixeira, M.A., Rodriguez, O., Gomes, P., Mata, V., and Rodrigues, A. (eds) (2013) *Perfume Engineering: Design, Performance & Classification*, Butterworth-Heinemann, Oxford.
81. Cortez, M.J., Rowe, H.G. (2012) Alternative response technologies: progressing learnings. Proceedings of Interspill 2012, Houston, TX.
82. Grau, E. and Mecking, S. (2013) *Green Chem.*, **15**, 1112–1115.
83. Zhang, J. (2012) *Rosin-Based Chemicals and Polymers*, Smithers Rapra Technology Ltd., Shawbury.
84. Monteiro, J.L.F. and Veloso, C.O. (2004) *Top. Catal.*, **27**, 169–179.
85. Swift, K.A.D. (2004) *Top. Catal.*, **27**, 143–155.
86. Ohnishi, R. and Tanabe, K. (1974) *Chem. Lett.*, **3**, 207–210.
87. Sheldon, R.A. and van Bekkum, H. (eds) (2001) *Fine Chemicals through Heterogeneous Catalysis*, Wiley-VCH Verlag GmbH, Berlin.
88. Allahuerdiev, A.I., Irandoust, S., and Murzin, D.Y. (1999) *J. Catal.*, **185**, 352–362.
89. Severino, A., Vital, J., and Lobo, L.S. (1993) *Stud. Surf. Sci. Catal.*, **78**, 685–692.
90. Lu, J., Kamigaito, M., Higashimura, T., Deng, Y., and Sawamoto, M. (1997) *Macromolecules*, **30**, 22–26.
91. Roberts, W.J. and Day, A.R. (1950) *J. Am. Chem. Soc.*, **72**, 1226–1230.
92. Kennedy, J.P. and Chou, T. (1976) *Adv. Polym. Sci.*, **21**, 1–39.
93. Lu, J., Kamigaito, M., Sawamoto, M., Higashimura, T., and Deng, Y. (1996) *J. Appl. Polym. Sci.*, **61**, 1011–1016.
94. Lu, J., Kamigaito, M., Higashimura, T., Deng, Y., and Sawamoto, M. (1997) *Macromolecules*, **30**, 27–31.
95. Lu, J., Liang, H., Zhang, R., and Li, B. (2001) *Polymer*, **42**, 4549–4553.
96. Keszler, B. and Kennedy, J. (1992) in *Macromolecules: Synthesis, Order, and Advanced Properties* (ed K.A. Armistead), Springer, Heidelberg, pp. 1–9.
97. Satoh, K., Sugiyama, H., and Kamigaito, M. (2006) *Green Chem.*, **8**, 878–882.
98. Yu, P., Li, A., Liang, H., and Lu, J. (2007) *J. Polym. Sci., Part A: Polym. Chem.*, **45**, 3739–3746.
99. Kobayashi, S., Lu, C., Hoye, T., and Hillmyer, M. (2009) *J. Am. Chem. Soc.*, **131**, 7960–7961.
100. Sarkar, P. and Bhowmick, A.K. (2014) *RSC Adv.*, **4**, 61343–61354.
101. Firdaus, M., Meier, M.A.R., Biermann, U., and Metzger, J.O. (2014) *Eur. J. Lipid Sci. Technol.*, **116**, 31–36.

102. Liu, B., Li, L., Sun, G., Liu, D., Li, S., and Cui, D. (2015) *Chem. Commun.*, **51**, 1039–1041.
103. Kolb, H.C., Finn, M.G., and Sharpless, K.B. (2001) *Angew. Chem. Int. Ed.*, **40**, 2004–2021.
104. Claudino, M., Mathevet, J.-M., Jonsson, M., and Johansson, M. (2013) *Polym. Chem.*, **5**, 3245–3260.
105. Janes, J.F., Marr, I.M., Unwin, N., Banthorpe, D.V., and Yusuf, A. (1993) *Flavour Fragance J.*, **8**, 289–294.
106. Sanchez-Vazquez, S.A., Sheppard, T.D., Evans, J.R.G., and Hailes, H.C. (2014) *RSC Adv.*, **4**, 61652–61655.
107. Colonna, M., Berti, C., Fiorini, M., Binassi, E., Mazzacurati, M., Vannini, M., and Karanam, S. (2011) *Green Chem.*, **13**, 2543–2548.
108. Firdaus, M., Montero de Espinosa, L., and Meier, M.A.R. (2011) *Macromolecules*, **44**, 7253–7262.
109. Firdaus, M. and Meier, M.A.R. (2013) *Green Chem.*, **15**, 370–380.
110. Byrne, C., Allen, S., Lobkovsky, E., and Coates, G. (2004) *J. Am. Chem. Soc.*, **126**, 11404–11405.
111. Roberge, D.M., Buhl, D., Niederer, J.P.M., and Holderich, W.F. (2001) *Appl. Catal., A-Gen.*, **215**, 111–124.
112. Prandi, J., Kagan, H.B., and Mimoun, H. (1986) *Tetrahedron Lett.*, **27**, 2617–2620.
113. Battioni, P., Renaud, J.P., Bartoli, J.F., Reina-Artiles, M., Fort, M., and Mansuy, D. (1988) *J. Am. Chem. Soc.*, **110**, 8462–8470.
114. Bahr, M., Bitto, A., and Mulhaupt, R. (2012) *Green Chem.*, **14**, 1447–1454.
115. Zhang, D., Hillmyer, M.A., and Tolman, W.B. (2005) *Biomacromolecules*, **6**, 2091–2095.
116. Shin, J., Martello, M.T., Shrestha, M., Wissinger, J.E., Tolman, W.B., and Hillmyer, M.A. (2011) *Macromolecules*, **44**, 87–94.
117. Mathers, R.T., Damodaran, K., Rendos, M.G., and Lavrich, M.S. (2009) *Macromolecules*, **42**, 1512–1518.
118. Lepoittevin, B., Wang, X., Baltaze, J.-P., Liu, H., Herry, J.-M., and Bellon-Fontaine, M.-N. (2011) *Eur. Polym. J.*, **47**, 1842–1851.
119. Lowe, J.R., Tolman, W.B., and Hillmyer, M.A. (2009) *Biomacromolecules*, **10**, 2003–2008.
120. Liu, X., Xin, W., and Zhang, J. (2009) *Green Chem.*, **11**, 1018–1025.
121. Martello, M.T., Burns, A., and Hillmyer, M. (2012) *ACS Macro Lett.*, **1**, 131–135.
122. Rudolphy, A. (1975) Modified natural resin binder and process for preparation. US Patent 3880788.
123. Wang, H., Liu, B., Liu, X., Zhang, J., and Xian, M. (2009) *Polym. Int.*, **58**, 1435–1441.
124. Kim, S.J., Kim, B.J., Jang, D.W., Kim, S.H., Park, S.Y., Lee, J.H., Lee, S.H., and Choi, D.H. (2001) *J. Appl. Polym. Sci.*, **79**, 687–695.
125. Bingham, J.F. and Marvel, C.S. (1972) *J. Polym. Sci., Part A: Polym. Chem.*, **10**, 921–929.
126. Zheng, Y., Yao, K., Lee, J., Chandler, D., Wang, C., Chu, F., and Tang, C. (2010) *Macromolecules*, **43**, 5922–5924.

13
Polymers from Monomers Derived from Biomass

Annamaria Celli, Martino Colonna, Alessandro Gandini, Claudio Gioia, Talita M. Lacerda, and Micaela Vannini

13.1
Polymers Derived from Furans

13.1.1
Introduction

Numerous reviews concerning furan-based monomers and polymers published in the last few years [1–7] emphasize the growing importance of these renewable resources following a comprehensive monograph in 1997 [8]. The explanation of their success does not lie in a fleeting trend but is founded instead on different and consistent reasons, including, first of all, their sustainability based on biomass sources that are not competitive with the food chain and additionally their aptitude to provide a wide range of original materials. The exciting challenge of academic and industrial researches is to optimize their preparation from polysaccharides (pentoses and hexoses) present in numerous lignocellulosic residues like sugarcane bagasse, rice lulls, and certain woods [9–16].

Uniquely, the furan ring is characterized by interesting peculiarities including its pronounced dienic character that makes it a perfect candidate for Diels–Alder (DA) reactions, that is, for click polymerizations with suitable dienophiles. Furans are moreover considered appropriate renewable substitutes for the corresponding benzene-based monomers, especially the 2,5-furandicarboxylic acid (FDCA) regarded as an effective alternative to terephthalic acid.

These and other premises justify the insertion of furans in the Bozell's "Top 10 Chemicals from Biomass," [17] particularly the furan derivatives more frequently exploited as building blocks depicted in Scheme 13.1. Their synthesis requires the exposure of lignocellulosic material derived from agricultural or forestry wastes to aqueous acid and fairly high temperatures to hydrolyze the polysaccharides and to cyclodehydrate the ensuing sugars [10, 12, 15]. While pentoses generate the monofunctional furfural (F), the hexoses produce the bifunctional hydroxymethylfurfural (HMF), as shown in Scheme 13.2. Whereas F has broad applications and has been produced on industrial scale for nearly

Chemicals and Fuels from Bio-Based Building Blocks, First Edition.
Edited by Fabrizio Cavani, Stefania Albonetti, Francesco Basile, and Alessandro Gandini.

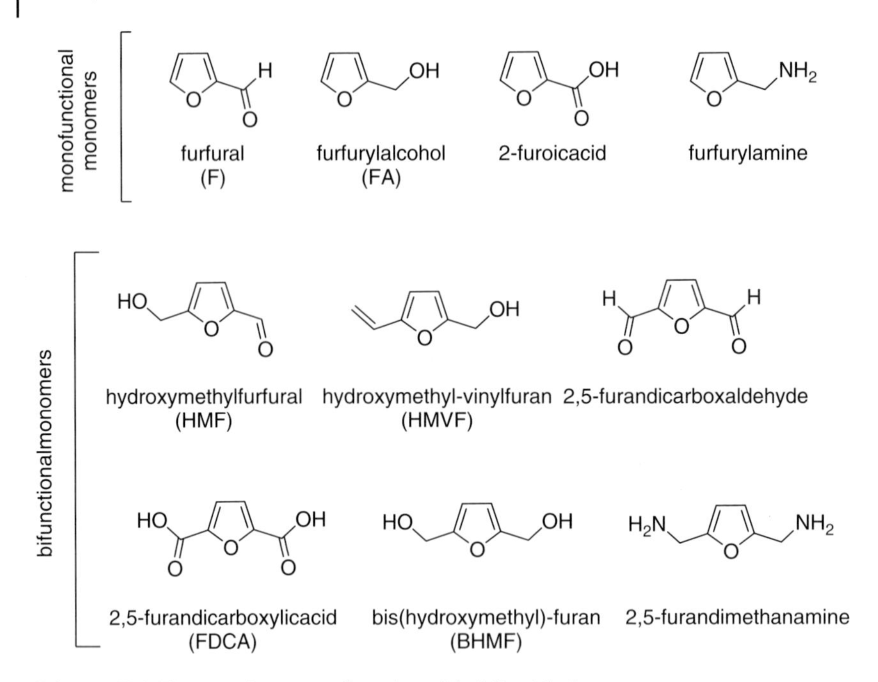

Scheme 13.1 The most important furan-based building blocks.

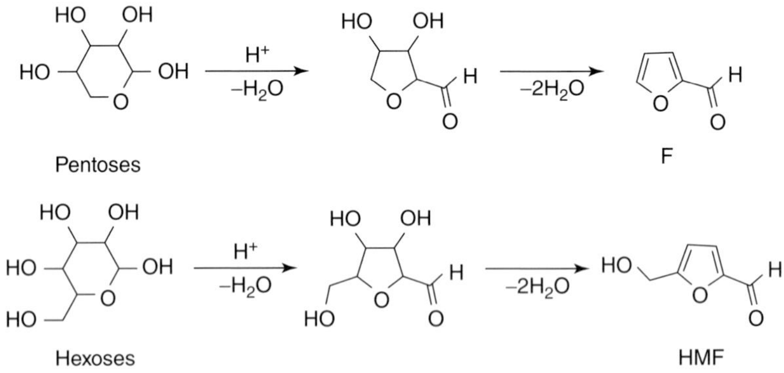

Scheme 13.2 Synthetic pathways to prepare the two basic furan derivatives from biomass.

a century, HMF, in contrast, has enormous potential as a raw material for the production of chemicals, polymers, and biofuels, but its production has only very recently reached an industrial stage [15].

The production of F and its derivatives is already inserted in strengthened processes mainly directed to furfuryl alcohol (FA) synthesis for the subsequent production of furan resins and fuels. The global annual production of F totalizes 300–400 kt [9, 11], and its selling price is lower than 1 € kg^{-1} [18]. In the Scheme 13.1 are reported the most exploited furan-based monomers, directly achievable from F or HMF: indeed F can be transformed into furfuryl alcohol by hydrogenation, into furfurylamine by reactive amination, into furoic acid by oxidation, and into furan by catalytic decarbonylation, following straightforward processes [18]. Similar modification reactions can be applied to HMF in order to prepare the analogous bifunctional structures (Scheme 13.1). Before HMF was readily available as a source of bifunctional monomers, Khrouf and coworkers [19] developed a smart and astute idea to convert monofunctional furans into difuran derivatives bearing two reactive groups by the acid-catalyzed condensation shown in Scheme 13.3. Whereas monomers derived from F, such as 2-alkenyl furans, 2-furyl oxyrane, and furfuryl acrylates and methacrylates [8, 20], are polymerized by chain-growth processes and generate macromolecular structures where the heterocycle is appended as a side moiety to the main chain, monomers from HMF (Scheme 13.1), or arising from the coupling reaction shown in Scheme 13.3, are suitable for step-growth mechanisms, and the ensuing polymers will bear the heterocycle within the macromolecular chain. It follows that the structure–properties relationships are quite different for these two types of architectures [8, 20]. Two relevant exception to this situation are the polycondensation of FA [8] and of 5-methylfurfural [21], which generates conjugated polymers, both giving the furan ring in the main chain. The choice of systems and materials discussed in the following reflects what we consider as the most relevant approaches to furan polymers in terms of originality and recent promising developments with regard to practical implementations.

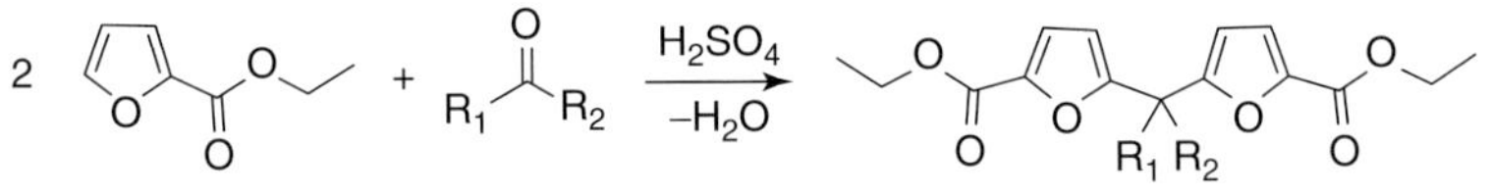

2 [ethyl 2-furoate] + $R_1C(O)R_2$ $\xrightarrow[-H_2O]{H_2SO_4}$ [difuran diester with R_1 R_2]

Scheme 13.3 Synthesis of difuran monomers.

13.1.2 Polyesters

After the first reports of furan-based polyesters, obtained by polycondensation of 2,5-furandicarbonyl chloride with various diols by applying interfacial or solution methods [22–24], Gandini and coworkers and, later, a Tunisian group begun a systematic study focused on the preparation of these materials by changing procedures (interfacial, solution, or melt polycondensations), monomers, catalysts, and temperatures [25–32]. They used both FDCA, which provided the first scientific report dealing with poly(ethylene-2,5-furandicarboxylate) (PEF) [27], that is, the heterocyclic homolog of PET, and the difuran diacid obtained according to Scheme 13.3, alone or in conjunction with terephthalic acid. The large difference in delocalization energy (22 mol kcal^{-1} vs 36 kcal mol^{-1} for furan and benzene, respectively) clearly indicates that the furan ring has a significant diene character, resulting in a different chemical reactivity [33, 34]. Moreover, terephthalic acid exhibits an interatomic distance of 5.731 Å between carboxylic acid groups, while in FDCA it is notably shorter (4.830 Å). The linear *p*-phenyl connection in terephthalic acid results in an angle of 180° between carboxylic acid carbons, while the nonlinear structure of FDCA yields an angle of 129.4°. All these differences result in different properties of the corresponding polyesters [33, 34].

In 2005, a patent application by Canon Kabushiki Kaisha [35] claimed the preparation of three FDCA-based polyesters (PEF, poly(propylene-2,5-furandicarboxylate) (PPF), and poly(butylene-2,5-furandicarboxylate) (PBF)). In 2007, a Japanese patent by Mitsubishi [36] claimed the preparation of two similar polyesters. More recently, the Dutch company Avantium developed a novel process for the synthesis of commercial furan polymers, the most studied being PEF, prepared by melt condensation of dimethyl ester of FDCA with ethylene glycol (EG) (1.3 equiv with respect the diacid) [37] in the presence of titanium isopropoxide. The polycondensation takes about 4 h for the stage at room pressure and other 2 h for the stage under vacuum.

Essentially the same conditions were reported by van Es's group [33] to prepare PEF, as well as the 1,3-propane glycol (PPF) and the 1,4-butane diol (PBF) counterparts, but the polycondensation time necessary to reach M_n values in the range 8000–12 000 exceeded 22 h, whereas a solid-state polymerization (SSP) of 3 days provided values up to eight times higher. Zhu *et al.* [38] described at the same time the synthesis of PBF directly from the FDCA and not the corresponding diester and also in this case, the time required for the melt polycondensation was very long, namely, more than 26 h, to achieve an M_w of 65 000. The simplest and industrially applicable synthetic procedure was described by Jiang and coworkers [39] who prepared PEF, PPF, PBF, and other homologs by melt polycondensation between FDCA and the diol in the presence of titanium tetrabutoxide with reaction times shorter than 10 h and M_n higher than 18 000. Then, a systematic in-depth study to optimize the polymerization conditions for PEF and PBF was carried out by Gruter and colleagues [40], who used small-scale parallel film reactors to compare the molecular weights and the yellowing of the ensuing polyesters by changing the type and amount of catalyst, reaction times, and temperatures.

Beyond these linear polyesters, other and more complex furan-based materials are prepared by using diols like 2,3-butane diol [41, 42] and various polyols, including glycerol [41, 43], with the purpose to extend the usual applications of polyesters as engineering plastics (in particular, bottle applications) and to move toward the coating field. They were thermally stable up to 270 °C (specifically the stability depended on the molecular weight) and were cross-linked by reacting the hydroxyl end groups with a polyfunctional isocyanate. The coatings showed good solvent resistance, proving the formation of a properly cured network.

Copolymers of FDCA were also investigated for different purposes. Poly(ethylene 2,5-furandicarboxylate-*co*-ethylene succinate) (PEFS) [44] was prepared by Yu *et al.*, who demonstrated that this material was a suitable alternative to poly(ethylene terephthalate-*co*-ethylene succinate (PETS), a well-known PET copolymer. Poly(ethylene terephthalate-*co*-ethylene 2,5-furandicarboxylate) PET/PEF [45] was synthesized by Sousa *et al.* in order to improve the biodegradability of PET, though maintaining its main properties. Wu and coworkers [46] developed copolymers based on PBF (specifically, poly(butylene 2,5-furandicarboxylate-*co*-butylene succinate) and poly(butylene 2,5-furandicarboxylate-*co*-butylene adipate) [47]) with tunable properties, ranging from crystalline polymers, possessing good tensile modulus (360–1800 MPa) and strength (20–35 MPa), to essentially amorphous polymers with low T_g and very high elongation at break (about 600%), useful as thermoplastics, elastomers, or impact modifiers. Poly(ethylene 2,5-furandicarboxylate-*co*-lactic acid) PEF/PLA [48] was synthesized as a nice example of bio-based and biodegradable materials. Finally, copolymers completely based on FDCA were prepared by Ma *et al.* [49] who discovered that the lower reactivity of ethylene glycol with FDCA versus 1,4-butanediol (BDO) directly influences the corresponding polymerization rates and therefore the relative diol incorporation and the final properties of the copolyesters.

Surprisingly, few papers report the synthesis of furan-based polyesters where the furan ring is not only coming from the dicarboxylic acid derivatives but from the 2,5-bis(hydroxymethyl)furan (BHMF) diol counterpart. Recently, Gandini *et al.* [28] illustrated the preparation of an all-furan-based polyester, obtained by interfacial polycondensation of BHMF and 2,5-furandicarbonyl chloride, but characterized by low molecular weight. Then, Ikezaki and Zeng prepared poly(2,5-furandimethylene succinate) (PFS) and its copolymers with butanediol by using *N*,*N*′-diisopropylcarbodiimide as coupling agent, with the aim to produce self-healing materials by exploiting DA reactions (see following paragraphs) with various maleimides (MIs) [50–52]. The condensation between BHMF and various diacid ethyl esters was carried out enzymatically using the *Candida antarctica* Lipase B [53] via a three-stage method.

Concerning the characterization of furan-based polyesters, it is interesting that only recently crystal structure, thermal behavior, phase transitions, and mechanical properties were reported for PEF [33, 34, 54, 55], PBF [38, 49, 56], and PHF (based on 1,6-hexanediol) [57]. These studies confirmed the potentially wide application fields for these novel polyesters. Table 13.1 gives a selection of thermal and mechanical data for the most popular and used polyesters (POF derives from FDCA and 1,8-octanediol).

One of the strongest potential of these polyesters seems to lie in the improved barrier properties, with PEF exhibiting a water diffusion coefficient 5 times lower than that of PET and oxygen permeability 10 times lower, despite its higher free volume (see Table 13.2) [34, 58–60]. These reductions are attributed to fundamental differences in segmental mobility due to the higher rigidity of the furan ring. Indeed, the nonlinear axis of ring rotation coupled with the dipolarity of the heterocycle in PEF significantly hinders the furan ring-flipping mechanism. This peculiarity could make the furan-based polyesters powerful and innovative materials for food packaging applications.

Table 13.1 Thermal and mechanical data of furan dicarboxylates.

	$M_n \times 10^{-3}$	$M_w \times 10^{-3}$	η_{sp} (dl g^{-1})	T_c[a] (°C)	T_g[a] (°C)	T_m[a] (°C)	T_d[b] (°C)	Young's modulus (MPa)	σ_{max} (MPa)	ε_{break} (%)
PEF references										
[33]	8.0	15.2	—	—	77	214	—	—	—	—
[49]	—	—	0.47 (intrinsic)		89	201	326	—	—	—
[27]	22.4	44.5	—	165	80	215	300	—	—	—
[35]	23	—	—	156	85	170	332	—	—	—
[39]	105	252	1.20	—	90	210	389	2070	66.7	4.2
PPF references										
[33]	10.1	16.3	—	—	40	171	—	—	—	—
[35]	15	—	—	102	39	150	335	—	—	—
[27]	21.6	27.6	—	127	50	174	295	—	—	—
[39]	60.2	89.8	1.21	—	58	—	375	1550	68.2	46
PBF references										
[49]	—	—	0.45 (intrinsic)	—	36	169	329	—	—	—
[33]	11.8	21.2	—	—	36	177	—	—	—	—
[39]	17.8	42.3	1.41	—	31	172	373	1110	19.8	2.8
[38]	23.2	65	—	—	39	172	—	959	31.8	1055
[35]	60	—	—	90	31	170	338 (5%)	—	—	—
PHF references										
[39]	32.1	66.7	1.04	—	28	148	375	493	35.5	210
POF references										
[39]	20.7	47.5	0.69	—	22	149	375	340	20.3	15

a) DSC data, measured at 10 °C min^{-1}.
b) TGA data.

Table 13.2 Comparison among some properties of PEF and PET [34, 58–60].

Polymer	Density[a] (g cm^{-3})	T_g[b] (°C)	T_m[b] (°C)	T_d, onset[c] (°C)	CO_2 permeation[d] (cm^3 mil/(100 in^2 day))	O_2 permeation[e] (cm^3 mil/(100 in^2 day))
PEF	1.430	85	211	389	4.449	0.095
PET	1.335	76	247	413	10.154	0.891

a) For amorphous samples.
b) DSC data.
c) Thermogravimetric data at 10 °C min^{-1} under nitrogen.
d) Measured by MOCON Permatran at 23 °C and 50% relative humidity.
e) Measured by MOCON Oxtran at 23 °C and 50% relative humidity.

13.1.3 Polyamides

The first furan-based polyamides were prepared by Gandini and Mitiakoudis [61] by bulk, interfacial, and/or solution polycondensation with various diamines, mostly bearing aromatic rings [62] (see Scheme 13.4). FDCA was used as is or converted into dichloride, and the 3,4-FDCA isomer was also used. Later, unusual polyamides were prepared by the reaction of diamines with the difuranic dicarboxylic acid (Scheme 13.2) [63–65].

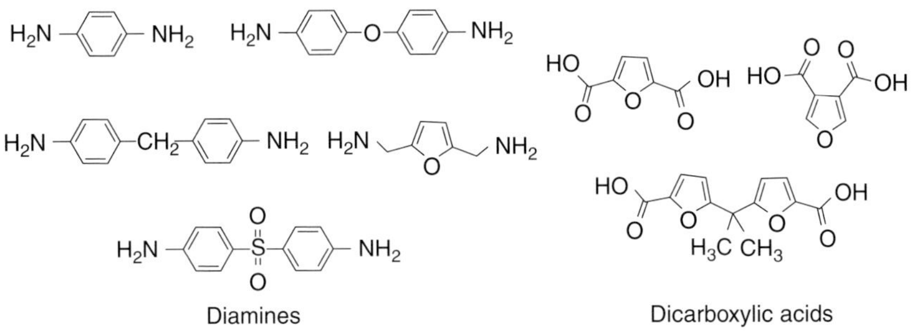

Diamines

Dicarboxylic acids

Scheme 13.4 Monomers used in furan-based polyamides.

A very original synthetic procedure was developed by Abid and Gandini [66] in which 2-furamide was oxymethylated and then self-polycondensed by reacting at the C_5 position of the heterocycle (see Scheme 13.5). These polyamides displayed an amorphous character, a very high thermal stability (over 400 °C), and high-T_g values (between 70 and 300 °C).

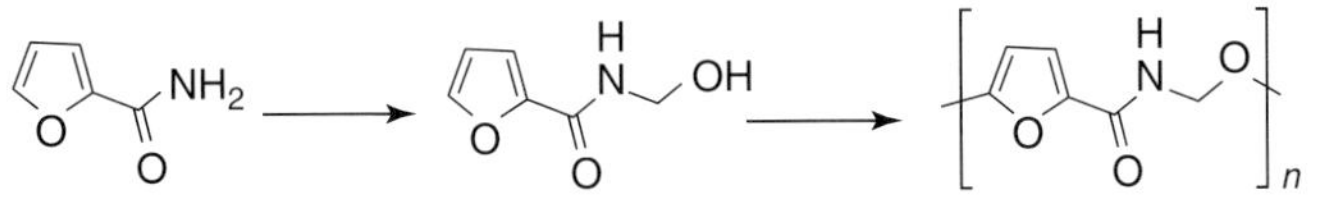

Scheme 13.5 2-Furamide self-condensation.

Recently, an interesting German paper reported a more "modern" synthetic approach by a two-stage bulk polycondensation [67] from the dimethyl ester of FDCA and various diamines. Specifically, poly(hexamethylene furandicarboxyamide) (PA6F), poly(octamethylene furandicarboxyamide) (PA8F), poly(decamethylene furandicarboxyamide) (PA10F), and poly(dodecamethylene furandicarboxyamide) (PA12F) were synthesized using dibutyl tin oxide as catalyst, a first stage under pressure at 140 °C and a second stage under vacuum increasing the temperature till 230 °C, over a total of 7 h. All the polyamides were amorphous with M_w in the range 11 000–19 000. Two patents deal with furan-based polyamides, one from DuPont [68] describing the synthesis in solution, the other [69] claiming that is not possible to prepare the PA6F by this approach and proposing, instead, the incorporation of the furan ring into the diamine monomer. Wilsens *et al.* [70] compared the properties of some polyesteramides containing isophthalic, terephthalic, or furandicarboxylic groups and found that the presence of furan ring in the polyesteramides perturbs the formation of intermolecular hydrogen bonds, resulting in a lowering of T_g and T_m with respect to the analogous materials bearing the other aromatic rings. Recently, Triki *et al.* [71] prepared amorphous polyesteramides by the melt copolycondensation of FDCA with hexamethylene diamine and ethylene glycol which produces materials with T_g values increasing with increasing amide content. Considering that the progress on synthesis of furan-based polyamides are quite recent, specific applications of these materials have not yet found an industrial development, but their peculiarities make them promising materials, especially in the packaging field.

13.1.4 Polyurethanes

The first examples of furan-based polyurethanes were reported in the early 1990s by Gandini and colleagues [72–75], who described the polycondensation of BHMF and 5,5′(oxy-bis(methylene))bis-2-furfural (OBMF) with various diisocyanates (see Scheme 13.6). It is interesting to observe that no polymers based on 2,5-furyl diisocyanate were prepared because of its excessive reactivity and also because the resulting polyurethanes were not stable. Therefore, the authors used its homolog bearing methylene groups between the ring and each NCO function (see Scheme 13.6) and the diisocyanate derived from difuranic acid (Scheme 13.3). With respect to commercial materials based on aromatic/aliphatic precursors, these furan polyurethanes tended to display lower moduli and thermal transitions but could fulfill important applications specific to their properties. The most original feature was however associated with a very high carbonaceous residue following thermal degradation, particularly with the fully furanic structures [74].

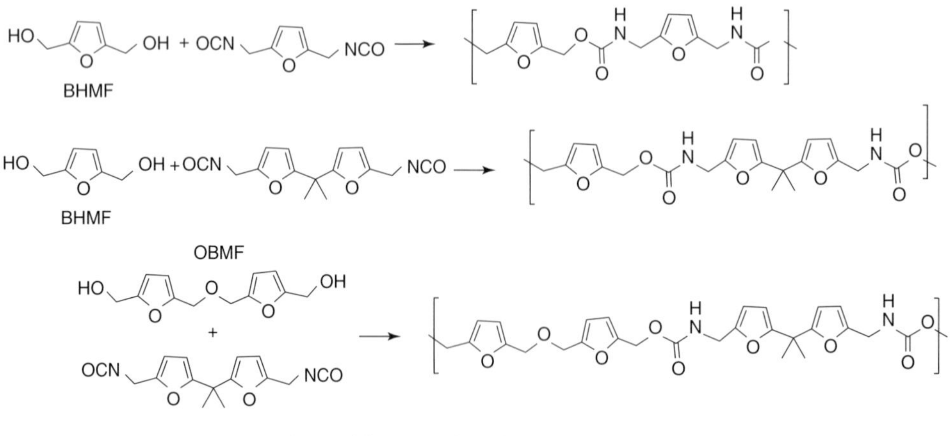

Scheme 13.6 Synthesis of furanic polyurethanes.

13.1.5 Polymers Based on the DA Reaction

The furan ring is able to react in DA cycloadditions, thanks to its pronounced dienic character. The reaction occurs easily if the furan dienic character is conveniently accentuated by placing, for example, electron donors as substituents on the heterocycle. Furthermore, the DA reaction is favored by the presence of alkyl substituents on the furan rings, if possible not sterically bulky. More specifically, Scheme 13.7 shows the reactivity scale of substituted furans toward MIs [3].

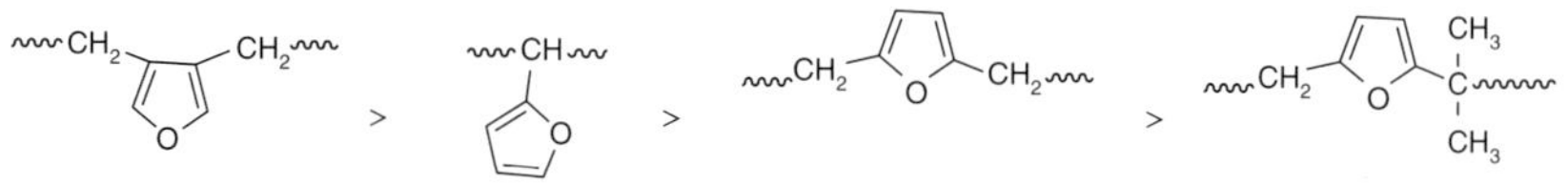

Scheme 13.7 Reactivity scale of substituted furans in their DA coupling with maleimides.

MIs are very reactive dienophiles and particularly suited as potential monomers in DA polycondensations because they are readily prepared as di- or trisubstituted structures. The DA condensation step in such a polymerization, depicted in Scheme 13.8, occurs by means of a concerted mechanism which is inherently thermoreversible. This peculiarity makes the reaction really attractive also because the temperatures up to ~60 °C for the click reaction (equilibrium shifted toward the adduct) and above ~110 °C for the unclick reaction (equilibrium shifted toward the reagents) fall within realistic values [3]. This reaction is not marred by side reactions, and no other degradation is envisaged within that temperature range. Moreover, even if these reactions produce two different stereoisomers (exo and endo), their relative abundance is not relevant in the polymers context, since both play the same role as chain links between monomer units. Only a Dutch research group speculated on the differences between the two isomers [76], concluding that the endo and exo DA stereoisomers show significantly different thermal responses, with the retro DA of the endo isomer typically taking place 20–40 °C below that of the corresponding exo DA adduct (with the exception of adduct derived from aromatic MIs).

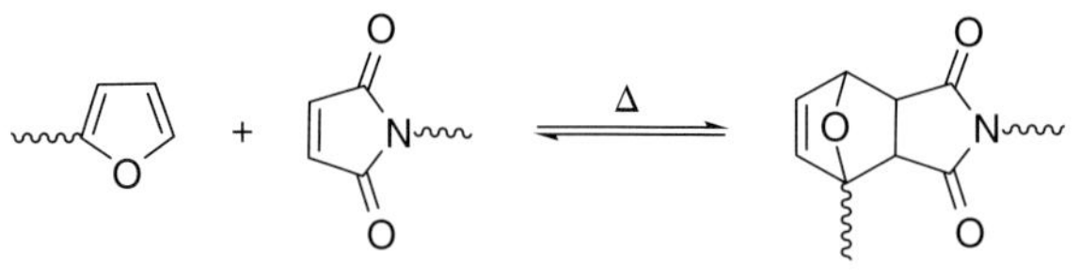

Scheme 13.8 DA equilibrium between furan and maleimide end groups in a macromolecular synthesis.

The exploitation of the DA reaction in polymer chemistry has gained a growing momentum in the last decade [3], and the trend is accelerating, as discussed here for the latest contributions. The most salient feature here is the thermoreversible character of this DA reaction [77–79] which provides the ensuing materials by limitless possibilities, such as removable thermosets [80], thermoresponsive optical polymers [81], novel materials for nanoscale probe lithography and data storage [82], modular polymeric color switches [83], and smart coatings [84]. The unique self-healing and recyclability properties of these systems are illustrated in Figure 13.1.

Since the polymers are materials susceptible to damage induced by mechanical, chemical, and thermal factors, the thermally reversible DA reactions may be applied to extend their life and safety, since damaged cross-linked materials can be simply healed by using a heating–cooling cycle. As the bond strength between the furan and MI moieties is at least three to four times lower than that of other covalent bonds, the retro DA reaction should be the main pathway for crack propagation [86], but healing is possible on the crack surfaces through the recovery of broken DA cross-links.

Multifunctional small molecules have been used as monomers for the construction of cross-linked polymers exhibiting a high efficiency of self-healing. Wudl's group [87] utilized a small molecule containing four furan groups and another possessing three MI groups as precursor to construct a dense macromolecular network which displayed excellent self-healing properties. Moreover, to further increase the self-healing efficiency of this kind of materials, it is preferable to use low melting point monomers, which have relatively high mobility in the cross-linked matrix so as to heal the crack effectively. It was reported [88] that

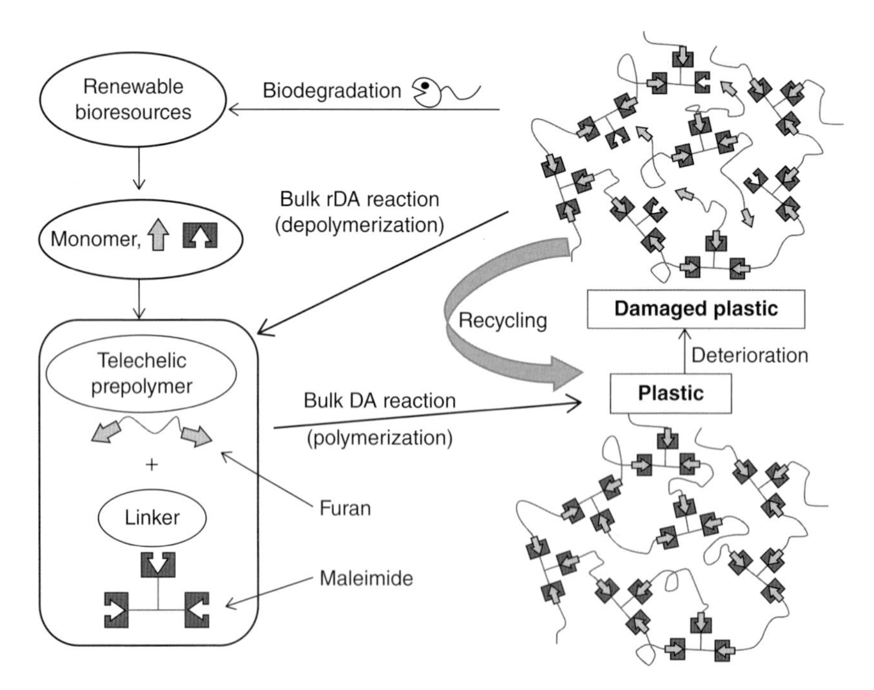

Figure 13.1 Schematic representation of a recyclable and self-mendable bio-based polymer system [85].

the self-healing efficiency of the polymers constructed with the low melting point furan and MI grew to about 80%. The quick healing procedure is attractive and indicates that the healing process occurs on a timescale controlled by the kinetics of the DA reaction rather than by a diffusion-controlled process.

Zeng *et al.* [50, 51] prepared polyesters by coupling bis(hydroxymethyl)furan with succinic acid (SA) and then studied their reversible DA cross-linking with different bismaleimides (BMIs) (Scheme 13.9). Variable amounts and different types of BMI led to materials displaying tunable mechanical properties with excellent self-healing ability, and their surface, when broken, could be rejoined without external stimulus.

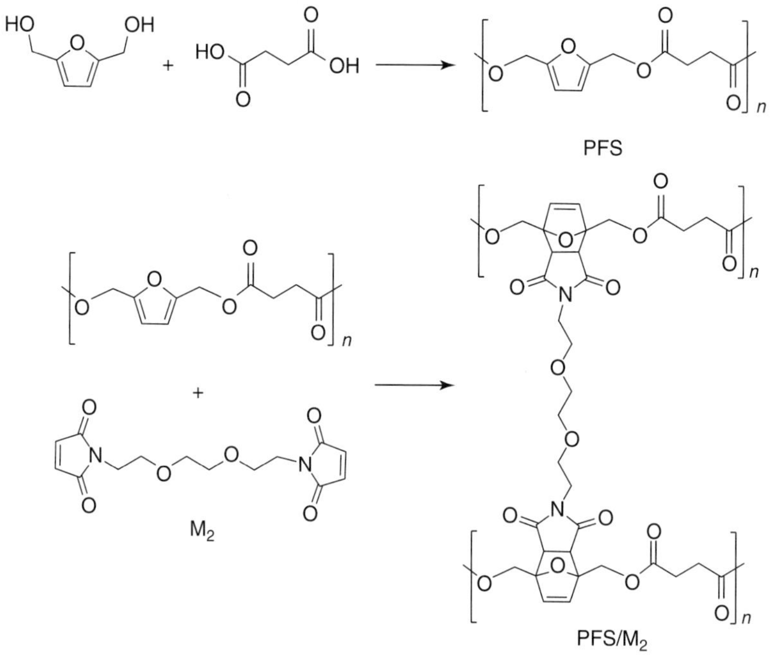

Scheme 13.9 Synthesis of poly(2,5-furandimethylene succinate) and reversible DA reaction between PFS and a bismaleimide (M_2) leading to the polymeric network.

Interestingly, the mendability is achievable even if the furans are present in modest proportions, as in the case of chain ends, as shown by Ishida and coworkers [89] who prepared recyclable plastics by the synthesis of furyl-telechelic PLA and poly(butylene succinate) (PBS) cross-linked with BMI and trismaleimides. Recently, Gandini *et al.* [90, 91] prepared bio-based polyesters by the DA reaction between BMI and novel monomers based on vegetable oils bearing furan ring appended through thiol–ene click chemistry.

Liu [92] prepared polyamides bearing pendants furan rings and studied the self-healing properties of the BMI cross-linked materials, which however had limited processability.

The application of these principles has proved very useful in processing, reshaping, thermal self-repairing, and extending the service life of polyurethanes [93]. The cross-linking can be carried out by three different approaches, namely, (i) by the DA reaction between pendant furans and furan telechelics with a multimaleimide linker [94–96], (ii) by the reaction between pendant MI with various difurans [97], or (iii) by coupling a difunctional DA adduct with the polyurethane chain [98]. Recently, Rivero and colleagues [99] prepared tough, flexible, and transparent polyurethane networks with healing capability at mild temperature conditions. Remendability was achieved by combining the occurrence of two processes at 50 °C, namely, a fast shape memory effect brought the free furan and MI moieties together, after which a progressive DA reaction could reform the covalent bonds on a longer timescale.

Kavitha and Singha [100] applied the DA reactions to copolymers of furfuryl methacrylate (FMA) and methyl methacrylate (MMA) prepared by the atom transfer radical technique (Scheme 13.10), and further DA reaction produced thermally mendable materials useful for coatings and adhesive.

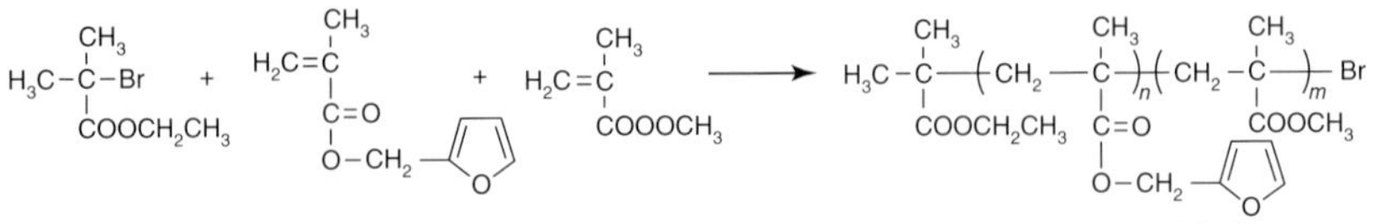

FMA MMA P(FMA-*co*-MMA)

Scheme 13.10 Atom transfer radical copolymerization of furfuryl methacrylate and methyl methacrylate.

Picchioni and coworkers [101] turned their attention to the modification of polyketones, via the Paal–Knoor reaction (Scheme 13.11) with furfurylamine followed by cross-linked with BMI, and, interestingly, the obtained networks, after various click–unclick cycles, maintained their mechanical properties, indicating the robustness of the furan-modified material and almost full mechanical reversibility. The same group [102] broadened this study by investigating the effect of hydrogen bonds on network properties.

Epoxy resins were subjected to DA studies carried out by Tian *et al.* [103, 104] in which new epoxy monomers containing furan rings were synthesized and then cured with BMI and anhydrides. The cured materials turned out to contain two types of intermonomer linkages, namely, thermally reversible DA bonds from the reaction between furan and MI groups and thermally irreversible bonds from the reaction between epoxide and anhydride groups. The high density of DA bonds in the cured epoxy resins guaranteed a recovery efficiency close to 100%.

Palmese's group [105] studied furan-functionalized epoxy resins cured with diamines and BMI with the intent to develop a polymer coating able to self-heal in solution. Indeed, for example, when a coating of metallic substrate fails, corrosion can seriously damage the underlying material. Solvent-induced swelling and softening of the crack surfaces allowed for mechanical interlocking, while compatible

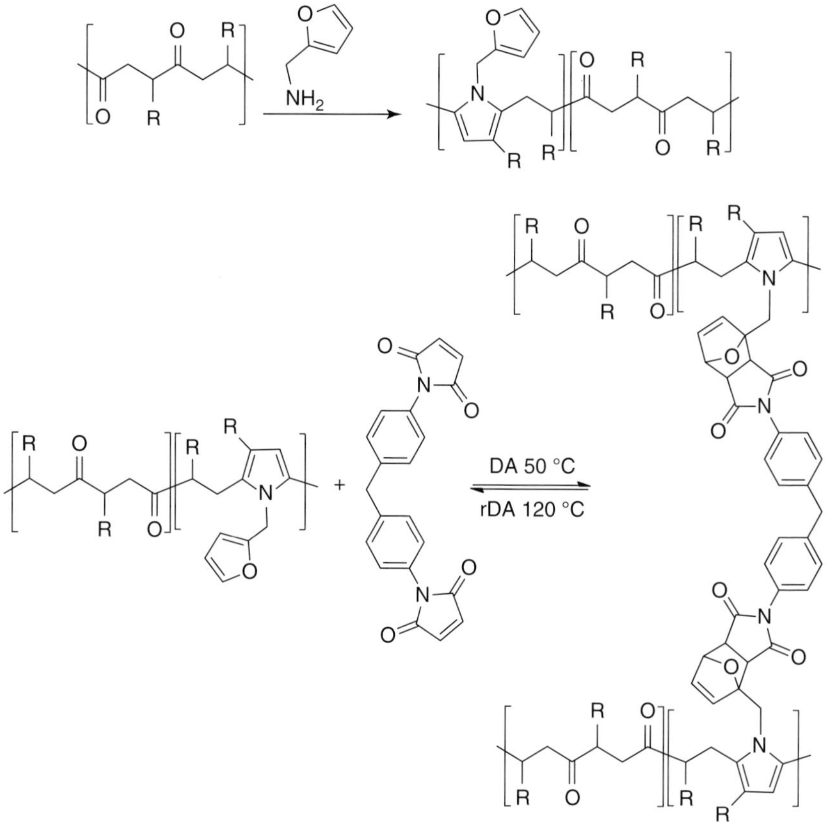

Scheme 13.11 Paal–Knorr reaction of alternating polyketone with furfurylamine and subsequent DA reaction with BMI.

functionalization of the polymer network and the healing agent caused covalent bonding through the DA reaction of furan and MI. Physical bonding resulted in 28% of recovery of initial strength, whereas covalent bonding resulted in an additional 42% healing efficiency [106].

DA chemistry was also attempted to furan-derivatized cellulose by the reaction of the 2-hydroxyethyl cellulose with 2-furoyl chloride [107] followed by cross-link with a BMI. Surprisingly, the authors do not consider that 2-carbonyl-substituted furans are DA unreactive because of the loss of dienic character, which explains, as should have been anticipated, that the system did not display the expected behavior.

A final important advance worth mentioning is the recent research into hydrogels prepared by the DA reaction [108, 109] in which water-soluble polyacrylates bearing pendant furan and MI moieties or fural-modified gelatin were cross-linked in an aqueous medium. These materials were fully characterized and showed promising properties in terms of applications, among others, in the biomedical realm.

13.1.6 Polyfurans

Apart the well-known conjugated polymers based on poly(2-furylene vinylene) based on 5-methyl furfural [21], polyfurans have scarcely been explored mostly because of the difficulty associated with their synthesis. The inherent structure of α-oligofurans (see Scheme 13.12) suggests however potential applications as organic materials in optoelectronic devices, such as solar cells, organic field-effect transistors (OFETs), and organic light-emitting diodes (OLEDs), as well as in highly efficient organic light-emitting transistors (OLETs), flexible displays, and specific chemical sensors. Since the main requirements for organic electronic materials are a low HOMO–LUMO gap (which also leads to absorption/emission in the visible or near infrared (NIR) range), good solid-state packing, stability, solubility (which is important for their processability), rigidity/planarity, and high luminescence (for light-emitting applications) [110], the polyfurans appear as good candidates. Indeed, the only oligofurans with six or more furan rings possess a HOMO–LUMO gap below 3 eV (which is the typical range for organic semiconductors). Moreover, the planarity prerequisite is totally satisfied by polyfurans, because they have a conjugated backbone stiffer than other systems, including oligothiophenes. This high rigidity allowed many different substituents to be introduced without distorting the planarity and conjugation of the polyfuran chain [111]. The excellent work of Bendikov's group [110–112] showed, among other features, that regioirregular alkyl-substituted polyfurans do not lose their conjugation and moreover exhibit reduced bandgaps.

They showed that the preparation of these materials must be carried out by electrochemical polymerization, because the chemically synthesized polyfurans are not highly conjugated owing to a significant degree of furan ring-opening defects. They were able to prepare stable polyfurans by polymerization of oligofurans using considerably lower potential than parent furans [112]. This is a very interesting discovery, because it contrasts the previous belief that polyfurans are intrinsically unstable and possess low conductivity. Recently, also the β-oligofurans (Scheme 13.12) have been the subject of few studies, but only with basic characterizations and no applications have yet been reported [113].

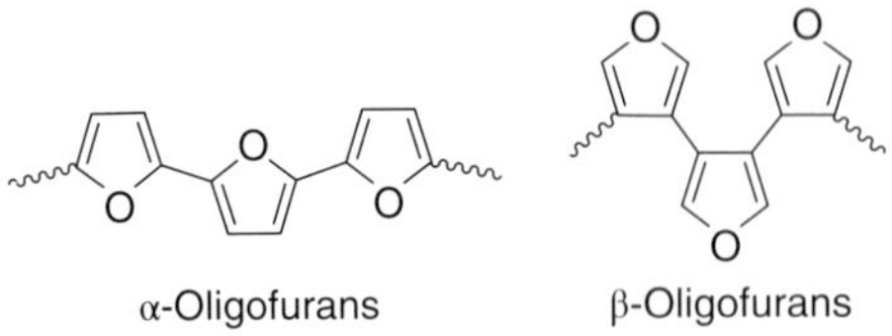

O O O O O O

α-Oligofurans β-Oligofurans

Scheme 13.12 Structures of oligofurans.

13.2
Polymers from Diacids, Hydroxyacids, Diols

13.2.1
Introduction

This chapter aims at discussing how some significant monomeric structures (diacids, hydroxyacids, and diols, shown in Scheme 13.13), all obtained from biomass, have been recently exploited to prepare bio-based polymers. Among these structures, the most studied and also industrially available bio-based monomers are aliphatic. They can derive, for instance, from cellulose, starch, or triglycerides and are characterized by relatively easy preparation and low cost [12, 17, 114]. In particular, succinic acid (SA) has been considered as one of the most important biomass-derived value-added compound, from which many other industrially significant chemicals, polymers, and copolymers can be prepared [115]. The classical problem of fully aliphatic polymer structures is however related to their poor properties, particularly, their generally very low T_g values, the highest being 55 °C for PLA. To overcome this drawback, a wide literature deals with the modification of the original macromolecules by introduction of alicyclic or aromatic units, through copolymerization, or by creating a complex network of cross-linkings to generate high-T_g biopolymers similar to that of important classical engineering plastics (such as poly(styrene), poly(vinyl chloride), poly(bisphenol-A carbonate), which are amorphous and have T_g values around or above 100 °C). Another notable aspect relates to the preparation of composites (for example, with vegetal fibers) and nanocomposites to improve the polymer matrix performances [116].

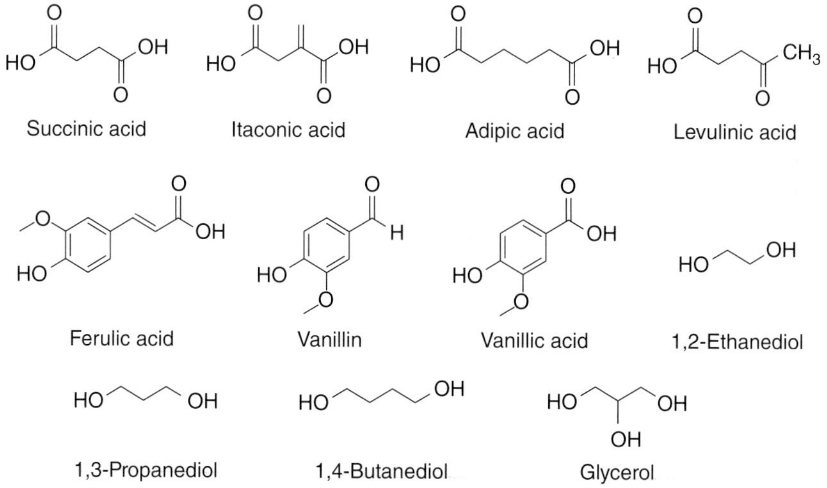

Succinic acid | Itaconic acid | Adipic acid | Levulinic acid

Ferulic acid | Vanillin | Vanillic acid | 1,2-Ethanediol

1,3-Propanediol | 1,4-Butanediol | Glycerol

Scheme 13.13 Examples of some important building blocks derived from renewable resources.

Many key commercial chemicals are aromatic and petrol derived. The access to bio-based and nonharmful aromatic monomers is one of the main challenges of the years to come. Lignin, an amorphous cross-linked polymer that gives structural integrity to plants, is the most abundant natural aromatic feedstock. Depolymerization of lignin should therefore be the way to access bio-based aromatics needed by the chemical industry. Unfortunately, this route is unmanageable and there are very few reports on efficient ways of recovering such aromatic products [117, 118]. The only monoaromatic compound currently produced industrially from lignin (from paper-processing by-products) is vanillin, which is, then, a highly promising bio-based building block for monomer synthesis. Some recent examples of exploitation of vanillin to prepare polymeric structures are described in Section 13.2.5.

13.2.2 SA and Its Polymers

SA, recognized as one of the most promising four-carbon 1,4-diacids [17, 119], is a key building block for preparing both commodity and specialty chemicals [115, 120] and can be polymerized with diols and diamines, including those obtained by conversion of SA itself, to produce bio-based polymers [121].

The current market of SA is around 30 000 tons per year and a sixfold increment of it is forecast by 2016 with the development of bio-SA [122]. Conventionally, SA is petrochemically produced on an industrial scale; however, in recent years, fermentation processes for bio-SA production have been developed [123–125]. In the following sections, the most significant classes of polymers that can be prepared from SA will be discussed, with peculiar attention to the recent developments.

13.2.2.1 Polyesters

Poly(butylene succinate) (PBS) is the most widely employed SA-based polymer. This poly(alkylene dicarboxylate) can be fully prepared from renewable resources through the polycondensation of bio-SA and bio-butanediol (bio-BDO). PBS is produced industrially from bio-SA by *ptt*, Mitsubishi Chemical Corporation, with the trade name of *BioPBS*, in Thailand with a capacity of 20 000 tons per year; moreover, Showa Denko produces PBS, starting from petro-SA and bio-SA (trade name Bionolle®) in Japan with a capacity of 6000 tons per year [126]. PBS is a white semicrystalline thermoplastic, with a melting temperature of about 115 °C and glass transition temperature of about −30 °C. It is characterized by a high thermal stability and very good mechanical properties: heat distortion temperature at 97 °C; tensile yield strength up to 30–35 MPa, similar to that of polypropylene; Young's modulus of 300–500 MPa; and flexibility [127]. It has excellent processability using conventional equipment at temperatures of 160–200 °C. Applications are expected in different fields like disposable packaging, flushable hygiene products, or in agriculture. It is biodegradable in compost, in moist soil, fresh water with activated sludge and sea water [128, 129], and biocompatible, hence suitable for biomedical applications.

Tuning the PBS properties is also needed for broadening its range of applications, and copolymerization appears to offer promising advantages. Different comonomers have been used to modulate PBS mechanical properties and degradability, for example, a dicarboxylic acid with a high number of methylene sequences, such as azelaic acid. Azelaic acid is a C_9 dicarboxylic acid, industrially produced from bio-based oleic acid. Fully bio-based poly(butylene succinate-*co*-butylene azelate) copolymers, characterized by low glass transition and melting temperatures, by a decrement of crystallinity, and by a reduction of thermal degradation [130–132], can find applications from elastomers to high-impact thermoplastics. The insertion of aromatic moieties into the main aliphatic backbone of PBS is strategic to increase thermal transition temperatures, improve mechanical properties, and modulate the biodegradability of the aliphatic polyesters. In poly(butylene succinate-*co*-butylene terephthalate) (PBST), tensile strength and elongation at break vary according to the degree of crystallinity and, as expected, the material biodegrades more slowly with an increase in aromatic comonomer content [133]. More complex structures, such as long-chain branched PBSTs, have also been prepared in order to notably improve rheological properties and melt strength, greatly expanding the range of applications of the material [134]. In order to avoid petro-derived terephthalate units, recently other rigid, cyclic, renewable monomers have been used in copolymerizations with PBS. Isosorbide or FDCA have been tested as comonomers to obtain materials suitable to film processing [46]. Although melt strength increases in the presence of rigid units, the lower crystallization rate does not favor the processing via extrusion blowing. Nevertheless, a large increase in elongation at break was obtained, which is an interesting feature for film application [135]. Tartaric acid-based comonomers have also been used to modulate PBS thermomechanical properties and biodegradability [136]. As previously described [52], PBS has also been copolymerized by using the rigid furan diol BHMF. These new copolymers have a more rigid macromolecular structure (T_g increases from −44 °C for PBS to 2 °C for the copolymer with 80 mol% of BHMF) and can form reversibly cross-linked network by the DA reaction with a BMI, thus providing self-healing properties.

With itaconic acid (Scheme 13.13) as comonomer, unsaturated polyesters can also be prepared, which belong to the class of cross-linkable thermoset. To avoid that the double bonds react during the synthesis, a selective enzymatic catalysis has been successfully used [137].

Linear and branched co- and terpolyesters based on SA and isosorbide, also in combination with other renewable monomers such as 2,3-butanediol and 1,3-propanediol (PDO), have been prepared for powder coating applications [138, 139]. Upon incorporation of 60–80% of isosorbide, the T_g is high enough to insure a good processing and storage stability. It was also found that incorporation of diols different from isosorbide has a beneficial effect on the color of the resin. Isosorbide can be substituted with isoidide [140], which is slightly more reactive in polycondensations, thermally stable, although not commercially available. Coatings from branched polyesters show improved mechanical and chemical resistance compared to those formulated from linear polymers.

Other high-T_g amorphous biodegradable polyesters have been recently synthesized, namely, copolyesters based on SA, isosorbide, and 1,4-cyclohexane dicarboxylic with T_g of 100–140 °C and copolyesters from SA, isosorbide, and isophthalic acid with T_g from 90 to 160 °C. These are significantly high values, compared with those of the most important classical engineering plastics [141].

13.2.2.2 **Poly(ester amide)s**

The introduction of amide groups into polyester chains has been used as a strategic approach to improve properties thanks to strong intermolecular hydrogen-bond interactions. Poly(ester amide)s of PBS, characterized by a periodic sequential structure of ester and amide groups, were obtained from succinate, BDO, and 1,4-butanediamine by two-step polycondensation reactions [142]. PBS oligoesters were first prepared and then fractionated by supercritical fluid chromatography. Then, equimolar amounts of monodispersed oligoester and butan-1,4-diamine were copolymerized. The ensuing periodic copolyesteramides displayed high thermal stability and a biodegradation rate which depended on the polyester unit length. Poly(ester amide)s prepared by insertion of naturally occurring α-amino acids (AAs) into a poly(alkylene succinate) backbone [143] provided significant advantages over biodegradable polyesters as biomedical materials. Indeed, the final polymers incorporated two of the most desirable heterolinks, namely, the ester bond providing biodegradation (hydrolysis), improving processability, and decreasing immunogenicity of AA-based polymers, and H-bonds providing desired mechanical properties at low molecular weights, increasing hydrophilicity, and promoting active interactions with the surrounding tissues after implantation. Pure polyamides made of AA-poly(α-amino acid)s are, by contrast, characterized by low rates of biodegradation and poor processability.

13.2.2.3 **Polyamides**

Several aliphatic polyamides can be prepared from bio-based building blocks. For example, the bio-based production of 1,4-butanediamine is possible via the chemical conversion of bio-SA. 1,4-Butanediamine is the basic monomer for polyamide 4,6, an engineering plastic with high crystallinity and high melting point, by reaction with adipic acid, and for polyamide 4,10, a high performance engineering plastic obtained from sebacic acid. Today, only partially bio-based PA4,10, derived from bio-based sebacic acid, is produced commercially [121].

A recent paper reports a novel water-soluble polyamide formed from diethyl succinate and hexamethylenediamine copolymerized with tributyl citrate according to Scheme 13.14 [144]. This kind of polyamide is structurally designed to mimic natural proteins and provide cryoprotection for living cells, biologically

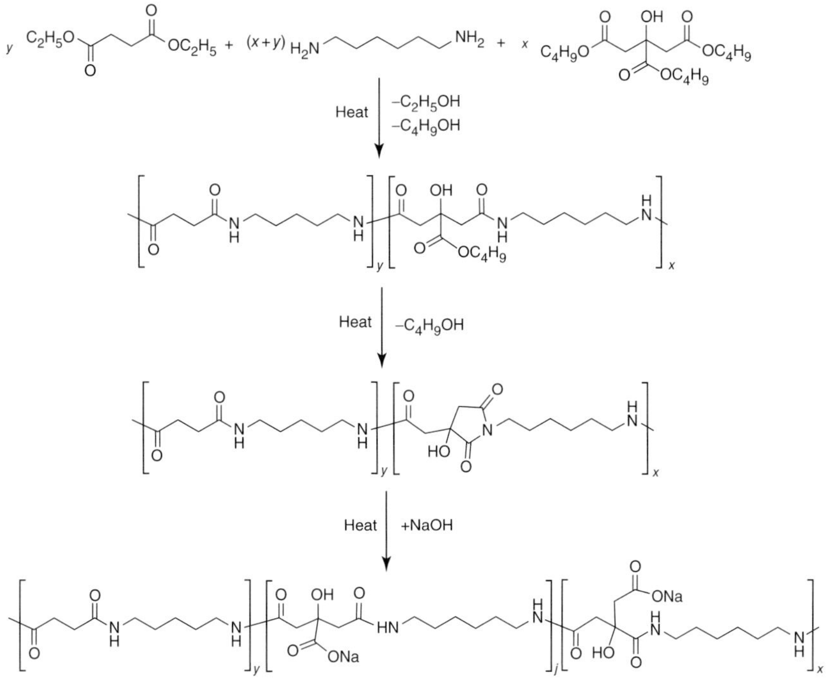

Scheme 13.14 Synthetic pathway to a polyamide formed from diethyl succinate and hexamethylenediamine copolymerized with tributyl citrate [144].

active substances, and food. In addition, it has an outstanding role in metal chelation, ion exchange, and drug delivery.

13.2.2.4 **Polyurethanes**

PBS polyols are building blocks for biodegradable thermoplastic polyurethane elastomers. For example, PBS oligomers with $M_n = 1000-2000$ and high contents of terminal hydroxyls are the soft diols and react with 4,4′-methylenebis diphenylisocyanate (MDI) and 1,4-butanediol, according to Scheme 13.15 [145]. With respect to the analogous polyurethane prepared from poly(butylene adipate), the greater density of H-bond-accepting carbonyl groups on the PBS chain results in increased hard segments–soft segment interactions. This behavior induces high T_g, low hard segment melting points and also influences mechanical properties, such as abrasion resistance. In similar structures, crystalline PBS and MDI constitute the hard segments, while amorphous poly(diethylene glycol succinate) forms the soft segments [146]. In the latter case, the composition is the parameter that mostly affects the physical properties, since the degree of crystallinity is strongly dependent on the PBS segment content. These materials are expected to find many applications as elastic materials.

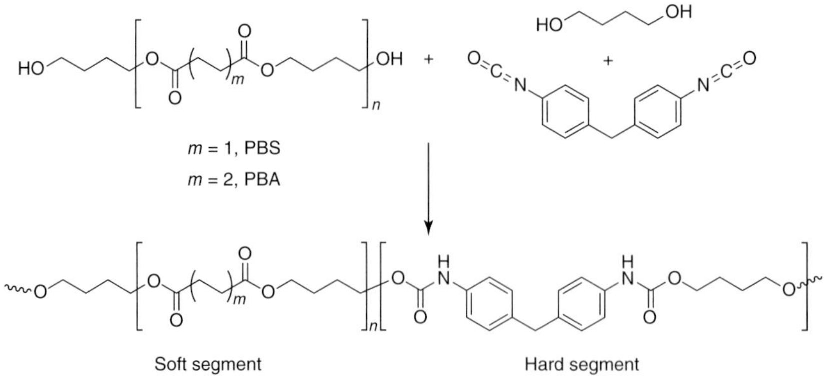

m = 1, PBS
m = 2, PBA

Soft segment Hard segment

Scheme 13.15 Synthesis of thermoplastic polyurethanes based on succinate polyesters [145].

13.2.3
Adipic Acid and Its Polymers

Adipic acid is another key aliphatic acid, with a global production of 4000 kt per year. Currently, its production is based on a chemical route from petrochemical cyclohexane, but bio-based processes have been developed at pilot plant scale and are under evaluation [121].

Adipic acid is a monomer for the production of polyamide 6,6 and polyamide 6, which are the most important nylons. At the end of 2013, Rennovia Inc. announced the production of the first 100% bio-based Nylon 6,6, made from Rennovia's renewable monomers [147]. From a commercial standpoint, adipic acid is also the monomer for the production of the BASF eco-friendly, compostable/biodegradable aliphatic–aromatic copolyester (poly(butylene adipate-*co*-terephthalate) and its blend with PLA, Ecoflex, and Ecovio, respectively) [148]. As already discussed, adipic acid is also involved in the preparation of bio-based polyester polyols for polyurethane production, for example, for coating applications [145, 149]. Finally, adipic acid has also been recently described as a monomer to prepare new hyperbranched polyesters (HBPEs) [150, 151], which have been synthesized starting from several multifunctional biobasic monomeric building blocks, multifunctional alcohols, and acids. In particular, Scheme 13.16 describes glycerol–adipic acid HBPEs obtained by starting with an excess of glycerol to obtain OH end groups. In these systems, the control of the end group functionality is a valuable attribute because many polymer properties depend on it, including solubility, melt viscosity, and thermal behavior. The final product characteristics can therefore be controlled by adjusting monomer stoichiometry and reaction time. The end groups are also available for subsequent cross-linking or for reactions with molecules that will be attached to the polymer skeleton, as in the case of the terminal polar groups reacting with active agents that can be encapsulated. Applications of these bio-based polymers

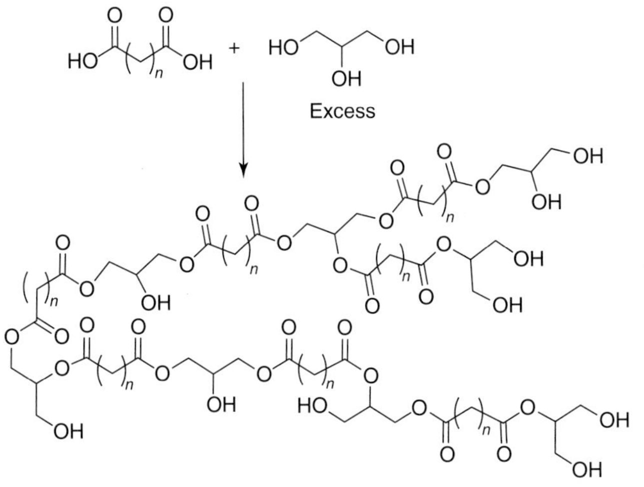

Scheme 13.16 Synthetic pathway to produce glycerol–adipic acid hyperbranched polyesters.

include delivery of active species, such as pharmaceuticals, pesticides, and antimicrobials.

13.2.4 Levulinic Acid and Its Polymers

Levulinic acid can be prepared by the acidic hydrolysis of lignocellulose, which causes the breakdown of polysaccharides. A degradation product of C_6-sugars is HMF, which can be converted to levulinic acid. This five-carbon keto-acid is a versatile building block for the synthesis of various organic chemicals, such as levulinate esters, γ-valerolactone, acrylic acid, 1,4-pentanediol, and β-acetylacrylic acid [152, 153].

The direct use of levulinic acid for the synthesis of polymers is rare. Bacterial cultures of *Burkholderia cepacia* was used as a biocatalyst for the preparation of a poly(β-hydroxyalkanoate) from levulinic acid [154]. Levulinic acid derivatives have instead been used as monomers. In particular, 5-hydroxylevulinic acid (5-HLA), synthesized via bromination and hydrolysis of levulinic acid, has been used for the preparation of poly(5-hydroxylevulinic acid) (PHLA) by direct polycondensation, according to Scheme 13.17 [155]. Although the polymerization is difficult (only a M_w of 1000 has been obtained), PHLA possesses an unordinary high glass transition temperature of about 120 °C. The high T_g was attributed to the formation of inter- and/or intramolecular hydrogen bonds due to a characteristic keto–enol tautomerism equilibria in the polymer structure. This polymer is far from fulfilling practical application but provides some information about the possibility of synthesizing aliphatic polyesters with high T_g. The combination of 5-HLA (ketone and enol forms) and diols gave rise to novel biodegradable

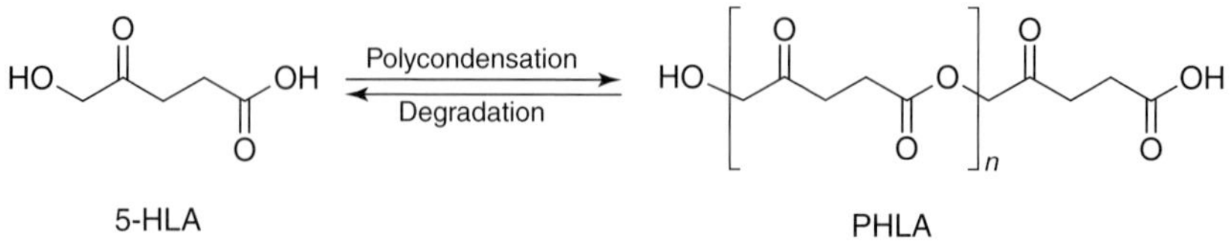

Polycondensation
Degradation
5-HLA
PHLA

Scheme 13.17 Schematic diagram of the synthesis and degradation of poly(5-hydroxylevulinic acid) [155].

cross-linked materials (Scheme 13.18) [156]. Interestingly in this case, T_g decreased notably, from 120 to −50\30 °C, according to the nature of the diol. The formation of cross-linking in the PHLA-diol destroys, at least partly, the hydrogen bonds thus reducing T_g more or less drastically.

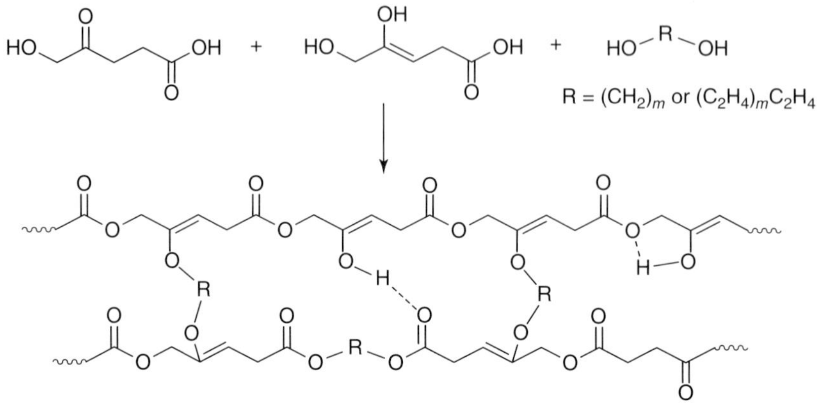

R = $(CH_2)_m$ or $(C_2H_4)_mC_2H_4$

Scheme 13.18 Possible cross-linking mechanism and microstructure of PHLA-diols [156].

In 2011, Leibig *et al.* [157] reported the selective ketalization of levulinic acid esters, providing an industrial pathway for developing bio-based building blocks for the synthesis of novel polymers, functional oligomers, polymer additives, and solvents. In these experiments, the authors observed that using a relatively low amount of moderate to strong protic acid catalyst resulted in the selective formation of ketal with respect to the product of transesterification, without sacrificing the reaction speed. Thus, levulinic acid can react with polyols like glycerol (Scheme 13.19) [158], since the keto group can form ketals with 1,2 or 1,3 hydroxyls, providing an opportunity for the synthesis of macromolecular structures.

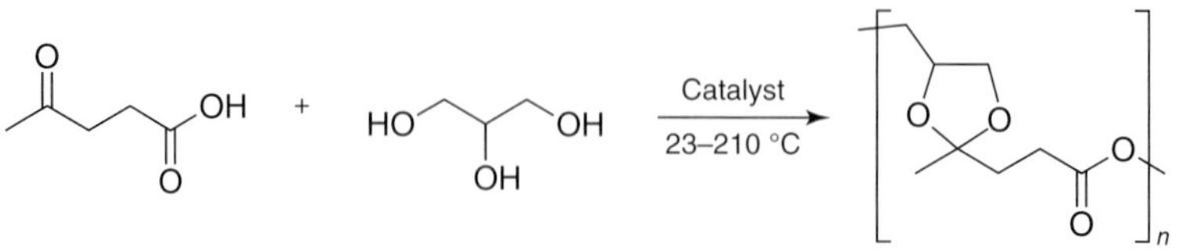

Catalyst
23–210 °C

Scheme 13.19 Synthesis of levulinic acid-*co*-glycerol oligomers [158].

13.2.5 Vanillin, Vanillic, and Ferulic Acids and Derived Polymers

As mentioned previously, the only molecule that is prepared industrially from lignin is vanillin (Scheme 13.13). Solvay and Borregaard dominate the vanillin market and, for both companies, part of their production is obtained from lignin. Recently, some attempts to utilize vanillin as monomer for bio-based polymers have been published. An electrochemical route to polymerize divanillin provides a functionalized polyvanillic polymer with potential applications like chelating metal ions from aqueous solutions [159]. Examples of the preparation of vanillic-based resins from vanillin or vanillic acid have been reported [160, 161]. Moreover, Fache *et al.* [162] investigated the functionalization of vanillin at different oxidation states, potentially available from lignin and directly available for polymerizations. Starting from three vanillin derivatives (methoxyhydroquinone, vanillic acid, and vanillyl alcohol), a platform of 22 bio-based compounds for polymer chemistry has been developed, which can be used in the synthesis of epoxy resins, polyesters, and polyurethanes. For example, epoxy thermosets with properties close to the bisphenol A-based industrial counterparts have been prepared, demonstrating that vanillin derivatives can be a viable bio-based alternative to bisphenol A [163].

Phenolic acids, such as coumaric, ferulic, and sinapic acids, are incorporated in lignin and presumably act as cross-linkers between lignin and polysaccharides to ensure rigidity of lignocellulosic composite. Although ferulic acid is not very abundant in lignins, according to the best estimates, prices could be as low as 1–3$ kg^{-1}, which is a reasonable target price for a commercial monomer. It exhibits antioxidant, antitumor, photoprotective, and antihypertensive activities, making it a promising additive to be used in food and cosmetic industries. However, despite its valuable nature, the use of ferulic acid in polymers is rather scarce, mainly due its poor availability. Alternatively, vanillin could be an attractive material for the preparation of ferulic acid and related derivatives, but for the time being, the proposed synthetic pathways require extra synthetic/purification steps leading to high production cost.

The design of sustainable PET mimics includes poly(dihydroferulic acid), PHFA, starting from vanillin, according to Scheme 13.20 [164]. The final polymer is characterized by a T_g of 73 °C, very similar to that of PET, whereas its melting temperature is lower (234 °C compared to 265 °C for PET) and its crystallization rate is higher than that of PET. It seems therefore that PHFA could be an alternative to the polyester derived from FDCA as a substitute to petro-derived PET for packaging applications.

Recently, a new class of alternating aliphatic–aromatic thermoplastic copolyesters has been prepared from renewable monomers through a metal catalyst-free process [165]. The authors developed a methodology for the production of renewable bisphenols incorporating ferulic acid moieties through an enzymatic process, according to Scheme 13.21. Then, the monomers based on bisphenols were polymerized by polycondensation with different aliphatic

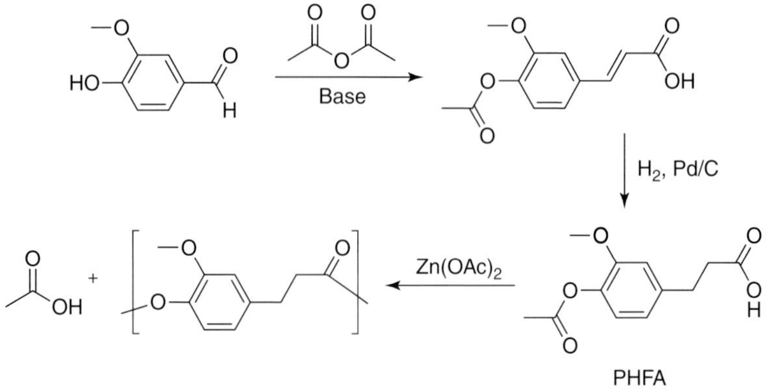

Scheme 13.20 Synthetic pathway leading to poly(dihydroferulic acid) from vanillin [164].

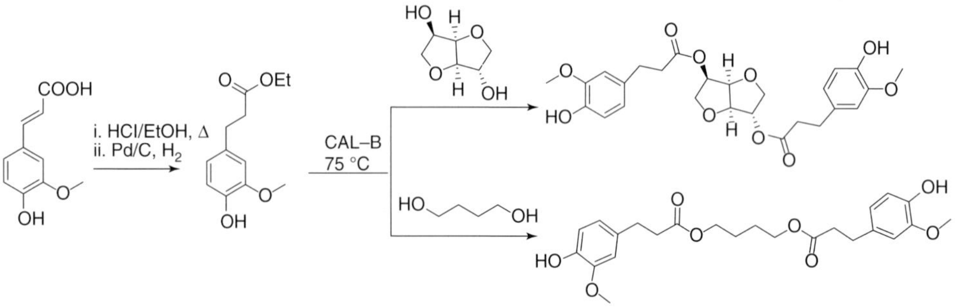

Scheme 13.21 Chemoenzymatic preparation of bio-based bisphenols [165].

diacids (Scheme 13.22). The final amorphous polymers had high thermal stability, with T_g ranging from 0 to 76 °C. Again, the properties of these new materials could be tuned to match those of fossil-based polyalkylene terephthalates, such as PET, or to prepare materials requiring lower T_g. Since these polymers are potentially biodegradable, they could be used in biocompatible applications, such as reconstructive bone surgery or sutures.

The same approach was applied to prepare poly(ester-urethane)s and poly(ester-alkenamer)s [166, 167], the latter from ferulic acid-based α,ω-diene monomers by a chemoenzymatic synthetic pathway, according to Scheme 13.23. Then, the monomers were successfully polymerized via ADMET in the presence of a second-generation Hoveyda–Grubbs catalyst (Scheme 13.24). The final amorphous polymers displayed low molecular weights, high thermal stability, and T_g that could be easily tuned by varying the chemical structure of the internal ester and the alkene length of the α,ω-diene monomers.

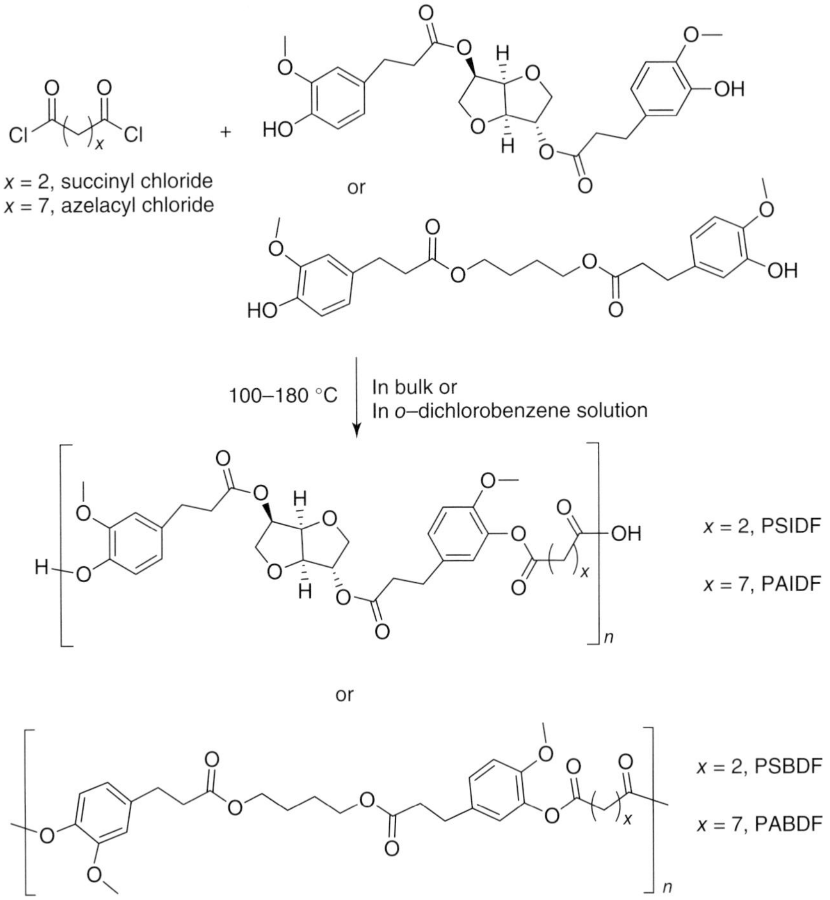

$x = 2$, succinyl chloride
$x = 7$, azelacyl chloride

or

100–180 °C
In bulk or
In *o*–dichlorobenzene solution

$x = 2$, PSIDF
$x = 7$, PAIDF

or

$x = 2$, PSBDF
$x = 7$, PABDF

Scheme 13.22 Preparation of aliphatic–aromatic polyesters containing ferulic acid moieties [165].

13.2.6 Diols and Their Polymers

Advances in biotechnology have helped in preparing several diols from renewable resources, which were traditionally obtained only from petroleum feedstocks. Bio-based 1,2-ethanediol (or ethylene glycol, EG) is prepared by chemical catalysis starting from ethanol whose biosynthesis has been developed through fermentation of various carbon sources. Novel fermentation processes are under development focusing on utilization of novel raw materials and microbial strains or development of integrated fermentation with ethanol separation [114]. Besides the ethanol route, also other possibilities for bio-EG production have been developed, such as via hydrogenolysis (i.e., covalent bond cleavage by hydrogen) of xylitol (originating from xylose), sorbitol (originating from glucose), or glycerol

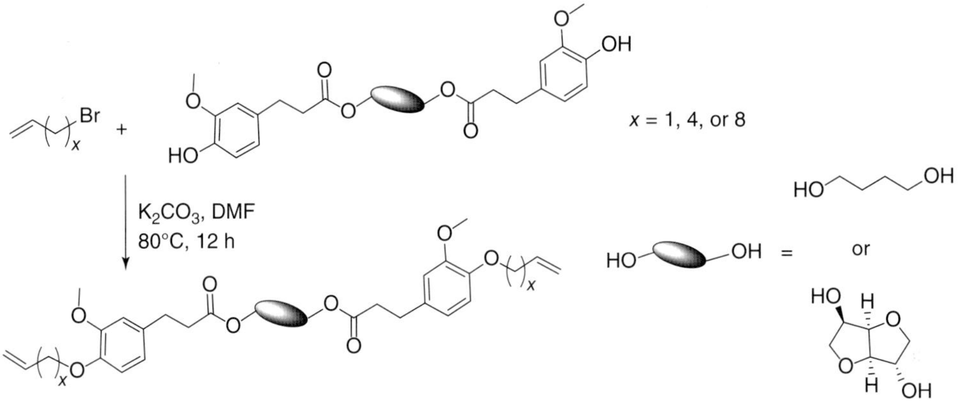

Scheme 13.23 Synthesis of ferulic acid-derived α,ω-diene monomers [167].

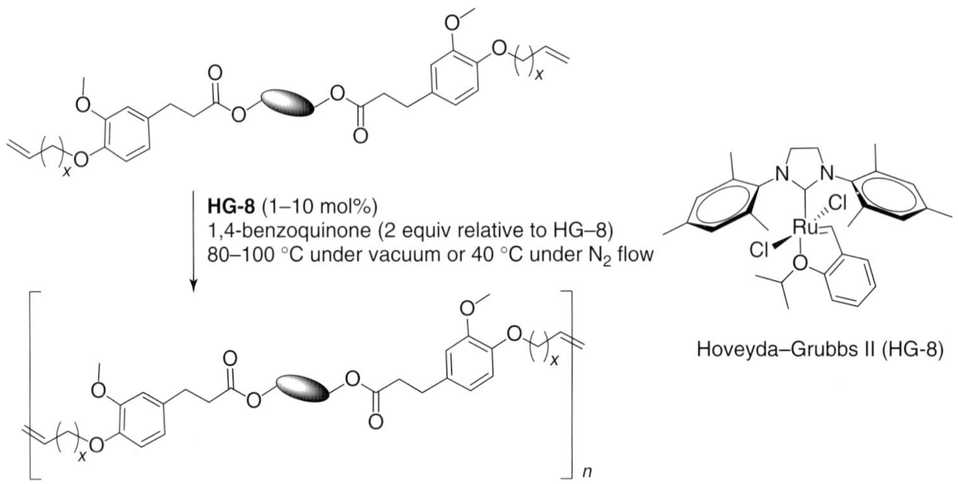

Scheme 13.24 Ferulic acid-derived poly(ester-alkenamer)s [167].

(a by-product of biodiesel production) [121]. EG is currently the object of a large interest from several companies because it is the monomer for poly(ethylene terephthalate) (PET), which is one of the largest volume industrial polymers (annual global production >50 million tons [12]), used for fibers and bottles for soft drinks and water. Nowadays considerable effort is being focused on the development of the so-called "plant bottle," that is, a PET bottle produced partially or entirely from renewable biomass. The simplest way to produce bio-PET is to use bio-EG by the existing technology to generate about 30% plant-based PET. This strategy has been followed by, for example, Coca Cola and Danone. Pepsi Cola, instead, used the strategy of producing 100% plant-based PET, where also terephthalic acid is obtained in sustainable way from bio-*p*-xylene [12].

1,3-propanediol (PDO), successfully prepared in aerobic bioprocesses starting from glucose from starch, has a market volume of 125 kt per year whose large part is bio-based [121]. These achievements have attracted companies to produce bio-poly(trimethylene terephthalate) (PTT), an original polyester bearing an odd number of methylene groups, which is characterized by good strength, stiffness, toughness, and heat resistance, for applications in packaging, carpets, textiles, films, and automotive parts. 1,4-butanediol (BDO) is currently produced in large volumes from petrochemical raw materials, but various companies and consortia are working on the development and upscaling of its bioproduction by direct fermentation of sugars to BDO (now at commercial scale), fermentation to SA and reduction to BDO, fermentation of ethanol, and starting from levulinic acid via a chemical route [121]. Examples of bio-based polymers derived from PDO or other diols have already been described in previous sections. Examples of new materials include a series of bio-based elastomeric copolyesters synthesized from PDO, sebacic acid, and itaconic acid. SA was also introduced to tailor the flexibility of these macromolecules, and zinc dimethacrylate (ZDA) was incorporated to improve mechanical strength. These composites show excellent shape memory properties [168].

13.3 Glycerol

13.3.1 Introduction

Glycerol is an abundant renewable resource, by-product of soap and biodiesel industries. Because of the biodiesel and fatty alcohol boom, about 2 million tons of renewable glycerol are produced annually [169], becoming a bio-based raw material of ample supply and low cost. It was also identified as one the 12 most interesting building blocks that can be produced from sugars via biological or chemical conversions and that can be converted into a number of high value chemicals or materials [119]. Indeed, although characterized by a very simple chemical structure, from a synthetic perspective, glycerol possesses a wealth of opportunities for selective chemical transformations, such as oxidation, dehydration, hydrogenolysis, selective protection, and esterification, providing an array of value-added small-molecule building blocks. Glycerol is also interesting as a nontoxic, edible, biodegradable compound, providing environmental benefits to the new platform products.

Last but not least, glycerol and glycerol derivatives can provide polymers which are attractive for their diversity in compositions and architectures, which can find opportunity of applications in different fields, including medicine. Various glycerol-based polymer structures, from linear to dendritic, are described in the literature [170, 171], and only some examples of the most recent developments are briefly described here.

13.3.2
Linear 1,3-Linked Glycerol Polymers

The main linear polymers containing 1,3-linked glycerol units, which are characterized by the presence of a protected or free secondary hydroxyl group, belong to the classes of polyethers, polyesters, and polycarbonates.

Oligomers of glycerol, also called polyglycerols (PGs), are obtained by homogeneously or heterogeneously catalyzed etherification of glycerol [172, 173]. They are highly biocompatible and thermally stable and therefore used as plasticizers. The secondary hydroxyl groups present along their polymer backbone can be esterified with carboxylic acid to produce a variety of amphiphilic polyglycerol esters (PGEs), characterized by a good hydrophilic–lipophilic balance. PG find numerous applications as antifogging and antistatic agents, lubricants, emulsifiers, and stabilizers. PGEs of fatty acids are used in food, cosmetics, and toiletries. Polyglycerol polyricinoleate, for example, is a powerful water-in-oil emulsifier, used in the food industry as an emulsifier for making chocolate.

PGs are also excellent polyols to prepare polyurethanes. Hu and Li describe the conversion of crude glycerol from biodiesel production to polyols with low contents of residual free fatty acids and fatty acid methyl esters. They were used to produce waterborne polyurethane dispersions that find applications in coating, showing excellent adhesion to steel panel surfaces and high pencil hardness [174].

Polyesters are obtained by conventional polycondensation of glycerol with a diacid, such as sebacic acid, adipic acid, and other diacids with different chain lengths, exploiting the higher reactivity of the primary hydroxyl groups compared with that of the secondary moiety. In order to prepare linear or, at least, controlled branched structures, the molar ratio of glycerol and diacid is critical, and protection–deprotection steps can be required. An example of a notable class of glycerol-based are poly(glycerol sebacates) (PGS), prepared by starting from mixtures of glycerol and sebacic acid at 120 °C in two steps, namely, a prepolycondensation for 24 h, followed by a cross-linking step at low pressure for 48 h (Scheme 13.25) [175].

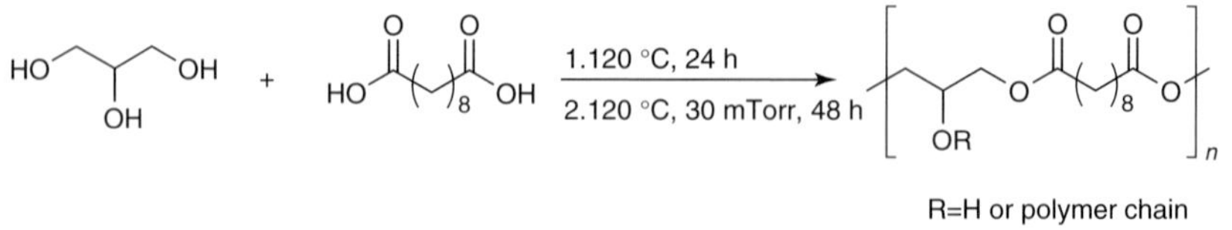

HO OH OH + HO (8) OH → 1.120 °C, 24 h; 2.120 °C, 30 mTorr, 48 h → [O (8) O OR]$_n$

R=H or polymer chain

Scheme 13.25 Synthetic pathway to prepare PGS polymers.

Since PGS are biodegradable, they are increasingly used for biomedical applications [176]. At 37 °C they are amorphous, characterized by elastomeric properties, due to cross-linking and hydrogen bonding interactions between the OH groups. Their mechanical properties can be tailored by changing the temperature of the second polymerization step (curing temperature) and the monomer molar ratio. By incorporating acrylic moieties into their structure, it is possible to have an additional control on the cross-link density and, hence, on the mechanical properties of the final materials. It is also notable that PGS also possess a shape memory behavior [177]. These materials have attractive properties for applications focused on soft tissue engineering, such as cardiac muscle, vascular tissue engineering, and cartilage, apart from drug delivery and tissue adhesive roles. Recently, PGS prepared with a three-step strategy and characterized by different cross-link density were also investigated for amphiphilic conetworks applications (APCNs), which require materials comprising hydrophilic and hydrophobic segments and sharing several attributes of hydrogels and surfactants [178].

Better controlled architectures can be obtained by enzyme-catalyzed condensation polymerization. Lipases can provide a good control over branching, avoiding cross-linking reactions, due to the steric constraints at the enzyme active sites [179]. By comparing the performance of a lipase (N435) and a conventional chemical catalyst (dibutyltin oxide, DBTO) during the bulk copolymerization of oleic diacid and glycerol, according to Scheme 13.26, it is notable that only N435 catalysis gave linear polymers with M_n varying from 4400 to 9000, whereas DBTO catalysis provide cross-linking systems, with gel formation and M_n values of only 1750. As already described, ketal–ester oligomers were obtained by acid-catalyzed condensation between levulinic acid and glycerol [158] to produce levulinic acid–glycerol oligomers. This work follows the results obtained by Mullen *et al.* [180] who studied the selective ketalization of levulinic acid esters, providing an industrial pathway for developing bio-based building blocks for the synthesis of novel polymers, functional oligomers, polymer additives, and solvents. Ketals showed similar hydrostability than esters and superior hydrostability relative to an anhydride in the absence of strong acid. The preparation of 1,3-linked glycerol carbonate exploits the ring-opening polymerization of six-member cyclic carbonates derived from glycerol. Scheme 13.27 shows a first example of the synthesis of poly(1,3-glycerol carbonate), where the monomer (5-benzyloxy-trimethylene carbonate) was synthesized from glycerol [181]. The same approach has been recently applied to prepare new poly(lactic acid-*co*-glycerol monostearate) copolymers, which in blend with poly(lactic

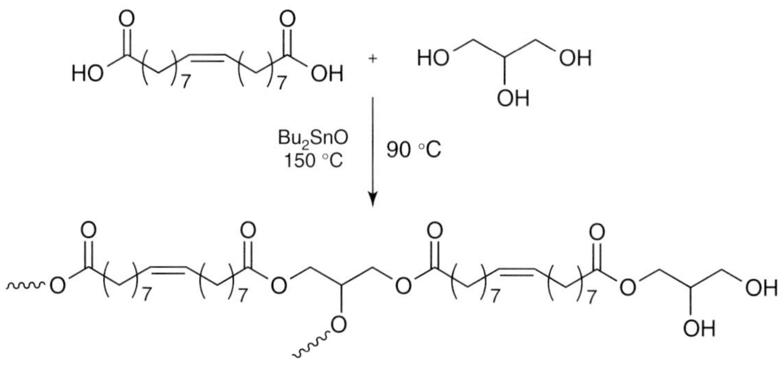

Scheme 13.26 Copolymerization of oleic diacid with glycerol [179].

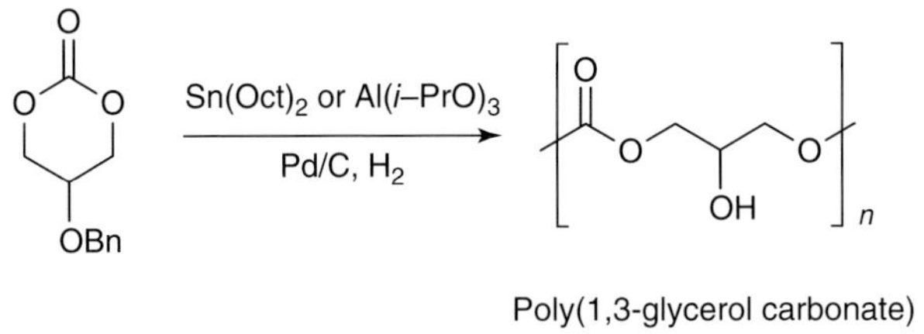

$Sn(Oct)_2$ or $Al(i–PrO)_3$

Pd/C, H_2

OBn

OH

Poly(1,3-glycerol carbonate)

Scheme 13.27 Synthesis of poly(1,3-glycerol carbonate) [181].

acid-*co*-glycolic acid) (PLGA) can tune the hydrophobicity of PLGA-based electrospun fibers for drug delivery applications [182].

13.4 Final Considerations

In the last decade notable research efforts have been spent to valorize biomass and biowaste into biochemical platforms. A wide range of aliphatic and aromatic building blocks (diacids, hydroxyacids, diols, etc.) can be produced, with a very large spectrum of properties and characteristics. Particularly, the furan-based monomers, due to their dienic character, can be exploited in the click chemistry polymerizations that are currently deeply investigated to develop self-healing materials or can be used in electrochemical polymerization to prepare conjugated polymers useful in the optoelectronic applications. Moreover, the furanic dicarboxylic derivative (FDCA) results to be a convincing alternative to terephthalic and isophthalic acid, largely used in the preparation of commodity plastics.

Furthermore, the obtained macromolecules (polyesters, polyamides, polyurethanes, copolymers, etc.) vary from flexible structures, with low T_g values (−30 °C for PBS), to more rigid architectures, mainly thanks to the presence of aromatic rings (T_g = 73 °C for PHFA derived from vanillin and T_g = 90 °C for poly(ethylene furandicarboxylate)). High T_g values can also be obtained thanks to the presence of hydrogen bonds that link polymeric chains, such as in PHLA (T_g = 120 °C). Besides thermoplastics, also bio-based elastomeric materials can be prepared by polyfunctional monomers, such as glycerol, that realize cross-linked systems. Therefore, biomass represents a sustainable feedstock of a large variety of valuable building blocks usable to expand the operative temperature range, the mechanical performance, and many other important properties of green polymers.

References

1. Gandini, A. (2011) in *Biopolymers-New Materials for Sustainable Films and Coatings* (ed. D. Plackett), John & Wiley Sons, Ltd., Chichester, pp. 179–209.
2. Gandini, A. (2010) *Polym. Chem.*, **1**, 245–251.
3. Gandini, A. (2013) *Prog. Polym. Sci.*, **38**, 1–29.

4. Gandini, A. and Belgacem, M.N. (2014) in *Handbook of Thermoset Plastics*, 3rd edn (eds H. Dodiuk and S.H. Goodman), Elsevier, Amsterdam, pp. 93–110.
5. Gandini, A. (2011) *Green Chem.*, **13**, 1061–1083.
6. González-Tejera, M.J., Sánchez de la Blanca, E., and Carrillo, I. (2008) *Synth. Met.*, **158**, 165–189.
7. Amarasekara, A.S. (2011) in *Renewable Polymers: Synthesis, Processing, and Technology* (ed. V. Mittal), John Wiley & Sons, Inc., Hoboken, NJ, pp. 381–428.
8. Gandini, A. and Belgacem, M.N. (1997) *Prog. Polym. Sci.*, **22**, 1203–1379.
9. Metkar, P.S., Till, E.J., Corbin, D.R., Pereira, C.J., Hutchenson, K.W., and Sengupta, S.K. (2015) *Green Chem.*, **17**, 1453–1466
10. Mamman, A.S., Lee, J.-M., Kim, Y.-C., Hwang, I.T., Park, N.-J., Hwang, Y.K., Chang, J.-S., and Hwang, J.-S. (2008) *Biofuels, Bioprod. Biorefin.*, **2**, 438–454.
11. Lange, J.P., van der Heide, E., van Buijtenen, J., and Price, R. (2012) *ChemSusChem*, **5**, 150–166.
12. Sheldon, R.A. (2014) *Green Chem.*, **16**, 950–963.
13. Triebl, C., Nikolakis, V., and Ierapetritou, M. (2013) *Comput. Chem. Eng.*, **52**, 26–34.
14. Pan, T., Deng, J., Xu, Q., Zuo, Y., Guo, Q.-X., and Fu, Y. (2013) *ChemSusChem*, **6**, 47–50.
15. van Putten, R.-J., van der Waal, J.C., de Jong, E., Rasrendra, C.B., Heeres, H.J., and de Vries, J.G. (2013) *Chem. Rev.*, **113**, 1499–1597.
16. Wang, S., Zhang, Z., Liu, B., and Li, J. (2014) *Ind. Eng. Chem. Res.*, **53**, 5820–5827.
17. Bozell, J.J. and Petersen, G.R. (2010) *Green Chem.*, **12**, 539–554.
18. Lichtenthaler, F.W. and Peters, S. (2004) *C.R. Chim.*, **7**, 65–90.
19. Khrouf, A., Boufi, S., El Gharbi, R., Belgacem, N.M., and Gandini, A. (1996) *Polym. Bull.*, **37**, 589–596.
20. Gandini, A. (2011) in *Green Polymerization Methods* (eds R.T. Mathers and M.A.R. Meier), Wiley-VCH Verlag GmbH, Weinheim, pp. 29–56.
21. Coutterez, C., Goussé, C., Genheim, R., Waig Fang, S., and Gandini, A. (2001) *ACS Symp. Ser.*, **784**, 98–109.
22. Moore, J.A. and Kelly, J.E. (1978) *Macromolecules*, **11**, 568–573.
23. Moore, J.A. and Kelly, J.E. (1979) *Polymer*, **20**, 627–628.
24. Storbeck, R. and Ballauff, M. (1993) *Polymer*, **34**, 5003–5006.
25. Khrouf, A., Abid, M., Boufi, S., El Gharbi, R., and Gandini, A. (1998) *Macromol. Chem. Phys.*, **199**, 2755–2765.
26. Gharbi, S., Andreolety, J.-P., and Gandini, A. (2000) *Eur. Polym. J.*, **36**, 463–472.
27. Gandini, A., Silvestre, A.J.D., Pascoal Neto, C., Sousa, A.F., and Gomes, M. (2009) *J. Polym. Sci., Part A: Polym. Chem.*, **47**, 295–298.
28. Gomes, M., Gandini, A., Silvestre, A.J.D., and Reis, B. (2011) *Polym. Chem.*, **49**, 3759–3768.
29. Abid, M., Kamoun, W., El Gharbi, R., and Fradet, A. (2008) *Macromol. Mater. Eng.*, **293**, 39–44.
30. Abid, M., Abid, S., and El Gharbi, R. (2012) *J. Macromol. Sci. Part A*, **49**, 758–763.
31. Bougarech, A., Abid, M., Gouanvé, F., Espuche, E., Abid, S., El Gharbi, R., and Fleury, E. (2013) *Polymer*, **54**, 5482–5489.
32. Bougarech, A., Abid, M., DaCruz-Boisson, F., Abid, S., El Gharbi, R., and Fleury, E. (2014) *Eur. Polym. J.*, **58**, 207–217.
33. Knoop, R.J.I., Vogelzang, W., van Haveren, J., and van Es, D.S. (2013) *J. Polym. Sci., Part A: Polym. Chem.*, **51**, 4191–4199.
34. Burgess, S.K., Leisen, J.E., Kraftschik, B.E., Mubarak, C.R., Kriegel, R.M., and Koros, W.J. (2014) *Macromolecules*, **47**, 1383–1391.
35. Matsuda, K., Matsuda, H., Horie, H., and Komuro, T. (2007) Polymer compound and method of synthesizing the same. PCT Patent WO2007/052847 A1, filed Nov. 6, 2006 and issued May 10, 2007.
36. Kato, S. and Kasai, A. (2008) Method for producing polyester resin including furan structure. Japanese Patent

JP2008/291244, filed Apr. 24, 2007, issued Dec. 04, 2008.

37. De Jong, E., Dam, M.A., Sipos, L., and Gruter, G.-J.M. (2012) in *Biobased Monomers, Polymers, and Materials*, ACS Symposium Series (eds P.B. Smith and R.A. Gross), American Chemical Society, Washington, DC, pp. 1–13.
38. Zhu, J., Cai, J., Xie, W., Chen, P.-H., Gazzano, M., and Scandola, M. (2013) *Macromolecules*, **46**, 796–804.
39. Jiang, M., Liu, Q., Zhang, Q., Ye, C., and Zhou, G. (2012) *J. Polym. Sci., Part A: Polym. Chem.*, **50**, 1026–1036.
40. Gruter, G.-J.M., Sipos, L., and Dam, M.A. (2012) *Comb. Chem. High Throughput Screening*, **15**, 180–188.
41. Gubbels, E., Jasinska-Walc, L., Noordover, B.A.J., and Koning, C.E. (2013) *Eur. Polym. J.*, **49**, 3188–3198.
42. Gubbels, E., Jasinska-Walc, L., and Koning, C.E. (2013) *J. Polym. Sci., Part A: Polym. Chem.*, **51**, 890–898.
43. Amarasekara, A.S., Razzaq, A., and Bonham, P. (2013) *ISRN Polym. Sci.*, 2013, ID 645169.
44. Yu, Z., Zhou, J., Cao, F., Wen, B., Zhu, X., and Wei, P. (2013) *J. Appl. Polym. Sci.*, **130**, 1415–1420.
45. Sousa, A.F., Matos, M., Freire, C.S.R., Silvestre, A.J.D., and Coelho, J.F.J. (2013) *Polymer*, **54**, 513–519.
46. Wu, L., Mincheva, R., Xu, Y., Raquez, J.-M., and Dubois, P. (2012) *Biomacromolecules*, **13**, 2973–2981.
47. Wu, B., Xu, Y., Bu, Z., Wu, L., Li, B.-G., and Dubois, P. (2014) *Polymer*, **55**, 3648–3655.
48. Matos, M., Sousa, A.F., Fonseca, A.C., Freire, C.S.R., Coelho, J.F.J., and Silvestre, A.J.D. (2014) *Macromol. Chem. Phys.*, **215**, 2175–2184.
49. Ma, J., Pang, Y., Wang, M., Xu, J., Ma, H., and Nie, X. (2012) *J. Mater. Chem.*, **22**, 3457–3461.
50. Zeng, C., Seino, H., Ren, J., Hatanaka, K., and Yoshie, N. (2013) *Polymer*, **54**, 5351–5357.
51. Zeng, C., Seino, H., Ren, J., Hatanaka, K., and Yoshie, N. (2013) *Macromolecules*, **46**, 1794–1802.
52. Ikezaki, T., Matsuoka, R., Hatanaka, K., and Yoshie, N. (2013) *J. Polym. Sci., Part A: Polym. Chem.*, **52**, 216–222.
53. Jiang, Y., Woortman, A.J.J., Alberda van Ekenstein, G.O.R., Petrović, D.M., and Loos, K. (2014) *Biomacromolecules*, **15**, 2482–2493.
54. Codou, A., Guigo, N., van Berkel, J., de Jong, E., and Sbirrazzuoli, N. (2014) *Macromol. Chem. Phys.*, **215**, 2065–2074.
55. Papageorgiou, G.Z., Tsanaktsis, V., and Bikiaris, D.N. (2014) *Phys. Chem. Chem. Phys.*, **16**, 7946–7958.
56. Papageorgiou, G.Z., Tsanaktsis, V., Papageorgiou, D.G., Exarhopoulos, S., Papageorgiou, M., and Bikiaris, D.N. (2014) *Polymer*, **55**, 3846–3858.
57. Papageorgiou, G.Z., Tsanaktsis, V., Papageorgiou, D.G., Chrissafis, K., Exarhopoulos, S., and Bikiaris, D.N. (2014) *Eur. Polym. J.*, **67**, 383–393
58. Burgess, S.K., Mikkilineni, D.S., Yu, D.B., Kim, D.J., Mubarak, C.R., Kriegel, R.M., and Koros, W.J. (2014) *Polymer*, **55**, 6861–6869.
59. Burgess, S.K., Mikkilineni, D.S., Yu, D.B., Kim, D.J., Mubarak, C.R., Kriegel, R.M., and Koros, W.J. (2014) *Polymer*, **55**, 6870–6882.
60. Burgess, S.K., Karvan, O., Johnson, J.R., Kriegel, R.M., and Koros, W.J. (2014) *Polymer*, **55**, 4748–4756.
61. Mitiakoudis, A., Gandini, A., and Cheradame, H. (1985) *Polym. Commun.*, **26**, 246–249.
62. Mitiakoudis, A. and Gandini, A. (1991) *Macromolecules*, **24**, 830–835.
63. Abid, S., El Gharbi, R., and Gandini, A. (2004) *Polymer*, **45**, 5793–5801.
64. Gharbi, S. and Gandini, A. (1999) *Acta Polym.*, **50**, 293–297.
65. Gharbi, S., Afli, A., El Gharbi, R., and Gandini, A. (2001) *Polym. Int.*, **50**, 509–514.
66. Abid, M., El Gharbi, R., and Gandini, A. (2000) *Polymer*, **41**, 3555–3560.
67. Grosshardt, O., Fehrenbacher, U., Kowollik, K., Tübke, B., Dingenouts, N., and Wilhelm, M. (2009) *Chem.-Ing.-Tech.*, **81**, 1829–1835.
68. Chan, J., Nederberg, F., Rajagopalan, B., Williams, S.R., and Cobb M.W. (2013) Furan based polyamides. PCT Patent WO2013/149180 A1, filed Mar. 29, 2013 and issued Oct. 03, 2013.

69. Jeol, S. (2013) Nouveau polyamide, procede de preparation et utilisations. WO Patent 2013/007585 A1, filed Jul. 05, 2012 and issued Jan. 17, 2013.
70. Wilsens, C.H.R.M., Deshmukh, Y.S., Noordover, B.A.J., and Rastogi, S. (2014) *Macromolecules*, **47**, 6196–6206.
71. Triki, R., Abid, M., Tessier, M., Abid, S., El Gharbi, R., and Fradet, A. (2013) *Eur. Polym. J.*, **49**, 1852–1860.
72. Belgacem, M.N., Quillerou, J., Gandini, A., Rivero, J., and Roux, G. (1989) *Eur. Polym. J.*, **25**, 1125–1130.
73. Belgacem, M.N., Quillerou, J., and Gandini, A. (1993) *Eur. Polym. J.*, **29**, 1217–1224.
74. Boufi, S., Belgacem, M.N., Quillerou, J., and Gandini, A. (1993) *Macromolecules*, **26**, 6706–6717.
75. Boufi, S., Gandini, A., and Belgacem, M.N. (1995) *Polymer*, **36**, 1689–1696.
76. Canadell, J., Fischer, H., de With, G., and van Benthem, R.A.T.M. (2010) *J. Polym. Sci., Part A: Polym. Chem.*, **48**, 3456–3467.
77. Gandini, A., Coelho, D., Gomes, M., Reis, B., and Silvestre, A. (2009) *J. Mater. Chem.*, **19**, 8656–8664.
78. Imai, Y., Itoh, H., Naka, K., and Chujo, Y. (2000) *Macromolecules*, **33**, 4343–4346.
79. Goiti, E., Huglin, M.B., and Rego, J.M. (2004) *Eur. Polym. J.*, **40**, 219–226.
80. Mcelhanon, J.R., Russick, E.M., Wheeler, D.R., Loy, D.A., and Aubert, J.H. (2002) *J. Appl. Polym. Sci.*, **85**, 1496–1502.
81. Adachi, K., Achimuthu, A.K., and Chujo, Y. (2004) *Macromolecules*, **37**, 9793–9797.
82. Gotsmann, B., Duerig, U., Frommer, J., and Hawker, C.J. (2006) *Adv. Funct. Mater.*, **16**, 1499–1505.
83. Paulöhrl, T., Inglis, A.J., and Barner-Kowollik, C. (2010) *Adv. Mater.*, **22**, 2788–2791.
84. Gevrek, T.N., Ozdeslik, R.N., Sahin, G.S., Yesilbag, G., Mutlu, S., and Sanyal, A. (2012) *Macromol. Chem. Phys.*, **213**, 166–172.
85. Ishida, K. and Yoshie, N. (2008) *Macromol. Biosci.*, **8**, 916–922.
86. Wu, D.Y., Meure, S., and Solomon, D. (2008) *Prog. Polym. Sci.*, **33**, 479–522.
87. Chen, X., Dam, M.A., Ono, K., Mal, A., Shen, H., Nutt, S.R., Sheran, K., and Wudl, F. (2002) *Science*, **295**, 1698–1702.
88. Chen, X., Wudl, F., Mal, A.K., Shen, H., and Nutt, S.R. (2003) *Macromolecules*, **36**, 1802–1807.
89. Ishida, K., Furuhashi, Y., and Yoshie, N. (2014) *Polym. Degrad. Stab.*, **110**, 149–155.
90. Vilela, C., Silvestre, A.J.D., and Gandini, A. (2013) *J. Polym. Sci., Part A: Polym. Chem.*, **51**, 2260–2270.
91. Vilela, C., Cruciani, L., Silvestre, A.J.D., and Gandini, A. (2012) *RSC Adv.*, **2**, 2966–2974.
92. Liu, Y.-L. and Chen, Y.-W. (2007) *Macromol. Chem. Phys.*, **208**, 224–232.
93. Yu, S., Zhang, R., Wu, Q., Chen, T., and Sun, P. (2013) *Adv. Mater.*, **25**, 4912–4917.
94. Du, P., Wu, M., Liu, X., Zheng, Z., Wang, X., Sun, P., Joncheray, T., and Zhang, Y. (2014) *New J. Chem.*, **38**, 770–776.
95. Ishida, K. and Yoshie, N. (2008) *Macromolecules*, **41**, 4753–4757.
96. Ishida, K., Weibel, V., and Yoshie, N. (2011) *Polymer*, **52**, 2877–2882.
97. Varganici, C.-D., Ursache, O., Gaina, C., Gaina, V., Rosu, D., and Simionescu, B.C. (2013) *Ind. Eng. Chem. Res.*, **52**, 5287–5295.
98. Gaina, V., Ursache, O., Gaina, C., and Buruiana, E. (2012) *Des. Monomers Polym.*, **15**, 63–73.
99. Rivero, G., Nguyen, L.-T.T., Hillewaere, X.K.D., and Du Prez, F.E. (2014) *Macromolecules*, **47**, 2010–2018.
100. Kavitha, A.A. and Singha, N.K. (2010) *Macromolecules*, **43**, 3193–3205.
101. Toncelli, C., De Reus, D.C., Picchioni, F., and Broekhuis, A.A. (2012) *Macromol. Chem. Phys.*, **213**, 157–165.
102. Araya-Hermosilla, R., Broekhuis, A.A., and Picchioni, F. (2014) *Eur. Polym. J.*, **50**, 127–134.
103. Tian, Q., Rong, M.Z., Zhang, M.Q., and Yuan, Y.C. (2010) *Polym. Int.*, **59**, 1339–1345.
104. Tian, Q., Yuan, Y.C., Rong, M.Z., and Zhang, M.Q. (2009) *J. Mater. Chem.*, **19**, 1289–1296.

105. Pratama, P.A., Peterson, A.M., and Palmese, G.R. (2012) *Macromol. Chem. Phys.*, **213**, 173–181.
106. Peterson, A., Jensen, R.E., and Palmese, G.R. (2010) *ACS Appl. Mater. Interfaces*, **2**, 1141–1149.
107. Ax, J. and Wenz, G. (2012) *Macromol. Chem. Phys.*, **213**, 182–186.
108. García-Astrain, C., Gandini, A., Coelho, D., Mondragon, I., Retegi, A., Eceiza, A., Corcuera, M.A., and Gabilondo, N. (2013) *Eur. Polym. J.*, **49**, 3998–4007.
109. García-Astrain, C., Gandini, A., Peña, C., Algar, I., Eceiza, A., Corcuera, M., and Gabilondo, N. (2014) *RSC Adv.*, **4**, 35578–35587.
110. Gidron, O. and Bendikov, M. (2014) *Angew. Chem. Int. Ed.*, **53**, 2546–2555.
111. Korshin, E.E., Leitus, G.M., and Bendikov, M. (2014) *Org. Biomol. Chem.*, **12**, 6661–6671.
112. Sheberla, D., Patra, S., Wijsboom, Y.H., Sharma, S., Sheynin, Y., Haj-Yahia, A.-E., Barak, A.H., Gidron, O., and Bendikov, M. (2015) *Chem. Sci.*, **6**, 360–371.
113. Fallon, T., Willis, A.C., Rae, A.D., Paddon-Row, M.N., and Sherburn, M.S. (2012) *Chem. Sci.*, **3**, 2133–2137.
114. Koutinas, A.A., Vlysidis, A., Pleissner, D., Kopsahelis, N., Garcia, I.L., Kookos, I.K., Papanikolaou, S., Kwan, T.H., and Lin, K.S.K. (2014) *Chem. Soc. Rev.*, **43**, 2587–2627.
115. Bechthold, I., Bretz, K., Kabasci, S., Kopitzky, R., and Springer, A. (2008) *Chem. Eng. Technol.*, **31** (5), 647–654.
116. Reddy, M.M., Vivekanandhan, S., Misra, M., Bhatia, S.K., and Mohanty, A.K. (2013) *Prog. Polym. Sci.*, **38**, 1653–1689.
117. Deepa, A.K. and Dhepe, P.L. (2014) *ACS Catal.*, **5**, 365–379
118. Laurichesse, S. and Avérous, L. (2014) *Prog. Polym. Sci.*, **39**, 1266–1290.
119. Werpy, T. and Peterson, G. (2004) *Top Value Added Chemicals from Biomass*, US Department of Energy.
120. Cukalovic, A. and Stevens, C.V. (2008) *Biofuels, Bioprod. Biorefin.*, **2**, 505–529.
121. Harmsen, P.F.H., Hackmann, M.M., and Bos, H.L. (2014) *Biofuels, Bioprod. Biorefin.*, **8**, 306–324.
122. Taylor, P. (2010) Biosuccinic Acid Ready to Take Off?, http://www.rsc.org/chemistryworld/News/2010/january/21011003.asp (accessed 30 September 2015).
123. Tan, J.P., Jahim, J.M., Wu, T.Y., Harun, S., Kim, B.H., and Mohammad, A.W. (2014) *Ind. Eng. Chem. Res.*, **53**, 16123–16134.
124. Leung, C.C.J., Cheung, A.S.Y., Zhang, A.Y.-Z., Lam, K.F., and Lin, C.S.K. (2012) *Biochem. Eng. J.*, **65**, 10–15.
125. Pinazo, J.M., Domine, M.E., Parvulescu, V., and Petru, F. (2014) *Catal. Today*, **239**, 17–24.
126. Pttmcc http://www.pttmcc.com (accessed 30 September 2015).
127. Xu, J. and Guo, B.-H. (2010) *Biotechnol. J.*, **5**, 1149–1163.
128. Fujimaki, T. (1998) *Polym. Degrad. Stab.*, **59**, 209–214.
129. Sasanuma, Y., Nonaka, Y., and Yamaguchi, Y. (2015) *Polymer*, **56**, 327–339.
130. Mincheva, R., Delangre, A., Raquez, J.-M., Narayan, R., and Dubois, P. (2013) *Biomacromolecules*, **14**, 890–899.
131. Díaz, A., Franco, L., and Puiggalí, J. (2014) *Thermochim. Acta*, **575**, 45–54.
132. Arandia, I., Mugica, A., Zubitur, M., Arbe, A., Liu, G., Wang, D., Mincheva, R., Dubois, P., and Müller, A.J. (2015) *Macromolecules*, **48**, 43–57.
133. Luo, S.L., Li, F.X., Yu, J.Y., and Cao, A.M. (2010) *J. Appl. Polym. Sci.*, **115**, 2203–2211.
134. Sun, Y., Wu, L., Bu, Z., Li, B.-G., Li, N., and Dai, J. (2014) *Ind. Eng. Chem. Res.*, **53**, 10380–10386.
135. Jacquel, N., Saint-Loup, R., Pascault, J.-P., Rousseau, A., and Fenouillot, F. (2015) *Polymer*, **59**, 234–242.
136. Zakharova, E., Lavilla, C., Alla, A., Martínez de Ilarduya, A., and Munoz-Guerra, S. (2014) *Eur. Polym. J.*, **61**, 263–273.
137. Jiang, Y., van Ekenstein, G.O.R.A., Woortman, A.J.J., and Loos, K. (2014) *Macromol. Chem. Phys.*, **215**, 2185–2197.
138. Noordover, B.A.J., van Staalduinen, V.G., Duchateau, R., Koning, C.E., van Benthem, R.A.T.M., Mak, M., Heise, A., Frissen, A.E., and

van Haveren, J. (2006) *Biomacromolecules*, **7**, 3406–3416.

139. van Haveren, J., Oostveen, E.A., Micciche, F., Noordover, B.A.J., Koning, C.E., van Benthem, R.A.T.M., Frissen, A.E., and Weijnen, J.G.J. (2007) *J. Coat. Technol. Res.*, **4**, 177–186.
140. Noordover, B.A.J., Heise, A., Malanowksi, P., Senatore, D., Mak, M., Molhoek, L., Duchateau, R., Koning, C.E., and van Benthem, R.A.T.M. (2009) *Prog. Org. Coat.*, **65**, 187–196.
141. Chatti, S., Weidner, S.M., Fildier, A., and Kricheldorf, H.R. (2013) *J. Polym. Sci., Part A: Polym. Chem.*, **51**, 2464–2471.
142. Abe, H. and Doi, Y. (2005) *Macromol. Rapid Commun.*, **25**, 1303–1308.
143. Díaz, A., Katsarava, R., and Puiggalí, J. (2014) *Int. J. Mol. Sci.*, **15**, 7064–7123.
144. Jiang, M., Chen, G., Lu, P., and Dong, J. (2014) *J. Appl. Polym. Sci.*, **131**. doi: 10.1002/app.39807.
145. Sonnenschein, M.F., Guillaudeu, S.J., Landes, B.G., and Wendt, B.L. (2010) *Polymer*, **51**, 3685–3692.
146. Li, S.-L., Zeng, J.-B., Wu, F., Yang, Y., and Wang, Y.-Z. (2014) *Ind. Eng. Chem. Res.*, **53**, 1404–1414.
147. Rennovia http://www.rennovia.com/wp-content/uploads/2014/12/Rennovia-produces-RENNLON-28TM-29-nylon-a-100-bio-based-nylon-6-polymer-Press-Release-10.1.2013.pdf (accessed 30 September 2015).
148. BASF http://www.plasticsportal.net/wa/plasticsEU~it_IT/portal/show/content/products/biodegradable_plastics/biodegradable_polymers (accessed 26 November 2015)
149. Xu, W., Zhou, L., Sun, W., Zhang, J., and Tu, W. (2015) *J. Appl. Polym. Sci.* **132**, doi: 10.1002/app.41246.
150. Zhang, T., Howell, B.A., Dumitrascu, A., and Martin, S.J. (2014) *Polymer*, **55**, 5065–5072.
151. Zhang, T., Howell, B.A., and Smith, P.B. (2014) *J. Therm. Anal. Calorim.*, **116**, 1369–1378.
152. Bozell, J.J., Moens, L., Elliott, D.C., Wang, Y., Neuenscwander, G.G., Fitzpatrick, S.W. *et al.* (2000) *Resour. Conserv. Recycl.*, **28**, 227–239.
153. Rackemann, D.W. and Doherty, W.O. (2011) *Biofuels, Bioprod. Biorefin.*, **5**, 198–214.
154. Keenan, T.M., Nakas, J.P., and Tanenbaum, S.W. (2006) *J. Ind. Microbiol. Biotechnol.*, **33**, 616–626.
155. Wu, L., Zhang, Y., Fan, H., Bu, Z., and Li, B.G. (2008) *J. Polym. Environ.*, **16**, 68–73.
156. Zhang, Y., Wu, L., and Li, B.-G. (2010) *J. Appl. Polym. Sci.*, **117**, 3315–3321.
157. Leibig, C., Mullen, B., Mullen, T., Rieth, L., and Badarinarayana, V. (2010) *Polym. Prepr.*, **51**, 763–764.
158. Amarasekara, A.S. and Hawkins, S.A. (2011) *Eur. Polym. J.*, **47**, 2451–2457.
159. Amarasekara, A.S., Wiredu, B., and Razzaq, A. (2012) *Green Chem.*, **14**, 2395–2397.
160. Stanzione, J.F. III,, Sadler, J.M., La Scala, J.L., Reno, K.H., and Wool, R.P. (2012) *Green Chem.*, **14**, 2346–2352.
161. Aouf, C., Lecomte, J., Villeneuve, P., Dubreucq, E., and Fulcrand, H. (2012) *Green Chem.*, **14**, 2328–2336.
162. Fache, M., Darroman, E., Besse, V., Auvergne, R., Caillol, S., and Boutevin, B. (2014) *Green Chem.*, **16**, 1987–1997.
163. Fache, M., Auvergne, R., Boutevin, B., and Caillol, S. (2014) *Eur. Polym. J.*, **67**, 527–538.
164. Mialon, L., Pemba, A.G., and Miller, S.A. (2010) *Green Chem.*, **12**, 1704–1706.
165. Pion, F., Ducrot, P.-H., and Allais, F. (2014) *Macromol. Chem. Phys.*, **215**, 431–439.
166. Oulame, M.Z., Pion, F., Allauddin, S., Raju, K.V.S.N., Ducrot, P.H., and Allais, F. (2015) *Eur. Polym. J.*, **63**, 186–193.
167. Barbara, I., Flourat, A.L., and Allais, F. (2015) *Eur. Polym. J.*, **62**, 236–243.
168. Guo, W., Shen, Z., Guo, B., Zhang, L., and Jia, D. (2014) *Polymer*, **55**, 4324–4331.
169. Ciriminna, R. (2014) *Eur. J. Lipid Sci. Technol.*, **116**, 1432–1439.
170. Gandini, A. and Lacerda, T.M. (2015) *Prog. Polym. Sci.*, **38**, 1–39.
171. Zhang, H. and Grinstaff, M.W. (2014) *Macromol. Rapid Commun.*, **35**, 1906–1924.

172. Ciriminna, R., Katryniok, B., Paul, S., Dumeignil, F., and Pagliaro, M. (2015) *Org. Process Res. Dev.*, **19**, 748–754.
173. Martin, A. and Richter, M. (2011) *Eur. J. Lipid Sci. Technol.*, **113**, 100–117.
174. Hu, S., Luo, X., and Li, Y. (2015) *J. Appl. Polym. Sci.* **132**, doi: 10.1002/APP.41425
175. Wang, Y., Ameer, G.A., Sheppard, B.J., and Langer, R. (2002) *Nat. Biotechnol.*, **20**, 602–606.
176. Rai, R., Tallawi, M., Grigore, A., and Boccaccini, A.R. (2012) *Prog. Polym. Sci.*, **37**, 1051–1078.
177. Cai, W. and Liu, C. (2008) *Mater. Lett.*, **62**, 2171–2173.
178. Kafouris, D., Kossivas, F., Constantinides, C., Nguyen, N.Q., Wesdemiotis, C., Patrickios, C.S., (2013) *Macromolecules*, **46**, 622-630.
179. Yang, Y., Lu, W., Cai, J., Hou, Y., Ouyang, S., Xie, W., and Gross, R.A.,(2011) *Macromolecules*, **44**, 1977–1985.
180. Mullen, B.D., Badarinarayana, V., Santos-Martinez, M., and Selifonov, S. (2010) *Top. Catal.*, **53**, 1235–1240.
181. Wang, X.L., Zhuo, R.X., Liu, R.J., He, F., and Liu, G. (2002) *J. Polym. Sci., Part A: Polym. Chem.*, **40**, 70–75.
182. Kaplan, J.A., Lei, H., Liu, R., Padera, R., Colson, Y.L., and Grinstaff, M.W. (2014) *Biomacromolecules*, **15**, 2548–2554.